C0-ARS-812

Topics in Applied Physics Volume 18

Topics in Applied Physics   Founded by Helmut K. V. Lotsch

Volume 1   **Dye Lasers**   Editor: F. P. Schäfer

Volume 2   **Laser Spectroscopy** of Atoms and Molecules
Editor: H. Walther

Volume 3   **Numerical and Asymptotic Techniques in Electromagnetics**
Editor: R. Mittra

Volume 4   **Interactions on Metal Surfaces**   Editor: R. Gomer

Volume 5   **Mössbauer Spectroscopy**   Editor: U. Gonser

Volume 6   **Picture Processing and Digital Filtering**   Editor: T. S. Huang

Volume 7   **Integrated Optics**   Editor: T. Tamir

Volume 8   **Light Scattering in Solids**   Editor: M. Cardona

Volume 9   **Laser Speckle** and Related Phenomena   Editor: J. C. Dainty

Volume 10   **Transient Electromagnetic Fields**   Editor: L. B. Felsen

Volume 11   **Digital Picture Analysis**   Editor: A. Rosenfeld

Volume 12   **Turbulence**   Editor: P. Bradshaw

Volume 13   **High-Resolution Laser Spectroscopy**   Editor: K. Shimoda

Volume 14   **Laser Monitoring of the Atmosphere**   Editor: E. D. Hinkley

Volume 15   **Radiationless Processes** in Molecules and Condensed Phases
Editor: F. K. Fong

Volume 16   **Nonlinear Infrared Generation**   Editor: Y.-R. Shen

Volume 17   **Electroluminescence**   Editor: J. I. Pankove

Volume 18   **Ultrashort Light Pulses,** Picosecond Techniques and Applications
Editor: S. L. Shapiro

Volume 19   **Optical and Infrared Detectors**   Editor: R. J. Keyes

---

Ultrashort Light Pulses

Picosecond Techniques and Applications

Edited by S. L. Shapiro

With Contributions by
D. H. Auston D. J. Bradley A. J. Campillo
K. B. Eisenthal E. P. Ippen D. von der Linde
C. V. Shank S. L. Shapiro

With 173 Figures

Springer-Verlag Berlin Heidelberg New York 1977

Stanley L. Shapiro, Ph. D.

University of California, Los Alamos Scientific Laboratory,
Los Alamos, NM 87545, USA

ISBN 3-540-08103-8 Springer-Verlag Berlin Heidelberg New York
ISBN 0-387-08103-8 Springer-Verlag New York Heidelberg Berlin

Library of Congress Cataloging in Publication Data. Main entry under title: Ultrashort light pulses. (Topics in applied physics; v. 18). Includes bibliographical references and index. 1. Laser pulses, Ultrashort—Addresses, essays, lectures. I. Auston, D. H. II. Shapiro, Stanley Leland, 1941-. QC689.5.L37U47. 535.5′8. 77-2402

Printed in Germany

Monophoto typesetting, offset printing, and bookbinding: Brühlsche Universitätsdruckerei, Lahn-Giessen.
2153/3130-543210

Preface

Soon after the invention of the laser, a brand-new area of endeavour emerged after the discovery that powerful ultrashort (picosecond) light pulses could be extracted from some lasers. Chemists, physicists, and engineers quickly recognized that such pulses would allow direct temporal studies of extremely rapid phenomena requiring, however, development of revolutionary ultrafast optical and electronic devices. For basic research the development of picosecond pulses was highly important because experimentalists were now able to measure directly the motions of atoms and molecules in liquids and solids: by disrupting a material from equilibrium with an intense picosecond pulse and then recording the time of return to the equilibrium state by picosecond techniques.

Studies of picosecond laser pulses—their generation and diagnostic techniques—are still undergoing a fairly rapid expansion, but a critical review of the state of the art by experienced workers in the field may be a timely help to new experimentalists. We shall review the sophisticated tools developed in the last ten years, including the modelocked picosecond-pulse-emitting lasers, the picosecond detection techniques, and picosecond devices. Moreover, we shall outline the basic foundations for the study of rapid events in chemistry and physics, which have emerged after many interesting experiments and which are now being applied in biology. An in-depth coverage of various aspects of the picosecond field should be helpful to scientists and engineers alike.

Because this volume had to be published quickly if it wanted to fill a present need, and because the material on picosecond pulses is both expanding and voluminous, this book could not be all-inclusive. Some overlap between chapters could not be avoided, but the reader may well benefit from a presentation that views the material from a different perspective.

The editor wishes to thank all contributors for their cooperation; Drs. *M. A. Duguay* and *H. Lotsch* for their advice and comments; and *F. Skoberne* for editorial assistance.

Los Alamos, New Mexico
November 1976

Stanley L. Shapiro

Contents

Contributors

Auston, David H.
Bell Laboratories, Murray Hill, NJ 07974, USA

Bradley, Daniel J.
Imperial College of Science and Technology, Optics Section, Department of Physics, London SW 7 2BZ, Great Britain

Campillo, Anthony J.
University of California, Los Alamos Scientific Laboratory, Los Alamos, NM 87545, USA

Eisenthal, Kenneth B.
Department of Chemistry, Columbia University, New York, NY 10027, USA

Ippen, Erich P.
Bell Laboratories, Holmdel, NJ 07733, USA

von der Linde, Dietrich
Max-Planck-Institut für Festkörperforschung, D-7000 Stuttgart 80, Fed. Rep. of Germany

Shank, Charles V.
Bell Laboratories, Holmdel, NJ 07733, USA

Shapiro, Stanley L.
University of California, Los Alamos Scientific Laboratory, Los Alamos, NM 87545, USA

1. Introduction–A Historical Overview

S. L. Shapiro

With 4 Figures

The rapid development of picosecond technology has allowed us to examine fundamental processes in materials. The following chapters will review these developments. Here we shall first investigate some historical concepts that underly the measurement of rapid phenomena. We shall see that some concepts are very old and that many are included in present technology, although in a most sophisticated form. We shall also attempt to discover why investigators have been interested in rapid phenomena in the past, and how some of their motivations and procedures differ from those of today. We shall try to answer the question why picosecond light pulses have become so attractive in such a short period of time, what some of the main research directions are, and what we can expect in the future. We hope to succeed in this general overview in our attempt of outlining the objectives and providing perspective into picosecond studies, although we shall only briefly survey that field; the new concepts are discussed in great detail in subsequent chapters.

1.1 Historical Concepts for Measuring Brief Time Intervals

To accurately determine an interval of short duration, one must first be able to measure time. One measuring technique is to identify events that recur regularly over and over again, such as the passing of a day. The ancients knew how to subdivide time by using such devices as sundials. For measuring still briefer time intervals, they used the hour glass or allowed a specific quantity of water to drip from a cistern. However, they apparently showed no interest in studying physical events with these methods.

1.1.1 Physiological Techniques

The study of brief time intervals really began in earnest with *Galileo Galilei* [1.1]. Celebrated for his theories of motion, many regard Galilei's experimental methods as the beginning of modern physics. What may not be as widely recognized is the fact that the key to many of Galilei's discoveries was his ability to measure short time intervals. Among Galilei's first methods of measuring time was the use of his pulse as a clock. His method undoubtedly reflected an instinctive feeling for physiological functions, because they appear to continually and uniformly flow irreversibly in one direction. He believed his methods were accurate to about one-tenth of a pulse beat [Ref. 1.1b, p. 171].

Galilei also discovered the isochronism of the pendulum in about 1580. Unfortunately, there are no records of his early experiments. A famous story relates how he timed with his pulse the swings back and forth of a lamp affixed to the ceiling of Pisa Cathedral. Although the story is now believed to be apocryphal, *Galilei* nonetheless did discover the isochronism. Later he devised a clock based on the pendulum, as well as a medical device to aid physicians in determining whether a patient's pulse rate was normal. If the displacement from equilibrium is small, the pendulum is in essence a harmonic oscillator with gravity playing the role of the spring constant. Harmonicity is the phenomenon behind the isochronism of the swings and behind the mechanism of similar clocks such as the balance wheel of a watch, electrical oscillations, and loaded tuning forks. *Galilei* [Ref. 1.1b, p. 172] also measured time by employing a large vessel of water placed at an elevated position; to the bottom of the vessel he soldered a small pipe that released a thin jet of water during the time interval to be measured; the water was collected in a glass and was then weighed on an accurate balance; because the clocks of Galilei's time were inaccurate, he cleverly turned the measurement of time into the measurement of weights.

Although a physiological function played a key role in Galilei's original measurements, it was soon recognized that physiological detectors were limited in measuring still briefer time intervals. *Von Segner* [1.2] performed an experiment in 1740 which showed that images persist on the retina for about 150 ms. He placed a glowing ember on the edge of a rotating disk and found that when the disk was rotated more quickly the apparent image on the retina resembled a curved comet. By recording the length of the perceived comet tail as a function of wheel velocity, *von Segner* determined the limitations of the eye for measuring brief time intervals.

To measure repetitive events that are shorter than a tenth of a second, investigators originally resorted to yet a third physiological function, sound. As we all know, sound represents a compression wave travelling in air, which humans detect because sound activates the small bones of the inner ear into oscillation. The range of human hearing is roughly between 24 and 24000 cps. This suggests the possibility of detecting periodic phenomena lasting between 10^{-2} and 10^{-4} s if the phenomena can be translated into a sound wave, because humans can discriminate between sound frequencies. Also, the sound can then be audibly compared against some standard such as a set of tuning forks, a plucked string, or similar device. An excellent standard of time based on sonic principles is the siren devised by *de la Tour* [1.3] in 1819. The siren consists of a disk with a series of equidistant holes drilled along its perimeter, and an orifice connected by tubing to a pressurized container. The disk is rotated at constant speed with the holes crossing the orifice, and, if the orifice is opened and closed with sufficient rapidity, a tone is produced whose frequencies can be related to the speed of rotation of the disk. The siren was quickly adopted as an instrument for measuring brief time intervals, as we shall shortly see in an experimental example in the next section. Frequency is often an indirect means by which to measure time.

1.1.2 Mechanical Techniques and Streak Concepts

In 1834 *Wheatstone* [1.4, 5], who was well aware of the limitation imposed by the finite time of visual response, noted that the path of a light-emitting body, if given a rapid transverse motion that would be combined with the velocity the body had originally, may be lengthened to any assignable extent and its velocity and duration could be measured. More conveniently, *Wheatstone* employed a rotating mirror to determine the duration of a spark as well as the velocity of electricity. As shown schematically in Fig. 1.1, he viewed the image of the spark reflected from a plane mirror that was rotated rapidly by means of a train of wheels. The essence of a streak-recording method is captured here: light emitted by an event at later times appears and is detected at a different spatial position, in this case, because the mechanical motion of the rotating mirror leads to a change of angular direction of the light with passing time. Using this mechanical technique, *Wheatstone* [1.4, 5] was able to demonstrate that some sparks last less than 10^{-6} s. To measure the speed of rotation of the mirror, he used two sonic methods—a siren and an arm striking a card—and found that the mirror was turning 800 times per second.

A second consideration is also realized here: to measure a short temporal event directly, one must provide a source whose temporal duration is comparable to the event itself. In Wheatstone's case, the ultrashort event was the electrical spark – controlled electrical methods were still fairly new in those days, as was the spark source. *Wheatstone* [1.4, 5] also used his apparatus to

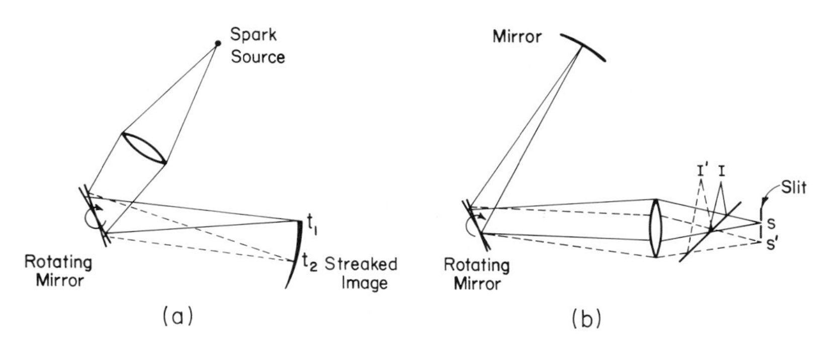

Fig. 1.1a and b. (a) Time history of a spark is displayed by Wheatstone's 1834 mechanical streak camera. Rotating mirror displaces light that is emitted at later times to a different spatial position. Length of image and knowledge of the angular velocity of the mirror give measure of lifetime. (b) Foucault's apparatus for measuring velocity of light. Light emitted at position S is reflected by a rotating mirror to a fixed mirror and then back again and is detected at I′ for clarity. Because of the finite velocity of light, image is displaced

show that certain flames vibrated rapidly, a fact suspected from a sound note emitted by the flames. Wheatstone's apparatus was applied to the study of fluorescence and phosphorescence of materials roughly a century after his discovery. Note, that the spark in Fig. 1.1, if simply focused so as to excite a sample in the precise position previously occupied by the spark source, allows one to study fast phenomena. Prior to this application of the Wheatstone-type device in the 1920's, *Becquerel* [1.6–8] was able to examine and measure phosphorescence lifetimes. He used a mechanical chopper to generate a short temporal light source, and a mechanical shutter and was able to measure phosphorescence events in the millisecond range.

A method of streaking was also used by *Foucault* [1.9–11] for his famous experiments in 1850 and 1862, which determined the velocity of light to fairly high precision. These experiments are shown schematically in Fig. 1.1. They involved a solar source, whose light passed through a slit and was imaged onto a curved mirror after reflection from a plane one. The return reflection, viewed at Point I for clarity, changes its position and elongates because the mirror rotates during the time light traverses back and forth between the rotating mirror and the curved mirror. By knowing the angular velocity of the mirror and the distance between mirrors (about 20 m in the original experiments) one may easily calculate the velocity of light, and *Foucault* obtained 2.980×10^{10} cm/s, a value which differs by about one part in one hundred and sixty from the best measurements today. By using similar techniques, Foucault was also able to show that light travelled slower in media with indices of refraction, a fact applied at the time to differentiate between an emission theory and a wave theory of light.

1.1.3 Studies in Motion with Spark Photography

Many concepts that eventually evolved into present-day probe and flash-photolysis experiments have, in some ways, been established long ago. We have already seen that microsecond sparks were available in 1834. In fact, by 1851, *Talbot* [1.12] had used such sparks successfully to obtain sharp photographs of a rapidly revolving newspaper page. Talbot's achievement emphasizes the point that high-speed spark photography of objects in motion is a very old concept. At the time, however, his accomplishment was overshadowed by the great advances in still photography. After all, the types of events people were most interested in a century ago were distant events and places, as related by photography. Eventually, the improvement in technique allowed *Muybridge* [1.13–18] in 1878 to photograph a large number of creatures in motion, particularly man and the horse. His early photographs were initiated by a sponsor, Governor *Leland Stanford* of California [1.15, 18]. *Stanford,* who was interested in thoroughbred horses, supposedly bet that at some time during a horse's gallop, all four legs were off the ground simultaneously. To prove this, Muybridge had a horse gallop past twelve cameras, spaced 21 inches apart, and, by breaking threads stretched across its path, the horse exposed the plates electronically. The exposure times were typically 1/500th of a second, and the photographs uncovered truths that the eye was incapable of perceiving. Fig. 1.2 shows some of Muybridge's photos; his photos, in fact, produced great surprise, because, while it is true that the horse did lift all four legs off the ground at one time, the horse never fully assumed the rocking-horse position of front legs stretched forward and hind legs backward, so widely depicted by artists of the time. *Muybridge* [1.16–17] later conducted an extensive research program on animals and humans in motion; his detailed investigations laid to rest many theories of animal motion and have remained a visual aid for artists and sculptors to this day. He also learned how to project his photographs so as to produce apparent motion; whether these could be considered the first true motion pictures is controversial, but his technique created the illusion of motion pictures and was shown by him throughout the world.

Fig. 1.2. High-speed photographs, such as these of a galloping horse taken by *Muybridge*, produced great excitement in the last century

1.1.4 Probe Technique Experiments in the Last Century

Although studies of animals in motion captured the interest and imagination of people in the last century, at least one set of photographic experiments, which measured short time intervals, was implemented in a way quite analogous to present-day experiments, and so deserves prominent mention. These are the schlieren experiments of *Töpler* [1.19–20]. His experiments followed the discovery by *Plateau* [1.21] in 1833 and *Stampfer* [1.22] in 1834 of the stroboscope and the design by *Foucault* [1.23] in 1859 of a sensitive method of detecting changes in the refractive index. *Töpler* combined these two techniques into a schlieren method and was able to take pictures of rapidly vibrating flames and of refractive-index changes in liquids. By 1867 he [1.19] was using a light spark of less than two microseconds to generate a sound wave, and then photographed the sound wave by using a second spark of equal duration as the first. Most importantly, he invented an electronic delay circuit that triggered the second spark at any arbitrary delay time. By taking pictures of the sound wave as a function of delay time, he obtained a complete history of the sound-wave phenomena, including motions, reflections, and amplitudes of the waves.

These techniques are, of course, very similar to present-day picosecond probe techniques where one pulse is used to generate a phenomenon, and a second pulse is used either to probe the phenomenon at adjusted delay times or to function as an analyzing device.

1.1.5 Popularization of Spark Photography – Chemical Applications

Edgerton [1.24–25] in 1931 popularized spark photogaphy by introducing packaged electronic spark units. Moreover, he [1.24] introduced a new type of lamp wherein a condenser charged to high voltage was discharged by means of a thyratron circuit into a tube containing a mixture of gases. These new lamps had a much-improved light-output efficiency and the emission was of short duration. Also, the frequency of the flash was controlled easily and accurately by means of a grid. From a scientific point of view these lamp techniques were applied most dramatically when *Porter* [1.26] in 1949 developed high-power flashlamps for studying fast photochemical reactions. With these powerful lamps he was able to produce a high concentration of intermediates in chemical reactions, such as free radicals and atoms, and was able to monitor the evolution of their spectra by means of a flash probe at later times. Only recently have these techniques been refined to the point of being useful also in the nanosecond and picosecond range by using lasers as the flash source for generating intermediates and using a fraction of the laser beam that was split off from the main beam as a probe beam. The synchronization problem is much alleviated in this fashion, because, a pulse derived from another one is easy to time. As described in Chapters 3, 4, 6 and 7, the most modern "flash-photolysis" techniques make use of probe beams generated by nonlinear optical processes such as stimulated Raman scattering, harmonic generation, stimulated emission from dyes, or, most popularly, by self-phase modulation which generates a very broad continuum. Nonlinear optical processes are only found at high optical field strengths, and very broad continua (~ 10000 cm^{-1}) could be demonstrated only with intense picosecond pulses. Many researches have now used these techniques for experiments in chemistry, physics, and biology.

1.1.6 Electrical Technique

A forerunner of the now popular picosecond optical-gate technique was the electrical gate. An important element in the gate is a Kerr cell, named after its discoverer. *Kerr* [1.27] in 1875 recognized that a dc electric field, when placed on certain isotropic substances, induces birefringence. It was later recognized that the molecular basis for the Kerr effect in substances composed of molecules that have permanent dipole moments, was the reorientation of the molecules so that their dipole moments were aligned along the field, which led to a change in refractive index. As early as 1899 *Abraham* and *Lemoine*

[1.28, 29] not only recognized that the Kerr effect could be used for measuring time intervals, but developed a device capable of measuring 10^{-9} s. A simplified schematic of their apparatus is shown in Fig. 1.3. The source, a spark discharge, is in series with a Kerr cell. The Kerr cell is composed of a polarizer and analyzer oriented 90° from each other, and in between is a cell of carbon disulfide upon which a dc field is placed at 45° with respect to the polarizers. When a dc voltage is placed on the cell, light is transmitted. By varying the optical path along which the light travelled to reach the Kerr cell, they observed light emitted by a spark as a function of delay time and found that they had to rotate their analyzer through smaller and smaller angles to cross out the light through the dc Kerr cell as the path length travelled by the light was increased. Equivalently, the intensity of the spark from the dc-triggered Kerr cell diminished if the distance was too large. From these measurements they concluded that the mechanism responsible for the Kerr effect lasted less than 10^{-8} s. The use of this type of device was extended to measure fluorescence lifetimes from dye molecules by *Gottling* [1.30] in 1923, and the resolution capability of the device eventually reached the stated capability of 1 ns in 1958 [1.31] when it

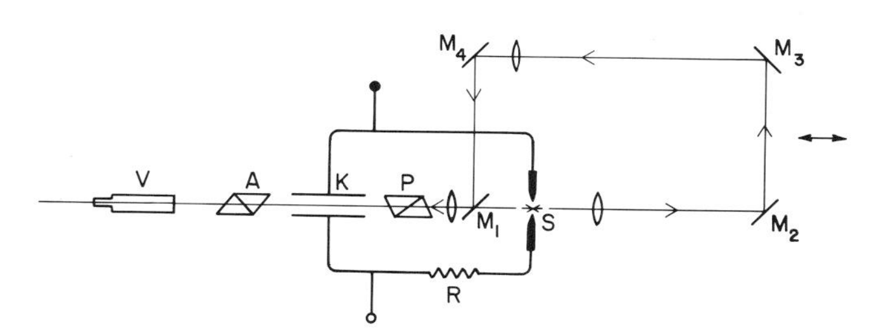

Fig. 1.3. Electrical pulse produces a spark S, and activates a Kerr dc shutter K. Light from the spark is collimated through a variable-delay path (mirrors M) and through the Kerr cell consisting of the polarizer P_1 of the CS_2 cell, and of the analyzer A. When the optical delay coincides with the triggering of the birefringence in the Kerr cell, an image is detected by the viewer V. However, when the delay is sufficiently long, no light is transmitted through the Kerr cell showing that the Kerr effect lasts less than 10^{-8} s. (After *Abraham* and *Lemoine*)

was used for high-speed photography. The experiment by *Abraham* and *Lemoine* illustrates a very important principle that is encountered time and again. To detect a phenomenon a device must be capable of being accurately synchronized with the phenomenon to be studied. In this particular case the time delay between origination of the spark and detection at the Kerr cell is set by varying the light path to compensate for the electrical delays. Synchronization becomes especially important in the time range of 10^{-9} s and less. The problem of jitter with picosecond events is usually solved by letting the picosecond pulses be an integral part of the devices themselves. The electrical gate just described is the ancestor of present-day optical gates.

1.1.7 Historical Summary

We have presented only a few examples to illustrate the links between the present and the past; many more examples could have easily been included. We have given this brief historical review to counteract a modern-day tendency to slight older learning, without any intent of diminishing present efforts. On the contrary, we wish to stress that it is the implementation of concepts which is hardest of all, and that improvements reach their highest sophistication in today's experiments. Even in early experiments, careful examination shows that creative ideas required the clever implementation of measurement techniques, which allowed investigators to overcome experimental difficulties and to translate their ideas into action.

1.2 Picosecond Techniques

1.2.1 Their Origins

The increasing (from about 1600 to date) ability to measure shorter and shorter time intervals is plotted in Fig. 1.4. Although the time durations plotted are only approximate, two points are worth stressing: First, that the plot is not a simple function of time, and second, that sudden dramatic progress was made between 1965 and 1975. That the plot is complex is not surprising; science proceeds by an endless sequence of related discoveries and sudden great advances, followed by periods of implementation and improvement. The sudden dip beginning in about 1820 is due to both the introduction of the

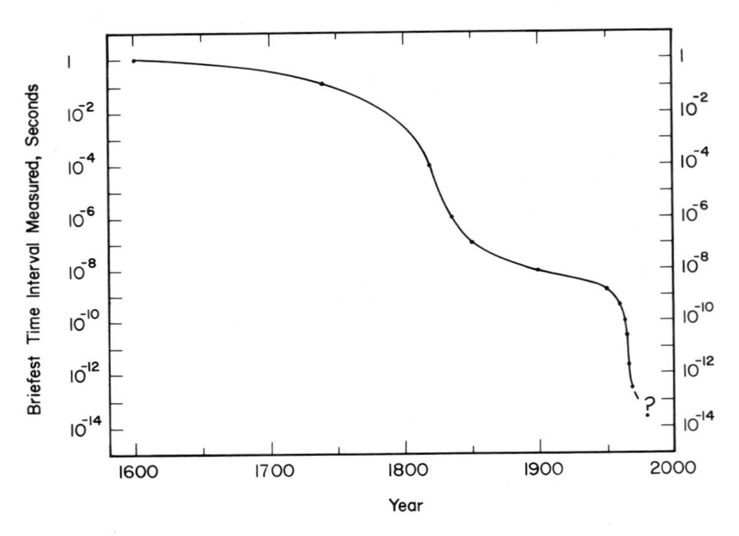

Fig. 1.4. Progress in measurement of brief time intervals. Note sharp improvement in last decade

streaking concept and to the first use of electrical sources and instruments. As electronic techniques improve, the capability for measuring brief time intervals improves. The spark replaces the flame as a source, the photomultiplier and oscilloscope replace the eye as a detector, cathode-ray tubes replace rotating mirrors, and even electronic tuning forks replace mechanical tuning forks. Around 1950 the capability improves, concurrent with the development of higher frequency circuitry. Nonetheless, the basic time interval which could be measured between 1900 and 1965 remained at about 1 ns despite the enormous growth of the electronics industry during this period. In fact, even the sources of excitation were no shorter than 1 ns. Quite evidently, the discovery that eventually resulted in greatly improved time measurements was the laser. Within five years picosecond pulses were a reality and within the next five years instruments were developed with a measurement capability three orders of magnitude better than ever before attainable.

In this summary, I have somewhat arbitrarily chosen not to include ultrashort lifetimes extrapolated by nuclear physicists by taking the inverse spectral linewidth of a nuclear transition or particular resonance [$\sim 10^{-24}$ s], or by microscopically measuring the length of a track produced by a high-energy particle in a nuclear emulsion, then obtaining a time interval by dividing these values by the particle velocity [$\sim 10^{-16}$ s]. Most points in Fig. 1.4 are produced by light techniques. Limited theoretical extrapolations or assumptions are needed, for example, for interpreting streak-camera results where an arbitrary and unknown waveform is faithfully displayed, regardless of theory.

Much of the ground work for picosecond light-pulse technology was laid by related discoveries during the five years preceding the use of high-power modelocked lasers. All main lasers now used for picosecond work were discovered during this period: the ruby, Nd: glass, Nd:YAG, and dye lasers. Moreover, several techniques for attaining high-power outputs were discovered, including the *Q*-switch and, especially, saturable absorbers as a nonlinear element in the cavity. Also, the first modelocked laser was a He:Ne gas laser whose pulsewidth of 2.5 ns, however, was no shorter than that from other sources. The bandwidth was too narrow to support ultrashort pulses, and the power too low for studying nonlinear optical effects. Nevertheless, this laser established many fundamental principles of modelocking and of pulse compression.

Besides the discovery of lasers in this five-year period, many nonlinear optical effects were detected, some of which were shortly to become an integral part of picosecond techniques, e.g., second- and third-harmonic generation, two- and three-photon absorption, stimulated Raman scattering, and the ac Kerr effect. The ac Kerr effect is a particularly fitting example to illustrate how laser technology contributes to new insights. The ac Kerr effect was discovered a full 90 years after the analogous dc effect had been identified, and was applied to chronoscopy studies 70 years after the dc effect had been applied for such studies; the power available in light beams prior to 1964 was simply too low to observe this minuscule nonlinear optical effect.

The first modelocked solid-state lasers built in 1965 and 1966 were the first devices emitting high-power ultrashort pulses. Many people date the picosecond era from that time. With these lasers came the introduction of the saturable filter for modelocking purposes. These nonlinear elements immediately became the most popular method for modelocking lasers, and they are used in nearly all picosecond experiments today. An early suggestion that the pulses from the Nd:glass laser might even be subpicosecond created great excitement. The fastest electronics at that time were unable to detect picosecond events. It immediately became apparent that new means of detection for picosecond phenomena were needed. For several years effort centered mostly on pulse-measuring techniques because accurate measurements of events would be difficult without this capacity. Because picosecond laser pulses were the only events on this time scale known at that time, it became an attractive possibility to incorporate such pulses in any detection device. It was no accident that many scientists who had made fundamental contributions to nonlinear optics quickly recognized the potential of nonlinear optical techniques for measuring picosecond pulse durations, and they were the ones who devised some of the first measurement techniques.

1.2.2 Advantages and Differences of the New Techniques

There are no springs, rotating mirrors, or extended sources in picosecond physics. Picosecond events are too fast for mechanical devices, and a careful track of distances is a necessity. Novel techniques have incorporated picosecond pulses so that they are part of the devices themselves.

Picosecond pulses do not deal with the motion of tangible objects, but rather with the motion of photons, electrons, and molecules. A simple example points out the futility of using picosecond pulses to look at large objects: a bullet moving at a velocity of 1000 m/s only moves 10 Å in 1 ps, less than the diameter of a good-sized molecule. So much for human events; on a picosecond time scale a speeding bullet moves like a dead ant. If one insisted upon a dichotomy between studies a century ago and now, it is probably this: Emphasis in the last century was placed on human perceptual events and/or the motion of large objects, whereas now, the picosecond techniques allow us to explore atomic and molecular events in depth.

Picosecond pulses are consequently used as a basic research tool in vast areas of physics and chemistry as well as in biology. Engineering applications now range from picosecond electronics to laser fusion. Picosecond pulses have two primary properties that make them particularly useful: ultrashort duration and, when needed, very high power. Because of their ultrashort duration a wealth of new phenomena has been uncovered. A picosecond, or 10^{-12} s, is only a trillionth of a second: in terms of present theoretical-physics concepts, which regard the speed of light as the upper limit on the propagation of energy, a picosecond is the time it takes light to travel just 0.3 mm. The shortest pulses obtained to date, a few tenths of a picosecond, are, according to Heisenberg's uncertainty principle, approaching within two to three orders of magnitude the shortest theoretically obtainable pulse durations in the ultraviolet-visible-infrared. This ultrashort temporal property allows the study of some of the most fundamental processes in materials, most which occur on a picosecond time scale. Among the fundamental processes that have been measured are the free decay of molecular vibrations and orientational fluctuations in liquids, phonon decay in solids, exciton decay and migration in solids, growth and decay of plasmas in gases and solids, charge-transfer processes and other nonradiative transfer processes between molecules, temperature-fluctuation processes, and much more. Picosecond pulses and their interaction with matter have therefore been and remain a frontier area in science and technology.

Even a small amount of energy, when compacted into a few picoseconds, can have enormous power. A terawatt (10^{12} W) can easily be produced with even a modest-size laser. This high-power property allows the investigation of many interesting nonlinear phenomena. Moreover, it is the property responsible for the ability to produce a suitably high concentration of particles or excited-state species within a brief interval, which proves useful for many measurement devices, e.g., operating a shutter, triggering a streak camera, or producing nonlinear optical effects.

Besides their ultrashort and high-power properties, picosecond techniques are different from previous methods in at least two aspects. The first is miniaturization. The finer the temporal resolution, the greater the care required to keep track of ever smaller distances. This care applies especially to the new electronic instruments. In streak cameras the distances between certain electrodes must be kept small to avoid the loss of temporal resolution because of the different velocities at which electrons spread. In the new electronic switches physical distances must be small for maximum switching capability. Another aspect that almost goes unnoticed because it is accomplished so easily is the production of pulses at regular intervals. The ability to produce repetitive pulses is useful for many types of measurements. Laser pulses are produced automatically at accurately spaced time intervals by the pulse traversing back and forth in the laser cavity. Of course, pulses can be set at any arbitrary time interval with respect to one another by etalons, echelons, and similar devices. The ability of the cw modelocked dye laser, in particular, to produce controlled pulse emission where each pulse has the same amplitude, the same duration, and occurs at regular time intervals, has resulted in a super measuring tool because repetitive sampling techniques are thus possible.

Of course, nearly all the traditional advantages of laser sources also accrue in the picosecond field. Beam divergence is small, some sources are tunable, the source is easy to handle, and the beam is monochromatic within limits.

1.3 Present Trends and Future Studies

Today, activity in the picosecond field is still intense. Interest covers a wide spectrum and involves many approaches, from the individual researcher at a university to large corporations, such as computer and telephone companies, and to the large national laser-fusion laboratories. There are many more researchers in the field than at any previous time, and the studies are much more diverse. Besides, the keen interest in picosecond phenomena stems largely from the fact that it is still not too expensive for private research. Moreover, a researcher in the field does not necessarily have to channel his efforts into one specific area. He can do research in physics, chemistry, biology, and engineering in a wide variety of materials, e.g., liquids, solids, plasmas, or gases. Doubtless this variety is a source of inspiration allowing the transfer of ideas from one area of endeavor to another.

As always, present trends are toward shorter and shorter pulses and toward higher and higher powers. These trends cannot go on indefinitely in the visible optical domain where the limit to temporal resolution, set by the uncertainty principle, is 10^{-15} s and dispersion problems intervene; we are already at the limit of propagating high power in materials as set by breakdown and nonlinear optical effects. Though the temporal resolution in the x-ray region can theoretically be still shorter (10^{-18} s at 1 keV), hard-x-ray lasers, which will presumably be pumped by powerful picosecond pulses, would probably have more impact in the biological area for holographic purposes and a less likely impact on chronoscopy. Because of experimental difficulties, it is doubtful that more esoteric but basic problems, such as photon-photon interactions in a vacuum, will be addressed successfully in the near future. History, however, teaches us that new ingenious approaches or surprises are inevitably found.

For the present we can expect continued application of standard techniques and can look forward to much broader application of the streak camera, of the cw modelocked dye laser, and of picosecond electronics. Subpicosecond cw dye lasers and picosecond electrical pulses have been developed only in the last two years, and the streak camera has only recently been extended into both the x-ray detection range and the subpicosecond resolution range for the visible domain. By the end of 1976 coherent pulses as short as 38 nm were generated.

In the past decade picosecond-pulse technology has seen continuous, steady development and improvement. Among the authors of this book are those entirely or partly responsible for some of the very recent developments including the subpicosecond dye laser and associated techniques, picosecond electronics, and the x-ray and subpicosecond streak cameras. These recent innovations testify to the activity in this vast field.

1.4 Organization of the Book

Following this overview, Chapter 2, by *Bradley,* introduces the subject of picosecond pulses and describes how to produce them. He outlines the basic concepts that underly the generation of picosecond pulses by modelocked lasers. The different kinds of modelocked lasers are described, and important measurements on the passive element responsible for modelocking, i.e., the saturable absorber, are analyzed in terms of laser operation. He describes the standard techniques for producing pulses at other frequencies by harmonic generation and discusses techniques for producing pulses that are tunable in wavelength in the visible, for the production of infrared and x-ray pulses, and for producing ultrashort pulses by nonlinear optical, and the cw modelocked dye laser techniques. Finally, he discusses some limitations of the detection techniques in the visible and x-ray regions for these pulses, and describes in detail the x-ray and subpicosecond streak camera.

When the modelocked laser was invented, no techniques existed for measuring ultrashort durations. These techniques are essential for measuring rapid events and even for studying the modelocked lasers. In Chapter 3 *Ippen* and *Shank* describe picosecond detection techniques in detail. They describe all the clever techniques used for detecting rapid phenomena and the advantages and disadvantages of each. They also describe new cw techniques they have developed for use with cw modelocked dye lasers. These techniques have allowed them to regularly produce the shortest pulse to date, 3×10^{-13} s or 300 fs. They tell how to use the recently developed techniques, and how they have applied them to study such phenomena as vibrational relaxation in dye molecules, and rotational relaxation times of molecules. The techniques are relatively inexpensive and will be powerful research tools in the future.

Two important topics covered in Chapter 4 by *Auston* are nonlinear optical effects and picosecond devices. He describes how several new nonlinear optical effects have been discovered with picosecond light pulses, and how many older effects have been greatly enhanced. He also tells how picosecond pulses have allowed investigators to study the transient response of materials, and how these pulses can be applied to study the many contributions to a nonlinear refractive index, by providing results free from ambiguity by separating the faster from slower contributions on the basis of temporal response. He also describes the electronic switches he recently invented, which use picosecond technology; they are easily the fastest electronic switches ever produced. The new capability for producing and manipulating picosecond electrical signals will prove useful in future scientific studies and may lead eventually to important technological advances.

The development of modelocked lasers and of associated measurement techniques forms the foundation that is vital to the pursuit of the study of rapid physical processes in materials. In Chapter 5, *von der Linde* gives an account of rapid relaxation measurements in liquids, solids, and plasmas. These explorations lie near the heart of picosecond activities, because many measurements of fundamental importance are becoming possible for the first time, and many relaxation phenomena can now be studied and extended in some depth. Included in von der Linde's chapter are descriptions of measurements of decays of excitons, phonons, polaritons, plasmas, molecular vibrations, and coherent phenomena. He describes how the molecular depopulation and dephasing times for vibrations in liquids have been measured directly for the first time. He discusses interesting new measurements on plasma decay in solids, and on interactions of excitations with each other and with impurities.

The division between Chapter 5 and 6 is somewhat arbitrary. The measurements described by *Eisenthal* in Chapter 6 are traditionally expected from chemists. Among the rapid phenomena studied and described in his chapter, are charge transfer, short-lived transient species, solvent relaxation, fluorescence, nonradiative transfer, and relaxation processes in small and large molecules. Many traditional detection schemes are described, such as flash photolysis, which are allowing the examination of singlet, triplet, and excited-state interactions on a picosecond time scale. He describes measurements that test models for molecular interactions in liquids.

In Chapter 7 *Campillo* and *Shapiro* describe an area relatively new to picosecond technology – biological measurements with ultrashort pulses. Interesting information has already been obtained about fundamental processes in photosynthesis: here the energy is collected by antenna pigment molecules and funneled to a reaction center. Information on the structure and on trapping within the photosynthetic units has been obtained recently. Also, information on new states within reaction centers is especially important and exciting. A state has been identified within the reaction center, which has been associated with an early step in photosynthesis, i.e., the separation of charges from one molecule to another. The energy stored by this charge separation begins the process of photosynthesis. Measurements of primary visual processes are also described, as well as measurements on DNA. An interesting measurement on picosecond transfer times in hemoglobin made with the powerful new cw technique is also described.

Laser fusion, an area where high-energy picosecond pulses play an important role, is not treated as a separate topic; laser fusion is an extensive topic and many of its elements fall outside the context of this book. However, some recent time-dependent measurements made in the last year are discussed as well as some of the design concepts for large laser-amplifier systems used for laser-fusion experiments. X-ray streak cameras with sufficient resolution should play an important role in laser-fusion target diagnostics in the future.

Modelocking for production of picosecond pulses is discussed in Chapter 2, but this subject is not covered as a separate topic per se in the book and the interested reader is referred to the excellent historical and tutorial survey of *Smith* [1.32], and to review papers on the theory of internal modulation by *Harris* [1.33], and by *Siegman* and *Kuizenga* [1.34].

Popular accounts on picosecond pulses may be found in [1.35–40].

References

1.1 Galileo Galilei: *Discorsi e Dimostrazioni Matematiche intorno a due nuoue Scienze*, (Elsevier, Leiden 1638). *Translated as:* Galileo Galilei: *Dialogues Concerning Two New Sciences*, ed. by H. Crew, A. de Salvio (Northwestern University Press, Evanston, Ill. 1950)
1.2 J. A. von Segner: *De raritate Luminis* (Göttingen 1740)
1.3 C. de la Tour: Ann. Chim. Phys. **12**, 167 (1819)
1.4 C. Wheatstone: Phil. Trans., **1834**, 583
1.5 C. Wheatstone: Phil. Mag. **6**, 61 (1835)
1.6 E. Bequerel: Ann. Chim. Phys. **55**, 5 (1859)
1.7 E. Bequerel: Ann. Chim. Phys. **57**, 40 (1859)
1.8 E. Bequerel: Ann. Chim. Phys. **62**, 5 (1861)
1.9 L. Foucault: Compt. Rend. **30**, 551 (1850)
1.10 L. Foucault: Compt. Rend. **55**, 501 (1862)
1.11 L. Foucault: Compt. Rend. **55**, 792 (1862)
1.12 H. F. Talbot: Phil. Mag. and J. Sci. **3**, 73 (1852)
1.13 E. Muybridge: Scientific American, October 1878
1.14 E. Muybridge: La Nature (Paris), December 1878
1.15 E. Muybridge: Nature **19**, 517 (1878)
1.16 E. Muybridge: In *Animals in Motion*, ed. by L. S. Brown (Dover Publ. Co., New York 1957)
1.17 E. Muybridge: In *Human Figure in Motion*, ed. by L. S. Brown (Dover Publ. Co., New York 1955)
1.18 E. Muybridge: In *Man in Motion*, ed. by R. B. Haas (University of California Press, Berkeley, Los Angeles, London 1976)
1.19 A. Töpler: Ann. Physik und Chemie **131**, 33, 180 (1867)
1.20 A. Töpler: Ann. Physik und Chemie **128**, 126 (1866)
1.21 J. A. F. Plateau: Ann. Chim. Phys. **53**, 304 (1833)
1.22 S. Stampfer: Jahrbücher k. k. Polytechnisches Institut, Wien **18**, 237 (1834)
1.23 L. Foucault: Monthly Notices Roy. Astron. Soc. **19**, 284 (1859)
1.24 H. E. Edgerton: J. SMPTE **16**, 735 (1931)
1.25 H. E. Edgerton: J. SMPTE **18**, 356 (1932)
1.26 G. Porter: Proc. Roy. Soc. (London) **A200**, 284 (1950)
1.27 J. Kerr: Phil. Mag. and J. Sci. **50**, 337 (1875)
1.28 H. Abraham, J. Lemoine: Compt. Rend. **129**, 206 (1899)
1.29 H. Abraham, J. Lemoine: Ann. Chim. **20**, 264 (1900)
1.30 P. F. Gottling: Phys. Rev. **22**, 566 (1923)
1.31 A. M. Zarem, F. R. Marshall, S. M. Hauser: Rev. Sci. Instr. **29**, 1041 (1958)
1.32 P. W. Smith: Proc. IEEE **58**, 1342 (1970)
1.33 S. E. Harris: Proc. IEEE **54**, 1401 (1966)
1.34 A. E. Siegman, D. J. Kuizenga: Opto-Electronics **6**, 43 (1974)
1.35 A. J. DeMaria, D. A. Stetser, W. H. Glenn: Science **156**, 1557 (1967)
1.36 N. Bloembergen: Comments on Solid State Physics **1**, 37 (1968)
1.37 P. M. Rentzepis: Science **169**, 239 (1970)
1.38 A. J. DeMaria, W. H. Glenn, M. E. Mack: Physics Today **24**, 19 (1971)
1.39 R. R. Alfano, S. L. Shapiro: Scientific American **228**, 42 (1973)
1.40 R. R. Alfano, S. L. Shapiro: Physics Today **28**, 30 (1975)

2. Methods of Generation

D. J. Bradley

With 46 Figures

It is over a decade since picosecond laser pulses were first produced by passive modelocking of a giant-pulse ruby laser by *Mocker* and *Collins* [2.1] in 1965 and then Nd : glass lasers by *DeMaria* et al. [2.2] in 1966. Since then the techniques for the generation of these pulses have been developed to the extent that it is possible to reliably produce pulses of bandwidth-limited durations of ~ 1 ps from both pulsed and cw lasers. In addition, theoretical models have been refined to the stage that there is excellent agreement with even the details of the experimental results, and there is now a very good understanding of the mechanisms by which ultrashort pulses evolve from the initial fluorescence intensity fluctuation patterns. These substantial advances in technology and physical understanding have been largely due to the simultaneous development of the methods of picosecond chronoscopy, particularly the direct linear measurement of pulse durations by electron-optical streak cameras. This pattern, of course, follows the historical pattern whereby developments in science and technology are almost always related to advances in measurement techniques. As a result the methods of picosecond laser pulse generation and measurement are now sufficiently refined and catalogued for them to be used with confidence for the investigation on a picosecond timescale of the interaction of coherent light with matter.

The purpose of this chapter is to summarize the present state of the art in picosecond pulse generation, dealing only with systems which are capable of producing pulses of durations ~ 10 ps or less. As a consequence, the methods of active modelocking of pulsed and continuous lasers are not considered, since such systems have to date not produced such short pulses. (Actively modelocked lasers have been extensively reviewed, [2.3, 4]). Also excluded are gas lasers, although pulses of durations ~ 100 ps have been obtained by actively modelocking the argon-ion laser [2.5] and the iodine photodissociation laser [2.6], and, from a passively modelocked high-pressure CO_2 laser [2.7], and it is likely that the newly developed broadband excimer and exciplex lasers based upon high-pressure gases (see Proceedings of Conference on High Power Gas Lasers 1975 [2.8]), will provide sources of picosecond pulses tunable throughout the uv.

The mathematical description of optical pulses is immediately followed by an account of the experimental methods of measuring temporal intensity profiles. Emphasis is placed upon the recently perfected electron-optical picosecond streak-camera techniques since these provide direct linear measurements of pulses over a wide range of photon energies from ~1 eV to several keV. The two main types of picosecond sources, giant-pulse solid state systems and dye lasers, are then discussed together with the extensive and detailed experimental studies of the temporal buildup of modelocking of these lasers, including measurements of the properties of saturable absorbers and the effects of the nonlinear index of refraction. Special attention is paid to the cw dye laser, since this is the only source, to date, of steady-state, bandwidth-limited subpicosecond pulses. Recent developments of dye lasers, synchronously pumped by modelocked gas and solid-state lasers, are also described. The fluctuation model of passive-modelocking is related to these experimental studies, in particular to the special properties of dye lasers. Methods of picosecond pulse amplification and the limitations set by the linear and nonlinear properties of the laser media are described. Finally some of the latest results obtained in frequency changing picosecond pulses into the uv, vuv, and ir spectral regions by harmonic-generation and by nonlinear optical mixing are briefly considered theoretically and experimentally.

2.1 Optical Pulse Properties and Methods of Measurement

Before discussing the details of the generation of picosecond pulses, it is necessary to establish an appropriate mathematical description of the optical pulses themselves and to define the terminology involved. It is also sensible to follow this with a discussion of the methods available for the measurement of the temporal profile, the essential property of an ultrashort pulse.

2.1.1 General Description of Modelocked Laser Pulses

Perfect modelocking represents one of the two completely organised states of operation of a laser. To each transverse mode of a laser cavity there corresponds a set of longitudinal modes (Fig. 2.1), each having the same form of spatial energy distribution in a transverse plane, but with different distributions along the axis of the resonator. These longitudinal modes are separated in frequency by $c/2L$, where L is the optical pathlength between the mirrors and c is the velocity of light [2.9]. By employing appropriately designed laser resonators, including the use of apertures, single transverse-mode operation can be readily obtained. The addition of intracavity frequency-selective components [2.10] can then ensure operation of the laser in a single longitudinal, single transverse mode. The second completely organised situation occurs when the set of longitudinal modes is maintained in fixed phase and fixed amplitude relationships. In this case, the output will be a well-defined function of time, the modelocked laser pulse train. With a lasing medium of adequate bandwidth a regularly spaced train of picosecond pulses can be produced in this manner.

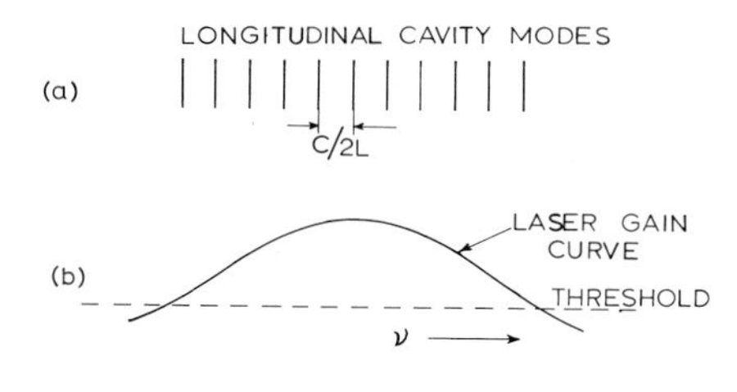

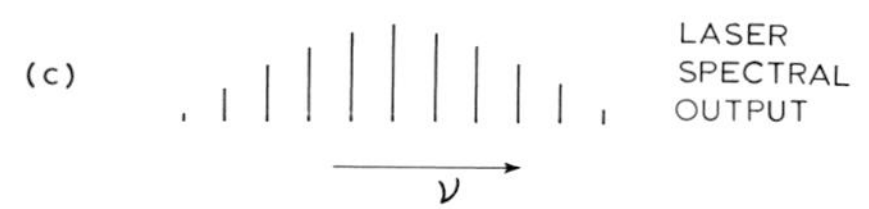

Fig. 2.1 a–c. Interaction of laser gain curve and longitudinal cavity modes resulting in oscillation at several resonator mode frequencies for which the laser gain is above the threshold value

Starting from the pair of Fourier integrals

$$E(t)=\frac{1}{\sqrt{2\pi}}\int_{-\infty}^{+\infty} e(\omega)\exp(-i\omega t)\,d\omega \tag{2.1}$$

and

$$e(\omega)=\frac{1}{\sqrt{2\pi}}\int_{-\infty}^{+\infty} E(t)\exp(i\omega t)dt \tag{2.2}$$

we define the complex signal $V(t)$ of a plane-wave optical pulse associated with the magnitude $E(t)$ of the electric field vector at a fixed point in space by writing [2.11, 12]

$$\begin{aligned} V(t)&=\frac{1}{\sqrt{2\pi}}\int_{0}^{\infty} 2e(\omega)\exp(-\mathrm{i}\omega t)d\omega \\ &=\frac{1}{\sqrt{2\pi}}\int_{-\infty}^{+\infty} v(\omega)\exp(-\mathrm{i}\omega t)\,d\omega\,. \end{aligned} \tag{2.3}$$

In obtaining (2.3) the amplitudes of the negative frequency components in the integral of (2.1) are suppressed and the amplitudes of the positive frequencies are doubled. It follows that

$$\begin{aligned} v(\omega)&=\frac{1}{\sqrt{2\pi}}\int_{-\infty}^{+\infty} V(t)\exp(+\mathrm{i}\omega t)dt \\ &=2e(\omega)\quad(\omega>0) \\ &=0\qquad(\omega<0). \end{aligned} \tag{2.4}$$

The complex functions $V(t)$ and $v(\omega)$ can be employed to define an optical pulse in the time and frequency domains, respectively. By writing

$$v(\omega)=a(\omega)\exp[-\mathrm{i}\phi(\omega)] \tag{2.5}$$

we define $a(\omega)$ the spectral amplitude and $\phi(\omega)$ the spectral phase [2.13]. When the pulse bandwidth $\Delta\omega$ is narrow compared with the mean optical frequency ω_0 we can write

$$V(t)=A(t)\exp\mathrm{i}[\Phi(t)-\omega_0 t]. \tag{2.6}$$

In this quasi-monochromatic case, the temporal amplitude $A(t)$ and the temporal phase $\Phi(t)$ of the optical-frequency wave are both slowly varying functions of time. The instantaneous intensity is then

$$I(t)=V(t)V^*(t)=A^2(t). \tag{2.7}$$

By analogy, the spectral intensity function (recorded by a spectroscope or spectral analyser) can be defined as

$$i(\omega)=v(\omega)v^*(\omega)=a^2(\omega) \tag{2.8}$$

Since, by Parseval's theorem,

$$\int_{-\infty}^{+\infty} I(t)dt \qquad \int_0^\infty a^2(\omega)d\omega, \tag{2.9}$$

the total energy of a pulse is proportional to the area under either of the, temporal or spectral, intensity profiles.

This notation clearly shows the symmetry between the temporal and frequency descriptions of an optical pulse. In each case, the structure of the pulse is completely defined by a phase and an intensity, and there is a one-to-one correspondence between the two intensity profiles, $I(t)$ and $i(\omega)$. The general relationship between the two functions arising from the Uncertainty Principle, is

$$(\Delta\omega\Delta t)/2\pi \geqslant K \tag{2.10}$$

where K is a constant of the order of unity, whose value depends upon the shapes of the intensity profiles. The shortest pulse obtainable with a given spectral bandwidth is described as being "transform-limited" or "bandwidth-limited". In this case the duration is

$$\Delta t=2\pi K(\Delta\omega)^{-1}. \tag{2.11}$$

The value of $(\Delta\omega\Delta t/2\pi)$, called the "time-bandwidth product" P, is a most important parameter in the study of ultrashort laser pulses, and is used as a measure of the extent to which the ideal modelocked situation has been achieved with a particular system. A "bandwidth-limited" pulse can also be defined physically as a pulse completely devoid of amplitude or frequency modulation.

A better understanding of the degree of organisation produced by mode-locking may be obtained by considering the pictorial representation of the time and frequency descriptions given in Fig. 2.2 for the two cases of a non-mode-locked laser and a perfectly modelocked laser, respectively. In generating these diagrams it was assumed [2.13] that the laser was operating with 101 discrete longitudinal modes, of equal frequency separation $\delta\omega$. In the time domain, the field pattern then repeats itself with a periodicity of $2\pi(\delta\omega)^{-1}$, corresponding to the double transit-time of the laser resonator, [2.14], even when the longitudinal modes are largely uncorrelated. In this quasi-periodical intensity fluctuation pattern, the duration of the shortest fluctuation is of the order of the inverse bandwidth. Below the lasing threshold, the amplitudes and phases of the cavity modes fluctuate randomly as a result of the independence of the different spontaneously emitting sources. While these fluctuations become progressively less marked as the threshold is passed and stimulated emission begins to dominate, each mode still remains largely uncorrelated with its neighbours and the temporal pattern is still very similar to that of thermal noise. In obtaining the computer simulation of Fig. 2.2a the 101 spectral phases were chosen randomly in the range $-\pi$ to $+\pi$ and the spectral intensities were generated with a Rayleigh distribution about a gaussian mean.

To achieve modelocked operation of the laser it is necessary to introduce some device that will correlate the spectral amplitudes and phases. Then the perfectly modelocked situation, also shown in Fig. 2.2, is obtained. For simplicity the spectral intensity was chosen to have a gaussian distribution

$$i(\omega)=\exp[-(\omega-\omega_0)^2/\alpha], \tag{2.12}$$

and the corresponding temporal intensity profile is then also gaussian

$$I(t)=\alpha\exp[-\alpha(t-t_0)^2]. \tag{2.13}$$

For this combination of intensity profiles $K=(2\ln 2)/\pi=0.441$ in (2.10) and (2.11). With care this ideal situation, resulting in a train of isolated, transform-limited pulses, can be achieved in practice with pulses as short as 0.3 ps (see *Modelocked* cw *Dye Lasers*).

While techniques and procedures for achieving good modelocking have evolved empirically over the years, it was only with the development of the picosecond electron-optical streak camera [2.13], with the capability of recording $I(t)$ directly with adequate time resolution, that a detailed understanding of the processes involved was obtained. The absence of this direct

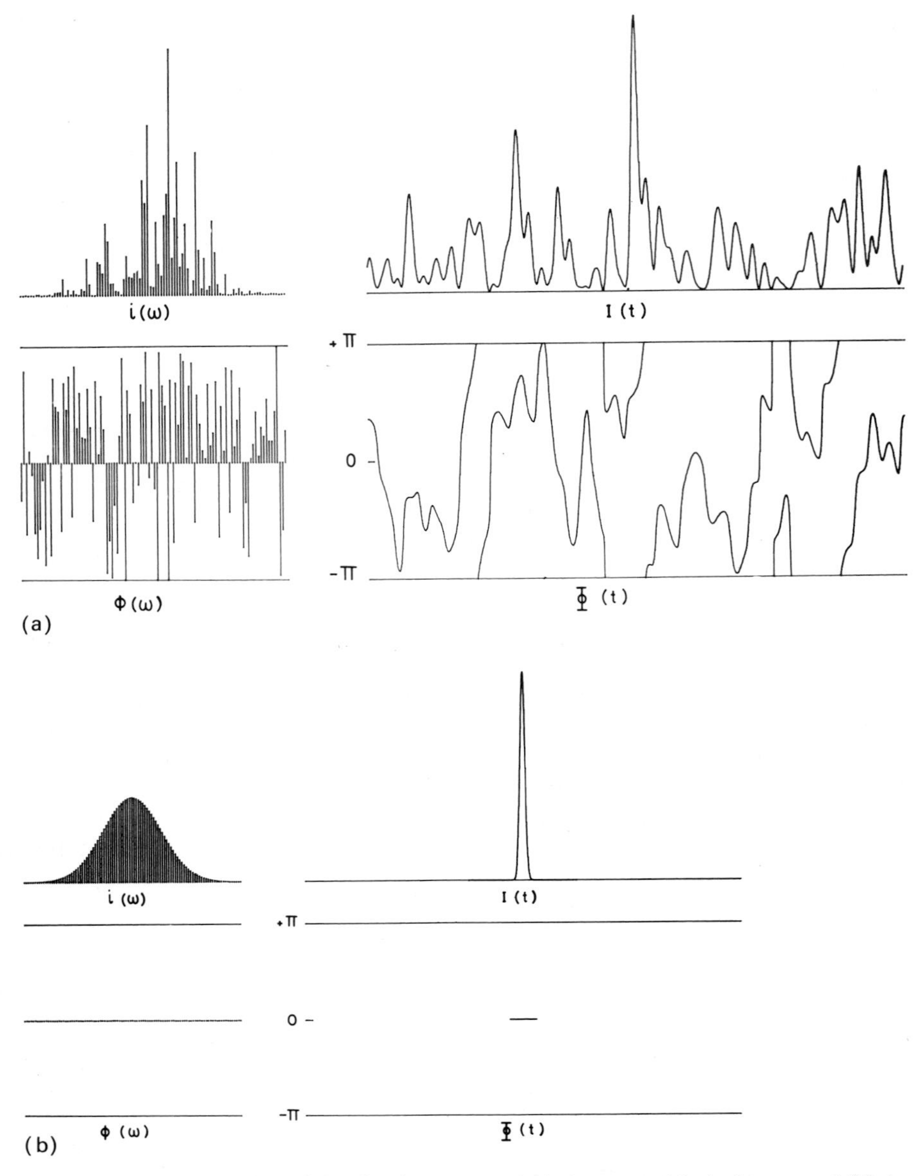

Fig. 2.2a and b. Simulation of the signal structure of (a) A non-modelocked laser, and (b) An ideally modelocked laser. In the frequency domain (left) in (a) the intensities $i(\omega)$ of the 101 discrete longitudinal modes have a Rayleigh distribution about a gaussian mean and the phases $\phi(\omega)$ are randomly distributed in the range $-\pi$ to $+\pi$. In the time domain (right) the phase $\Phi(t)$ fluctuates randomly, while the intensity $I(t)$ has the characteristics of thermal noise. In (b) the 101 spectral intensities have a gaussian distribution while the spectral phases are identically zero. In the time domain the signal is a bandwidth-limited gaussian pulse and the temporal intensity should be scaled up by $\times 20$ to correspond quantitatively with (a) (from *Bradley* and *New* [2.13])

measurement capability had resulted in erroneous reports of the generation of isolated picosecond and subpicosecond pulses by early experimenters, because they did not fully understand the precautions to be taken in applying the nonlinear correlation techniques then employed for pulse duration measurements. Because measurement of the temporal intensity profile $I(t)$ with picosecond time resolution has been the key to the understanding and development of modelocking techniques over the last decade, the state of the art of picosecond chronoscopy will be summarized before proceeding to a discussion of laser experiments.

2.1.2 Measurement of Pulse Intensity Profile I(t)

The simplest and most direct method of recording the temporal intensity profile of a light pulse is provided by the photodiode-oscilloscope combination. While the response time of the photoelectric effect is thought to be $\sim 10^{-14}$ s [2.15] the fastest real-time oscilloscopes have bandwidths of ~ 5 GHz, corresponding to a risetime of ~ 70 ps. Photodiodes with risetimes of ~ 100 ps and with current characteristics capable of driving such an oscilloscope are commercially available. As an example, a 10 GHz IMPATT silicon photodiode mounted in a coaxial cable has been employed to drive a 5 GHz oscilloscope with a sensitivity of 3 V cm^{-1} [2.16]. If the pulses to be measured are repetitive, as in cw modelocked lasers, and if real-time measurement of a single pulse is not required, pulse sampling techniques can be employed. In this case, the photodiode risetime limits the time resolution to 100 ps, since the oscilloscope sampling unit can have a risetime as short as 25 ps. Thus direct pulse measurements down to a time resolution of ~ 100 ps can be readily achieved. However, for experiments with modelocked lasers this performance is inadequate, and subpicosecond time resolution is required to diagnose presently available laser systems. This time resolution can only be achieved by nonlinear correlation methods or by electron-optical chronoscopy.

Two-Photon Fluorescence (TPF) Measurements

The nonlinear techniques for measuring the intensities of ultrashort pulses based upon the recording of the various correlation functions of $I(t)$ have been reviewed extensively. [2.13]. While the rapid development of ultrafast streak cameras has provided means for the direct recording of $I(t)$ with subpicosecond time resolution, nonlinear measurements are still widely employed, and most of the significant early results were obtained through their use. Since time measurements employing second-harmonic generation, especially of cw lasers, and the use of the optical Kerr-effect shutter are described in later chapters, it will be necessary to consider here only the two-photon fluorescence technique.

The widely employed two-photon fluorescence method [2.17] of ultrashort pulse duration measurement uses the experimental arrangement shown in

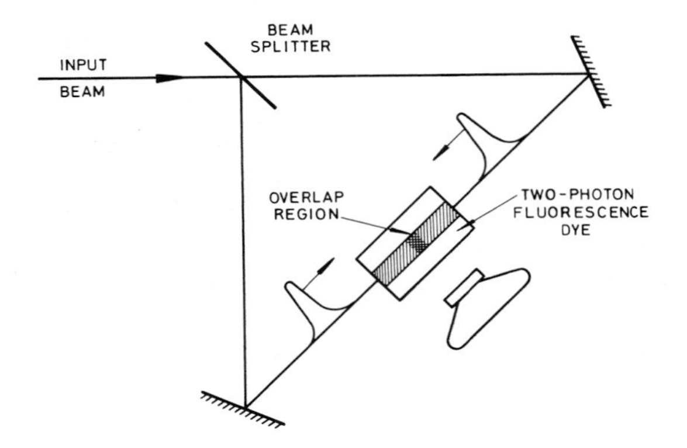

Fig. 2.3. The triangular configuration for the two-photon fluorescence (TPF) method of picosecond pulse duration measurement. The alignment of the optical triangle is critical if good results are to be obtained

Fig. 2.3. Each pulse of a train, or a single switched-out pulse, is split into two pulses of equal intensity which, in the usual triangular arrangement, overlap in a dye cell. The dye is chosen so that it can be excited to fluorescence by light of the laser wavelength only by two-photon absorption. The intensity of the fluorescence subsequently emitted is then proportional to the square of the intensity of the exciting radiation. At the centre of the dye cell the intensities of the two pulses add to enhance the fluorescence. The fluorescence intensity photographed at a distance Z from the midpoint of the dye cell is given by [2.18]

$$I(\tau)=A\left[\int_{-\infty}^{+\infty} I(t)^2 dt+2\int_{-\infty}^{+\infty} I(t)I(t+\tau)dt\right] \tag{2.14}$$

where A is a constant of proportionality and $\tau=2Z/c$. In deriving this relation, time averaging by the photographic process and spatial averaging over several optical wavelengths are assumed. This pulse duration measurement technique has the advantage of directly relating the spatial distribution of the fluorescence to the second-order correlation function $G^2(\tau)$. Eq. (2.14) can be rewritten as

$$I(\tau)=A\left[G^2(0)+2G^2(\tau)\right] \tag{2.15}$$

with $G^2(\tau)=\int_{-\infty}^{+\infty} I(t)I(t+\tau)dt.$ (2.16)

The complete pulse autocorrelation profile is therefore displayed simultaneously as a function of distance. Since $G^2(\tau)$ becomes zero away from the region of overlap, it is easy to see that the peak to background contrast ratio R of the two-photon fluorescence (TPF) pattern, which is given by

$$R\equiv I(0)/I(\infty) \tag{2.17}$$

has the value 3. When the laser radiation is not a series of single, isolated pulses, but is simply a burst of narrow-bandwidth radiation noise, there will still be a peak at $\tau=0$ in the TPF pattern, due to the overlap of individual fluctuations with themselves. It has been shown [2.19, 20] that for noise fluctuations the contrast ratio has the value 1.5. For a short burst of noise (*not* a single transform limited pulse) while the peak to background contrast ratio is 3:1, the TPF pattern has a broad base representing the total duration of the burst, with a fine "noise spike" superimposed. Fig. 2.4 was obtained from the first photographs [2.21] of TPF profiles showing this feature. In these experiments the width of the fine spike varied from 0.2 ps to 1 ps but was always correlated with the simultaneously recorded neodymium:glass laser spectral widths of 150 Å to 30 Å. TPF records of single pulses of durations 3 ps, with the theoretical contrast ratio of 3:1, were also recorded (see also Chapt. 3).

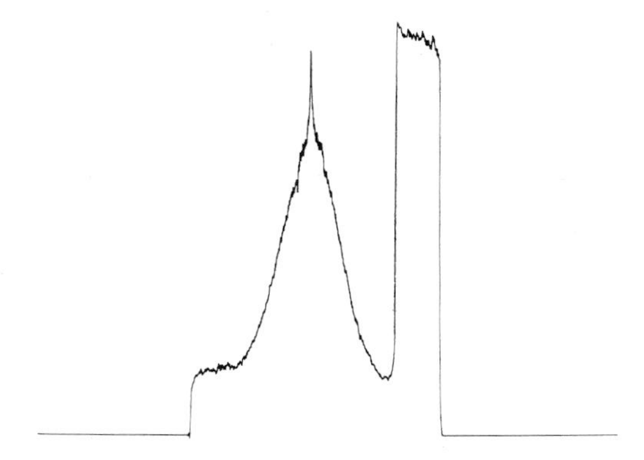

Fig. 2.4. Microdensitometer trace of the TPF pattern obtained from a mode-locked Nd:glass laser (*Bradley* et al. [2.21]). The step on the right indicates a contrast ratio of 2.6. The width of the fine spectral spike is ~0.2 ps

Early reports of subpicosecond pulses from neodymium:glass lasers and of ~5 ps pulses from ruby lasers were based upon TPF patterns, in which only the noise spike was measured. Unfortunately, this led to overoptimistic expectations for the use of these lasers in experiments involving the measurement of picosecond phenomena. The situation was further complicated by the effect of fluorescence quenching under excitation by intense laser radiation [2.22, 23], which leads to a reversal of the TPF "noise spike". Moreover, the nonlinear TPF display technique does not uniquely define the shape of the laser pulse, and the presence of a substantial proportion of the laser energy outside the ultrashort pulses may not be detected. Finally, low intensity pulses cannot be measured. These defects have been overcome by the development of direct photoelectric picosecond chronoscopy with electron-optical streak cameras.

Electron-Optical Picosecond Chronoscopy

The method of studying rapidly varying luminous phenomena by electron-optical chronoscopy was first proposed in 1956 by *Zavoiskii* and *Fanchenko* [2.15], who pointed out that the time resolution of an image tube streak camera is ultimately limited by the spread of the photoelectron transit times in the first image tube. This spread arises from variations in the initial velocities of the photoelectrons and is mainly developed close to the photocathode, where the photoelectrons are moving slowly, but for x-ray and xuv tubes, and in subpicosecond operation at longer wavelengths, significant time dispersion occurs throughout the photoelectron path (see below).

Photoelectron time dispersion in the image tubes available for use in high-speed streak cameras up to 1970 limited the time resolution to greater than 50 ps. Problems also arose from image distortion and loss of spatial resolution resulting from high photocurrent densities inside the streak tubes [2.24] and it was not clear if picosecond imaging could be obtained. In 1969, picosecond exposures without image distortion were obtained with a gated four-stage cascade image intensifier tube by *Bradley*, et al. [2.25] and the same year it was realised by *Bradley* [2.26] that this time-dispersion limitation could be overcome by the simple expedient of locating a planar, fine-mesh, high-potential electrode close to the tube photocathode. With an otherwise standard streak tube altered in this manner, time resolution was improved to ~5 ps, employing second-harmonic pulses generated by a modelocked Nd:glass laser as test sources [2.27]. Soon after, by applying 20 ns high voltage pulses to the shutter grid of a conventional RCA streak image tube, with a S-1 photocathode, *Schelev* et al. [2.28] measured the durations of the ~10 ps pulses from a Nd:glass laser.

Photoelectron Time Dispersion. Time dispersion occurs over the whole of the photoelectron pathlength in an electron-optical streak tube (Fig. 2.5). The time taken for a photoelectron of initial energy eV_0 in electron volts, emitted normal to the photocathode, to travel a distance x along the tube axis is given by

$$t=(m/2e)^{\frac{1}{2}}\int[V_0+V(x)]^{-\frac{1}{2}}dx. \tag{2.18}$$

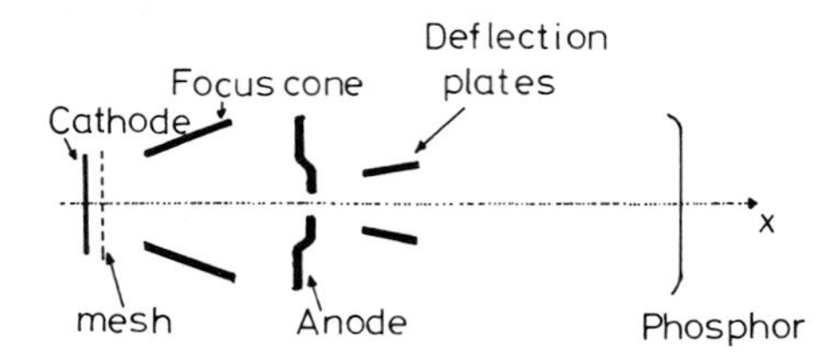

Fig. 2.5. Electrode arrangement and photoelectron transit-time dispersion in electron-optical streak image tube. Photoelectron transit-time to deflection plates is

$$t=\left(\frac{m}{2e}\right)^{\frac{1}{2}}\int_{\text{cathode}}^{\text{def.}}\{V_0+V(x)\}^{-\frac{1}{2}}dx$$

where

V_0 initial photoelectron energy
$V(x)$ axial potential distribution of tube

$V(x)$ is the axial potential distribution for the tube and e and m are the electronic charge and mass, respectively. If $V(x)$ varies linearly with distance x from the photocathode, then (2.18) can be written as

$$t=(2m/e)^{\frac{1}{2}}[(V_0+V)^{\frac{1}{2}}-V_0^{\frac{1}{2}}]E^{-1}. \tag{2.19}$$

E is the uniform axial electric field strength and t is now the time taken to reach a point of potential V. When $V_0 \simeq \Delta V_0 \ll V$ (which is true for all systems employed to date) the well-known result for the dispersion spread of the transit times Δt_{D} for a spread in initial energies $e\Delta V_0$

$$\Delta t_{\mathrm{D}} \cong (m\Delta V_0/2e)^{\frac{1}{2}}E^{-1}, \tag{2.20a}$$

is obtained [2.13]. Expressing the initial photoelectron energy spread in terms of velocity ΔU, (2.20a) can also be written as

$$\Delta t_{\mathrm{D}}=m\Delta U/eE. \tag{2.20b}$$

The electric field strength E near the photocathode must clearly be maximized so that the transit time spread is minimized, since the quicker the electrons

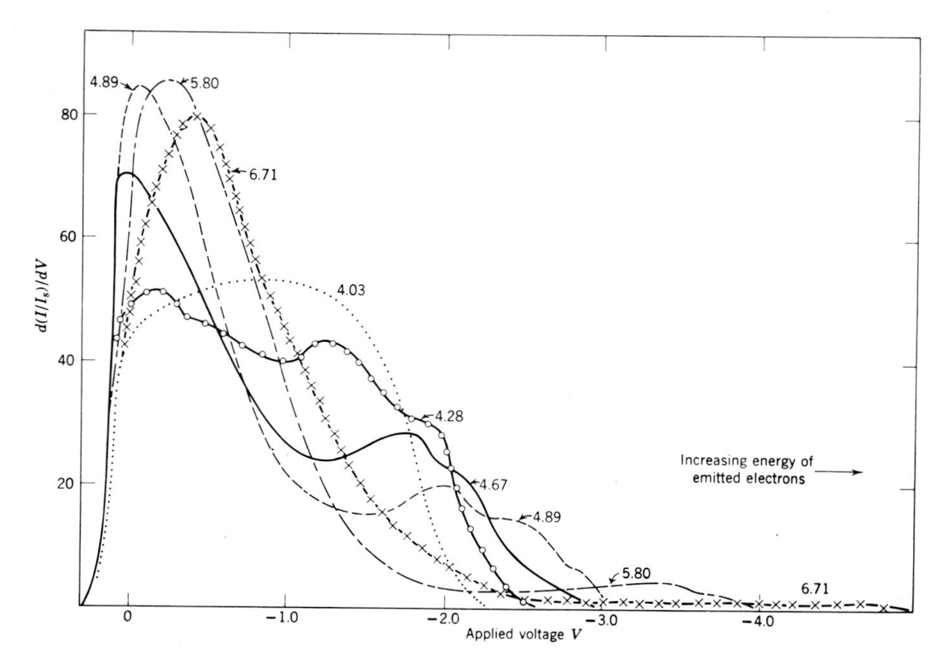

Fig. 2.6. Energy distribution of photoelectrons emmitted by Cs_3Sb photocathode for different photon energies (numbers labelling the curves are in eV) (from *Bradley* [2.56])

are accelerated to a high velocity, the faster they "forget" their small initial differences in energy. The value of ΔU for a particular photocathode depends on the wavelength of the illuminating light. As an example, for an $S1$ photocathode and 1060 nm light (the wavelength of the Nd:glass laser) the spread in photoelectron energies has a half-width of ~ 0.3 eV, leading to a value of 2.3×10^7 cm s^{-1} for ΔU. To obtain a time resolution of 1 ps in this case requires a photocathode extraction field $E = 13$ kV cm^{-1}. For light of wavelength 600 nm the energy spread is ~ 1 eV so that the time resolution limit at this wavelength is increased to ~ 2 ps for the same photocathode extraction field. Energy spreads of up to ~ 2 eV are produced by other photocathodes with uv illumination (Fig. 2.6). The ultimate time resolution of a streak camera then depends both on the type of photocathode and the wavelength of the light, as well as on the extraction field strength. Thus to construct a camera with picosecond time resolution throughout the spectrum from the vuv to the near ir, extraction fields in the region of 20 kV cm^{-1} were needed. This performance was first demonstrated with the second-generation Photochron II streak tube [2.29, 30] which also has subpicosecond resolution in selected wavelength regions depending upon the photocathode type.

Streak-Camera Systems. The essential components of a streak-camera system and the experimental arrangement used for testing its performance are shown in Figs. 2.7 and 2.8, respectively. The availability of frequency-tunable mode-locked dye lasers [2.31] capable of reliably producing pulses of durations $\leqslant 2$ ps permitted the direct measurement of the resolution of the new streak

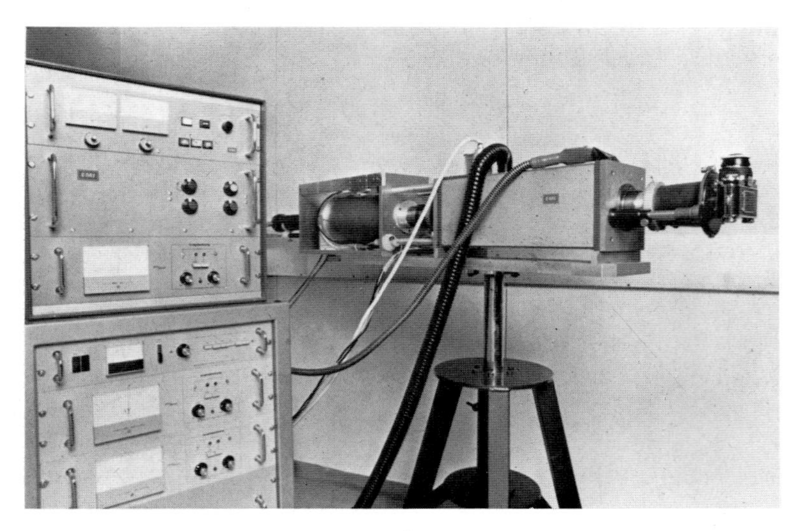

Fig. 2.7. Photograph of Photocron streak camera showing the power supplies, the streak-tube mounting, the four-stage magnetically focused image intensifier and the recording camera

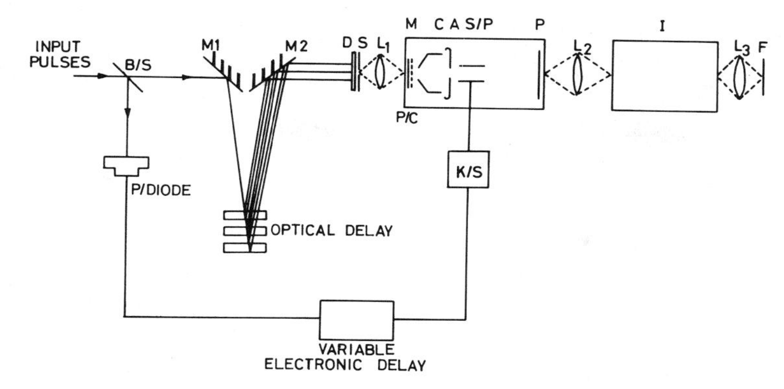

Fig. 2.8. Experimental arrangement for measuring picosecond pulse durations with the Photochron streak camera (from *Bradley* [2.56])

tubes [2.32]. An optical delay line arranged (Fig. 2.8) to produce, from a single laser pulse, a series of identical pulses with appropriate separations greatly facilitates calibration of the streak timescale. After an appropriate, variable, electronic delay, the photodiode pulse triggers the krytron circuit, which generates the streak deflection voltage ramp. For a streak length of 5 cm a deflection voltage of 1.5 kV is required. Writing speeds (at the streak-tube phosphor) of $>2\times10^{10}$ cm^{-1} with a jitter of ± 50 ps can be obtained (Fig. 2.9).

Fig. 2.10 shows a typical streak photograph. From microdensitometer traces, such as that of Fig. 2.11, recorded pulsewidths, incorporating both camera response time and laser pulsewidth, as short as 1.5 ps were obtained, thus confirming the subpicosecond resolution capability of the camera. To a good approximation the recorded pulsewidth may be expressed as

$$\Delta t_R = \sqrt{(\Delta t_s)^2 + (\Delta t_D)^2 + (\Delta t_P)^2} \tag{2.21}$$

where Δt_s is the time resolution limit arising from the finite spatial resolution of the overall streak-camera electron-optical and optical systems at a particular streak writing speed, and Δt_P is the laser pulse duration. Normally, the slit-width is set so that its effect on the spatial resolution is negligible. Substituting (2.20b) in (2.21) yields

$$\Delta t_R = [(\Delta t_s)^2 + (\Delta t_P)^2 + k^2 E^{-2}]^{\frac{1}{2}} \tag{2.22}$$

where $k = m\Delta U/e$. In the first experiments with a picosecond streak camera it was confirmed that Δt_R depended linearly on E^{-2} [2.27] and that Δt_D was a linear function of E^{-1}.

As an example the use of (2.21), in the first measurements to directly demonstrate subpicosecond time resolution [2.30] with the experimental arrange-

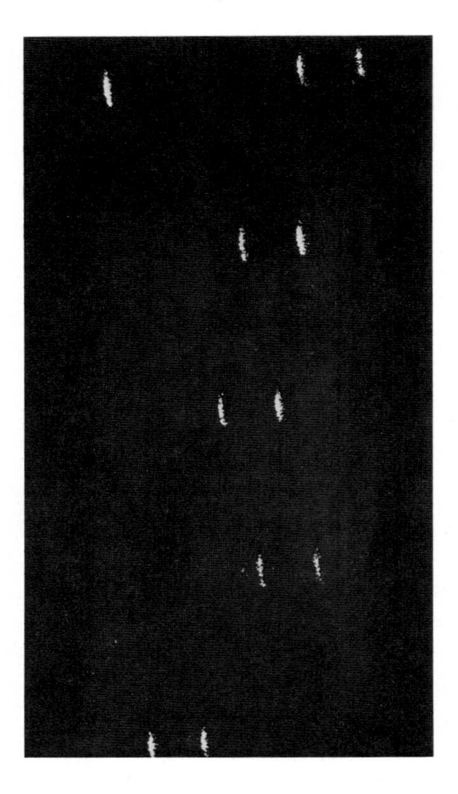

Fig. 2.9. Successive recordings of streaks of pulses from flashlamp-pumped modelocked dye laser showing jitter of ~50 ps. In each streak record the pair of pulses (generated from a single pulse by reflection from the two faces of an optical flat of appropriate thickness) has a separation of 60 ps

ment shown in Fig. 2.12, will be described in some detail. Six pulses from the centre of the pulse train from a rhodamine 6*G* dye laser, modelocked by DQOCI, were selected out by a Pockels cell optical-switch and amplified to peak powers of ~300 MW. After amplification the pulses were focused into a Raman cell containing ethanol. Transient stimulated Raman scattering from the C–H stretching vibration generates a Stokes frequency at 733.7 nm. The transverse relaxation time of this vibration is ~0.3 ps, and for a laser pulse of duration ~1.5 ps the Stokes pulse generated close to threshold power conditions is shortened [2.34]. The Raman Stokes pulses were transmitted through filters to remove the dye-laser pumping pulses, and two pulses separated by 60 ps were generated from each Raman pulse by reflection from a quartz plate. From the microdensitometer trace (Fig. 2.13) a total recorded duration of 900 fs (900×10^{-15} s) was measured. For pulses of durations as short as this, significant time dispersion occurs outside the photocathode to the mesh region of the streak tube and it is no longer sufficient to employ the approximation of (2.20b) in deriving the time resolution of the camera. Instead (2.20a) must be used to calculate the dispersion over the total photoelectron pathlength from photocathode to anode [2.35]. At a wavelength of 733.7 nm the time-

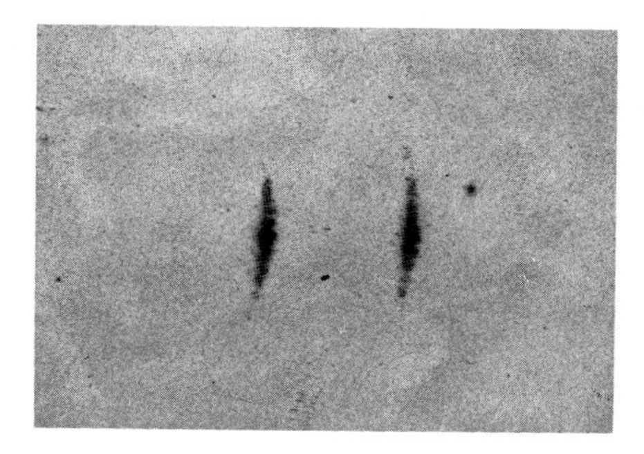

Fig. 2.10. Streak photograph, taken with Photocron II camera, of two pulses separated by 60 ps generated from a pulse (λ 605 nm) from a rhodamine 6G dye laser modelocked by DODCI (from *Bradley* [2.56])

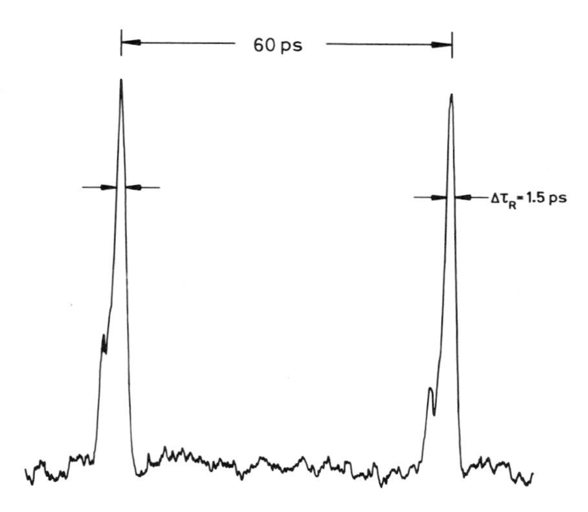

Fig. 2.11. Microdensitometer trace (linear optical density) of a pair of streak images of a pulse (λ 605 nm) from a flashlamp-pumped rhodamine 6G dye laser modelocked by DQOCI, demonstrating a total recorded pulse duration of 1.5 ps and confirming the subpicosecond time-resolution limit of the streak camera (from *Bradley* [2.56])

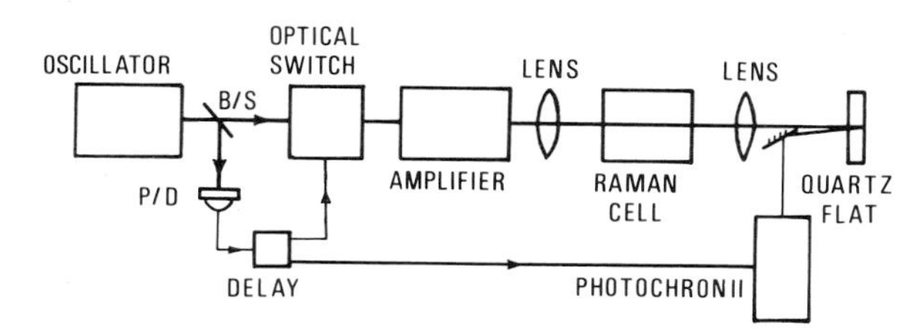

Fig. 2.12. Block diagram of arrangement for generation of subpicosecond pulses by transient Raman scattering in ethanol of 300 MW picosecond pulses from rhodamine 6G dye laser oscillator-amplifier system

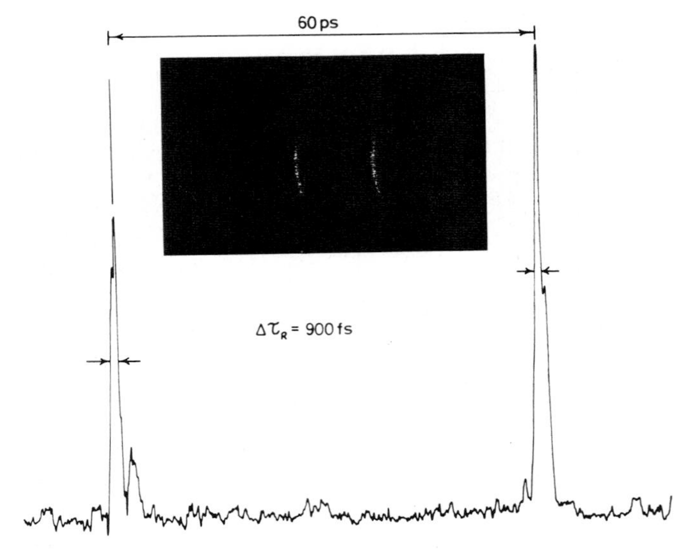

Fig. 2.13. Streak photograph and corresponding microdensitometer trace (arbitrary linear density scale) of Raman-Stokes pulses at 733.7 nm showing recorded half-width of 900 fs

dispersion spread between the photocathode and the mesh is 380 fs, and time dispersion in the mesh to anode region of the tube adds 120 fs, to give a total time-dispersion resolution limit of 500 fs. At a phosphor-screen writing speed of 2×10^{10} cm s^{-1} and with a dynamic spatial resolution of better than 10 line pairs/mm a total camera instrumental resolution of 700 fs is obtained. Using (2.21) to deconvolve this value from the measured pulsewidth of 900 fs gives a Raman pulse duration of 570 fs.

Other picosecond streak-camera systems have recently been developed in the USSR. *Basov* et al. [2.36] and *Fanchenko* and *Frolov* [2.37] employed the newly developed "Picochron" streak camera also to study the structure of the pulses from a modelocked Nd:glass laser. By adding an extra electrostatic lens, an electric field strength of up to 6 kV cm^{-1} at the photocathode is produced in the "Picochron" image tube, giving a limiting time resolution of $\sim$2 ps. An x-band resonance-cavity deflection system continuously sweeps a point image along an elliptical path on the output phosphor. While the resulting total recording time of 15 to 20 ns simplifies sweep timing and synchronisation problems, the use of a point image instead of a slit image seriously reduces the information content per time-resolution element. The facility for picosecond time-resolved spectroscopy or source spatial variation studies is also obviously lost. Another design of slit-streak image tube based on the fine-mesh accelerating electrode principle of the Photochron tubes was introduced in 1972 [2.38] with an electric field strength near the photocathode up to 60 kV cm^{-1} and an experimentally demonstrated time resolution of $\sim$1 ps.

With a new deflection electrode arrangement, the deflection sensitivity was improved to become comparable with that of the Photochron tubes so as to achieve subpicosecond resolution [2.39].

xuv and x-Ray Streak Cameras. The study of high-density, high-temperature plasmas for laser fusion [2.40] and the development of coherent light sources in the vuv and xuv [2.41–43] spectral regions require the extension of electron-optical chronoscopy to these shorter wavelengths, with temporal resolution in the picosecond range. This has been achieved by employing gold photocathodes in tubes developed for operation at longer wavelengths. The first camera systems had photocathodes operating with front-surface emission [2.45] and in the transmission mode [2.46–48]. The measured time-resolution limits of these devices were restricted to 60–150 ps by the durations of the laser plasma x-ray pulses employed as test sources. Later experiments [2.35] with short-pulse x-ray emissions produced by 10 ps pulses from a Nd:glass laser demonstrated structures as short as 20 ps. The experimental arrangement is shown in Fig. 2.14. A single pulse from a modelocked laser oscillator was intensity divided by reflection from a double-mirror configuration to give multiple pulses of variable but known separation. When amplified up to energies of ~100 mJ these pulses were focused successively on to a plane copper target to generate a plasma of diameter 100 μm, emitting recombination continua and line radiation up to photon energies of ~1 keV. X-rays in a broad band of energy around 1 keV, selected by an Al foil filter, passed through a tapered slit of width varying from 7 μm to 57 μm to project a magnified shadowgraph image of the slit on to the gold photocathode of the streak tube, at a glancing

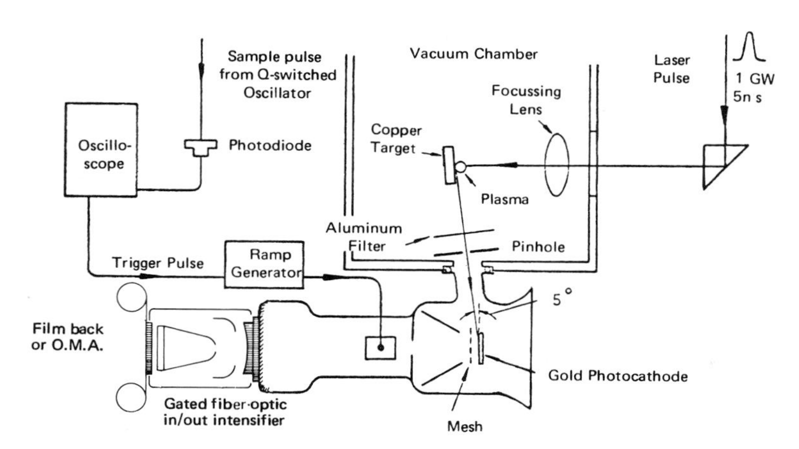

Fig. 2.14. Schematic diagram of x-ray streak camera and laser plasma target chamber. (1) Focusing lens, 12 cm *f*/3 (2) plane Cu target (3) plasma x-ray source (4) Al filter (5) tapered slit (6) gold photocathode (7) accelerating mesh (8) phosphor screen (9) photodiode (10) streak deflection plates (from *Bradley* et al. [2.35])

angle of 5°. The use of the tapered slit gave a continuously varying photographic exposure. The photocathode field strength was 6 kV cm^{-1} and streak writing speeds of up to 2×10^9 cm s^{-1} were used. Fig. 2.15 is a x-ray streak photograph when the incident laser pulses were separated by 66 ps and the clear resolution is evident from the 70% depth of modulation of the corresponding microdensitometer trace. The shortest recorded pulsewidths were 22 ps. Because of the relatively large energies of the photoelectrons ejected by x-rays there is an associated large spread in photoelectron energy. Using (2.18) the total transit time spread for photoelectrons emitted with different initial energies up to 1 keV was calculated, employing an approximate axial

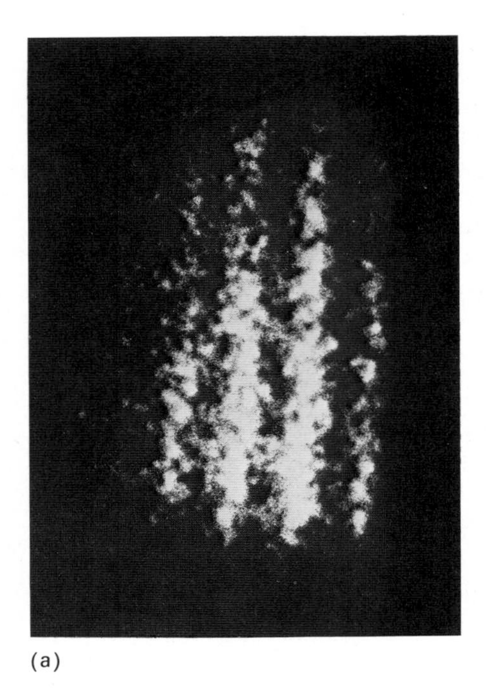
(a)

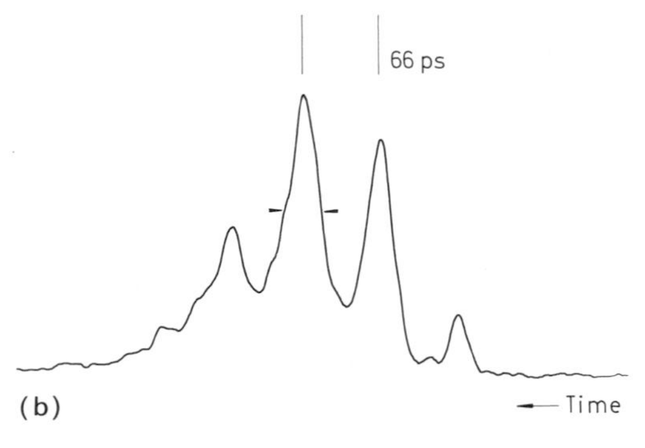

(b)

Fig. 2.15a and b. (a) Streak photograph of x-ray pulses generated by the arrangement of Fig. 2.14 when the laser separation was 66 ps. (b) Microdensitometer trace (linear density scale) of (a). Arrows indicate the half peak intensity level

distribution of the streak-tube potential. Since, from (2.19), the recorded pulse duration will depend upon the value of E and hence on the mesh potential, streak records were obtained for various values of this potential. Knowing the spatial resolution contribution to the recorded pulsewidth, the residual pulse duration, $[(\Delta t_p)^2+(\Delta t_s)^2]^{\frac{1}{2}}$, was calculated and the best fit to the experimental points was obtained for $eV_0=60$ eV, $e\Delta V_0=30$ eV and $\Delta t_p=25$ ps. Then the time dispersion in the photocathode-mesh region amounts to ~ 8 ps, whereas the total dispersion in the tube is 20 ps. Recent direct measurements [2.44] of the photoelectron energy distribution for a gold photocathode irradiated by 1.5 keV x-rays gave the values $eV_0=1.1$ eV and $e\Delta V_0=3.8$ eV for the secondary photoelectrons. These values would indicate a dispersion-limited time resolution of ~ 6 ps for the tube of [2.35].

These results emphasize the need to properly evaluate all contributions to time dispersion in x-ray and xuv streak cameras when estimating the time resolution limit. Otherwise a too optimistic estimate may be obtained [2.47, 48]. In a recently developed x-ray streak camera the buildup of time dispersion has been considerably reduced by employing a proximity-focused image tube [2.49]. A microchannel plate operates both as an accelerating mesh electrode (to eliminate time dispersion close to the photocathode as in the Photochron tubes) and as a passive collimator, (to replace the usual electrostatic focusing lens). The overall tube length is considerably reduced and the calculated time-resolution limit is ~ 3 ps. Similar results, but with higher spatial resolution should be achievable with magnetically focused tubes [2.50].

Recent technical improvements in picosecond streak cameras include the use of microchannel image intensifiers and fibre-optical coupling (Fig. 2.16) to increase the overall system gain and to provide a more compact and versatile instrument [2.51, 52]. The replacement of photography by optical multi-

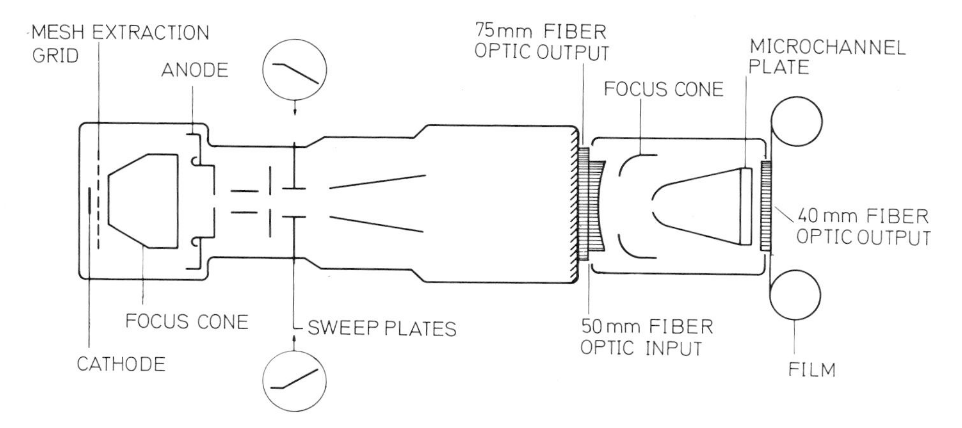

Fig. 2.16. Details of picosecond streak camera system employing fibre-optical coupling (Imacon 675 and 675/II models, Hadland Photonics Ltd.), based upon the Photocron streak tube

channel analysers gives an immediate linear digital record of $I(t)$ [2.53, 54]. The shortest subpicosecond pulses are obtained from cw modelocked dye lasers (see *Modelocked* cw *Dye Lasers*, see p. 54) with peak powers of 10^2 to 10^3 W. While streak cameras are capable of recording these pulses directly, for studies, for example, of the fluorescence generated repeatedly by such pulses, synchroscan operation of the streak camera [2.55] permits the accumulation of time-resolved data. In this mode of operation the streak plate deflecting voltage is driven synchronously at the repetition frequency ($\sim$100 MHz) of the modelocked pulse train, so that successive streaks are superimposed on the tube phosphor. The use of digital recording and storage then promises a very convenient system for the study of repetitive phenomena at low light levels. Instrumentation for electron-optical picosecond chronoscopy is still developing very rapidly [2.56]. However with direct linear recording, not only of laser pulses but also of luminous phenomena, over a wide range of photon energies from 1 eV to 10 keV with picosecond time resolution or better, it is clear that a major revolution in chronoscopy has already taken place.

2.2 Types of Modelocked Lasers

The various laser systems with which picosecond pulses have been reliably produced can be divided into two main classes: *i*) giant pulse lasers, such as Nd:glass and ruby lasers, and *ii*) continuous, or quasi-continuous systems, of which the dye laser is the outstanding example.

2.2.1 Giant-Pulse Lasers

Picosecond pulses were first obtained from ruby [2.1] and Nd:glass [2.2] lasers, Q-switched and modelocked by saturable-absorber dye solutions inside the laser resonators. The evolution of ultrashort pulses in these lasers has since been extensively studied both theoretically and experimentally. A typical experimental arrangement is shown in Fig. 2.17. Investigations have shown that the following conditions are necessary to achieve reliable modelocking in giant-pulse lasers.

i) Thermal distortion in the laser rod should be minimized or corrected for, and the laser should operate in a low order transverse mode. These conditions are achieved by employing a spherical mirror with an ancillary corrector lens to produce a generalised confocal cavity [2.57–59] or by filtering the pump light to reduce thermal distortion [2.60]. In either case an intracavity aperture is used to control the transverse mode structure.

ii) Spurious reflections inside the laser resonator and feedback from outside surfaces should be avoided. This is achieved either with a Brewster-angled laser rod and wedged resonator components or with antireflection coatings [2.18, 21, 61]. Outside the cavity lenses of long focal length are useful in isolating other experimental equipment from the laser oscillator [2.62].

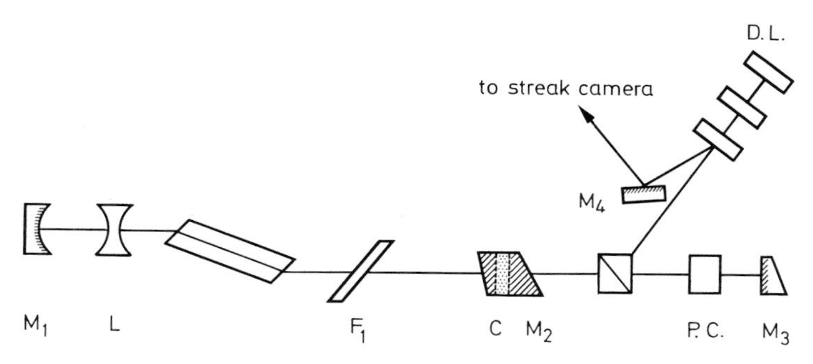

Fig. 2.17. Passively modelocked Nd:glass oscillator. L = thermal distortion corrector lens. C = optically contacted saturable-absorber cell. PC = pockels cell. D.L. = optical delay line. F_1 = flashlamp light filter (from *Bradley* and *Sibett* [2.58])

iii) The saturable absorber dye should be contained in a thin cell in optical contact with one of the laser mirror surfaces [2.60, 62].

iv) The laser should be operated close to threshold and at a high value ($\gtrsim 70\%$) for the low light level transmission of the absorber cell [2.58, 59, 63].

Because of the intrinsic randomness of the passive modelocking process, a completely modelocked pulse train is not produced with a success rate much in excess of $\sim 80\%$ even with these precautions, particularly in the case of the broad-bandwidth, inhomogeneously broadened Nd:glass laser.

It is now well known [2.58, 59, 60, 64] how the shape and duration of the ultrashort pulses generated by a modelocked Nd:glass laser vary along a pulse train. At the beginning of a train the pulses have smooth intensity profiles and durations of 4 to 10 ps. When special care is taken [2.58, 60] bandwidth-limited pulses of 3 to 4 ps are obtained at the beginning of the train. As an example, Fig. 2.18 shows the distributions of pulse durations at two points of the trains emitted by the arrangement of Fig. 2.17. In this system the saturable absorber dye (Eastman Kodak 9860 in dichlorethane) was contained in a cell in contact with the 70% reflectivity plane mirror M_2. (Low light level transmission of the cell was 70%). About ten pulses were switched out of the train by a Pockels cell shutter PC and particular pulses within this group were studied with the streak camera. When the fifth pulse of the train was streaked, pulse durations of 3 to 5 ps were consistently obtained for an absorber dye cell thickness of 50 μm. The streak photograph of Fig. 2.19 shows two images of a single pulse obtained by employing an optical delay line (D.L. of Fig. 2.17) to subdivide each pulse into multiple pulses of predetermined separations. The microdensitometer trace shows a clear separation, as expected from the ~ 2 ps time resolution of the camera. The limiting value of the time-bandwidth product, $\Delta\nu\Delta t = 0.5$, was obtained for several 3 ps pulses. Similar results were obtained by *von der Linde* [2.60] who used careful TPF measurements of the pulse durations. Both investigations demonstrate the advantage of using an absorber dye cell in contact with a laser mirror for reliable generation of

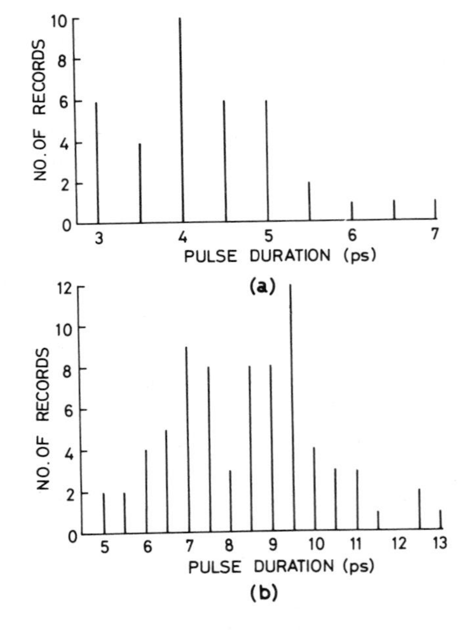

Fig. 2.18a and b. Histogram giving the statistical distribution of the duration of the pulses generated by the modelocked Nd:glass oscillator arrangement of Fig. 2.17 (a) 90 ns from start (10 pulses along train), (b) 500 ns from the start

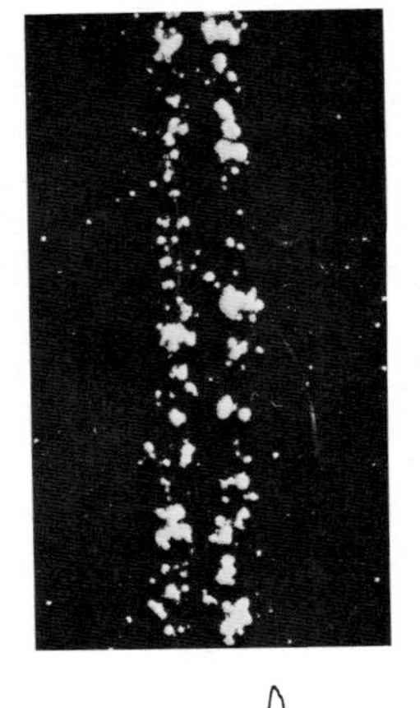

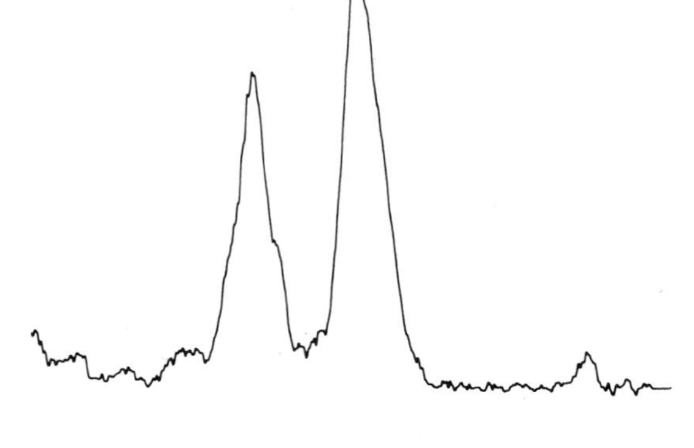

Fig. 2.19. *Top:* Streak photograph of two 3 ps pulses, separated by 10 ps, generated in the optical delay line of Fig. 2.17 from a single pulse. *Bottom:* Microdensitometer trace

ultrashort pulses. The streak-camera measurements also confirmed an earlier report [2.62] that in this mode of operation the shortest pulses are obtained with the shortest dye cell.

Further along the pulse train the spectral bandwidths of the pulses increase rapidly and substructures develop in both the pulse spectra and in the temporal intensity envelopes. This spectral broadening arises from self-phase modulation [2.61, 65–67] caused by the nonlinear interaction of the intense light field with the glass rod. With increasing self-phase modulation the spectral broadening becomes so large that the pulse spectral bandwidth exceeds the amplification bandwidth of the laser material. This results in breakup of the pulse into multiple components, quickly followed by degradation into a burst of noise fluctuations [2.68]. Earlier TPF measurements [2.69] of the cross-correlation of each pulse of a train with its two nearest neighbours had shown that the fluctuation structure of $I(t)$ changed from one round-trip to the next, while the time-integrated autocorrelation function of the whole train exhibited the fine central spike typical of a pulse of random fluctuations (Fig. 2.20). Self-focusing, also arising from the nonlinear refractive index, deviates the radiation out of the laser resonator so that a dip appears in the ultrashort pulse profile [2.70]. Substructure continues to build up as a result of repeated passages through the active medium [2.63, 71]. The effect of self-focusing can also be seen in the oscillogram of the pulse train by a dip in the train envelope, which is characteristic of a well modelocked laser operating at high pulse energies (Fig. 2.21). From the practical point of view these nonlinear effects can be avoided by switching out pulses from the beginning of the pulse train. However, they seriously limit the power density that can be propagated in ultrashort pulse amplifiers. The introduction of a two-photon absorber [2.57, 72] dye cell into the laser cavity greatly reduces the amount of self-focusing. A similar effect has also been obtained by employing intracavity second-harmonic generation [2.73]. In both cases the induced nonlinear absorption limits the

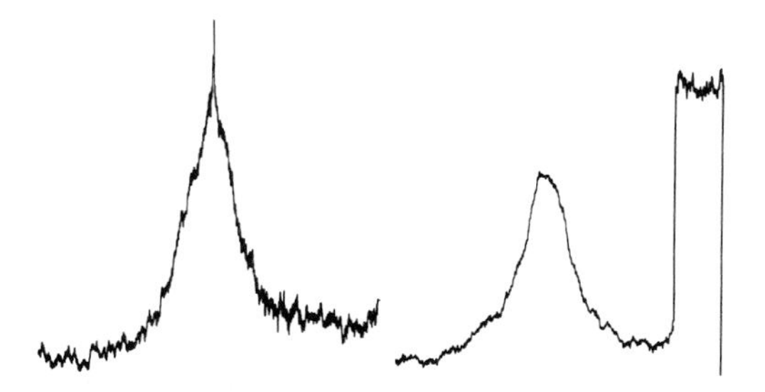

Fig. 2.20. Microdensitometer traces (linear density scale) of the TPF profiles obtained with a modelocked Nd:glass laser. The time-integrated autocorrelation profile (left) of the whole pulse train exhibits the fine central spike similar to that of Fig. 2.4. The spike does not appear in the cross-correlation profile (right) indicating that the pulse temporal substructure changes from one round trip to the next (from *Bradley* et al. [2.62])

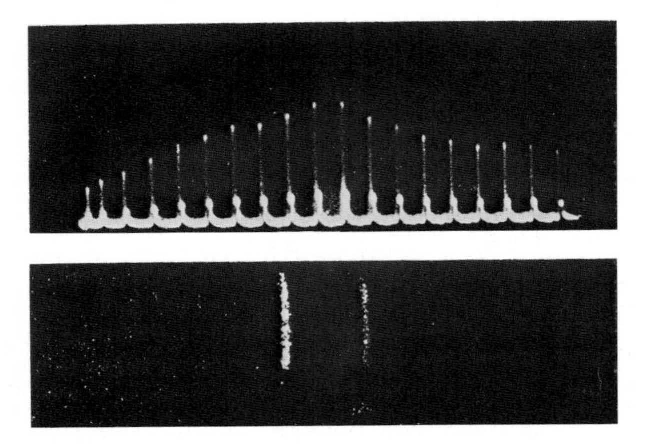

Fig. 2.21. *Top* Oscillogram of modelocked pulse train from Nd:glass laser showing dip in intensity due to self-focusing effects. *Bottom* Streak-camera record of pulse from the beginning of the train before self-action effects arise

peak pulse intensity, leading to more reliable modelocking with better spatial properties in the beams.

Similar pulse trains have been obtained by employing the same experimental arrangements with passively modelocked ruby [2.23, 57, 74, 75] and Nd:YAG [2.54, 71, 76, 77] lasers. Because of the narrower fluorescence bandwidths of these materials the pulse durations lie in the range of 15 to 50 ps and, hence, are too long for many interesting scientific applications of ultrashort laser pulses (see later chapters).

Because of variations in pulse duration throughout the train and because of transient effects in many materials, a single pulse must often be selected for measuring accurately rapid processes in materials [2.60]. Single pulse selection [2.57, 78–84] may be accomplished by passing a train of pulses through a Kerr cell or a Pockels cell that has been activated electronically for a short time, in order to change the polarization of only one of the pulses. To generate the high voltages necessary to rotate the polarization in the Pockels cells, a high voltage spark gap or an all electronic circuit may be used [2.30, 84]. To trigger the circuits, a small fraction of the laser beam is diverted, and by adjusting the optical and electronic delay times, a pulse may be selected from any portion of the train. Experimenters usually choose earlier pulses in the train because of their shorter durations, and cleaner envelopes.

Before discussing the application of the fluctuation model of a passively modelocked laser to giant-pulse systems, the corresponding results obtained with modelocked dye lasers will be described. Measurements on dye laser systems have elucidated more clearly the significance of absorber recovery time and gain saturation in the generation of picosecond pulses by passive modelocking. Comparison between the ruby laser and the cresyl-violet dye laser, operating at the same wavelength and modelocked by the same saturable absorbers, has proved to be particularly illuminating.

2.2.2 Dye Lasers

Flashlamp Pumped Systems

While ultrashort pulses were first generated in dye lasers by synchronous pumping with pulse trains from modelocked ruby lasers [2.85, 86] and, after second-harmonic generation, Nd:glass lasers [2.87, 88], the shortest pulses were obtained from passively modelocked systems [2.89]. *Schmidt* and *Schafer* [2.90] had shown that the saturable absorber DODCI[1] could be employed to modulate the output intensity profile of a flashlamp-pumped rhodamine 6G dye laser. This result encouraged other workers to attempt to generate picosecond pulses in this manner, particularly since these pulses could be frequency tuned across the broad spectral bandwidths of the laser dyes. Tunable picosecond pulses were obtained with both rhodamine 6G and rhodamine B lasers [2.91] by employing a diffraction grating as one laser resonator reflector, and immersing the output mirror in the DODCI solution. Two-photon fluorescence measurements showed that in these initial experiments pulse durations of ~ 5 ps were reliably produced. The introduction of interferometric tuning [2.86] permitted the generation of pulses of bandwidth-limited durations, frequency tunable from 580 to 700 nm [2.92, 93] by employing rhodamine and cresyl-violet laser dyes with the appropriate polymethine saturable absorbers. About the same time the development of picosecond streak cameras permitted direct linear measurement of pulse shapes [2.94, 95].

As a result of extensive streak-camera investigations of ultrashort pulse generation in flashlamp-pumped dye lasers [2.96–99] it became apparent that the mechanism of modelocking dye lasers was quite different from that occurring in giant-pulse lasers. Investigations of the modelocking dyes [2.96, 97, 100, 101] also had shown that the saturable absorber recovery time was at least two orders of magnitude longer than the duration of the modelocked dye laser pulses. These experimental results were incorporated in a simple rate-equation theory [2.102, 103] which showed that the combined actions of amplifier and absorber saturation produced the observed rapid pulse compression.

The experimental arrangement now used for modelocked pulsed dye lasers is shown in Fig. 2.22. The laser dye is pumped uniformly by two Xenon flashlamps in a double-elliptical pumping reflector arrangement. About 100 J of stored electrical energy is dissipated by each flashlamp in ~ 1 μs. The windows of the quartz cell are wedged at 1° and the laser dye is circulated continually through a fine filter and a heat exchanger for cooling. The saturable absorber solution is contained in a cell and is in optical contact with the 100% reflectivity

[1] DODCI = 3,3′-diethyloxadicarbocyanine iodide.

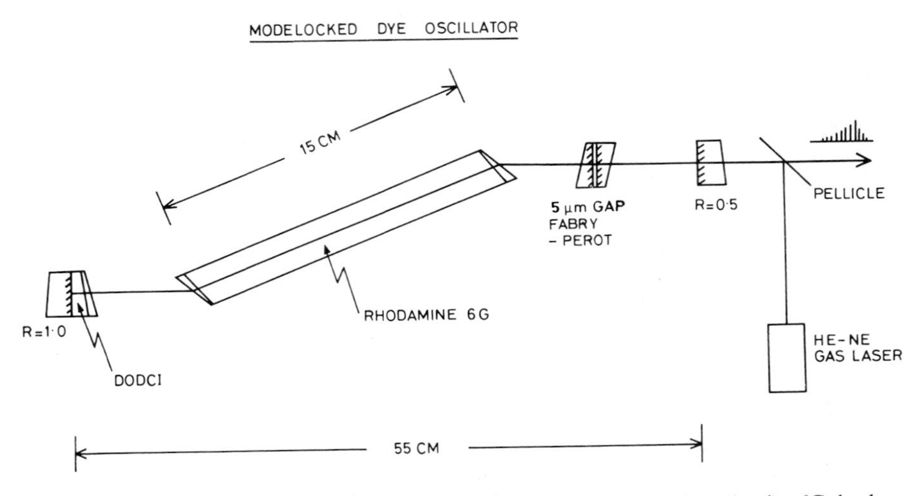

Fig. 2.22. Experimental arrangement for modelocking flashlamp-pumped rhodamine 6G dye laser

mirror. The absorber dye cell window is also wedged at 1°. The lasing dyes are dissolved in water or alcohols, depending upon the operating spectral range required. The laser bandwidth is frequency narrowed and tuned with an intracavity, optically contacted Fabry-Perot etalon of plate separation ~5 μm [2.92, 93]. In this manner transform-limited pulses were first obtained [2.92] over the spectral range 603 to 625 nm (Fig. 2.23). Because of variations in the gain of the active medium and in the absorption of the modelocking dye as the operating wavelength is altered, it is necessary to adjust the saturable absorber concentration for each wavelength to maintain the laser operating just above threshold. The pulse-train envelope also varies with wavelength as can be seen in Fig. 2.24. At the short wavelength end of the tuning range the ultrashort pulse forms very rapidly, whereas at longer wavelengths the output is similar to that of an untuned laser with an initial long pulse buildup section. (Without interferometer tuning the output spectrum of a rhodamine 6G laser modelocked by DODCI shifts to the red (615 nm) and has a bandwidth of ~4 nm [2.92].) Microdensitometer traces of typical, simultaneously recorded, TPF profiles and laser spectra near the extremes of the tuning range are shown in Fig. 2.25. In the absence of frequency modulation (chirping) the laser time-bandwidth product should have the value 0.441 for a gaussian pulse or 0.60 for a lorentzian-shaped pulse. If the modelocked dye laser is operated untuned, or slightly above threshold when tuned, the pulse spectra broaden and develop periodic structures typical of self-phase modulation, as in the case of the solid state lasers. Spectra of pulses selected out from different parts of the modelocked train again showed [2.92] that the spectral broadening increased monotonically along the trains. From these measurements a value for the nonlinear refractive index of the ethanol solvent was deduced.

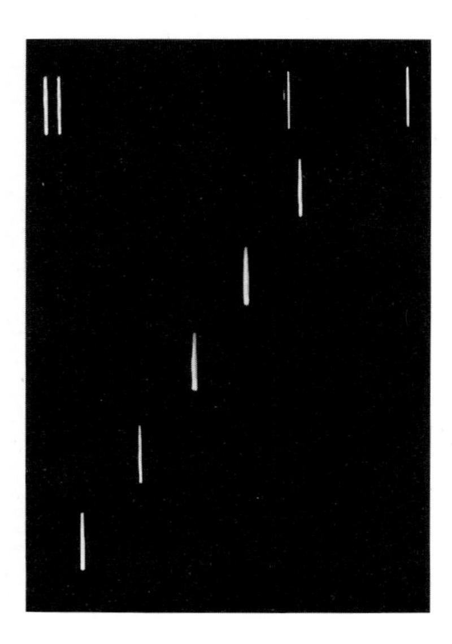

Fig. 2.23. Spectra showing tuning of bandwidth-limited picosecond pulses. Plate of spectrograph was moved vertically as laser was tuned, by rotation of intracavity Fabry-Perot etalon, from 603 nm to 625 nm. Mercury calibration spectrum at top (from *Arthurs* et al. [2.92])

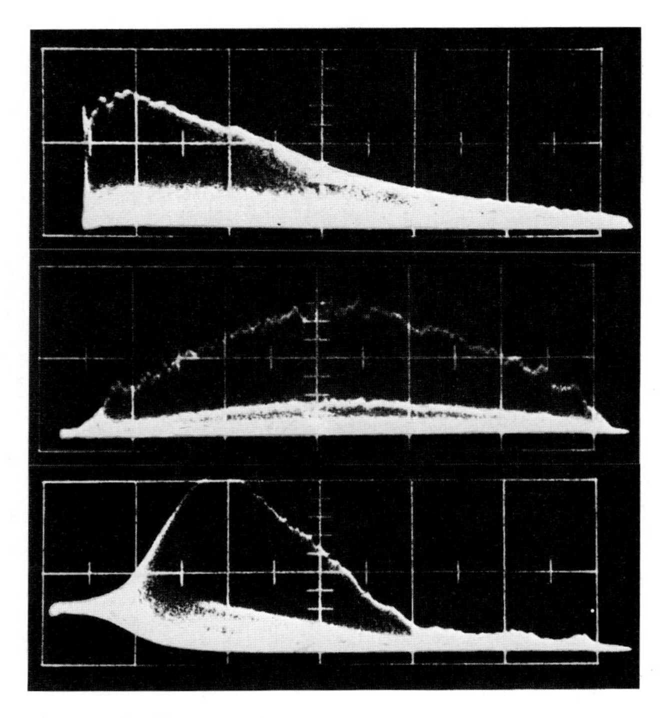

Fig. 2.24. Oscillograms of modelocked pulse trains from etalon tuned Rh 6G laser operating at, *Top* 605 nm, *Middle* 615 nm and *Bottom* 625 nm (from *Roddie* [2.104])

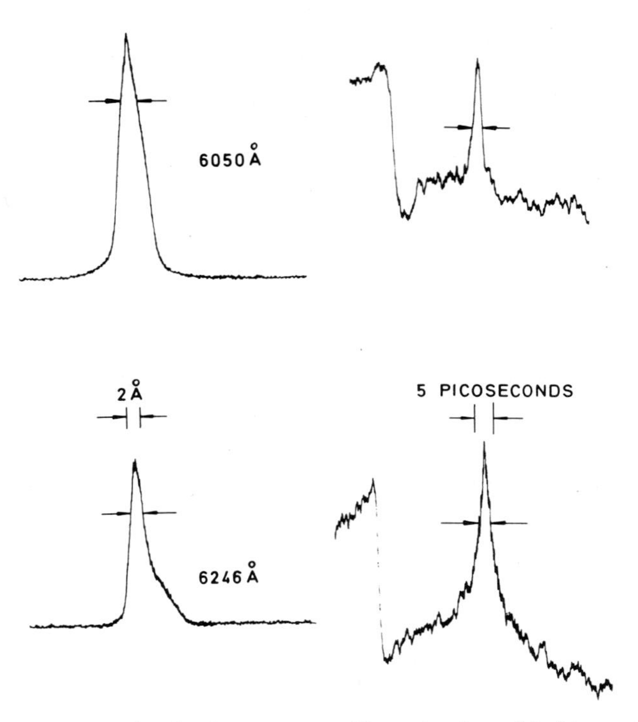

Fig. 2.25. Microdensitometer traces (linear density scale) of two pairs of simultaneously recorded TPF pulses and spectra of bandwidth-limited Rh 6G picosecond pulses (from *Arthurs* et al. [2.92])

The frequency range covered by flashlamp-pumped modelocked dye lasers has since been extended to both shorter and longer wavelengths with the laser dyes and corresponding saturable absorbers listed in Table 2.1. General technical improvements in the dye flow systems, in uniformity of optical pumping, and in laser optical components now permit the routine generation of pulses of a few picoseconds durations or less over the wide range of wavelengths given in the table. Operation of modelocked dye lasers is much less sensitive to variations in pumping energy or to thermal effects in the laser media than is operation of solid state systems. Consequently, the lasers can be used at higher repetition rates, mainly because two strongly nonlinear pulse shortening mechanisms are operating simultaneously and partly because the liquid laser media are more easily kept at uniform temperatures.

Temporal Buildup of Modelocking in Dye Lasers

The paradox that the shortest laser pulses were generated in dye lasers passively modelocked by saturable absorbers of relaxation times of hundreds of picoseconds was cleared up by detailed experimental investigations of the buildup of picosecond pulses [2.96–99, 104, 105]. The nature of the initial buildup

Table 2.1. Picosecond pulse generation in flashlamp-pumped dye lasers

Laser dye	Saturable absorber	Tuning range (nm)	Pulse durations, ps	References
Esculin monohydrate	DASPI	465–480		[1]
Rhodamine 6G	DQOCI	575–600	1.5–3	[2, 3]
	DODCI	600–625	2–3	[3, 4]
Rhodamine B	DQTCI	605–630	3–4	[5]
	DODCI	615–645	2–3	[5, 6]
Cresyl Violet	DTDCI	660–704	3–5	[5, 6, 7]
+Rhodamine 6G	DDCI	645–680	~3	[5, 6, 7]
	DOTCI	645–680	~4	[5]
DOTCI	HITCCI	795–805		[8]

Dye formulae

DASPI = 2-(P-Dimethylaminostryryl)-pyridylethyl iodide
DQTCI = 1,3′-diethyl-4,5′-quinolythia-carbocyanine iodide
DTDCI = 3,3′-diethyl-2,2′-thiadicarbocyanine iodide
DDCI = 1,1′-diethyl-2,2′-dicarbocyanine iodide
DODCI = 3,3′-diethyl-oxadicarbocyanine iodide
DOTCI = 3,3′-diethyl-2,2′-oxatricarbocyanine iodide
HITCI = 1,3,3,1′,3′,3′ Hexamethyl-2,2′-indotricarbocyanine iodide
DQOCI = 1,3′-diethyl-4,2′-quinolyoxacarbocyanide iodide

References

1 E. G. Arthurs, D. J. Bradley, T. G. Glynn, A. G. Roddie, W. Sibbett: (unpublished)
2 R. S. Adrain, E. G. Arthurs, D. J. Bradley, A. G. Roddie, J. R. Taylor: Opt. Commun. **12**, 136 (1974)
3 P. R. Bird, D. J. Bradley, W. Sibbett: *Proceedings of* 11*th International Congress on High Speed Photography* (Chapman and Hall, London 1975), p. 112
4 E. G. Arthurs, D. J. Bradley, A. G. Roddie: Appl. Phys. Lett. **19**, 480 (1971)
5 E. G. Arthurs, D. J. Bradley, A. G. Roddie: Appl. Phys. Lett. **20**, 125 (1972)
6 E. G. Arthurs: PhD Thesis, The Queen's University of Belfast (1972)
7 E. G. Arthurs, D. J. Bradley, P. N. Puntambekar, I. S. Ruddock: Opt. Commun. **12**, 360 (1974)
8 A. Hirth, K. Vollrath, D. J. Lougnot: Opt. Commun. **8**, 318 (1973)

phase was found to depend upon the operating wavelength and the particular saturable absorber employed. For rhodamine 6G modelocked by DODCI and tuned to operate at 605 nm, a very fast buildup of modelocking occurs, and streaks such as that of Fig. 2.26 showed that within a few round trips (20 ns) from the start of the laser action, a fluctuation noise burst of ~100 ps total duration existed in the laser cavity. The subsequent history of evolution to a single picosecond pulse is shown in Fig. 2.27. After ~25 round trips (120 ns) the noise burst has been substantially compressed and then very quickly evolves into a single 2 ps pulse after ~35 round trips. (For calibration purposes two images of each pulse separated by 57 ps were generated in an optical delay line). The very rapid pulse shortening at this wavelength can be clearly seen in the microdensitometer traces of Fig. 2.28 obtained with the experimental arrangement of Fig. 2.29. By using the pulse delay system two pulses, separated by two round trips, were imaged at different points of the streak-camera slit so as to produce simultaneously separate streak images on a single

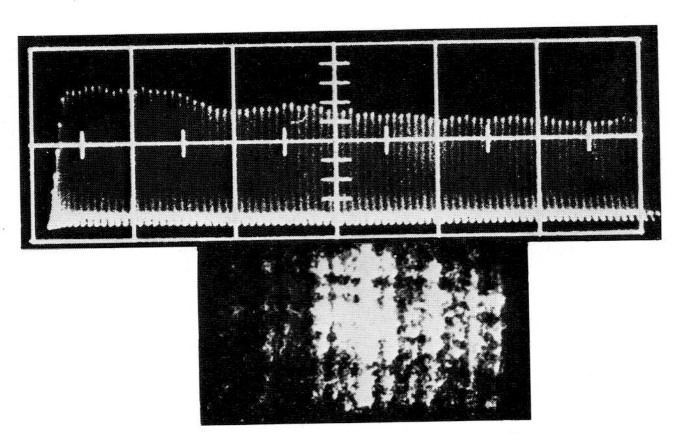

Fig. 2.26. *Top* Oscillogram of the modelocked pulse train of a rhodamine 6G dye laser tuned to operate at 605 nm. Time scale: 50 ns per major division. *Bottom* Streak photograph of fluctuation noise burst "pulse" 20 ns from the start of the pulse train (from *Arthurs* et al. [2.96])

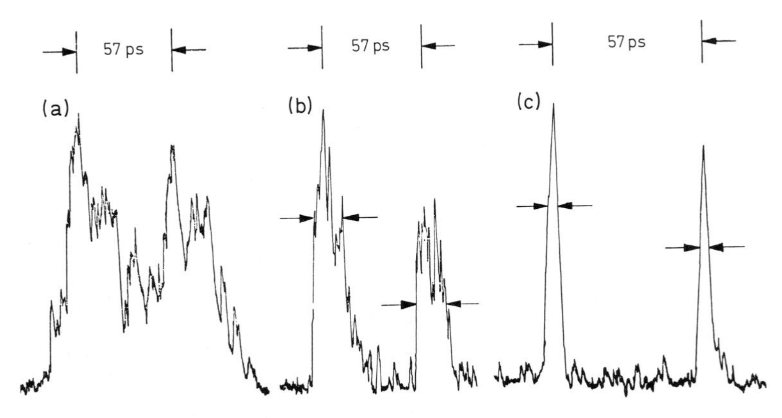

Fig. 2.27 a–c. Microdensitometer trace of streak-camera records showing the history of evolution to a single ultrashort pulse in a rhodamine 6G laser modelocked by DODCI. (a) Noise fluctuations at the start of pulse train (b) Halfwidth of envelope of noise pattern reduced to 17 ps after 30 round trips (c) Single 2 ps pulse after 35 round trips. In each case the streak record shows a double pattern generated in the optical delay line from the original pulse so as to provide a known time calibration (from *Adrain* [2.105])

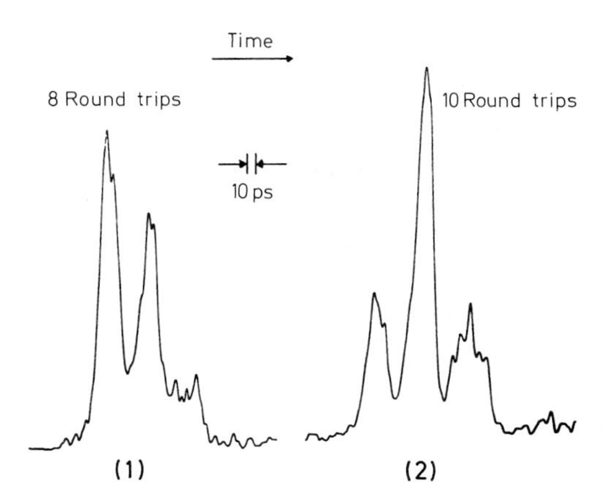

Fig. 2.28. Microdensitometer trace of two simultaneously recorded streak photographs showing pulse evolution in Rh 6G laser. The number of round trips is counted from the start of the pulse train (from *Arthurs* et al. [2.96])

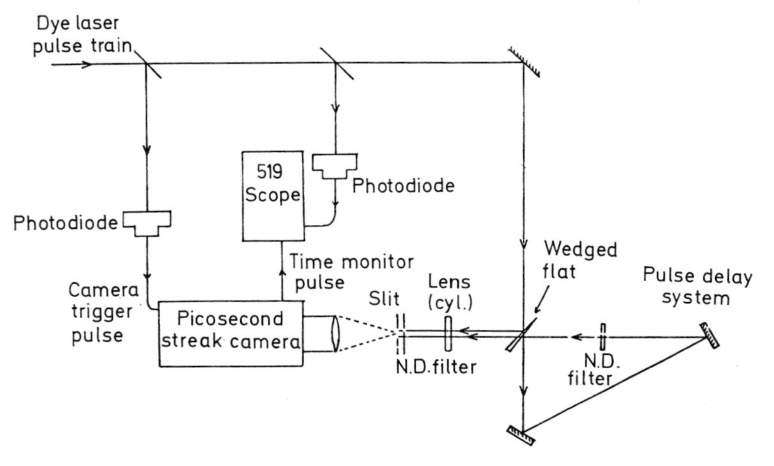

Fig. 2.29. Arrangements for the simultaneous recording of pulse profiles at two different times along the pulse train (see Fig. 2.28)

photograph. The microdensitometer traces[2] show clearly the selection of a single fluctuation after only two round trips. The fact that the pulse profiles are attenuated at the rear of the envelope as well as at the front demonstrated the important role that saturable amplification played in the passively mode-locked dye laser. When the rhodamine 6G laser is tuned to longer wavelengths the pulse buildup takes longer. This behaviour was explained by *New* [2.102] in terms of the wavelength variation of the ratio of the cross sections of the

[2] Since the vertical scale is linear optical density, the smaller fluctuations appear much more intense relative to the pulse peaks.

laser and modelocking dyes. Later studies with the Photochron II streak camera, with subpicosecond time resolution, confirmed that the final pulse duration obtainable was ~ 2 ps (Fig. 2.10) and that the pulse to background ratio was $>10^4$ implying that $>95\%$ of the laser energy was contained within the final isolated picosecond pulse.

Saturable Absorber Recovery Time and Photoisomer Generation. The fact that DODCI is effective as a saturable absorber for modelocking dye lasers operating in the spectral range 584 to 645 nm, although its absorption maximum is at 580 nm, led to the suggestion [2.93] that at longer wavelengths modelocking is due to the generation of a photoisomer, with a ground state lifetime of $\sim 3 \times 10^{-4}$ s and an absorption maximum at 620 nm [2.100]. This photoisomer is generated when DODCI is excited by single-photon, or two-photon [2.57] absorption, giving rise to strong fluorescence peaked at 646 nm. Also, conventional laser-pumped [2.106] and travelling-wave [2.57] DODCI dye lasers operate at ~ 650 nm. Detailed measurements [2.97] of the fluorescence spectra of both the normal and photoisomer forms of DODCI, with microsecond and picosecond excitation, gave the fluorescence spectra of the photoisomer species (Fig. 2.30).

When the temporal behaviour of the fluorescence from the intracavity absorber cell was monitored at 600 nm and 650 nm with the laser tuned to operate at various wavelengths, the fluorescence at 650 nm initially grew rapidly, relative to the rise of the laser intensity, but then levelled off to follow the laser intensity envelope shape. Likewise the fluorescence at 600 nm fell initially and then also stabilized. The time required for this initial stabilization period was a function of the lasing wavelength. At 607 nm, 100 ns was required by which time a well-defined single modelocked pulse was circulating inside the resonator. For lasing wavelengths longer than 610 nm the attainment of the balance took ~ 200 ns. It had earlier been found [2.93] that at these wavelengths the modelocked pulses also began to appear after ~ 200 ns. At wavelengths shorter than 605 nm, where the normal form of DODCI has a large absorption

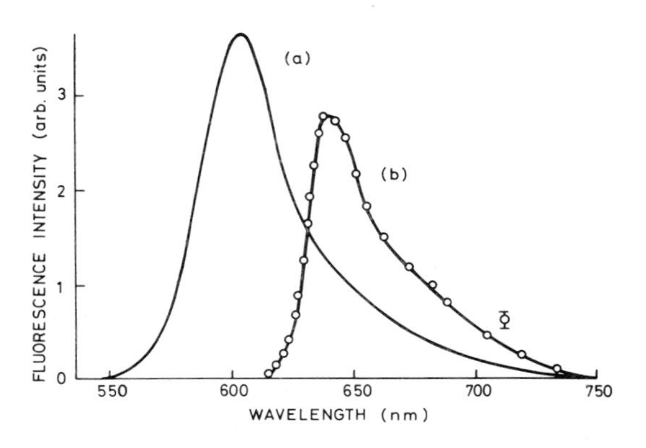

Fig. 2.30. Corrected fluorescence spectra of (a) DODCI recorded on a spectrophotofluorimeter and (b) DODCI photoisomer obtained experimentally (from *Arthurs* et al. [2.97])

cross section, the formation of the photoisomer effectively reduces the dye concentration and this explains why modelocking by DODCI is difficult at these wavelengths.

The excited-state fluorescence lifetimes of the two species of DODCI were measured when the pulse train from a modelocked, rhodamine 6G laser, tuned to 605 nm, was focused into a cell containing the absorber solution. A microdensitometer trace of a typical streak-camera fluorescence time profile is shown in Fig. 2.31. From a large number of such profiles the fluorescence decay time was measured to be 330 ± 40 ps for both the normal and photoisomer species. The decay of the sidelight fluorescence from the intracavity cell was also investigated with the streak camera and the results obtained were identical. However, in the case of the intracavity cell the photoisomer fluorescence was dominant.

The recovery time of the DODCI absorption following intense picosecond excitation was investigated with the experimental arrangements of Fig. 2.32a, b [2.97]. For studying the dye while operating as a saturable absorber *inside* the laser resonator, the modelocked pulse train (Fig. 2.32a) was telescoped up to reduce beam divergence, apertured, and then attenuated by neutral density filters to produce a low intensity uniform beam. After passing through a delay line, a polarizing prism, and further apertures, the beam was arranged to overlap again with the resonator beam inside the DODCI modelocking cell. A micrometer control of the position of the delay line prism permitted timing of the arrival of the pulses of the probe beam to 0.5 ps. The probing pulse train was monitored, before and after transmission through the cell.

Fig. 2.33 shows the transmission at a particular time during the train plotted against the delay of the probing pulse, for the laser operating at ~610 nm. The 1/e recovery time τ_A of the absorption is ~225 ps. Similar

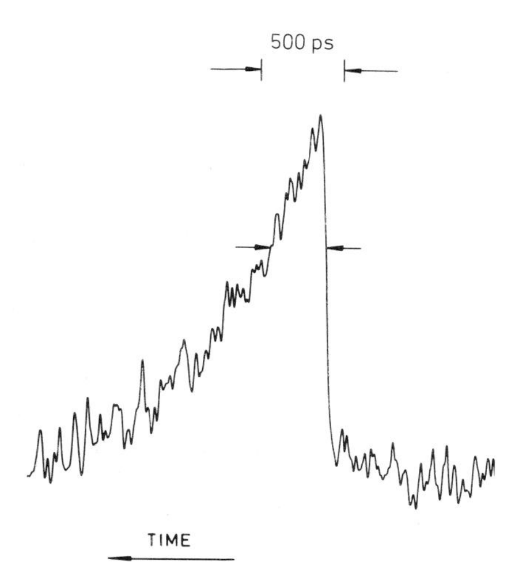

Fig. 2.31. Microdensitometer trace of streak-camera record of DODCI fluorescence time profile under same conditions as inside a mode-locking cell in the laser cavity. Corrected e^{-1} decay time is indicated. (Ordinate is linear density scale and the recorded signal to noise ratio is 20:1)

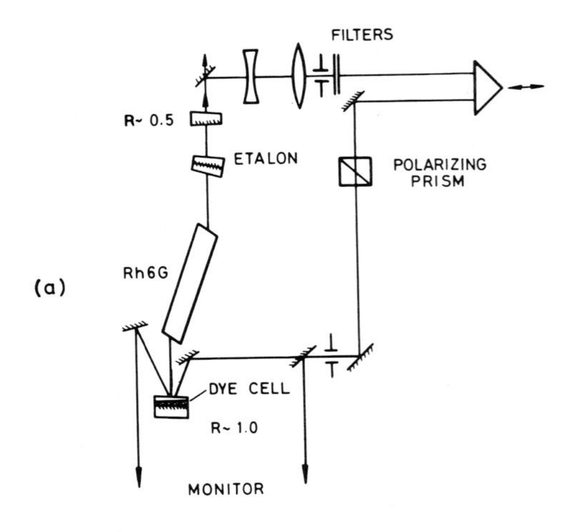

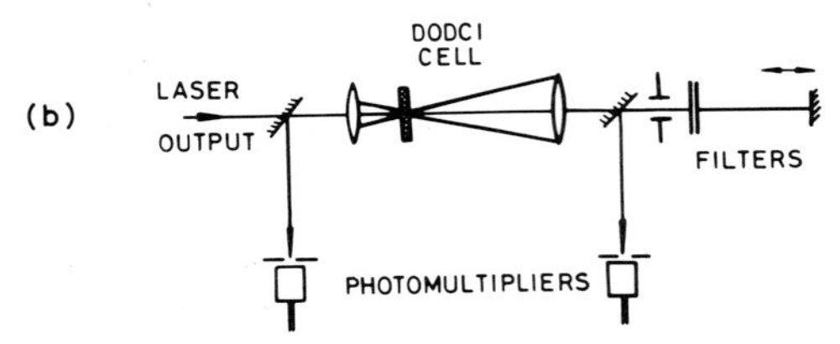

Fig. 2.32a and b. Methods of measuring saturable-absorber recovery time. (a) Inside modelocked dye laser cavity, (b) outside the laser resonator

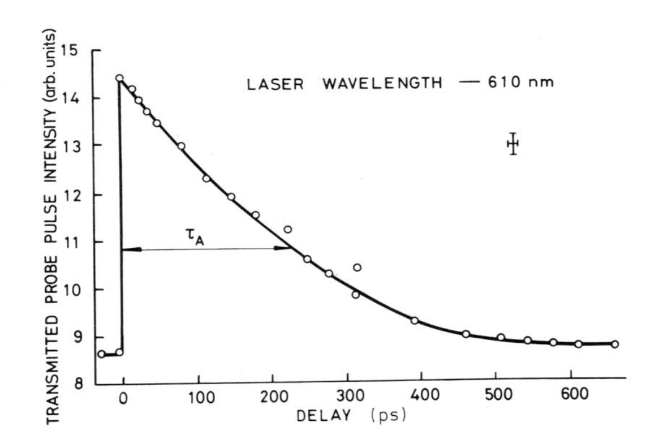

Fig. 2.33. Transmission of intracavity DODCI cell after excitation by a picosecond pulse as measured by arrangement (a) of Fig. 2.32 (from *Arthurs* et al. [2.97])

recovery profiles were obtained for all times during the pulse train, after the evolution of ultrashort pulses, for the laser operating in the range 595 to 620 nm. In all cases, the transmission remained constant over the 3.5 ns period following the absorption recovery until the next exciting pulse entered the cell. The value of this interpulse transmission $T(\infty)$ relative to the low-level transmission T_L of the unexcited DODCI depended on the operating wavelength of the laser. At wavelengths shorter than ~605 nm, where the photoisomer absorption cross section is larger [2.100] $T(\infty)$ was less than T_L.

The absorption recovery was also measured outside the laser resonator with the arrangement of Fig. 2.25b. The results obtained gave typical values of ~240 ps for τ_A. Assuming an optically thin sample cell, containing a two-level absorber, with spatially uniform excitation, by an intense, delta function pulse, the transmission $T(t)$ of the medium at time t after excitation is given by

$$\mathrm{Ln}\,[T(t)/T(0)] = [1 - \exp(-t/\tau_1)]\,\mathrm{Ln}\,[T(\infty)/T(0)] \tag{2.23}$$

where τ_1 is the lifetime of the upper state. This relation shows [2.107] that the aperture time τ_A is strongly dependent upon both $[T(\infty)/T(0)]$ and τ_1. Knowing τ_A, the value $\tau_1 \sim 300$ ps was calculated for particular values of $[T(\infty)/T(0)]$.

Comparison of Modelocked Ruby and Cresyl-Violet Lasers. With the development of the modelocked cresyl-violet laser it was possible [2.98, 99] to directly compare the mechanism of pulse generation in the ruby laser and in a dye laser operating at the same wavelength and each modelocked by saturable absorbers of different molecular relaxation times (Table 2.2).

The ruby laser, operated in a single transverse mode of beam divergence <1 mrad, produced regular pulse trains of total duration ~250 ns (~100 mJ total energy) when modelocked by DDCI or DTDCI. Both absorbers were contained in a cell of 200 μm thickness in optical contact with the 100% reflectivity spherical mirror [2.108]. With DDCI the pulse durations were in the range 15 to 30 ps but the pulses produced by DTDCI were longer, 90 to 105 ps. In both cases the fluctuation structure near the end of the linear phase of laser amplification develops into a train of isolated single pulses at the start of the giant pulse. More details can be seen in the streak-camera traces. The average duration of the fluctuations in Fig. 2.34a is ~120 ps, in good agreement with the ruby laser bandwidths of ~0.2 Å [2.108] at the end of the linear amplification stage. The selection of the most intense fluctuations and their gradual compression is clearly shown in the later streaks. This pattern of development is typical of solid state lasers modelocked by absorbers of very short relaxation times. The pattern obtained with DTDCI is, on the other hand, very similar to that seen in dye lasers, except that in the case of the ruby laser the final pulse duration is determined by the aperture time of the modelocking DTDCI dye cell, which was measured to be ~120 ps. (The corresponding aperture time for the DDCI cell was 13 ps.) It is interesting to note that the DTDCI cell aperture time equals the shortest fluctuation time in the trace of Fig. 2.34. Since in this case the dye relaxation time is longer, bleaching of the absorption occurs by energy saturation and the absorber tends to select a broad group of fluctuations. In keeping with this picture the final pulse shape measured with the streak camera was asymmetrical with a sharp leading edge.

These results with the ruby laser, showing that the absorber cell aperture time determined the pattern of pulse generation, demonstrated that the two

Table 2.2. Lifetimes (relaxation times) of saturable absorber molecules

Dye	Molecular lifetimes		References
	Absorption	Fluorescence	
DDCI			
In methanol	25 ps	—	[1]
	—	14 ps	[2]
In ethanol	—	11 ps	[3]
DCI (cryptocyanine)			
In methanol	80 ps	—	[4]
	16 ps	—	[5]
	—	110 ps	[6]
	—	20–40 ps	[7]
In ethanol	—	37 ps	[3]
DCI′			
In ethanol	10 ps	—	[8]
	—	20 ps	[1]
DTDCI			
In ethanol	175 ps	185 ps	[1]
	—	1.2 ns	[9]
DODCI			
In ethanol	10–250 ps	—	[10, 11]
	1.15 ns	—	[12]
	1.2 ns	—	[13]
	300 ps	—	[14]
	—	420 ps; 1.2 ns	[15]
	—	330 ps	[16]

References

1 E.G. Arthurs, D.J. Bradley, P.N. Puntambekar, I.S. Ruddock, T.J. Glynn: Opt. Commun. **12**, 360 (1974)
2 M.A. Duguay, J.W. Hansen: Opt. Commun. **1**, 254 (1969)
3 D.N. Dempster, T. Morrow, R. Rankin, G.F. Thompson: Chem. Phys. Lett. **22**, 222 (1973)
4 G. Mourou, G. Busca, M.M. Denariez-Roberge: Opt. Commun. **4**, 40 (1971)
5 J.P. Fouassier, D.J. Lougnot, J. Faure: Chem. Phys. Lett. **30**, 448, (1975)
6 G. Mourou, G. Busca, M.M. Denariez-Roberge: IEEE J. **QE-9**, 745 (1973)
7 M.A. Duguay, J.W. Hansen: Opt. Commun. **1**, 254 (1969)
8 M.W. McGeoch: Opt. Commun. **7**, 116 (1973)
9 D.N. Dempster, T. Morrow, R. Rankin, G.F. Thompson: J. Chem. Soc., Faraday II **68**, 1479 (1972)
10 G.E. Busch, R.P. Jones, P.M. Rentzepis: Chem. Phys. Lett. **18**, 178 (1973)
11 G.E. Busch, K.S. Greve, G.L. Olson, R.P. Jones, P.M. Rentzepis: Chem. Phys. Lett. **33**, 412 (1975)
12 D. Magde, M.W. Windsor: Chem. Phys. Lett. **27**, 31 (1974)
13 C.V. Shank, E.P. Ippen: Appl. Phys. Lett. **26**, 62 (1975)
14 E.G. Arthurs, D.J. Bradley, A.G. Roddie: Opt. Commun. **8**, 118 (1973)
15 J.C. Mialocq, A.W. Boyd, J. Jaraudias, J. Sutton: Chem. Phys. Lett. **37**, 236 (1976)
16 E.G. Arthurs, D.J. Bradley, A.G. Roddie: Chem. Phys. Lett. **22**, 230 (1973)

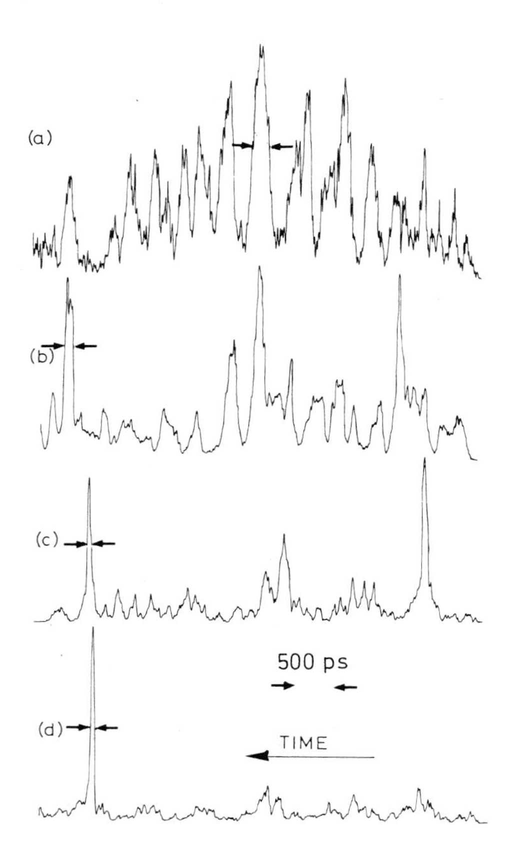

Fig. 2.34 a–d. Sequence of streak camera microdensitometer traces (time resolution ~30 ps) of pulse evolution in ruby laser modelocked by DDCI. (a) 3 μs (b) 2.5 μs (c) 1.5 μs (d) 1 μs before peak of giant pulse, respectively

cases of modelocking dye lasers modelocked commonly met dye lasers modelocked by absorbers of long lifetimes and solid state lasers with short duration absorbers, are special cases. In ruby and Nd : glass lasers the absorber cell aperture time sets a lower limit to the final pulse generated. In the case of dye lasers the very short storage times of the laser media result in the domination of saturable amplification so that picosecond pulses can be generated with saturable absorbers of a very wide range of relaxation times. This analysis was confirmed by studies of modelocking of the cresyl-violet laser [2.98]. The absorption spectra of the modelocking polymethine dyes used are shown in Fig. 2.35. The absorber-cell aperture times were measured by the methods used for DODCI, employing picosecond pulses from the modelocked laser tuned to operate at the appropriate wavelengths. The results obtained are given in Table 2.2 together with the corresponding results obtained by other researchers. The ability of DTDCI to modelock at 694.3 nm arises from the generation of a photoisomer with an absorption spectrum shifted to the red [2.100] as in the case of rhodamine 6 G modelocked by DODCI at 625 nm. While there has been controversy [2.101, 109] over the value of the molecular lifetime of DODCI, and

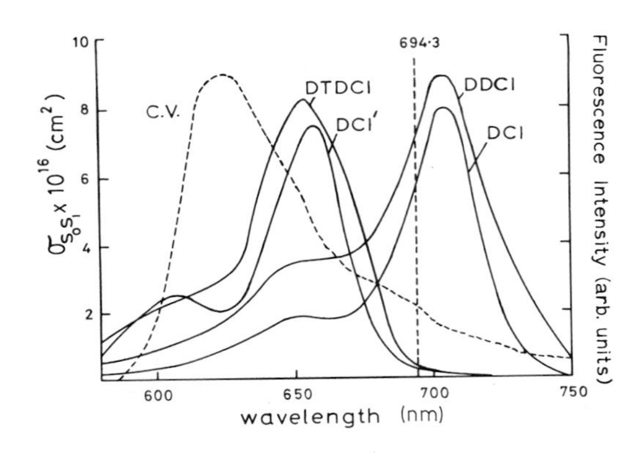

Fig. 2.35. Variation of absorption cross sections of saturable absorbers as a function of wavelength. The fluorescence spectrum of cresyl-violet is also shown

even about the existence of the photoisomer [2.110], from the point of view of understanding the modelocking process the relevant recovery time is the shortest *aperture time* of the intracavity saturable absorber dye cell, *under modelocking conditions*. As pointed out by *Mourou* et al. [2.111] care has to be taken in determining the molecular lifetime from absorption relaxation measurements, to correct the aperture time for the particular low light-level transmission used and the induced absorption change. Also stimulated emission can affect the measurements [2.112]. In the cases of DODCI and DTDCI, photoisomer generation adds further complications.

The efficiency of lasing is improved by employing a mixed solution, in ethanol, of rhodamine 6G and cresyl-violet. The longest pulse trains are generated with DTDCI and, in general, the rate of picosecond pulse buildup increases with the gain of the active medium, i.e., as the laser is tuned to shorter wavelengths. Pulse durations of ~4 to 7 ps were consistently obtained at the middle of the train with all of the four modelocking dyes (DTDCI, DDCI, DCI and DCI′). The effect of absorber relaxation time is clearly seen for the two cases in the streak camera records of Fig. 2.36. Traces (b)–(d) for DTDCI modelocking correspond to times 150, 350 and 500 ns, respectively, after lasing threshold was achieved. (Double transit time was 4 ns.) With the broad lasing bandwidth (~3 nm) of the untuned laser the shortest initial fluctuations would have durations of ~200 fs (200×10^{-13} s), so the saturation of the DTDCI absorber would therefore be determined by the integrated energy density. With DDCI and an initial bandwidth of 0.5 nm the streak traces (at times corresponding to 85, 90 and 95 round trips, respectively, after threshold) show an evolutionary process similar to that occurring in the ruby laser modelocked by the same dye (see Fig. 2.34). In the dye laser case a considerably more rapid selection rate arises from the much greater effect of gain depletion.

Modelocked cw *Dye Lasers*

While the continuous working dye laser has been actively modelocked [2.113, 114] the shortest pulses are obtained in passively modelocked systems. Pulses

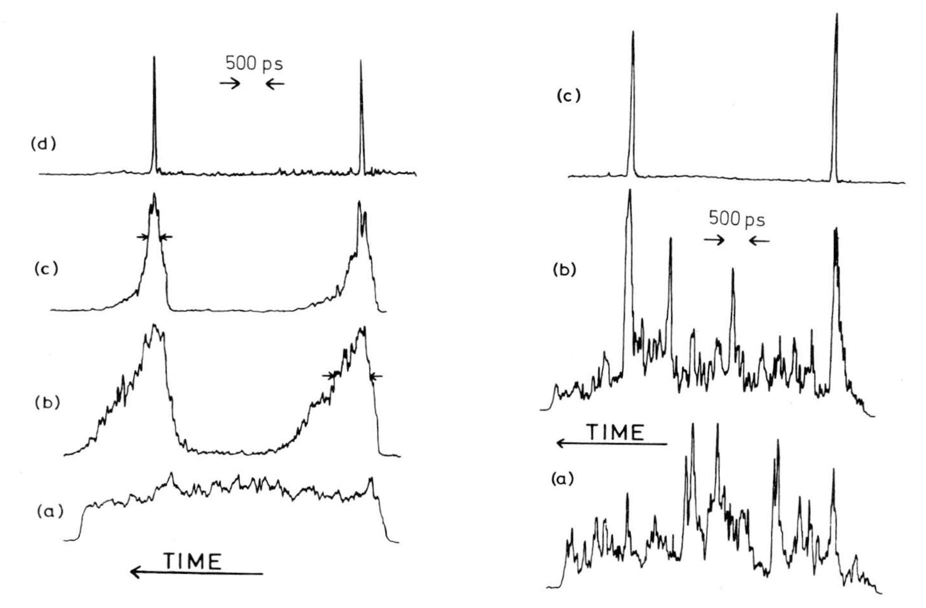

Fig. 2.36. Sequence of streak-camera microdensitometer traces (time resolution ~30 ps) showing ultrashort pulse development in cresyl-violet laser modelocked by (L.H.S.) DTDCI (untuned.) and (R.H.S.) DDCI (tuned to 695 nm) (see text for details)

of durations of a few picoseconds were first obtained from the cw rhodamine 6G dye laser [2.115, 32, 116] passively modelocked by DODCI. Since then, subpicosecond pulses have been obtained by compressing, with a grating pair, the frequency-chirped pulses generated in a system employing two free-flowing dye streams, one for the active medium and one for the absorber, respectively [2.117] (see also Chap. 3). Transform-limited pulses of durations as short as 0.3 ps have been produced [2.118] in the cavity configuration shown in Fig. 2.37, which allows the use of all lines of the Argon-ion laser [2.98]. The saturable-absorber dye flows, in contact with the ~100% broadband reflectivity mirror, in a narrow channel of thickness variable from 200 μm to 500 μm. Both the window and the channel of the absorber dye cell are wedged to 1° to eliminate subcavity resonances. The rhodamine 6G dye laser beam is focused into the absorber solution by a lens so as to increase the power density fourfold. The lens and the cell window are both antireflection coated. The laser dye flows in a jet near the centre of the resonator. Output mirror transmissions from 1 to 6% at 600 nm are employed and the round-trip time varies from 10 ns to 12 ns. With an absorber cell thickness of 0.5 mm pulses of ~1 ps durations are readily obtained. Pulse durations are further reduced to 0.3 ps (at 610 nm) when the DODCI cell length is shortened to 200 μm. Under these conditions the subpicosecond pulses are tunable over the range 598 to 615 nm. Pulse

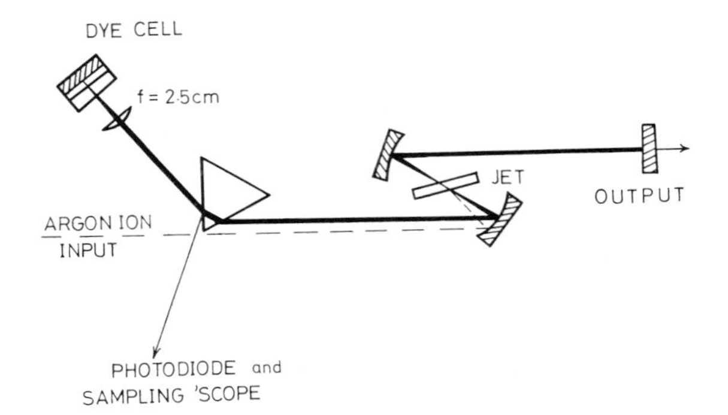

Fig. 2.37. Passively modelocked cw dye laser configuration

duration and stability are not dependent upon pumping level and the dye laser can operate up to 20% above threshold to give an average output power of 15 mW for 0.5 ps pulses (peak power 300 W). When the optically contacted absorber cell was replaced by a second free-flowing dye stream, single pulses were obtained only close to threshold as has been observed by other workers [2.119]. At higher pumping powers structured pulses were produced, as detected both by second-harmonic autocorrelation and streak-camera measurements. Another striking difference lies in the pulse shapes produced by the contacted and free-flowing absorber arrangements. In the former, the best fit to the autocorrelation traces gives a sech^2 pulse shape in good agreement with a recent theoretical model [2.120]. For the noncontacted cell the pulses are characterized by exponential trailing edges. Similar single-sided exponential pulse shapes have been reported by *Ippen* and *Shank* [2.117].

Confirmation of the generation of bandwidth-limited subpicosecond pulses by the narrow-gap contacted absorber cell was given by simultaneous temporal and spectral intensity recordings, such as Fig. 2.38. These pulses, having the high degree of organisation represented in Fig. 2.2b, are particularly valuable for experiments in nonlinear optics because the interpretation of the results is then relatively simple.

As with the flashlamp-pumped rhodamine 6G laser modelocked by DODCI, at wavelengths less than 600 nm modelocking of the cw laser deteriorates due to the decrease in the DODCI photoisomer absorption cross section at these wavelengths. Again, the use of DQOCI extends the range of ultrashort pulse generation in the cw laser to cover the spectral region 580 to 613 nm. Subpicosecond pulses have also been obtained with DQOCI, although detailed investigations of the tuning range have not been carried out. However, it is anticipated that bandwidth-limited subpicosecond pulses will be available over most of the DQOCI modelocking range.

Employing the various combinations of laser dyes and saturable absorbers given in Table 2.3 continuous tuning of the modelocked cw dye laser has been achieved over the range 580 to 630 nm with the optically contacted absorber dye cell arrangement. It is likely that bandwidth-limited pulses of $\sim$100 fs

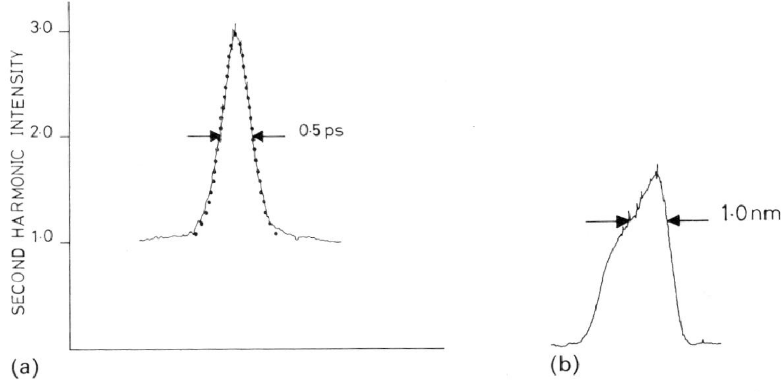

Fig. 2.38a and b. (a) Second-harmonic autocorrelation trace for laser operating at wavelength 607 nm. The discrete points are calculated for a $sech^2$ laser pulse intensity profile. (b) Microdensitometer trace of the spectrum of modelocked laser recorded simultaneously with the pulse duration measurement of (a). Arrows indicate the spectral halfwidth (linear density scale)

Table 2.3. Continuous wave Dye Lasers Modelocked by Contacted Absorber Dye Cell

Laser dye	Absorber dye	Pulse duration (ps)	Tuning range (nm)
Rhodamine 6G	DODCI	Sub-picosecond	598–615
		0.3[a]–1.5	592–617
	DQOCI	0.6[a]–2	580–613
Rhodamine B	DODCI	3–4	610–630
	DQOCI	4–5	600–620
	Cresyl-violet	3–4	610–620
Sodium fluorescein	Rhodamine 6G	5–7	546

[a] 200 μm cell thickness
(Data from I. S. Ruddock: PhD Thesis, University of London (1976))

durations could be obtained with narrower dye cells and general engineering improvements. There must be a minimum number of optical cycles to define the stationary-pulse $sech^2$ intensity profile, and because there would be only ~50 cycles in a 100 fs pulse, it is evident that we are approaching the limit for ultrashort pulse generation in the visible region.

Synchronously Pumped Dye Lasers

As mentioned in *Flashlamp Pumped Systems* (see p. 41), the generation of ultrashort pulses in dye lasers was first achieved by pumping with pulse trains from high power modelocked ruby [2.85, 86] and from the second-harmonic of modelocked Nd : glass lasers [2.87, 88]. By setting the cavity length of the dye lasers equal to, or a submultiple of, that of the pumping lasers, the gain of the dye laser was impulsively driven in synchronism with the cavity round-trip repetition rate, thus producing a train of short duration pulses. The pulses of shortest duration (15 to 20 ps) were obtained from the second-harmonic pulse train of a Nd:glass laser longitudinally pumping rhodamine *B* in a matched 1 m cavity [2.89]. This technique of synchronous pumping by modelocked solid state lasers has recently been taken up again [2.121]. Bandwidth-limited pulses of durations as short as 12 ps have been obtained [2.122, 123] by pumping with second-harmonic pulses from a repetitively pulsed (50 Hz), passively modelocked Nd:YAG laser. The dye laser was spectrally narrowed and tuned by two intracavity Fabry-Perot etalons. When the dye laser cavity length was matched to the pulse separation of the pumping laser train, the dye pulses were generated in time coincidence with, and had durations (~30 ps) little longer than the pump pulses. By slightly lengthening the dye laser cavity from the matched length (by ~0.8 mm) the resulting gain depletion arising from the increasing delay, with respect to the pumping pulse, of the circulating dye laser pulse, results in a shortening of this pulse by a half. An extensive tuning range of 549 to 727 nm was obtained for synchronous modelocked operation with rhodamine 6G, rhodamine B, cresyl-violet, and carbazine 122 dyes in a variety of solvents. Because of the reduced ground-state absorption in the highly inverted dyes under synchronous pumping, it is possible to tune to shorter wavelengths than in the case of flashlamp-pumped and cw dye lasers. With rhodamine 6G the bandwidth-limited pulses had durations of 12 to 16 ps with peak powers of ~1 MW. These results are to be compared with bandwidth-limited pulse durations of ~2 ps and peak powers of also ~1 MW obtainable from passively modelocked flashlamp-pumped systems [2.92].

Synchronous pumping has also been employed with modelocked argon-ion lasers [2.124–127]. Stable modelocking, to produce bandwidth-limited pulses of ~7 ps duration at a repetition rate of 10^8 Hz, and frequency tunable from 565 to 625 nm, has been obtained [2.128]. The dye laser is pumped by a modelocked argon-ion laser of ~800 mW average power and pulsewidths of 200 ps. The conversion efficiency is ~40% and the resulting dye laser pulses have peak powers of ~500 W (average power 350 mW).

Earlier, *Runge* [2.129] had reported an arrangement in which cresyl-violet, Nile Blue *A* and DTDCI dyes had been employed to passively modelock a He-Ne laser, and were found to lase and to emit a train of pulses with the same repetition rate as the He-Ne laser. By separating out the dye laser and gas laser cavities, DTDCI could be wavelength tuned over a broad spectral region when intracavity pumped in this way by the He-Ne laser [2.130]. Modelocked pulse trains at two frequencies have also been produced when a mixture of rhodamine 6G and cresyl-violet was pumped in the usual cw dye laser configuration by an argon-ion laser [2.131]. At a pump power of 1 W, simultaneously modelocked lasing was observed at the two wavelengths of 574 and 644 nm. However, no pulse duration measurements were reported and at higher pump powers simultaneous cw lasing at the two wavelengths occurred.

Fan and *Gustafson* [2.132] have generated narrow-band picosecond pulses from a rhodamine B dye laser with a cavity length of 50 to 125 μm pumped by frequency-doubled pulses ($\sim$5 ps) from a passively modelocked Nd:glass laser. Pulses of $\sim$3 ps duration with a spectral bandwidth of $\sim$5 Å (time-bandwidth product value of 1.25) were produced with an energy conversion efficiency of up to 40%. The emission wavelength varied from 0.60 to 0.63 μm depending upon dye concentration and cavity length. As pointed out by *Roess* [2.133] a high ratio of pumping pulse duration to cavity lifetime results in laser pulse shortening.

2.3 The Fluctuation Model of Modelocked Lasers

The experimental results, described and discussed in Section 2.2, can be explained for both giant-pulse lasers and dye lasers in terms of the fluctuation model of the passively modelocked laser. This model was first formulated by *Letokhov* [2.134] and, independently, by *Fleck* [2.135, 136], to show how ultrashort pulses are generated in giant-pulse systems from the initial fluorescence intensity fluctuation patterns of the laser media.

2.3.1 Passive Modelocking of Giant-Pulse Lasers

The various stages of evolution of a modelocked pulse are shown in Fig. 2.39. Spontaneous fluorescence emission begins at the start of the flashlamp pumping pulse. As the pumping continues, the gain of the laser medium builds up and the intracavity radiation pattern develops a structure periodic in the round-trip time $T=2L/c$. When the gain is sufficient to overcome the linear and nonlinear losses in the cavity, laser threshold is achieved and the period of linear amplification begins (at time t_1 in Fig. 2.39). Because of the higher gain at the centre of the lasing bandwidth, spectral narrowing occurs during this stage with consequential increases in the durations of the fluctuations in each round-trip period. In the nonlinear stage of ultrashort pulse development, which begins at time t_2, the random pulse structure is transformed by the combined actions of nonlinear saturation of the absorber and laser gain saturation. As a result, one of the fluctuation spikes grows in intensity until it dominates. This surviving pulse also experiences some shortening in duration and an increase in spectral bandwidth. The important role of laser gain depletion during the nonlinear stage of amplification in producing a strong discrimination of pulse intensities was pointed out by *Glenn* [2.137, 138]. Earlier treatments [2.14] assumed negligible gain saturation. This assumption is in keeping neither with the values of the laser parameters nor with the results of the experimental investigations.

In the nonlinear amplification phase the net gain experienced by a pulse per round trip is given by [2.138]

$$dI_k/dk = I_k[A_k - \Gamma - B_0/(1+I_k)] \qquad (2.24)$$

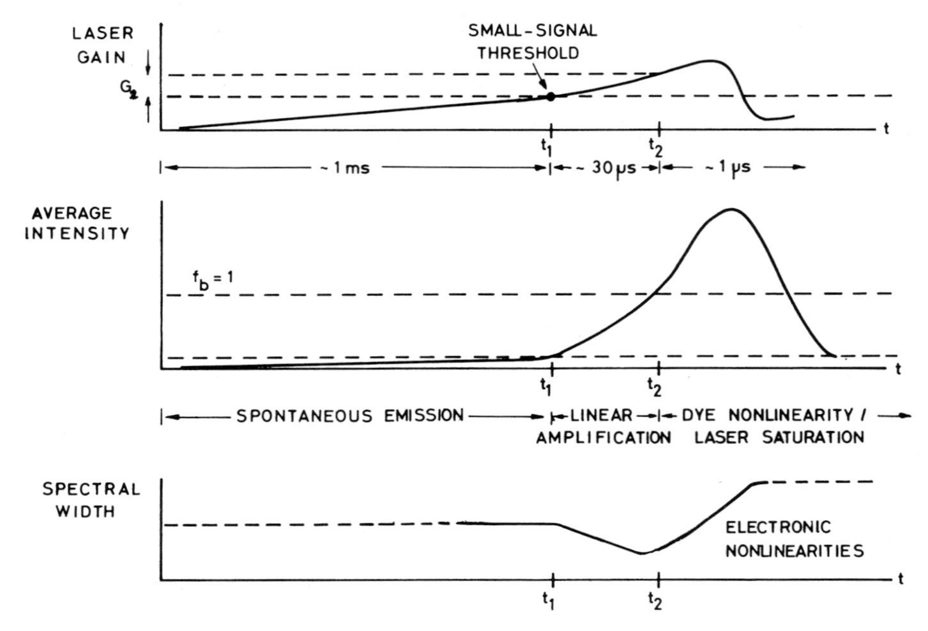

Fig. 2.39. Behaviour of gain, intensity and spectral width of passively modelocked Nd:glass laser (From *G. H. C. New*, unpublished)

where I_k is the pulse intensity at the k-th pass, normalized to the absorber saturation intensity $I_s=(2\sigma_b T_{1b})^{-1}$, A_k is the gain per pass, and Γ represents the cavity losses arising from scattering and the mirror transmissions, σ_b and T_{1b} are the cross section and the relaxation time of the saturable absorber dye, respectively, and B_0 is the initial absorption coefficient. The gain coefficient A_k obeys the equation

$$dA_k/dk = -A_k(2\sigma_a I_s T)I_k + fT \tag{2.25}$$

where f is the rate of flashlamp pumping and σ_a is the gain cross section. In deriving (2.24) and (2.25) it is assumed that the nonlinear absorber has instantaneous response, and that the flashlamp pumping produces a linearly increasing value of A, as shown in Fig. 2.39. The effective gain coefficient is

$$G = A - \Gamma - B_0/(1+I_k). \tag{2.26}$$

At time t_2, G is positive for all the pulses of the circulating intensity pattern. After that time only pulses above the critical intensity, given by

$$I_c = [B_0/(A-\Gamma)] - 1 = [B_0/(B_0 + \Delta A)] - 1 \tag{2.27}$$

for which $G=O$, will continue to grow since as the gain becomes depleted, $\Delta A = A - \Gamma - B_0$, becomes negative.

In the computer simulation model employed by Glenn, the gain was assumed to be constant during a pass through the laser and was recalculated at the end of each pass, taking into account the total energy extracted from the laser medium. The set of M starting pulses (corresponding to M longitudinal cavity modes) was assumed to have the probability density $W(I)=\exp(-I)$, for which the most probable, normalized, intensity of the N-th largest pulse has the value $\log(M/N)$. Clearly, the most important property of the fluctuation pattern is the ratio of the intensities of the two largest fluctuations at the end of the linear amplification stage. The probability that the largest pulse is more than R times as intense as the next largest can be shown to be, approximately, $R/(M+1)^{R-1}$. If each pulse is assigned its most probable value, the ratio between the two largest pulses is ~ 1.18 and for $M=100$, the ratio will exceed this value for $\sim 50\%$ of the time.

Figure 2.40 illustrates the basic type of behaviour predicted by the fluctuation model of Glenn for the case of the modelocked Nd:glass laser, operating in a cavity of 5 ns round-trip time. The upper family of curves shows the development of the five largest pulses out of an ensemble of 100, when gain

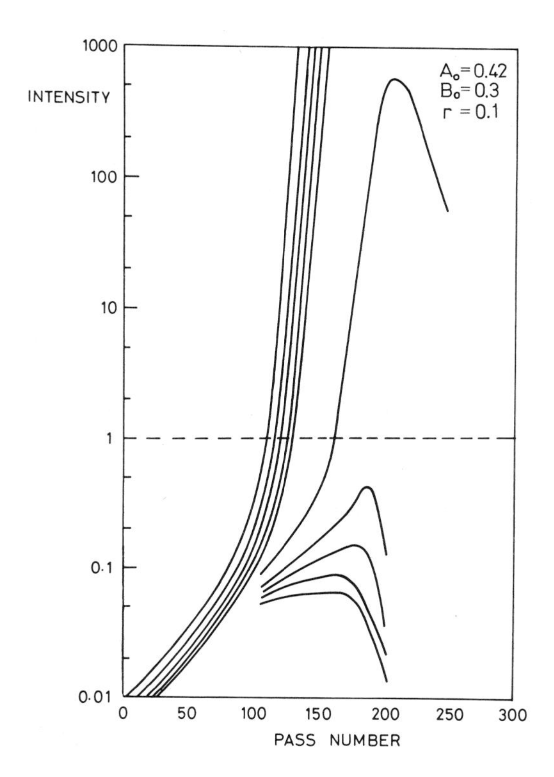

Fig. 2.40. Development of five largest pulses out of an ensemble of 100 for $B_0=0.3$, $\Gamma=0.1$ and $\Delta A=0.02$ at the start of the process. Upper curves show development when gain saturation is neglected and lower family of curves shows the effect of gain saturation (see text). (From *Glenn* [2.138])

saturation is neglected and $B_0 = 0.3$, $\Gamma = 0.1$ and $\Delta A = 0.02$ at the start of the process. It is seen that discrimination is very weak and that the pulses grow without limit, which obviously cannot hold in practice. When saturable amplification is taken into account, the behaviour shown in the lower set of curves is predicted. The discrimination in favour of the most intense pulses is greatly increased by the slowing down of the rate of growth in the region of nonlinear amplification. As a result, a final intensity ratio of 2500 near the peak of the laser giant pulse is produced. The computations also showed that when B_0 is reduced to 0.2, the final ratio exceeds 9000, for an initial value of the intensity ratio of the two strongest pulses of only 1.02. This ratio would be exceeded 93% of the time.

These result clearly show that the most important parameter governing the behaviour of the laser is the excess gain $\Delta A = A - \Gamma - B_0$ existing at the start of the nonlinear phase of passive modelocking. For a small range of values of $\Delta A > 0$ the combination of the absorber nonlinearity and laser gain saturation provides sufficient discrimination to select the largest of the initial fluctuations, so as to produce a train of isolated single pulses. The important experimental result that the best mode locking of the Nd:glass laser is obtained by operating just above threshold ($\Delta A \leqslant 0.03$) is also obtained from the fluctuation model with gain saturation taken into account. When the excess gain is too low the intensity never breaks through the dye, but for higher gain discrimination is poor.

When the relaxation time T_{1b} of the absorber is longer than the mean duration of the fluctuation peaks at the end of the linear amplification stage, the fluctuation pulses of duration $< T_{1b}$ are poorly selected and can remain as noise spikes inside the final pulse intensity envelope. The optimum case occurs obviously [2.14] when $M \cong T/T_{1b}$ and then the most intense peak will be selected very effectively.

2.3.2 Passively Modelocked Dye Lasers

As we have seen in subsection 2.2.2, the generation of ultrashort pulses occurs very rapidly in modelocked dye lasers, and the final pulse durations are much shorter than the recovery times of the saturable absorbers employed. This very different behaviour of dye lasers, compared with solid state lasers, derives from the short (nanosecond) excited state lifetimes (T_{1a}) of the amplifying dye media. Because of the resulting short energy storage times, even flashlamp pumped dye lasers operate in a quasi-cw manner.

Employing a rate-equation analysis, *New* [2.102] first showed that in these circumstances amplifier and absorber saturation can combine to produce rapid pulse shortening, even when the pulse duration is already considerably shorter than both the gain and the absorber relaxation times. Later, the early stages of modelocking were analytically treated [2.139] and conditions were derived for the growth of oscillating perturbations. *Garside* and *Lim* [2.140, 141] employed a coherent field analysis and obtained similar numerical solutions for both the buildup of the modelocking processes, and the eventual generation of ultrashort pulses. More recently *Haus* [2.120] has obtained a closed-form solution for the pulse intensity profile for the conditions under which cw modelocked dye lasers operate when a bandwidth-limiting element is used. Also *Yasa* et al. [2.142] have applied to dye lasers the computational model of *Glenn* [2.138] for the initial laser fluorescence fluctuation pattern to show numerically how these initial fluctuations evolve into a single picosecond pulses. *Yasa* and *Teschke* [2.143] have applied the same initial fluctuation pattern model to ultrashort pulse evolution in synchronously pumped dye lasers to obtain a stationary pulse solution. Reasonably good agreement with the experiments has been obtained with these various theoretical models, which are still being actively developed. It will be sufficient here to pick out the main points of the analyses and to show how they relate to the experimental results.

Rate-Equation Analysis

Frequency-dependent phenomena are explicitly left out in the rate-equation analysis, and as a result the ultimate pulse durations cannot be predicted quantitatively. However, the rate-equation approach is simple and provides particularly useful insight into pulse-compression mechanisms.

For pulses substantially shorter than either of the recovery times T_{1a} or T_{1b}, saturation is energy dependent and the pulse energy $j=\sigma_a\int_{-\infty}^{+\infty}F(\tau)d\tau$ [$F(\tau)$ is the photon flux at the local time $\tau=t-(x/c)$] obeys the energy balance equation [2.144]

$$A_L[1-\exp(-j)]-B_L s^{-1}[1-\exp(-sj)]-\Gamma j=0 \tag{2.28 (a)}$$

where A_L and B_L are the values of the respective coefficients at the leading edge of the pulse. The corresponding values in the trailing edge after the passage of a pulse are given by

$$A_T=A_L\exp(-j);\; B_T=B_L\exp(-sj). \tag{2.28 (b)}$$

Γ is the linear loss and s is the important parameter introduced by *New* [2.103] the value of which is given by $s=kA_a\sigma_b/A_b\sigma_a$. A_a and A_b are the laser beam cross-sectional areas at the positions of the amplifier and absorber dyes, respectively. For an absorber dye cell in optical contact with a laser mirror, k has the value 2 [2.103] when the cell is thin compared with the pulse length.

New traced out the complete time development of a pulse, showing that the rate of pulse shortening is extremely rapid, and derived the regions of pulse compression in terms of the small signal laser gain, the round-trip time T, expressed in units of T_{1a}, and the value of s. The conclusions of the analysis were in good agreement with the experimental results, particularly that the best modelocking is achieved with high values of s and that the cavity length has to be adjusted to the correct value for optimum performance.

When the rate-equation analysis is applied to the early growth of modelocking in a ring dye laser the initial rate of buildup is found to be a strong function of the absorber relaxation times T_{1b}, being slow when T_{1b} is short, in agreement with the results obtained with cresyl-violet modelocked by DTDCI and DIC, described in *Temporal Buildup of Modelocking in Dye Lasers* (see p. 44). The minimum condition for modelocking

$$s(T_{1b}/T_{1a}) > A_u/B_u \tag{2.29}$$

is also obtained, where A_u and B_u are the unsaturated (small signal) amplification and absorption coefficients, respectively, per cavity transit.

Steady-State Pulse Solutions

Garside and *Lim* [2.140] applied standard density matrix formalism, with the rotating wave and the slowly varying envelope approximations, also to a ring dye laser to study modelocking conditions by a perturbation analysis. This analysis again showed that the recovery time of the absorber does not restrain the durations of the modelocked pulses. The results obtained by this approach were compared in detail [2.141] with those of *New* for both cw and flashlamp-pumped dye lasers. These authors further showed [2.141] that under typical experimental conditions flashlamp-pumped dye lasers can generate pulses that are close to being steady state. The wavelength dependence of the buildup of modelocking by DODCI in the rhodamine 6G dye laser (Fig. 2.24) also came out of the computations.

Haus [2.120] and *Haus* et al. [2.146] considered the situation when both the gain and loss per transit are relatively small ($\sim 20\%$). This situation is a good approximation to the cw dye laser case. These authors assumed that the saturable-absorber loss and the laser-medium gain coefficients could be expanded to second order in their pulse-energy dependence, and that the dispersion of the system, as a function of frequency, could be expanded to the second order in frequency (i.e., the laser pulse spectrum is narrow compared with either the laser dye fluorescence spectrum or the saturable-absorber bandwidth).

Closed-form analytic solutions were obtained to the steady-state equation

$$[1 + l(t) - g(t) - (\Delta\omega)^{-2} d^2/dt^2] E(t) = -(\delta/\Delta\omega) dE(t)/dt \tag{2.30}$$

where $l(t)$, $g(t)$ are the loss and gain on transit through the absorbing and active media respectively; δ is a time delay (or advance) parameter proportional to any deviation δT of the pulse repetition period from the round-trip time T. The presence of a bandwidth-limiting element (tuning prism or etalon of bandwidth $\Delta\omega$) results in the introduction of the operator $-(\Delta\omega)^{-2} d^2/dt^2$, which leads to spreading of the pulse temporal profile. For the equilibrium situation this pulse broadening is balanced by the pulse compression processes.

The conditions required to achieve modelocking with a saturable absorber of long relaxation time when expressed in terms of the same parameters, were similar to those of *New*, with quantitative discrepancies of only $\sim 10\%$. By approximating the excess-gain versus pulse energy curve by a parabola the steady-state pulse shape was shown to be a hyperbolic secant, with intensity $I(t)$ proportional to $\mathrm{sech}^2(t/\Delta t_p)$. The pulse width Δt_p was found to be inversely proportional to the disperser element bandwidth $\Delta\omega$, to the pulse energy, and to the square root of the small-signal value of the absorber loss $l(t)$. These predictions are in good agreement with the properties of the bandwidth-limited subpicosecond pulses obtained from the cw dye laser arrangement described in *Modelocked* cw *Dye Lasers* (see p. 54).

2.4 Picosecond Pulse Amplification

In Section 2.2 it was shown that, for a modelocked laser oscillator producing bandwidth-limited pulses, the energy content and peak powers of individual pulses were limited to the values 10^{-10} J (300 W) for the cw dye laser, 50 μJ (25 MW) from the flashlamp-pumped dye-laser and 1 mJ (200 MW) for giant-pulse solid-state lasers. For several applications of picosecond pulses amplifiers must be used to obtain greater energies, without producing either amplitude or frequency modulation of the oscillator pulses. The physical parameters affecting the design of picosecond pulse amplifiers, and limiting the performance, are the stimulated emission cross section σ, the lifetime of the upper laser level T_1, and the nonlinear refractive index n_2 of the host material.

Consider the rate equations, for a pulse travelling in the positive x direction through an amplifying medium [2.102, 147]

$$\dot{n}=(n_0-n)T_1^{-1}-n\sigma F \tag{2.31}$$

and $$\partial F/\partial x=n\sigma F \tag{2.32}$$

where F is the photon flux, n_0 is the population inversion at $F=0$, and the differentiation in (2.31) is with respect to the local time $\tau=t-(x/v)$, where v is the velocity of light in the medium. For picosecond pulses $T_1 \gg$ the pulse duration Δt_p and (2.31) gives

$$n(\tau)=n_i \exp(-\sigma J_\tau) \tag{2.33}$$

where

$$J_\tau=\int_{-\infty}^{\tau} F(\tau)d\tau. \tag{2.34}$$

n_i is the initial population inversion before the arrival of the pulse, and J_∞ represents the total pulse energy per unit area. Combining (2.32) and (2.33) yields

$$\partial F(\tau)/\partial x = n_i \sigma F \exp(-\sigma J_\tau). \tag{2.35}$$

Far out on the leading edge of a short pulse, travelling through an amplifier of length l, J_τ has a negligible value. The small signal gain for this portion of the pulse envelope is given by

$$g_{ss} = \exp(n_i \sigma l). \tag{2.36}$$

The total amplified energy density at the end of the amplifier is

$$E(l) = h\nu J_\infty(l) = h\nu\sigma^{-1} \ln\{1 + g_{ss}[\exp \sigma J_\infty(0) - 1]\}. \tag{2.37}$$

From (2.33), the saturation energy density (the energy density required to deplete the population inversion to e^{-1}) is

$$E_s = h\nu/\sigma. \tag{2.38}$$

When $E(0) \ll E_s$, the population inversion is not significantly changed by the passage of the pulse and the pulse profile is unaltered. When $E(0) \sim E_s$ amplification becomes nonlinear and the front of the pulse is more strongly amplified, whereas for $E(0) > E_s$ the stored energy is completely swept out by the pulse and the gain in energy depends only on the value of the total inversion. The size of the stimulated emission cross section affects amplifier design in other ways. If σ is too large, parasitic oscillations and amplified spontaneous emission (ASE) can, if suitable precautions are not taken, build up to a level where amplifier stored energy is seriously depleted. This problem arises particularly with dye laser systems where $\sigma \sim 10^{-16}$ cm^2. On the other hand, for efficient energy extraction at low pulse energies, a high value of σ is advantageous.

When an amplified picosecond pulse approaches a high enough power density, nonlinear refractive index effects begin to seriously affect the pulse properties. Whole beam self-focusing [2.148–150] occurs when the nonlinear phase shift builds up to 2π in a distance of $(2\pi/\lambda)\,a^2$, where a is the beam radius [2.151] and λ is the laser wavelength. The phase shift built up in a propagation distance l is given by

$$\Delta\phi \equiv B = (2\pi/\lambda)\,(n_2) \int_0^l I(x)dx. \tag{2.39}$$

I is the light intensity. For picosecond pulses the nonlinear refractive index, n_2, can be regarded [2.152, 153] as arising mostly from the nonlinear electronic polarizability (distortion of the electronic clouds). For glass the critical power for self-focusing of a uniform beam of gaussian cross section is $\sim 5\times 10^6$ W. Small scale spatial fluctuations can cause self-focusing of portions of the beam at power levels well below those for total beam collapse [2.154, 155]. The most unstable spatial frequency grows according to the relation $I = I_0 \exp B$ [2.156]. The cumulative value of the B integral is thus a measure of the significance of nonlinear effects.

The intensity dependent index of refraction also produces self-phase modulation, as we have seen in the discussion of modelocked oscillators in Section 2.2. This frequency modulation is converted to modulation of the pulse temporal profile in the presence of linear dispersion [2.157]. Thus knowledge of the values of the nonlinear index n_2, and of the effects arising from n_2, is of great importance for the design of picosecond amplifiers.

2.4.1 Neodymium : Glass Amplifiers

The stimulated emission cross sections for the host materials used in Nd : glass lasers are $\sim 3-5\times 10^{-20}$ cm^2 [2.158]. The saturation energy density is ~ 5 J cm^{-2}, equivalent to a power density of 1 TW cm^{-2} for a pulse of 5 ps duration. However when the laser power density exceeds a few gigawatts per square centimetre nonlinear effects in the glass become apparent. Thus, any oscillator-amplifier system has to be operated under small-signal gain conditions, and the power density has to be maintained below the threshold levels for small-scale self-focusing and self-phase modulation. To reduce the possibility of small-scale self-focusing it is desirable to eliminate any small particles of dust or dirt on the optical surfaces and to reduce residual aberrations as much as possible. The most unstable spatial frequency corresponds to a structure of 100 μm dimension at a laser intensity of 10 GW cm^{-2} [2.156]. At the focus of a lens this structure falls outside the diffraction focus of the main beam. Thus, placing a limiting aperture (spatial filter) at the focus blocks the higher spatial frequencies and prevents the cumulative amplification of beam structures. The use of apodizing apertures [2.159] prevents the beam from developing diffraction fringes.

A picosecond Nd:glass laser system capable of amplifying a 5 ps pulse to an energy of 3 J (0.6 TW) is shown in Figure 2.41 [2.160]. This system employs a new phosphate glass LHG5 (Hoya) possessing a low value for n_2, zero refractive index temperature gradient (dn/dT) and a high stimulated emission cross section. The resulting increase in small-signal gain means that the amplifier pathlength can be shorter. Combined with the smaller value of n_2, this results in a reduced value of the B integral for a given overall amplification. Highly stable, TEM_{00} single-transverse mode, pulses of durations 3 to 10 ps and 1 mJ energy are obtained from the passively modelocked oscillator by switching out with a Kerr cell driven by a laser triggered spark gap. These pulses are passed through short length (15 to 20 cm) rod amplifiers to prevent self-focusing, self-phase modulation, amplified spontaneous emission and

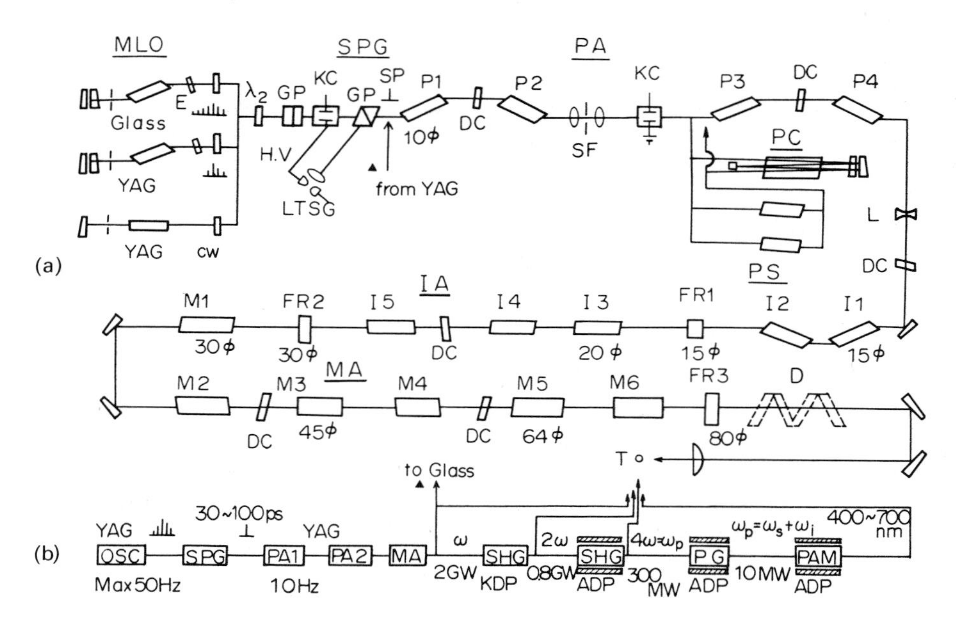

Fig. 2.41a and b. Experimental arrangement for (a) amplification and (b) frequency conversion of ultrashort pulses from Nd:glass and Nd:YAG lasers (from *Kuroda* et al. [2.160])

parasitic oscillation. The amplifiers are also decoupled and isolated by saturable absorber dye cells and by Faraday rotators with dielectric polarizers. The beam profile is cleaned up by spatial filtering. The final rod amplifier has a diameter of 64 mm and the power density is ~ 18 GW cm^{-2}.

New fluoride laser glasses with even smaller values of n_2 are under development [2.158] with gain coefficients and saturation parameters comparable to those of silicate glasses. These glasses should permit greater flexibility in the construction of picosecond amplifier systems.

Amplifier systems may also be used in conjunction with saturable absorbers to produce subpicosecond pulses [2.161, 162].

2.4.2 Amplification of Dye Laser Pulses

Dye laser amplifiers differ from Nd:glass systems in their disparate storage times of ~ 5 ns and ~ 300 μs, respectively. As a result, in a flashlamp-pumped dye laser amplifier it is only the energy stored during the 5 ns or so before the arrival of a picosecond pulse that contributes. Because of the large stimulated emission cross section ($\sigma \sim 10^{-16}$ cm^2) dye amplifier dimensions are limited by the need to avoid ASE losses, which are significant even with low stored energy densities. Also, as already discussed, the large cross section results in a high small-signal gain and a low saturation energy density (~ 3 mJ cm^{-2}).

Amplification to peak powers of >3 GW have been achieved with the arrangement shown in Fig. 2.42 [2.33]. Modelocking of an aqueous solution of rhodamine 6G (Rh6G) by an ethanolic solution of DODCI covered the spectral range 595 to 625 nm while the region 575 to 600 nm was covered by modelocking an ethanolic solution of Rh6G by DQOCI. The pulses employed for amplification studies had durations of ~ 2 ps, as measured with a streak camera, and the corresponding spectral bandwidths (FWHM) were ~ 2 nm. The first amplifier employed a 5 mm internal diameter dye cell, and the second amplifier a 1 cm diameter dye cell. Firing of the two amplifiers was synchronized to ± 100 ns and maximum gain always occurred at the peak of the 5 μs flash-lamp pulse. With the system aligned and firing only the amplifiers, up to 500 mJ of ASE was emitted from the second amplifier. This arose, in part, from reflection from the oscillator output mirror and could be overcome by employing ethanolic solutions of Rh6G. Then ASE occurred mainly at wavelengths shorter than 585 nm. When amplifying pulses of wavelengths >600 nm a solution of 3,3′-dimethylthiacyanine iodide placed between the oscillator and the first amplifier selectively absorbed the ASE.

Alternatively, when the ASE was not spectrally different from the picosecond pulses, a Fabry-Perot filter was used as isolator. Small signal gain measurements were carried out, for each amplifier independently, by attenuating the oscillator output one-hundred fold. Fig. 2.43 gives the variation of peak small-signal amplifier gain with wavelength, for different dye solutions and shows the importance of choosing the correct amplifying solution for optimising small-signal gain at a particular wavelength. To obtain the wavelength dependence of gain in the saturated regime (σ varies with wavelength (Fig. 2.43) and hence the value of the saturation energy density E_s) the unattenuated oscillator beam was passed through the first amplifier and the beam area telescoped down to 3 mm^2 before the second amplifier. From Fig. 2.44 it can be seen that, with the energy densities employed, the gain at each wavelength diminishes rapidly from its small-signal value to a wavelength in-

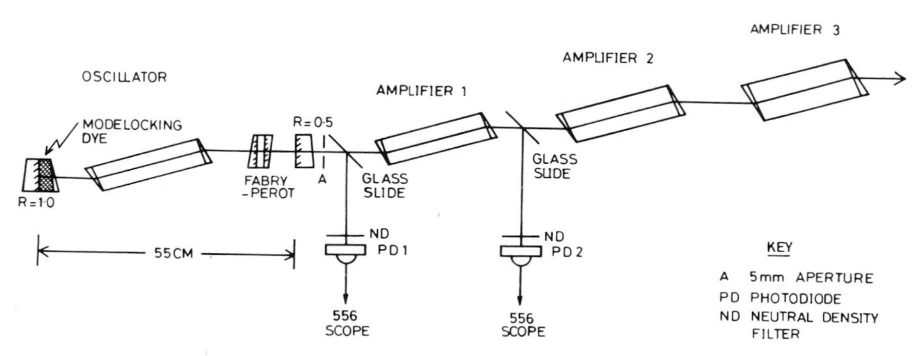

Fig. 2.42. Schematic diagram of modelocked dye-laser oscillator and amplifier chain for amplification of picosecond pulses to powers of >3 GW (from *Adrain* et al. [2.33])

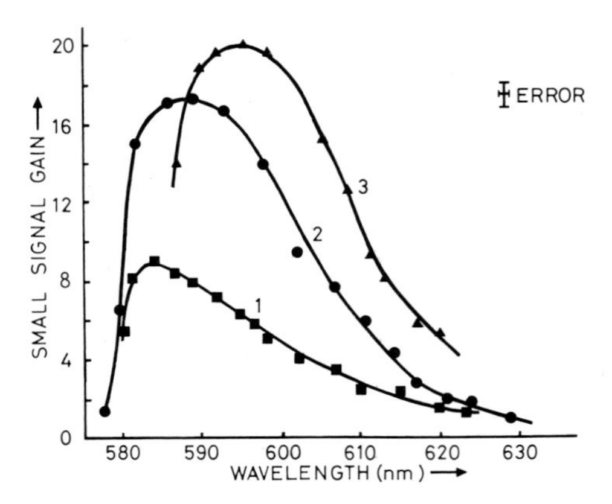

Fig. 2.43. Variation of small signal gain with wavelength for, (1) 10 mm diameter amplifier (9×10^{-5} M Rh 6G in ethanol), (2) 5 mm diameter amplifier (1.5×10^{-4} M Rh 6G + 10^{-5} M Rh B in ethanol), (3) 5 mm diameter amplifier (2.5×10^{-4} M Rh 6G in water)

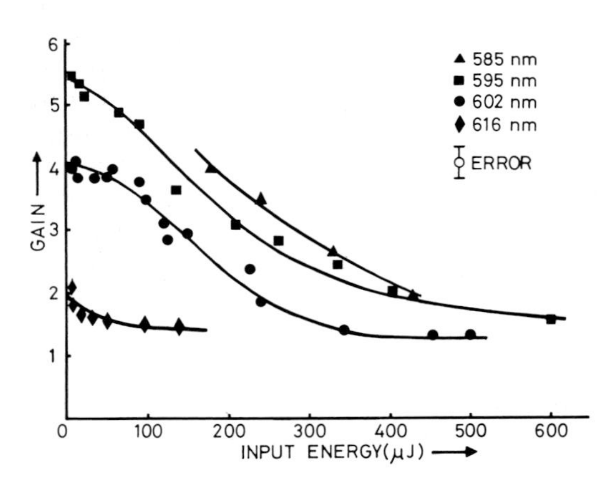

Fig. 2.44. Gain of 10 mm diameter dye laser amplifier at four wavelength as a function of input picosecond pulse energy

dependent saturated gain of ~1.4. These results clearly show how the amplification of picosecond pulses in a dye laser amplifier varies with both wavelength and input pulse energy density. As an example, a 70 μJ pulse, at 605 nm, can be amplified to ~1 mJ in the first amplifier and to ~4 mJ in the second amplifier. A third amplifier of 15 mm internal diameter pumped by 4 ablative lamps further increases the pulse energy to ~7 mJ, corresponding to a peak power of 3.5 GW. Streak-camera measurements of the amplified pulses showed no shortening of the pulse durations. Any self-steepening effect [2.157] would be masked by nonlinear self-action.

At these power densities self-phase modulation produced spectral broadening of the pulses by ~ 0.1 nm in the first (water solution) amplifier and by ~ 0.4 nm in the second (ethanolic solution) amplifier. Knowing the laser power density, the values $n_2 = 1.4 \times 10^{-13}$ esu for water and $n_2 = 4.5 \times 10^{-13}$ for ethanol can be calculated [2.150], in good relative agreement with the measured ratio of 4:1 for the values of n_2 for ethanol and water [2.163].

A nitrogen laser pumped amplifier is a better match to the storage time of a laser dye, since the pumping pulses last only a few nanoseconds. It is also possible to arrange the relative timing of the N_2 laser pulse so that one selected pulse of the modelocked train is amplified. Single-pulse selector electro-optical switches are then not required. *Schmidt* [2.164] has used this approach to amplify ~ 2.5 μJ pulses, from a modelocked flashlamp-pumped rhodamine 6G laser, in a cell containing a rhodamine B solution transversely pumped by the beam from an N_2 laser (3 mJ in 8 ns). The entrance and exit windows of the amplifier cell were anti-reflection coated and tilted to avoid feedback. The pumped region had dimensions 10 mm by 0.1 mm by 0.1 mm, and the oscillator beam was telescoped down to an area of 0.25 mm^2 to give an input energy density of ~ 1 mJ cm^{-2}. The amplified pulse of ~ 100 μJ was thus well above the saturation energy density. Further stages of amplification or a more powerful N_2 pumping laser should provide greater pulse energies. Since the nitrogen laser is an effective pump for a large number of dyes this method could be employed for the amplification of synchronously modelocked dye lasers, with their wider spectral coverage.

An alternative approach to amplification of the pulses from a modelocked oscillator is the modelocking of high-power laser oscillators by injection locking to a low-power modelocked signal from a cw laser, as proposed by *Moses* et al. [2.165]. This method has the advantage of locking the spectral and directional properties of the high-powered pulsed laser to those of the cw beam. Gains of $\sim 3 \times 10^4$ have been demonstrated [2.165] with a cw Rh6G dye laser, modelocked by synchronous pumping with a modelocked argon-ion laser operating at 5145 Å. The flashlamp-pumped Rh6G laser shared a common mirror (reflectivity 96%) with the cw laser. The optical lengths of the three lasers were precisely matched, employing a second-harmonic autocorrelation interferometer. When all the cavity lengths were matched (without the flashlamp being fired) the amplitude of the second-harmonic signal increased suddenly accompanied by a corresponding increase in the width of the cw lasing spectrum. The flashlamp-pumped system was spectrally controlled by a birefringent electro-optical tuner [2.166]. A birefringent Lyot filter in the cw laser gave a ~ 0.2 nm bandwidth, and tuning over the entire range of the dye for both lasers was obtained without realignment. The bandwidth of the flashlamp-pumped system was ~ 0.1 nm and good pulses were obtained near threshold. The pulse durations lengthened from 15 ps (of the cw laser) to 20 ps, corresponding to a time-bandwidth product, $\Delta\nu\Delta t$, value of ~ 2. Provided that bandwidth-limited pulse durations were maintained, this technique could be useful for the amplification of ~ 300 W peak power subpicosecond pulses from the passively modelocked cw dye laser [2.118, 168]. Also several high-power flashlamp-pumped lasers could be modelocked, synchronised and aligned by a single cw oscillator [2.168].

2.5 Frequency Changing

Harmonic generation provides the easiest means of producing picosecond pulses at short wavelengths. Apart from satisfying the normal phase-matching conditions (see *Akhmanov* et al. [2.169] for an excellent review of optical harmonic generation and optical frequency multipliers) group velocity mismatch has to be taken into account. For efficient harmonic generation it is necessary to have the optical path length L in the nonlinear material shorter than the group velocity characteristic length $L_{\Delta t} = \Delta t / \Delta u^{-1}$ where $\Delta u^{-1} (\equiv u_N^{-1} - u_1^{-1})$ is the group velocity mismatch parameter and u_1, u_N are the group velocities of the fundamental and N-th harmonic frequencies, respectively. The group velocity mismatch for the commonly used nonlinear crystals is given in Table 2.4. It can be seen that the effect is large in $LiNbO_3$ and $LiIO_3$ for 1.06 μm pulses and is even larger for ultraviolet frequency multipliers. When $L < L_{\Delta t}$ (the quasi-static regime), if the fundamental pulse has a gaussian intensity temporal profile, the harmonic pulse is also gaussian with the duration reduced by a factor $\sqrt{N}$ and the bandwidth increased by the same factor. When $L > L_{\Delta t}$ the width of the harmonic spectrum narrows [2.170] and the pulse duration increases during propagation through the crystal to the value $L\, \Delta u^{-1}$. This effect has been demonstrated for second-harmonic generation of Nd:glass laser picosecond pulses in KDP and in $LiNbO_3$ [2.171]. The spectral width and the harmonic pulse duration also depend strongly on the beam divergence [2.172]. On the positive side, the threshold intensities for optical breakdown in crystals increase for ultrashort pulses as the inverse of the laser pulse duration [2.173] and reach values of 10^{10} to 10^{11} W cm^{-2} for picosecond pulses.

Akhmanov et al. [2.174] generated the fourth-harmonic frequency of picosecond pulses from a Nd:glass laser by successive second-harmonic generation in KDP crystals, with an overall energy conversion efficiency of 2%. The laser was operating in many transverse modes and self-phase modulation probably

Table 2.4. Characteristic group-delay length $L_{\Delta t}$ for second-harmonic generation of a 1 ps pulse

Fundamental wavelength	$L_{\Delta t}$ cm			
	KDP	$LiNbO_3$	$LiIO_3$	CDA
1.06 μm	3.7	0.2	0.5	1.4
0.53 μm	0.3	—	—	—

(From Akhmanov et al. [2.169])

was affecting the temporal profiles of the pulses. Phase modulation can result in a deviation from the optimal phase-matching relationship when there is group velocity mismatch. Energy is then pumped back into the fundamental frequency [2.175]. With breakup of the temporal profile due to self-action effects, group velocity mismatch will certainly come into play. *Kung* et al. [2.176] obtained 80% efficiency for frequency doubling, and 10% efficiency for cascade tripling of bandwith-limited 50 ps pulses from a Nd:YAG laser. Employing an intracavity cesium dihydrogen arsenate (CDA), 90° phase-matched, second-harmonic generator *Weisman* and *Rice* [2.73] generated trains of pulses of nearly equal energies at the fundamental and harmonic frequencies of a Nd:glass laser. The second-harmonic bandwidth was narrowest when phase matching was optimized by temperature tuning the CDA crystal and operation of the modelocked laser was then at its most stable (pulse duration ~9 ps). This indicated that self-focusing and self-phase modulation were being suppressed by the nonlinear loss introduced by the harmonic generator [2.177].

Second-harmonic generation of pulses from synchronously pumped cw dye lasers over the uv spectral region from 265 nm to 335 nm has been obtained [2.5] using RhB, Rh6G and Na-fluorescein dyes. The pulse duration of the Rh6G laser was 7.5 ps at 600 nm and was bandwidth limited. With angle-tuned KDP crystals the energy conversion efficiency was ~0.05% giving uv average powers of 50 to 250 μW. With 90° phase-matched ADP a conversion efficiency of 2% was measured at 270 nm. The authors of [2.5] expect to achieve an average (modelocked) uv output power of 5 mW from Rh6G with a 90° phase-matched ADA harmonic generator. *Ippen* and *Shank* [2.178] reported conversion to the uv of the pulses from a passively modelocked dye laser by frequency doubling in a $LiIO_3$ crystal, of thickness 0.25 mm. The large nonlinear coefficient of this material more than compensates for the shorter characteristic group-delay length. Visible pulses with a peak power of ~5 kW can be converted to produce ~1 ps, 307.5 nm pulses with ~15% efficiency.

Pulses from dye lasers synchronously pumped by a modelocked Nd:YAG laser have been sum and difference mixed with both the fundamental and the second-harmonic pulses of the pumping laser, to generate tunable ultrashort pulses in the uv from 270 to 432 nm, and in the ir from 1.13 to 5.6 μm. [2.123] This technique has the advantages over optical parametric generators [2.179, 180] in that large beam divergences and broad spectral bandwidths are avoided. Also lower input power levels are required since the mixing process does not start from parametric noise. The modelocked Nd:YAG oscillator trains of 30 ps pulses (in a TEM_{00} beam of ~3 mJ total energy and a transform-limited bandwidth of ~0.1 nm) were amplified (sevenfold energy gain) and frequency doubled in a KDP crystal (energy conversion 30%). The dye laser was frequency narrowed and tuned by two Fabry-Perot etalons, of 6 μm and 100 μm spacings, to produce 12 ps bandwidth-limited pulses in a linearly polarized TEM_{00} mode from Rh6G. With RhB, cresyl-violet perchlorate, and carbazine 122 dyes a continuous tuning range of 549 to 727 nm was covered. Sum-mixing of the dye laser pulses with 1.064 m pulses in KDP generated pulses from 362 to 432 nm with an efficiency of 20% at 369 nm, the efficiency being primarily proportional to the 1.064 μm power. Sum-mixing the dye laser pulses with the 532 nm pulse train in ADP covered the spectral range from 270 to 307 nm. Difference mixing with 1.064 m radiation in $LiIO_3$ produced ir pulses tunable from 1.1 to 2.3 μm with ~2% efficiency and with 532 nm pulses the range 1.98 to 5.6 μm was covered. In all cases phase matching by angle tuning the crystals at room temperature was employed.

Nonlinear crystals capable of phase-matched harmonic generation are not transparent at wavelengths shorter than ~200 nm. Thus to reach the vacuum ultraviolet (vuv) gases and metal vapours must be employed as the nonlinear media for phase-matched harmonic generation. Third-order nonlinear processes in isotropic media [2.181] have been employed for the production of tunable [2.182] and fixed frequency [2.183] vuv radiation. Employing 30 ps pulses (peak powers $\sim 3 \times 10^8$ W) from a modelocked Nd:YAG laser, *Bloom* et al. [2.184] generated the third-harmonic frequency in a phase-matched mixture of rubidium and xenon with 10% energy conversion efficiency. Using the same laser system *Bloom* et al. [2.185] obtained frequency tripling in a homogeneous mixture of sodium and magnesium vapour with ~4% energy efficiency. Tripling in Rb-Xe mixtures has also been achieved with 7 ps pulses from a Nd:glass laser (2×10^8 W) with 2% energy conversion [2.186]. A further stage of frequency tripling to 118.2 nm radiation had been obtained earlier [2.187, 188]. Starting with a single pulse (25 ps, 3×10^8 W) at 1.064 μm frequency doubled in ADP to 532 nm and mixed, also in ADP, to give 354.7 nm radiation (peak power 10^7 W), the uv pulse was then frequency tripled in a phase-matched mixture of Xe and Ar to produce a vuv pulse at 118.2 nm. (Later, more careful measurements of the conversion efficiency of this stage revealed that the original measurement of 2.8% efficiency was too high by a factor of 20 [2.188].)

Employing a modelocked dye laser [2.189] it was possible to tune close to a two-photon resonance in calcium so as to enhance the nonlinear susceptibility [2.190] for third-harmonic generation. Pulses from a flashlamp-pumped Rh6G laser (2 ps, 50 μJ) were amplified to peak powers of ~300 MW and focused into calcium vapour, phase matched by xenon. The durations of the 200 nm pulses, measured by a streak camera, were the same as those of the fundamental frequency pulses (~4 ps). With two simultaneously driven dye lasers, synchronously pumped and modelocked by a common Nd:glass laser, *Royt* et al. [2.191] generated frequency-tunable 9 ps pulses in the vuv (190 to 195 nm) by four-wave mixing in strontium vapour. Resonant enhancement for both the two-photon and the single-photon frequencies was achieved. The conversion efficiency was 10^{-5} for power densities of 5×10^8 W cm^{-2} at the two-photon transition wavelength and 5×10^7 W cm^{-2} at the second dye-laser wavelength.

Picosecond pulses at 173.6 nm for amplification by the Xe_2 vuv laser have been obtained with the arrangement of Fig. 2.45 [2.192, 193] by mixing fundamental frequency (50 MW) and second-harmonic pulses from a modelocked ruby laser in magnesium vapour phase matched by xenon. There is a near two-photon resonance in magnesium for one ruby second-harmonic photon plus one fundamental-frequency photon (Fig. 2.46). A peak power of 200 W at 173.4 nm corresponded to a power conversion efficiency of 4×10^{-6}. Amplification of the oscillator pulses could provide vuv megawatt, picosecond pulses if saturation or optical Stark shifts do not affect the conversion efficiency [2.194]. Further frequency tripling in argon to the xuv (at 57.8 nm) could then be produced, after amplification if necessary by an excimer Xe_2 laser. This step has recently been achieved [2.43, 195] for the megawatt, nanosecond pulses

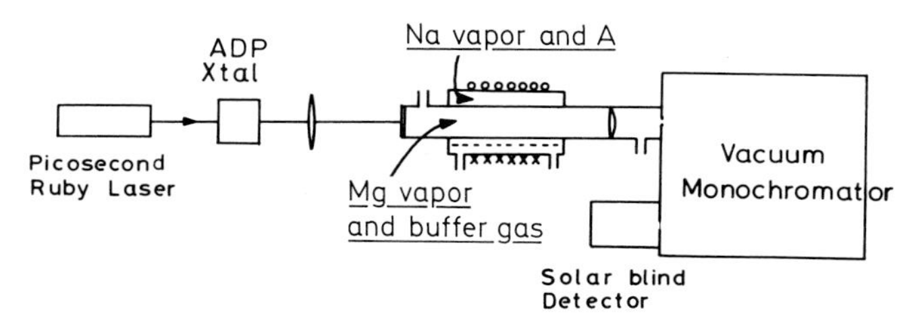

Fig. 2.45. Experimental arrangement for the generation of the fourth-harmonic frequency of a modelocked ruby laser. After the second-harmonic generation in ADP, one second-harmonic photon is sum-mixed with two fundamental photons in magnesium vapour phase matched by xenon (from *Bradley* [2.192])

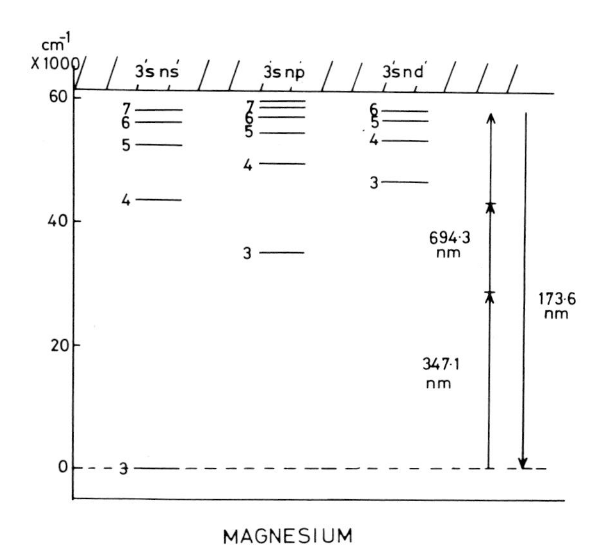

Fig. 2.46. Energy level diagram of MgI showing the transitions involved in two-photon resonance-enhanced four-wave mixing

from a frequency tunable Xe_2 laser [2.193, 195]. In this way xuv picosecond pulses could be obtained at useful powers.

In this chapter we have reviewed the methods of generating picosecond pulses by modelocked lasers. In the xuv region coherent generation has just recently been obtained [2.195], and new techniques are also being developed for the ir region. For example, ultrashort pulses can also be generated by an optical analogue of NMR free induction decay (FID) [2.197, 198]. Subnanosecond 10.6 μm pulses have been produced by sharply terminating a long duration CO_2 pulse by optical breakdown, and then passing it through a narrow CO_2 absorber [2.199]. Nearly total absorption occurred until the absorber reradiated a short FID pulse upon input pulse termination. Direct pulse observation as well as autocorrelation techniques [2.200, 201] have inferred pulses as short as 30 to 50 ps.

References

2.1 H.W. Mocker, R.J. Collins: Appl. Phys. Lett. **7**, 270 (1965)
2.2 A. J. DeMaria, D. A. Stetser, H. Heynau: Appl. Phys. Lett. **8**, 174 (1966)
2.3 A.E. Siegman, D.J. Kuizenga: Opto. Electronics, **6**, 43 (1974)
2.4 P.W. Smith: Proc. IEEE **58**, 1342 (1970)
2.5 J. de Vries, D. Bebelaar, J. Langelaar: Opt. Commun. **18**, 24 (1976)
2.6 E.D. Jones, M.A. Palmer, F.R. Franklin: Opt. Quant. Elect. **8**, 231 (1976)
2.7 A.J. Alcock, A.C. Walker: Appl. Phys. Lett. **25**, 299 (1974)
2.8 E. R. Pike (Ed.): *High Power Gas Lasers*, 1975, Proceedings of Summer School on Physics and Technology of High-Power Gas Lasers, Capri, 1975, Conference Series No. 29, (Institute of Physics, Bristol and London 1975)
2.9 H. Kogelnik, T. Li: Appl. Opt. **5**, 1550 (1966); Proc. IEEE **54**, 1312 (1966)
2.10 P.W. Smith: Proc. IEEE **60**, 422 (1972)
2.11 M. Born, E. Wolf: *Principles of Optics*, 3rd Edition (Pergamon Press, New York 1965) p. 494
2.12 L. Mandel, E. Wolf: Rev. Mod. Phys. **37**, 231 (1965)
2.13 D.J. Bradley, G.H.C. New: Proc. IEEE **62**, 313 (1974)
2.14 P.G. Kryukov, V.S. Letokhov: IEEE J. QE-**8**, 766 (1972)
2.15 E.K. Zavoiskii, S.D. Fanchenko: Sov. Phys. Doklady **1**, 285 (1956)
2.16 T. Ohmi, S. Hasuo, S. Hari: J. Appl. Phys. **43**, 3773 (1972)
2.17 J.A. Giordmaine, P.M. Rentzepis, S.L. Shapiro, K.W. Wecht: Appl. Phys. Lett. **11**, 218 (1967)
2.18 A. J. DeMaria, W. H. Glenn, M. J. Brienza, M. E. Mack: Proc. IEEE **57**, 2 (1969)
2.19 H.P. Weber: Phys. Lett. **27** A, 321 (1968)
2.20 J.R. Klauder, M.A. Duguay, J.A. Giordmaine, S.L. Shapiro: Appl. Phys. Lett. **13**, 174 (1968)
2.21 D.J. Bradley, G.H.C. New, S.J. Caughey: Phys. Lett **30** A, 78 (1969)
2.22 D.J. Bradley, T. Morrow, M.S. Petty: Opt. Commun. **2**, 1 (1970)
2.23 D.J. Bradley, M.H.R. Hutchinson, H. Koetser, T. Morrow, G.H.C. New, M.S. Petty: Proc. Roy. Soc. (London) **328** A, 97 (1972)
2.24 N. Ahmad, B.C. Gale, M.H. Key: Advan. Electron. Phys. **28** B, 999 (1969)
2.25 D.J. Bradley, J.F. Higgins, M.H. Key: Appl. Phys. Lett. **16**, 53 (1970)
2.26 D.J. Bradley: UK Patent Spec. 31167/70 (1970); U.S. Patent 3761614 (1973)
2.27 D.J. Bradley, B. Liddy, W.E. Sleat: Opt. Commun. **2**, 391 (1971)
2.28 M. Ya. Schelev, M.C. Richardson, A.J. Alcock: Appl. Phys. Lett. **18**, 354 (1971)

2.29 P. R. Bird, D. J. Bradley, W. Sibbett: Proc. 11th Intern. Congr. High Speed Photography, ed. by P.J. Rolls (Chapman and Hall, London 1974) p. 112
2.30 D.J. Bradley, W. Sibbett: Appl. Phys. Lett. **27**, 382 (1975)
2.31 D.J. Bradley: Opto-Electronics **6**, 25 (1974)
2.32 E. G. Arthurs, D. J. Bradley, B. Liddy, F. O'Neill, A. G. Roddie, W. Sibbett, W. E. Sleat: Proc. 10th Intern. Congr. High Speed Photography (A.N.R.T. Paris 1972) p. 117
2.33 R.S. Adrain, E.G. Arthurs, D.J. Bradley, A.G. Roddie, J.R. Taylor: Opt. Commun. **12**, 136 (1974)
2.34 R.S. Adrain, E.G. Arthus, W. Sibbett: Opt. Commun. **15**, 290 (1975)
2.35 D.J. Bradley, A.G. Roddie, W. Sibbett, M.H. Key, M.J. Lamb, C.L.S. Lewis, P. Sachsenmaier: Opt. Commun. **15**, 231 (1975)
2.36 N.G. Basov, M.M. Butslov, P.G. Kriukov, Yu. A. Matveets, E.A. Smirnova, S.D. Fanchenko, S.V. Chekalin, R.V. Chikin: Preprint 82, Lebedev Phys. Inst. Moscow, USSR (1972)
2.37 S.D. Fanchenko, B.A. Frolov: Sov. Phys. JETP Lett. **16**, 101 (1972)
2.38 M.M. Butslov, A.A. Malyutin, A.M. Prokhorov, B.M. Stepanov, M.Ya. Schelev: Proc. 10th Intern. Congr. High-Speed Photography (A.N.R.T. Paris 1972) p. 122
2.39 M.Ya. Schelev: Proc. 11th Congr. High-Speed Photography, ed. by P.J. Rolls (Chapman and Hall, London 1974) p. 32
2.40 J. Nuckolls, L. Wood, A. Thiessen, G. Zimmerman: Nature **239**, 139 (1972)
2.41 S.E. Harris, J.F. Young, A.H. Kung, D.M. Bloom, G.C. Bjorklund: In *Laser Spectroscopy*, ed. by R.G. Brewer, A. Mooradian (Plenum Press, New York and London 1974) p. 59
2.42 D.J. Bradley: In *Lecture notes in Physics*, Vol. 43: *Laser Spectroscopy* (Springer, Berlin, Heidelberg, New York 1975) p. 55
2.43 D.J. Bradley, M.H.R. Hutchinson, C.C. Ling: In *Springer Series in Optical Sciences*, Vol. 3: Tunable Lasers and Applications, Proceedings of the Leon Conference, Norway 1976, ed. by A. Mooradian, T. Jaeger, P. Stokseth (Springer, Berlin, Heidelberg, New York 1976) p. 40
2.44 B.L. Henke, J.A. Smith, D.T. Attwood: 1976 (unpublished)
2.45 P.R. Bird, D.J. Bradley, A.G. Roddie, W. Sibbett, M.H. Key, M.J. Lamb, C.L.S. Lewis: Proc. 11th Congr. High Speed Photography, ed. by P.J. Rolls (Chapman and Hall, London 1974) p. 118
2.46 G.I. Bryukhnevitch, Yu.S. Kasyanov, V.V. Korobkin, A.M. Prokhorov, B.M. Stepanov, V.K. Chevokin, M.Ya. Schelev: Proc. 11th Intern. Congr. High Speed Photography, ed. by P.J. Rolls (Chapman and Hall, London 1974) p. 554
2.47 Yu.S. Kasyanov, A.A. Malyutin, M.C. Richardson, V.K. Chevokin: Proc. 11th Intern. Congr. High Speed Photography, ed. by P.J. Rolls (Chapman and Hall, London 1974) p. 561
2.48 C.F. McConaghy, L.W. Coleman: Appl. Phys. Lett. **25**, 268 (1974)
2.49 A. J. Lieber, H. D. Sutphin, C. B. Webb: Abstract of paper at Intern. Conf. Physics of x-ray spectra, Gaithersburg, August 1976 (unpublished)
2.50 D.J. Bradley: 1976 (unpublished)
2.51 S.W. Thomas, J.W. Houghton, G. Tripp, L.W. Coleman: Proc. 11th Intern. Congr. High Speed Photography, ed. by P.J. Rolls (Chapman and Hall, London 1974) p. 101
2.52 G. Jeah-Francois, P. Nodenot, V.V. Korobkin, Yu.V. Korobkin, A.V. Prokhindeev, M.Ya. Schelev: Proc. 11th Intern. Congr. High Speed Photography, ed. by P.J. Rolls (Chapman and Hall, London 1974) p. 190
2.53 G. Porter: Opt. Commun. **18**, 141 (1976); private communication (1975)
2.54 L.A. Lompre, G. Mainfray, J. Thebault: Appl. Phys. Lett. **26**, 500 (1975)
2.55 R. Hadland, K. Helbrough, A.E. Huston: Proc. 11th Intern. Congr. High Speed Photography, ed. by P.J. Rolls (Chapman and Hall, London 1974) p. 107; K. Helbrough, M.C. Adams, W. Sibbett, T.H. Williams: Abstract of paper read at 12th Intern. Congr. High Speed Photography, Toronto, August 1976
2.56 D.J. Bradley: Proc. 11th Intern. Congr. High Speed Photography, ed. by P.J. Rolls (Chapman and Hall, London 1974) p. 23
2.57 D.J. Bradley, M.H.R. Hutchinson, H. Koetser: Proc. Roy. Soc. London **A329**, 105 (1972)

2.58 D.J. Bradley, W. Sibbett: Opt. Commun. **9**, 17 (1973)
2.59 N.G. Basov, M.M. Butslov, P.G. Krykov, Yu.A. Matveets, E.A. Smirnova, B.M. Stepanov, S.D. Franchenko, S.V. Chekalin, R.V. Chikin: Soviet Phys. JETP **38**, 449 (1974)
2.60 D. von der Linde: IEEE J. QE-**8**, 328 (1972)
2.61 M.A. Duguay, J.W. Hansen, S.L. Shapiro: IEEE J. QE-**6**, 725 (1970)
2.62 D.J. Bradley, G.H.C. New, S.J. Caughey: Opt. Commun. **2**, 41 (1970)
2.63 E.M. Gordeev, P.G. Kryukov, Yu.A. Matveets, B.M. Stepanov, S.D. Fanchenko, S.V. Chekalin, A.V. Sharkov: Sov. J. Quantum. Electron. **5**, 129 (1975)
2.64 M.C. Richardson: IEEE J. QE-**9**, 768 (1973)
2.65 F. De Martini, C.H. Townes, T.K. Gustafson, P.L. Kelley: Phys. Rev. **164**, 312 (1967)
2.66 R.A. Fisher, P.L. Kelley, T.K. Gustafson: Appl. Phys. Lett. **14**, 140 (1969)
2.67 V.V. Korobkin, A.A. Malyutin, M.Ya. Schelev: Sov. Phys. JETP Lett. **11**, 103 (1970)
2.68 R. C. Eckardt, C. H. Lee, J. N. Bradford: Opto-Electronics **6**, 67 (1974)
2.69 D.J. Bradley, G.H.C. New, S. Caughey: Phys. Lett. **33A**, 313 (1970)
2.70 N.E. Bykovskii, V. Kan, P.G. Kryukov, Yu.A. Matveets, N.L. Ni, Yu.V. Senatskii, S.V. Chekalin: Sov. J. Quant. Electron. **2**, 56 (1972)
2.71 S.V. Chekalin, P.G. Mryukov, Yu.A. Matveets, O.B. Shatkerashvilii: Opto-Electronics **6**, 249 (1974)
2.72 M.H.R. Hutchinson: PhD Thesis, The Queen's University of Belfast (1971)
2.73 R.B. Weisman, S.A. Rice: Spectros. Lett. **8**, 329 (1975)
2.74 N.G. Basov, Yu.A. Drozhbin, P.G. Kryukov, V.B. Lebedev, V.S. Letokhov, Yu.A. Matveets: Sov. Phys. JETP Lett. **9**, 256 (1969)
2.75 N. Nakashima, N. Mataga: Chem. Phys. Lett. **35**, 487 (1975)
2.76 E.D. Jones, M.A. Palmer: Opt. Quant. Electron. **7**, 520 (1975)
2.77 A.N. Zherikhin, V.A. Kovalenko, P.G. Kryukov, Yu.A. Matveets, S.V. Chekalin, O.B. Shatberashivili: Sov. J. Quant. Electron. **4**, 210 (1974)
2.78 A.W. Penney, H.A. Heynau: Appl. Phys. Lett. **9**, 257 (1966)
2.79 W.R. Hook, R.H. Dishington, R.P. Hilberg: Appl. Phys. Lett. **9**, 125 (1966)
2.80 G. Kachen, L. Steinmetz, J. Kysilka: Appl. Phys. Lett. **13**, 229 (1968)
2.81 M. Michon, H. Guillet, D. Le Goff, S. Raynaud: Rev. Sci. Instr. **40**, 263 (1969)
2.82 A.J. Alcock, M.C. Richardson: Opt. Commun. **2**, 65 (1970)
2.83 D. von der Linde, O. Bernecker, A. Laubereau: Opt. Commun. **2**, 215 (1970)
2.84 R.C. Hyer, H.D. Sutphin, K.R. Winn: Rev. Sci. Instr. **46**, 1333 (1975)
2.85 D.J. Bradley, A.J.F. Durrant: Phys. Lett. **27A**, 73 (1968)
2.86 D.J. Bradley, A.J.F. Durrant, G.M. Gale, M. Moore, P.D. Smith, IEEE J. QE-**4**, 707 (1968)
2.87 W.H. Glenn, M.J. Brienza, A.J. De Maria: Appl. Phys. Lett. **12**, 54 (1968)
2.88 B.H. Soffer, J.W. Linn: J. Appl. Phys. **39**, 5859 (1968)
2.89 D.J. Bradley, A.J.F. Durrant, F. O'Neill, B. Sutherland: Phys. Lett. **30A**, 535 (1969)
2.90 W. Schmidt, F.P. Schafer: Phys. Lett. **26A**, 558 (1968)
2.91 D.J. Bradley, F. O'Neill: Opto-Electronics **1**, 69 (1969)
2.92 E.G. Arthurs, D.J. Bradley, A.G. Roddie: Appl. Phys. Lett. **19**, 480 (1971)
2.93 E.G. Arthurs, D.J. Bradley, A.G. Roddie: Appl. Phys. Lett. **20**, 125 (1972)
2.94 D.J. Bradley, B. Liddy, A.G. Roddie, W. Sibbett, W.E. Sleat: Opt. Commun. **3**, 426 (1971)
2.95 D.J. Bradley, B. Liddy, W. Sibbett, W.E. Sleat: Appl. Phys. Lett. **20**, 219 (1972)
2.96 E.G. Arthurs, D.J. Bradley, A.G. Roddie: Appl. Phys. Lett. **23**, 88 (1973)
2.97 E.G. Arthurs, D.J. Bradley, A.G. Roddie: Chem. Phys. Lett. **22**, 230 (1973); Opt. Commun. **8**, 118 (1973)
2.98 E.G. Arthurs, D.J. Bradley, P.N. Puntambekar, I.S. Ruddock, T.J. Glynn: Opt. Commun. **12**, 360 (1974)
2.99 E.G. Arthurs, D.J. Bradley, T.J. Glynn: Opt. Commun. **12**, 136 (1974)
2.100 D.N. Dempster, T. Morrow, R. Rankin, G.F. Thompson: J. Chem. Soc. Faraday **II 68**, 1479 (1972)
2.101 D. Magde, M.W. Windsor: Chem. Phys. Lett. **27**, 31 (1974)
2.102 G.H.C. New: Opt. Commun. **6**, 188 (1972)

2.103 G.H.C. New: IEEE J. QE-**10**, 115 (1974)
2.104 A.G. Roddie: PhD Thesis, The Queen's University of Belfast (1072)
2.105 R.S. Adrain: PhD Thesis, The Queen's University of Belfast (1974)
2.106 F.P. Schafer, W. Schmidt, W. Marth: Phys. Lett. **24A**, 280 (1967)
2.107 G. Mourou, G. Busca, M.M. Denariez-Roberge: Opt. Commun. **4**, 40 (1971)
2.108 D.J. Bradley, T. Morrow, M.S. Petty: Opt. Commun. **2**, 1 (1970); M.S. Petty: PhD Thesis, The Queen's University of Belfast (1970)
2.109 J.C. Mialocq, A.W. Boyd, J. Jaraudias, J. Sutton: Chem. Phys. Lett. **37**, 236 (1976)
2.110 G.E. Busch, R.P. Jones, P.M. Rentzepis: Chem. Phys. Lett. **18**, 178 (1973)
2.111 G. Mourou, B. Drouin, M.M. Denariez-Roberge: Appl. Phys. Lett. **20**, 453 (1972)
2.112 H.E. Lessing, E. Lippert, W. Rapp: Chem. Phys. Lett. **7**, 227 (1970)
2.113 D. Kuizenga: Appl. Phys. Lett. **19**, 260 (1971)
2.114 A. Dienes, E.P. Ippen, C.V. Shank: Appl. Phys. Lett. **19**, 258 (1971)
2.115 E.P. Ippen, C.V. Shank, D. Dienes: Appl. Phys. Lett. **21**, 348 (1972)
2.116 F. O'Neill: Opt. Commun. **6**, 360 (1972)
2.117 E.P. Ippen, C.V. Shank: Appl. Phys. Lett. **27**, 488 (1975)
2.118 I.S. Ruddock, D.J. Bradley: Appl. Phys. Lett. **29**, 296 (1976)
2.119 C.V. Shank, E.P. Ippen: Appl. Phys. Lett. **24**, 373 (1974)
2.120 H.A. Haus: IEEE J. QE-**11**, 736 (1975); Opt. Commun. **15**, 29 (1975)
2.121 T.R. Royt, W.L. Faust, L.S. Goldberg, C.H. Lee: Appl. Phys. Lett. **25**, 514 (1974)
2.122 L.S. Goldberg, C.A. Moore: Appl. Phys. Lett. **27**, 217 (1975)
2.123 L.S. Goldberg, C.A. Moore: Lecture Notes in Physics, Vol. 43: *Laser Spectroscopy* (Springer, Berlin, Heidelberg, New York 1975) p. 248; Opt. Commun. **16**, 21 (1976)
2.124 C.K. Chan, S.O. Sari: Appl. Phys. Lett. **25**, 403 (1974)
2.125 C.K. Chan, S.O. Sari, R.E. Foster: J. Appl. Phys. **47**, 1139 (1976)
2.126 H. Mahr, M.D. Hirsch: Opt. Commun. **13**, 96 (1975)
2.127 J.M. Harris, R.M. Chrisman, F.E. Lytle: Appl. Phys. Lett. **26**, 16 (1975)
2.128 H. Mahr: private communication (1976)
2.129 P.K. Runge: Opt. Commun. **4**, 195 (1971)
2.130 P.K. Runge: Opt. Commun. **5**, 311 (1972)
2.131 Z.A. Yasa, O. Teschke: Appl. Phys. Lett. **27**, 446 (1975)
2.132 B. Fan, T.K. Gustafson: Appl. Phys. Lett. **28**, 202 (1976)
2.133 D. Roess: J. Appl. Phys. **37**, 2004 (1966)
2.134 V.S. Letokhov: Soviet Phys. JETP **28**, 562 (1969); Sov. Phys. JETP, **28**, 1026 (1969)
2.135 J.A. Fleck: Appl. Phys. Lett. **12**, 178 (1968); J. Appl. Phys. **39**, 3318 (1968)
2.136 J.A. Fleck: Phys. Rev. B. **1**, 84 (1970)
2.137 W.H. Glenn: Rep. No. L920935-8, United Aircraft Corp. (1972)
2.138 W.H. Glenn: IEEE J. QE-**11**, 8 (1975)
2.139 G.H.C. New, D.H. Rea: J. Appl. Phys. **47**, 3107 (1976)
2.140 B.K. Garside, T.K. Lim: J. Appl. Phys. **44**, 2335 (1973); Opt. Commun. **12**, 8 (1974)
2.141 B.K. Garside, T.K. Lim: Opt. Commun. **12**, 240 (1974)
2.142 Z.A. Yasa, O. Teschke, L.W. Bravermans, A. Dienes: Opt. Commun. **15**, 354 (1975)
2.143 Z.A. Yasa, O. Teschke: Opt. Commun. **15**, 169 (1975)
2.144 G.H.C. New, K.E. Orkney, M.J.W. Nock: Optical and Quant. Electron. **8**, 425 (1976)
2.145 P.G. Kryukov, V.S. Letokhov: Sov. Phys. Usp. **12**, 641 (1971)
2.146 H.A. Haus, C.V. Shank, E.P. Ippen: Opt. Commun. **15**, 29 (1975)
2.147 A. Icsevgi, W.E. Lamb: Phys. Rev. **185**, 517 (1969)
2.148 R.Y. Chiao, E. Garmire, C.H. Townes: Phys. Rev. Lett. **13**, 479 (1964)
2.149 P.L. Kelley: Phys. Rev. Lett. **15**, 1005 (1965)
2.150 S.A. Akhmanov, R.V. Khoklov, A.P. Sukhorukov: In *Laser Handbook*, ed. by F.T. Arecchi, E.O. Schultz-Dubois (North Holland Publ. Co., Amsterdam 1972) p. 1151
2.151 E.L. Dawes, J.H. Marburger: Phys. Rev. **179**, 862 (1969)
2.152 R.C. Brewer, C.H. Lee: Phys. Rev. Lett. **21**, 267 (1968)
2.153 R.R. Alfano, S.L. Shapiro: Phys. Rev. Lett. **24**, 592 (1970); **24**, 1217 (1970)
2.154 A.J. Campillo, S.L. Shapiro, B.R. Suydam: Appl. Phys. Lett. **23**, 628 (1973); **24**, 178 (1974)

2.155 E.S. Bliss, D.R. Speck, J.F. Holzrichter, J.H. Erkkila, A.J. Glass: Appl. Phys. Lett. **25**, 448 (1974)
2.156 J. Trenholme: Lawrence Livermore Lab. Rep., U.C.R.L. 50021-74 (1975) p. 178
2.157 R.A. Fisher, W.K. Bischel: J. Appl. Phys. **46**, 4921 (1975)
2.158 M.J. Weber, C.B. Layne, R.A. Saroyan, D. Milam: Opt. Commun. **18**, 171 (1976)
2.159 P. Jacquinot, B. Roizen-Dossier: In *Progress in Optics*, ed. by E. Wolf (North Holland Publ. Co., Amsterdam 1964) Vol. III, p. 130
2.160 H. Kuroda, H. Masuko, S. Maekawa: Opt. Commun. **18**, 169 (1976)
2.161 A. Penzkofer, D. von der Linde, A. Laubereau, W. Kaiser: Appl. Phys. Lett. **20**, 351 (1972)
2.162 A. Penzkofer: Opto-Electronics, **6**, 87 (1974)
2.163 W. Werncke, A. Law, M. Pfeiffer, K. Lenz, H.J. Weigmann, C.D. Thuy: Opt. Commun. **4**, 413 (1972)
2.164 A.J. Schmidt: Opt. Commun. **14**, 287 (1975)
2.165 E.I. Moses, J.J. Turner, C.L. Tang: Appl. Phys. Lett. **28**, 258 (1976); Opt. Commun. **18**, 24 (1976); Appl. Phys. Lett. **27**, 441 (1975)
2.166 J.M. Telle, C.L. Tang: Appl. Phys. Lett. **24**, 85 (1974); **26**, 572 (1975)
2.167 I.S. Ruddock, W. Sibbett, D.J. Bradley: Opt. Commun. **18**, 26 (1976)
2.168 I.S. Ruddock: PhD Thesis, University of London (1976)
2.169 S.A. Akhmanov, A.I. Kovrygin, A.P. Sukhorukov: In *Nonlinear Optics*, ed. by H. Rabin, C.C. Tang (Academic Press, New York, London 1975) p. 475 and references therein
2.170 S.A. Akhmanov, A.P.S. Sukhorukov, A.S. Chirkin: Sov. Phys. JETP **55**, 1480 (1968)
2.171 S.L. Shapiro: Appl. Phys. Lett. **13**, 19 (1968)
2.172 R.Yu. Orlov, A.S. Chirkin, T. Usmanov: Sov. Phys. JETP **57**, 1069 (1969)
2.173 D.F. Fradin, N. Bloembergen, J. Letellier: Appl. Phys. Lett. **22**, 635 (1973)
2.174 S.A. Akhmanov, I.B. Skidan, R.Yu. Orlov, L.I. Telegrin: Sov. Phys. JETP Lett. **16**, 471 (1972)
2.175 Yu.N. Karamzin, A.P. Sukhorukov: Sov. J. Quant. Electron. **5**, 496 (1975)
2.176 A. Kung, J.F. Young, G. Bjorklund, S.E. Harris: Phys. Rev. Lett. **29**, 985 (1972)
2.177 J.E. Murray, S.E. Harris: J. Appl. Phys. **41**, 609 (1970)
2.178 E.P. Ippen, C.V. Shank: Opt. Commun. **18**, 27 (1976)
2.179 R.L. Byer: In *Nonlinear Optics*, ed. by H. Rabin, C.L. Tang (Academic Press, New York, London 1975) p. 587 and references therein
2.180 S.E. Harris: Proc. IEEE **57**, 2096 (1969)
2.181 J.F. Ward, G.H.C. New: Phys. Rev. **185**, 57 (1969)
2.182 R.T. Hodgson, P.P. Sorokin, J.J. Wynne: Phys. Rev. Lett. **32**, 343, (1974)
2.183 S.E. Harris, J.F. Young, A.H. Kung, D.M. Bloom, G.C. Bjorklund: In *Laser Spectroscopy*, ed. by R.G. Brewer, A. Mooradian (Plenum Press, New York and London 1974) p. 59
2.184 D.M. Bloom, G.W. Bekkers, J.F. Young, S.E. Harris: Appl. Phys. Lett. **26**, 687 (1975)
2.185 D.M. Bloom, J.F. Young, S.E. Harris: Appl. Phys. Lett. **27**, 390 (1975)
2.186 H. Puell, C.R. Vidall: Opt. Commun. **18**, 107 (1976)
2.187 A.H. Kung, J.F. Young, S.E. Harris: Appl. Phys. Lett. **22**, 301 (1973)
2.188 A.H. Kung, J.F. Young, S.E. Harris, Erratum: Appl. Phys. Lett. **28**, 239 (1975)
2.189 W. Sibbett, D.J. Bradley, S.F. Bryant: Opt. Commun. **18**, 107 (1976)
2.190 P.D. Maker, R.W. Terhune, C.M. Savage: Proc. III Int. Conf. Quantum Electron, Paris 1964, p. 1559
2.191 T.R. Royt, C.H. Lee, W.L. Faust: Opt. Commun. **18**, 108 (1976)
2.192 D.J. Bradley: In *Laser Spectroscopy*, ed. by S. Haroche, J.C. Pebay-Peyroula, T.W. Hansch, S.E. Harris (Springer, Berlin, Heidelberg, New York 1975) p. 55
2.193 E.G. Arthurs, D.J. Bradley, C.B. Edwards, S. Domanski, D.R. Hull, C.C. Ling, M.H.R. Hutchinson: In *Proc. Intern. Conf. Electron Beam Research and Technology*, ed. by G. Yonas (Sandia Laboratories, SAND 76-5122, 1976) p. 193
2.194 J.N. Elgin, G.H.C. New: Opt. Commun. **16**, 242 (1976)
2.195 M.H.R. Hutchinson, C.C. Ling, D.J. Bradley: Opt. Commun. **18**, 203 (1976)
2.196 D.J. Bradley, D.R. Hull, M.H.R. Hutchinson, M.W. McGeoch: Opt. Commun. **7**, 187 (1975)

2.197 S.L. McCall: PhD Thesis, Univ. of Calif., Berkeley (1968)
2.198 D.C. Burnham, R.Y. Chiao: Phys. Rev. **188**, 667 (1969)
2.199 E. Yablonovitch, J. Goldhar: Appl. Phys. Lett. **25**, 580 (1974)
2.200 B.J. Feldman, R.A. Fisher, E.J. McLellan, S.J. Thomas: Opt. Commun. **18**, 72 (1976)
2.201 E. Yablonovitch, H.S. Kwok: Opt. Commun. **18**, 103 (1976)

3. Techniques for Measurement

E. P. Ippen and C. V. Shank

With 28 Figures

Picosecond optical pulses provide a unique means for studying ultrafast processes associated with the interaction of light with matter. Implementation of these studies has required the development of new measurement techniques capable of picosecond time resolution. In this chapter, we describe the various methods that are now available for characterizing picosecond laser pulses and for detecting rapid events created by them. Our emphasis here is on the relative advantages and limitations of the techniques themselves and less on the results of particular experiments. An understanding of these techniques is necessary for proper evaluation of any picosecond experiment. Many inconsistencies in early work have been due not only to the variability of pulsed laser sources but to improper interpretation of experimental results. As the various pitfalls of picosecond measurement become better understood, experimental studies become more reliable. At the same time, a better understanding of ultrafast process will undoubtedly lead to new and better measurement techniques.

3.1 Pulsewidth Measurements

The invention of the passively modelocked Nd:glass laser in 1965 [3.1] provided a pressing need for new techniques to measure the duration of ultrashort optical pulses. Direct measurement by the combined use of photodetectors and oscilloscopes was no longer adequate to temporally resolve the pulses being produced. Within a year, however, an indirect technique with subpicosecond time resolution had been proposed and demonstrated [3.2, 3, 4]. This technique, based on the nonlinear process of second-harmonic generation (SHG), is illustrated diagramatically in Fig. 3.1. The optical pulse is divided into two beams which travel different paths before being recombined in a nonlinear crystal. By polarizing the two beams differently [3.2, 4] or by making them noncollinear [3.3], it can be arranged that no SHG is detected when either beam is blocked or when the two pulses arrive at the crystal at sufficiently different times. Temporal overlap of the two pulses at the crystal can be varied by mechanically changing one of the path lengths. The amount of SHG detected is a maximum when the pulses are coincident and decreases as one is delayed with respect to the other.

The primary experimental difficulty in using the SHG method in conjunction with pulsed lasers is that it requires plotting the pulse correlation point by point with successive firings of the laser. Although the development of cw modelocked lasers has greatly revived interest in this technique, its use with pulsed lasers was effectively ended within one year by the invention of the two-photon-fluorescence (TPF) method [3.5]. The TPF technique in its most commonly used form is illustrated in Fig. 3.2. An input pulse is divided

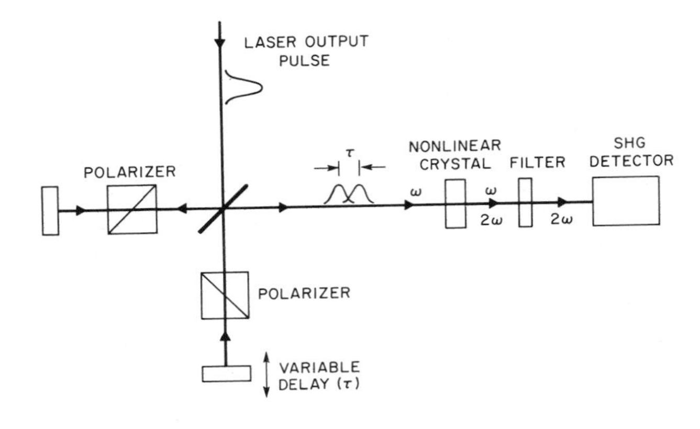

Fig. 3.1. Interferometric arrangement for pulse correlation measurements by SHG. The polarizers allow the two pulses to have different polarizations

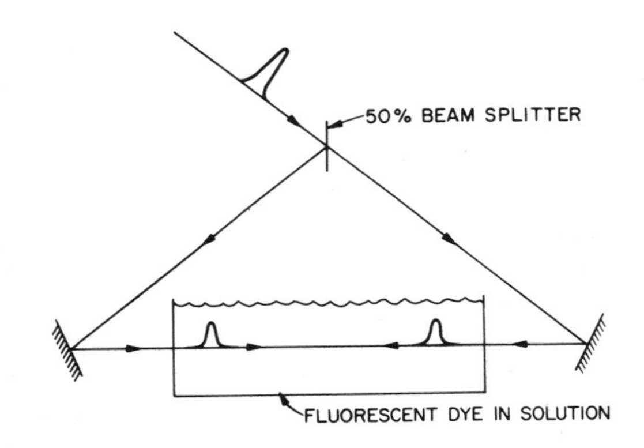

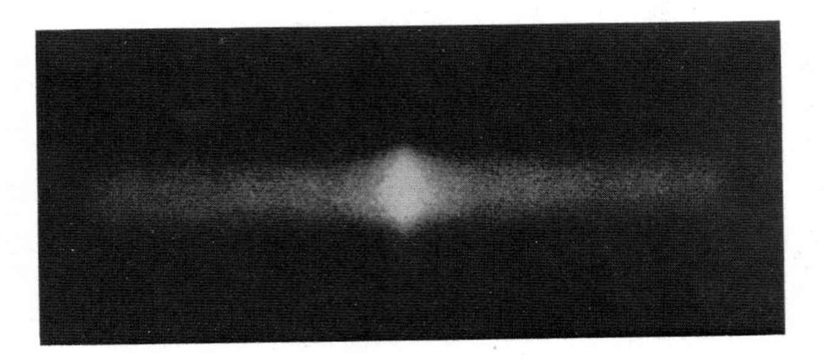

Fig. 3.2. Triangular arrangement for TPF measurement and photograph of an experimental fluorescence trace [3.158]

into two beams which then travel in opposite directions in an organic dye solution. Fluorescence from the dye is proportional to two-photon absorption which is a maximum at the point where the two pulses are coincident in time. With this scheme a single photograph of the fluorescence track provides a complete measurement of pulse correlation. If carefully performed, SHG and TPF measurements can provide a reliable estimate of pulse duration. A critical review of the use of both techniques is given below. They do not give a direct display of the pulse shape. It has been necessary to use different techniques to determine such pulse characteristics as temporal asymmetry and dynamic spectral behavior. In the sections that follow, we also review and try to assess some of these other techniques.

We restrict ourselves in this section to nonlinear optical methods. Direct linear measurements with subpicosecond resolution [3.6] have recently been made possible by advances in streak-camera technology. These are discussed in Chapter 2. Techniques for the generation of picosecond electrical pulses are described in Chapter 4.

3.1.1 Correlation Functions

Nonlinear optical techniques for pulse measurement do not provide a direct display of pulse shape but give instead measurements of correlation functions. It is important therefore to consider in some detail the theoretical relationship between a signal $I(t)$ and its correlation functions. This is a subject that has been treated extensively in the literature [3.7–25], and we will only review here some of the more important features of this theory.

The second-order autocorrelation function of the intensity $I(t)$ is given in normalized form by

$$G^2(\tau)=\frac{\langle I(t)I(t+\tau)\rangle}{\langle I^2(t)\rangle} \tag{3.1}$$

where the brackets indicate an average over a sufficiently long interval of time. This is the function one can obtain by SHG or TPF. If $I(t)$ is a single isolated pulse, $G^2(\tau)$ vanishes for large relative delay τ and its half-width provides a measure of the duration of $I(t)$. Precise determination of a pulsewidth requires further knowledge about the shape of $I(t)$. It is obvious from (3.1) that $G^2(\tau)$ is always symmetric regardless of any asymmetry in $I(t)$. This fact is the fundamental limitation on the use of $G^2(\tau)$ to determine pulse shape. If $I(t)$ were known to be symmetric, its shape could in fact be deduced from $G^2(\tau)$.

In general, higher-order correlation functions must be used in addition to $G^2(\tau)$ to determine $I(t)$ uniquely. The n-th order correlation function is given by

$$G^n(\tau_1, \tau_2 \ldots \tau_{n-1})=\frac{\langle I(t)I(t+\tau_1)\ldots I(t+\tau_{n-1})\rangle}{\langle I^n(t)\rangle}. \tag{3.2}$$

It has been shown mathematically [3.11, 23] that exact knowledge of G^2 and G^3 is sufficient to describe all higher orders and hence the pulse itself. In practice, where experimental errors limit one's accuracy in determining G^3, higher-order measurements are more sensitive to pulse asymmetry and may be helpful. Consider for example the special case

$$G^n(0, 0 \ldots \tau) = \frac{\langle I^{(n-1)}(t) I(t+\tau) \rangle}{\langle I^n(t) \rangle}. \tag{3.3}$$

As the order increases, $I^{n-1}(t)$ becomes a sharper function of time and therefore a better probe of the shape of $I(t)$. With picosecond laser pulses, correlation functions up to order five have been obtained [3.26].

The relationship between $G^2(\tau)$ and different types of optical signals is most easily illustrated with specific examples. Consider first the case where $I(t)$ is a gaussian random variable, the intensity produced by a continuous source of thermal noise. A laser operating in large number of randomly phased modes approximates such a source. The intensity can be described by

$$p(I)dI = \frac{1}{\langle I \rangle} e^{-I/\langle I \rangle} dI \tag{3.4}$$

where $p(I)dI$ is the probability of observing the intensity I in an interval dI, and $\langle I \rangle$ is the average intensity. For large τ, where $I(t+\tau)$ and $I(t)$ are independent random variables, (3.4) yields

$$G^2(\tau \to \infty) = \frac{\langle I(t) \rangle^2}{\langle I^2(t) \rangle} = \frac{1}{2}. \tag{3.5}$$

Since $G^2(0)$ is by definition unity, we see that even a random signal produces a peak in its autocorrelation with $G^2(0)/G^2(\infty) = 2$. The width of this peak is a measure of the temporal coherence of the signal, and is just related to the inverse of the spectral bandwidth of the source. The only information about the actual temporal behavior is contained in the contrast ratio, $G^2(0)/G^2(\infty)$. Contrast ratios different from 2 imply some deviation from purely random behavior. Smoother signals give lower contrast and enhancement of intensity peaks increases contrast.

Several higher-order correlations of gaussian noise are also of interest. The third-order correlation function

$$G^3(0, \tau) = \frac{\langle I^2(t) I(t+\tau) \rangle}{\langle I^3(t) \rangle} \tag{3.6}$$

has a contrast ratio of 3. With nonlinear harmonic generation followed by second-order correlation one can also measure

$$G^{2n}(0\ldots,\tau\ldots)=\frac{\langle I^n(t)I^n(t+\tau)\rangle}{\langle I^{2n}(t)\rangle} \tag{3.7}$$

which results [3.22, 23] in a contrast ratio of $(2n)!/(n!)^2$. In the limit of large n one obtains the contrast of a single, isolated pulse (produced by nonlinear selection of the largest noise spike in $I(t)$).

The next case we consider is that of an isolated burst of gaussian noise. This has its practical manifestation in a laser that is partially modelocked [3.13]. We describe this by $I(t)=I_1(t)\cdot I_2(t)$ where $I_1(t)$ is a random variable as above and $I_2(t)$ is a more slowly varying envelope function. It can be shown [3.19] that

$$G^2(t)=G_1^2(\tau)\cdot G_2^2(\tau) \tag{3.8}$$

where $G_1^2(\tau)$ and $G_2^2(\tau)$ are the autocorrelation functions of $I_1(t)$ and $I_2(t)$, respectively. From an initial ($\tau=0$) value of 1, $G^2(\tau)$ falls to a value of $(1/2)G_2^2(\tau)$ for τ longer than the coherence time of the random variable $I_1(t)$ and then to zero as $G_2^2(\tau)\sim\langle I_2(t)I_2(t+\tau)\rangle\to 0$. In this case the temporal isolation produced by the envelope of the signal results in a high contrast ratio overall. The contrast between noise and envelope contributions is unity. A careful determination of $G^2(\tau)$ yields information about the envelope of the signal as well as its coherence. Note that higher-order correlations of a noise burst tend to accentuate the relative amplitude of the coherence spike [3.15]. It should be mentioned that in the case of an isolated pulse of noise, the actual shape of the correlation function may differ from the expectation value. Examples of such deviations have been given in the literature [3.17, 22].

The second-order correlations one expects from these different types of optical intensity signals are given in Fig. 3.3. For comparison with experiment, the form that one measures in SHG and TPF experiments with inherent background is also shown. Higher-order correlation measurements produce similar shapes but with different contrast ratios. It has been common in the literature to estimate the duration of a pulse Δt by simply measuring the width $\Delta\tau$ of that part of the correlation due to pulse envelope. Of course the actual $\Delta\tau/\Delta t$ depends upon the precise pulse shape, but the uncertainty in such an estimate can be less than a factor of two. Table 3.1 gives $\Delta\tau/\Delta t$ for several theoretical pulse shapes. Also shown for each case is the predicted spectral width. Comparison of experimental bandwidth with the predicted width is an important test of the assumed pulse shape. It is interesting to note that the gaussian pulse envelope which is known to have the minimum time-bandwidth uncertainty product does not have the minimum half-width product. The Lorentzian intensity profile $I(t)=\tau^2/(t^2+\tau^2)$ without chirp has a singular spectrum [3.27]. If the measured bandwidth is much greater than that expected for a reasonable pulse shape, two possible causes are appararent. Either the experiment has not resolved the noise correlation spike indicative of amplitude substructure, or the actual pulse contains frequency modulation which does not show up in a correlation measurement [3.19]. If a frequency sweep is present, it is possible to compress the pulse in time. Methods for doing this are discussed below in the section on dynamic spectroscopy.

A final important point is that correlation measurements are not particularly sensitive to low-level background signals which may be present in addition to a short intense pulse [3.10, 11, 16, 17]. Other types of measurement have been necessary to estimate the fraction of the total energy contained in the short-pulse component [3.26, 28, 29].

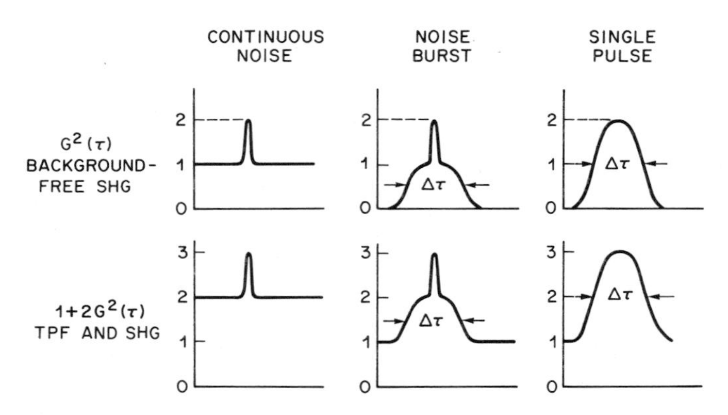

Fig. 3.3. Theoretical correlation traces for SHG and TPF measurements

Table 3.1. Correlation widths and spectral bandwidths for four different transform-limited pulse shapes. Δt and $\Delta\tau$ are full widths at half maximum of $I(t)$ and $G^2(\tau)$, respectively. $\Delta\nu$ is FWHM of the measurable frequency spectrum

$I(t)$	$\Delta\tau/\Delta t$	$\Delta t\Delta\nu$
$1\ (0 \leq t \leq \Delta t)$	1	0.886
$\exp\left\{-\frac{(4\ln 2)t^2}{\Delta t^2}\right\}$	$\sqrt{2}$	0.441
$\mathrm{sech}^2\left\{\frac{1.76\,t}{\Delta t}\right\}$	1.55	0.315
$\exp\left\{-\frac{(\ln 2)t}{\Delta t}\right\}\ (t \geq 0)$	2	0.11

3.1.2 The Two-Photon Fluorescence Method

Since pulse measurement by two-photon fluorescence was first demonstrated by *Giordmaine* et al. [3.5], the TPF method has been the most widely used technique for studying pulses produced by high-power modelocked lasers. Its popularity has been due to its inherent simplicity and its ability to provide a complete measurement of $G^2(\tau)$ with a single pulse. Correlation takes place in a liquid dye solution that has no absorption at the pulse wavelength but fluoresces in the presence of high intensity by virtue of two-photon absorption. The integrated intensity of fluorescence at any point is proportional to the time averaged fourth power of the optical electric field. For two identical pulses moving in opposite direction, the fluorescence distribution is proportional to

$$f(\tau)=1+2G^2(\tau)+r_1(\tau) \tag{3.9}$$

where $r_1(\tau)$ is a rapidly varying fringe-like component [3.15, 25] that averages to zero over several optical periods and is generally not observed experimentally. If c/n is the pulse velocity in the medium and z is the distance from perfect overlap, then $\tau=2nz/c$. Thus, one measures the second-order correlation function in the presence of constant background.

Early experiments [3.5, 8, 30–32] were performed simply by reflecting pulses back upon themselves with a mirror immersed in the two-photon absorbing dye solution. This technique had the disadvantage that autocorrelation had to be measured adjacent to the mirror. Because of this difficulty it has become more common to use the triangular configuration [3.33, 34] shown in Fig. 3.2. This arrangement also facilitates the TPF correlation of two different pulses. When pulses of significantly different frequency are used (i.e., 1.06 μm fundamental and 0.53 μm harmonic) it becomes possible [3.35] to greatly increase the contrast by making the lower-frequency pulse (for which no two-photon absorption is possible) much more intense than the other. With pulses of sufficiently different wavelengths, it is also possible to geometrically expand the correlation trace by passing both pulses through the TPF medium in the same direction [3.35]. The delay τ as a function of distance is then determined by the group velocity difference between the pulses in the medium. In theory, cross correlation of a pulse with its second harmonic yields a measurement of $G^3(0,\tau)$, but this is true only if the second harmonic has been generated with sufficient care to ensure a faithful $I^2(t)$ copy of the initial substructure.

In early TPF experiments little attention was paid to the observed contrast between pulse and background. This led at first to the belief [3.30, 32, 36–39] that regular trains of pulses were being generated in lasers that had not intentionally been modelocked. When it was shown [3.7, 8, 14, 40, 41] that correlation peaks result from noise spikes as well as isolated pulses, careful experiments were performed to verify the theoretical contrast ratios. Experiments with free-running lasers [3.8, 22, 35, 42] verified the 3:2 contrast for noise. Correlation measurements with the second harmonic of such a laser [3.30] have shown the higher contrast of fourth-order correlation. Proper 3:1 contrast for the modelocked Nd:glass laser was first demonstrated by using photoelectric detection and a thin dye cell to scan the correlation region [3.22, 42, 43]. Similar results have since been obtained photographically in careful experiments using selected single pulses [3.27, 44].

As simple as the TPF method appears at first glance, there are a number of reasons why it is difficult to obtain precise measurements with TPF. Experimental contrast ratios can depend critically on alignment of the two optical beams [3.44–46] and to a lesser extent on difference between the two intensities [3.15, 22] and absorption in the medium [3.47]. In certain experiments, intensity dependent quenching of the dye fluorescence has also been observed to drastically alter the measured contrast [3.48, 49]. In photographs of a TPF trace, camera misalignment, film resolution and dynamic range may distort the result. A stepped calibration of film exposure to pulse-induced fluorescence intensity should accompany every TPF photograph.

At best, a TPF measurement only determines $G^2(\tau)$. As discussed in the previous section, this provides limited information about the actual shape of an optical pulse. The background inherent to the TPF method also makes such measurements very insensitive to low-level signals in the presence of an intense short pulse [3.10, 11, 16, 17, 28]. Additional measurements are required to ensure that a major fraction of the overall signal energy is confined to the short pulse.

3.1.3 Second-Harmonic Generation Methods

The very first measurements of picosecond duration optical pulses were made using second-harmonic generation. In 1966 three different techniques were reported almost simultaneously. *Weber* proposed [3.2] and later demonstrated [3.50] the technique shown in Fig. 3.1 in which two orthogonally polarized pulses pass collinearly through a nonlinear crystal. The crystal (KDP) is oriented such that phase-matched SHG is produced only when both polarizations are present. Measurement of the integrated SHG intensity versus relative delay between the two pulses results in a direct measurement of $G^2(\tau)$. *Armstrong* [3.4] also made use of orthogonally polarized beams to obtain background-free correlation by SHG at the surface of a GaAs crystal. *Maier* et al. [3.3] demonstrated that a background-free measurement of $G^2(\tau)$ may be obtained with parallel polarized pulses traveling noncollinearly through a nonlinear crystal. SHG is produced in a third direction determined by phase matching. This technique has proven especially valuable when phase matching with orthogonally polarized beams is not possible [3.51, 52].

In the background-free SHG experiments the measured function is

$$f(\tau)=G^2(\tau)+r_2(\tau) \tag{3.10}$$

where $r_2(\tau)$ is a rapidly varying term, due to phase interference of the two beams, that averages to zero [3.15, 25]. The time delay τ is varied by mechanically changing the path length traversed by one of the pulses. If τ is varied in a step-like manner, special care must be taken to resolve $G^2(\tau)$ near $\tau=0$ and to average $r_2(\tau)$. Perhaps for this reason early measurements with Nd:glass lasers [3.4, 50, 53] failed to detect the presence of pulse substructure. Background-free exeriments with free-running lasers [3.54, 55] and noisy pulses [3.51] have since demonstrated the expected 2:1 contrast in $G^2(\tau)$ for such

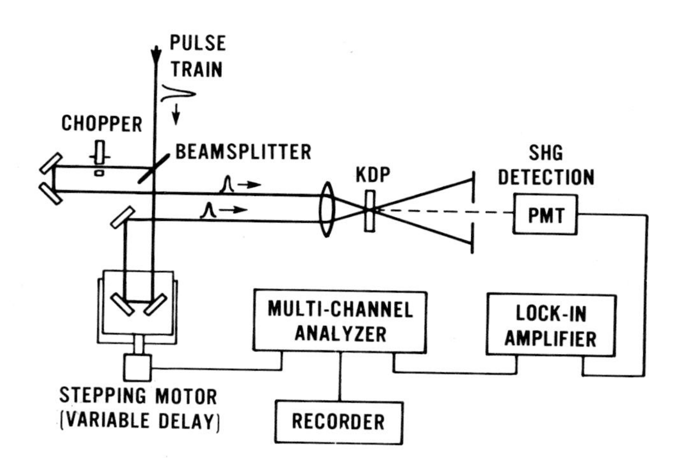

Fig. 3.4. Experimental setup for noncollinear, background-free correlation using SHG. The detection system allows averaging of the measurement by repeated scanning of the delay interval [3.52]

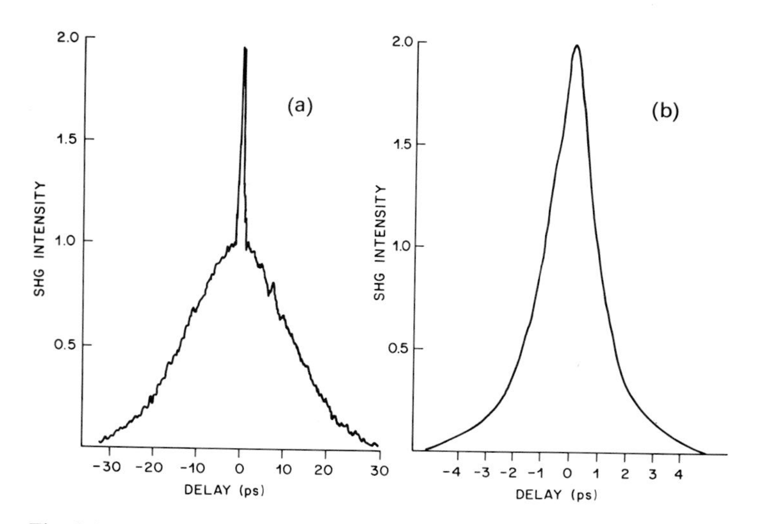

Fig. 3.5a and b. Background-free SHG correlation traces obtained with a modelocked cw dye laser. (a) Incomplete modelocking results in pulses with considerable substructure. (b) Good modelocking is evidenced by much shorter pulses with no apparent substructure. Note the difference in time scale between (a) and (b)

cases. Bandwidth-limited pulses as short as 0.3 ps have been measured [3.52]. An experimental arrangement for noncollinear correlation that has been found to be convenient for use with modelocked cw lasers is shown in Fig. 3.4. A stepper-motor allows digitally controllable scan over the desired delay interval. Repeated scanning and accumulation of the correlation signal in a multi-channel analyzer guards against the possibility of curve distortion due to changing pulse characteristics. Experimental curves obtained for two types of modelocked dye laser operation are shown in Fig. 3.5. Incomplete mode-locking produces a noisy pulse trace whose envelope is shown to fit the auto-correlation of a $\mathrm{sech}^2\,(t/t_\mathrm{p})$ shape. Good modelocking results in the isolated pulse trace of Fig. 3.5(b). In the latter case the entire pulse envelope is of sub-picosecond duration and the delay axis has been expanded to show the exponential behavior of the pulse wings.

In some experiments with low-power lasers [3.56–59], SHG pulse correlations have been obtained with collinear, identically polarized beams. The measured function in such an arrangement is equivalent to that in a TPF experiment and is given by (3.9). Contrast ratios of 3:2:1 for noisy pulses [3.58] and 3:1 for well modelocked pulses [3.59] have been observed. Fringes due to the interference term $r_1(\tau)$ in (3.9) can also be observed [3.59]. One advantage of measurements with background is that proper contrast provides assurance that the entral peak in $G^2(\tau)$ has been obtained.

As with TPF measurements, a proper SHG correlation measurement requires careful beam alignment [3.44–46]; but problems associated with photographic interpretation are eliminated. The required point by point scan of $G^2(\tau)$ has proven unwieldy in experiments with pulsed lasers but is well suited for use with modelocked cw lasers. The particular advantage of SHG to lower power lasers is its relatively high nonlinear conversion efficiency under phase-matched conditions. It must be remembered, however, that phase matching in dispersive materials imposes a restriction on the bandwidth over which efficient generation occurs and can limit the resolution of a correlation measurement if too thick a crystal is used [3.60–63]. When orthogonally polarized pulses are used, the velocity difference between the two may require the use of thin crystals [3.2]. Even with like pulses, group velocity mismatch between the pulses and their second harmonic can affect the resolution of a measurement. Background-free SHG is more sensitive than TPF to low-level background intensity, but shares with TPF the basic insensitivity of $G^2(\tau)$ to the actual pulse shape.

3.1.4 The Optical Kerr Shutter

The optical Kerr shutter is a device with wide-ranging application to pico-second measurement. In this section, we analyze operation of the shutter in terms of its ability to measure pulses by correlation. *Mayer* and *Gires* [3.64] first showed that powerful, polarized light pulses could be used in lieu of electrical pulses to induce birefringence in traditional Kerr cell liquids. The optically induced birefringence arises from a partial alignment of molecules in the Kerr liquid. The anisotropic molecules experience a torque which tends to rotate the axis with greatest polarizability into the plane of the optical electrical field. This birefringence was used by *Shimizu* and *Stoicheff* [3.65] to study small-scale filaments produced by self-focusing of laser pulses in CS_2. A travelling-wave optical Kerr shutter for picosecond pulse measurement based on this principle was invented by *Duguay* and *Hansen* [3.66]. Basic operation of the optical Kerr shutter is illustrated schematically in Fig. 3.6. An intense, polarized light pulse $I_1(t)$ induces a short-lived birefringence in the Kerr medium (3.64). A signal beam $I_2(t)$, initially polarized 45° with respect to the intense pulse, has its polarization effectively rotated by this birefringence. This allows transmission of the signal through the analyzing polarizer which is initially set for null. The induced difference between indices of refraction parallel and perpendicular to $I_1(t)$ is given by

$$\Delta n(t) = n_{2B} \int_{-\infty}^{t} I_1(t') \mathrm{e}^{-(t-t')/\tau_r} dt'/\tau_r \tag{3.11}$$

where n_{2B} is the nonlinear birefringence coefficient and τ_r is the time constant for relaxation of the effect in the medium. The intensity $I_1(t) = \overline{E_1^2(t)}$, the average value of the electric field squared in the medium. Integrated transmitted intensity through the analyzer as a function of delay τ between I_1 and I_2 is

$$I_T(\tau) = \int_{-\infty}^{\infty} I_2(t+\tau) \sin^2 [\phi(t)/2] dt \tag{3.12}$$

where $\phi(t) = 2\pi \Delta n(t) L/\lambda$. L is the length of the interaction and λ is the signal wavelength in vacuum.

A simple result is obtained if the response time τ_r of the medium is much shorter than the duration of $I_1(t)$ and the birefringence is small, $\phi(t) \ll 1$. Then

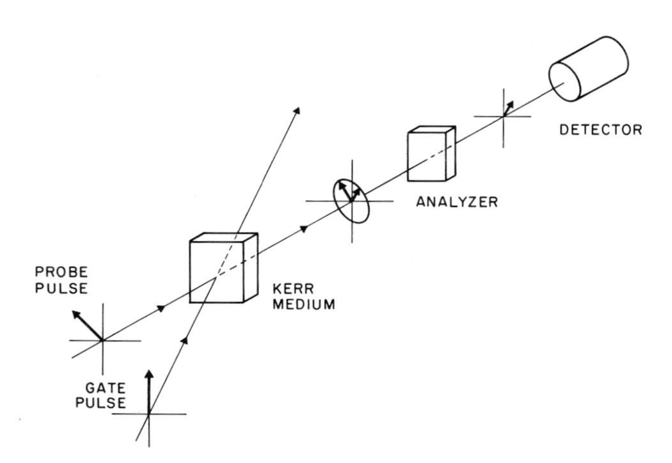

Fig. 3.6. Schematic of an optical Kerr shutter in the null configuration. A linear gate can be constructed by inserting a $\lambda/4$ plate in front of the analyzer (see text)

one gets the proportionality

$$I_T(\tau) \propto \int_{-\infty}^{\infty} I_2(t+\tau) I_1^2(t) dt \tag{3.13}$$

which we recognize as a third-order correlation measurement [3.67]. In the original experiment by *Duguay* and *Hansen* [3.66], where $I_2(t)$ was the second harmonic of $I_1(t)$, the measurement is a fourth-order correlation

$$I_T(\tau) \propto \int_{-\infty}^{\infty} I_1^2(t+\tau) I_1^2(t) dt. \tag{3.14}$$

In each of the above cases, the correlation function must be obtained by a sequence of measurements. An experimental arrangement that offers the possibility of a single shot correlation measurement is shown in Fig. 3.7 [3.68–70]. The signal pulse $I_2(t)$ passes through a scattering medium that makes it visible from the side. This side view is photographed through the shutter operated by $I_1(t)$. Fig. 3.8 shows the result. A single photograph like this of light in flight can provide a high contrast measurement of $I_T(\tau)$.

Yet another mode of Kerr shutter operation has been described by *Ippen* and *Shank* [3.71]. By circularly polarizing the incident signal beam one obtains

$$I_T(\tau) = \tfrac{1}{2} \int_{-\infty}^{\infty} I_2(t+\tau)[1 + \sin\phi(t)] dt. \tag{3.15}$$

For small $\phi(t)$ and fast recovery of the Kerr medium

$$I_T(\tau) = \frac{1}{2} \int_{-\infty}^{\infty} I_2(t) dt + \frac{\pi n_2 L}{\lambda} \int_{-\infty}^{\infty} I_2(t+\tau) I_1(t) dt. \tag{3.16}$$

Because of the large background this scheme is not suited for use with pulsed lasers. With a modelocked cw laser, however, the background can be removed electronically by chopping the gating beam $I_1(t)$ which only modulates the second term in (3.16). Thus, second-order correlation measurements are also possible with a Kerr shutter. Because of the inherent symmetry of $G^2(\tau)$ if I_1 and I_2 are identical, this scheme can also be used to make an unambiguous measurement of short Kerr effect lifetimes [3.71].

The most widely used Kerr medium has been CS_2. Its large Kerr coefficient n_2 [3.72, 73] depends on orientation of the asymmetric molecule. The relaxation time of this process which limits temporal resolution of the shutter has been shown to be about 2 ps [3.71, 74]. Faster gating may be achieved by the use of optical glass in which the Kerr effect is primarily electronic [3.73]. The disadvantage of glass is its relatively low coefficient n_2 [3.73, 75]. Larger values of n_2 and fast response may be obtained with certain conjugated-chain molecules [3.76, 77]. The time response of a Kerr shutter can also be limited by geometrical factors unless collinear beams are used. If the gate and signal pulses are at different wavelengths, group velocity difference in the Kerr

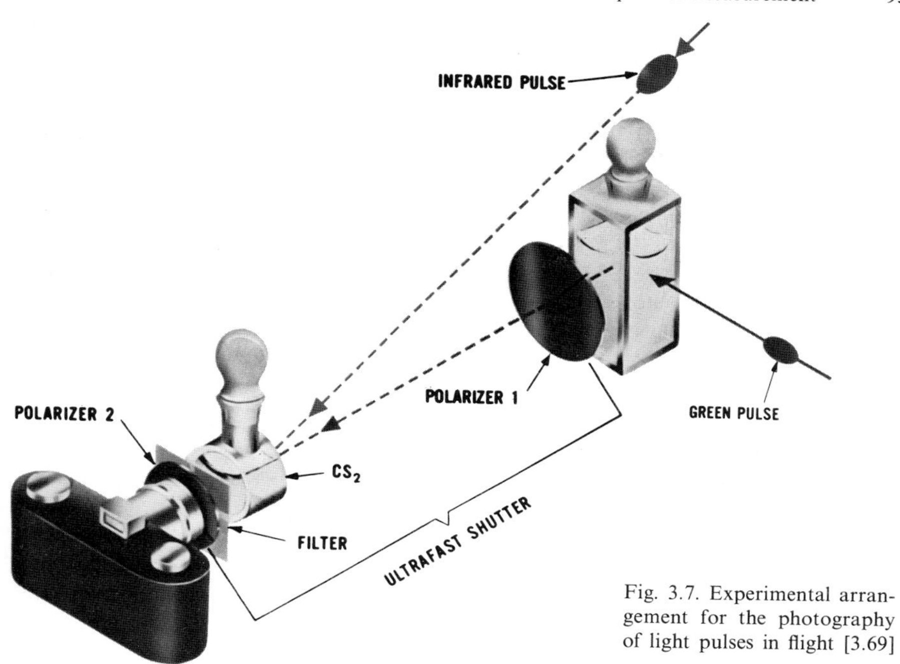

Fig. 3.7. Experimental arrangement for the photography of light pulses in flight [3.69]

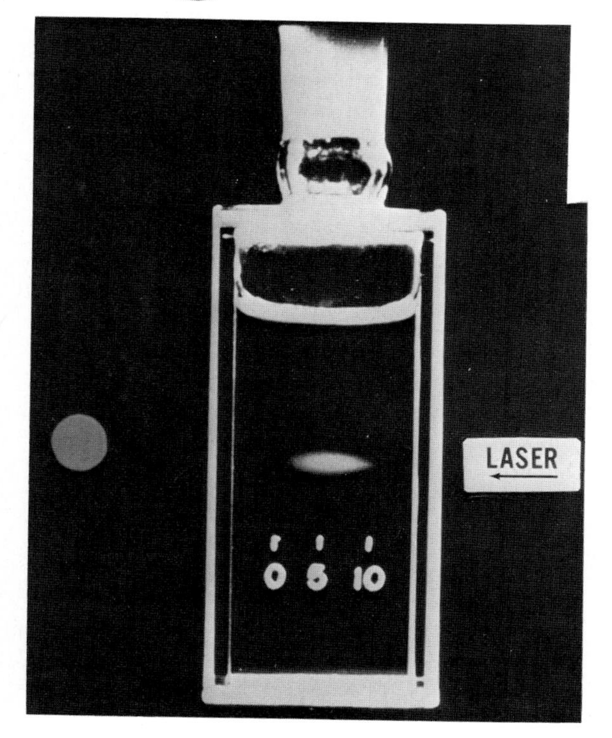

Fig. 3.8. Kerr-shutter photograph of a green light pulse in flight (from right to left) through a cell containing a colloidal suspension of milk in water [3.69]

medium must be considered [3.66]. In the photographic arrangement of Fig. 3.7, resolution may be limited by the angular aperture and the depth of the beam in scattering medium [3.70].

3.1.5 Pulse Compression and Dynamic Spectroscopy

The pulse measurement techniques described thus far are designed to provide information about the amplitude structure of optical signals. They are not sensitive to phase (frequency) modulation. Simultaneous measurement of the optical frequency spectrum serves as an initial test for the possibility of such modulation. Other experiments are necessary to elucidate the actual spectral behavior. One of the easiest modulations to test for is that of a monatomic frequency sweep (chirp). If a pulse has a negative chirp (higher frequencies leading lower frequencies), passage through normally dispersive media can compress the pulse in time. The more important case for laser pulses is that of a positive chirp. Such a chirp can be produced either by the linear dispersion of elements in a laser [3.78] or by the nonlinear process of self-phase modulation [3.79]. Compression in this case requires the use of anomalous dispersion. A simple device, which provides enough anomalous dispersion to compress picosecond pulses, is the grating pair devised by *Treacy* [3.80–82]. Operation of the grating pair is illustrated in Fig. 3.9. The rate of compression can be varied simply by changing the distance between gratings. In his experiments with Nd:glass laser pulses, *Treacy* [3.80] observed narrowing in TPF traces following a grating pair and concluded that the pulses had a large positive chirp. Similar experiments were later performed by *Bradley*, et al. [3.83]. Their results indicated the presence of both chirp and amplitude substructure. Pulsewidth measurements following compression must be made with care since random phase fluctuations can also be converted into amplitude substructure and produce a noise-like peaking of the correlation trace [3.84].

Recent experiments with the modelocked cw dye laser [3.52] have demonstrated the compression of chirped pulses into the subpicosecond region.

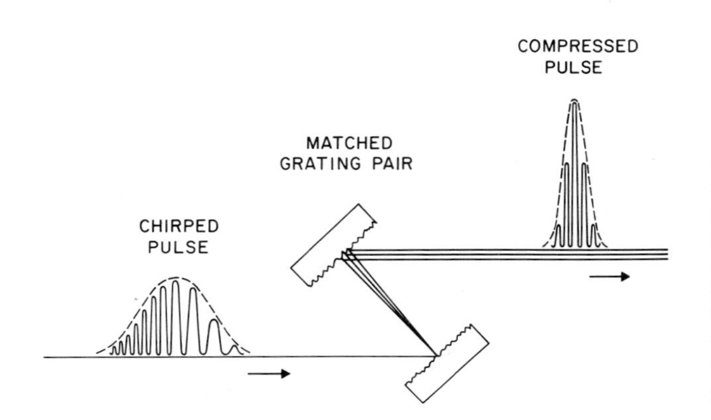

Fig. 3.9. A positively chirped pulse is compressed in time by reflection of a pair of gratings which delays the longer wavelength components in the front part of the pulse with respect to the shorter wavelength components in rear

SHG correlation traces made before and after grating-pair compression are shown in Fig. 3.10 (a, b). The additional shortening evident in Fig. 3.10 (c) was achieved by spectral filtering, indicating that the chirp was not entirely linear. Confirmation of this nonlinear chirp has been obtained by complete dynamic spectroscopic characterization of these pulses (see below). The shortest pulses obtained are estimated to be 0.3 ps in duration.

If optical pulses are not naturally chirped, it is possible to induce such a frequency sweep by active phase modulation [3.85]. This technique is impractical with ultrashort pulses, but has been used [3.86] to compress 500 ps pulses by a factor of two. In these experiments, *Duguay* and *Hansen* [3.86] demonstrated the use of a Gires-Tournois interferometer [3.87] as an alternative to the grating pair. The interferometer can be highly dispersive and virtually lossless but suffers the disadvantage that it can be used only over narrow frequency intervals near resonance. A more attractive chirping technique for intense pulses is self-phase modulation which can produce sufficient chirp to allow the generation of subpicosecond pulses [3.88]. Experimentally, *Laubereau* [3.89] has demonstrated compression of pulses from 20 ps to 2 ps using self-phase modulation in CS_2 and a grating pair. In a similar manner, *Lehmberg* and *McMahon* [3.90] were able to compress powerful 100 ps pulses to 7 ps with a grating pair separation of 23 m.

A direct application of compressed pulses to temporal profile measurements was demonstrated by *Treacy* [3.81]. Using compressed pulses to probe the initial pulse shape (a cross correlation by TPF), he found that the initial pulses from the Nd:glass laser were asymmetric with the trailing edge sharper than the leading edge. Although similar asymmetry was also observed by *Shelton*

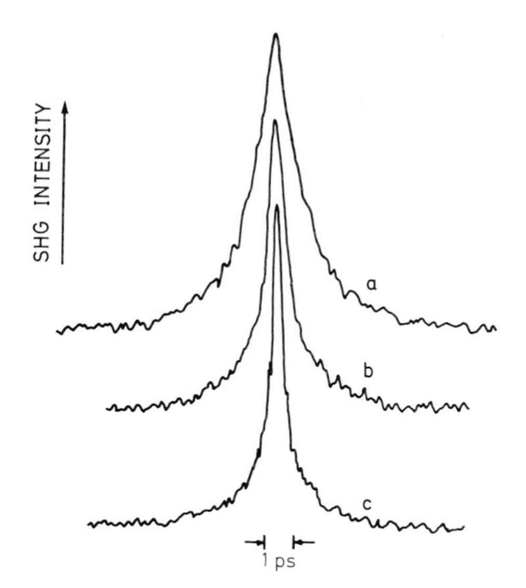

Fig. 3.10a–c. SHG correlation traces of pulses from the modelocked cw dye laser. (a) Laser output pulses, (b) after grating-pair compression, (c) after compression and spectral filtering [3.52]

and *Shen* [3.91], these results cannot be regarded as characteristic of Nd:glass laser pulses. In different experiments *Auston* [3.26] saw no strong asymmetry, while *von der Linde* and *Laubereau* [3.92] found asymmetry in the low amplitude wings with the leading edge sharper than the trailing. Correlation between compressed and uncompressed pulses from the modelocked cw dye laser [3.52] has revealed clear asymmetry. Pulses from this laser are found to have rapidly rising leading edges and considerably longer, exponential tails (see Fig. 3.13(a)).

Although pulse compressibility may be used as an indicator of frequency modulation, unambiguous characterization of a pulse requires an actual display of amplitude and frequency versus time. Techniques for providing such a display with picosecond resolution have been discussed by *Treacy* [3.93]. In an elegant experiment [3.93, 94] he was able to observe directly the frequency sweep in Nd:glass laser pulses. The experimental arrangement is shown in Fig. 3.11. In the focal plane of a simple grating spectrometer, different frequency components are dispersed spatially. They also arrive in the focal plane with the relative delays they have in the pulse. To obtain picosecond time resolution of the arrival times, however, it is necessary to "straighten out" the canted wavefronts produced by the grating. Treacy accomplished this with a properly designed echelon (stepped mirror). Two focal plane images (one inverted with respect to the other) are combined in a TPF cell. A slant in the

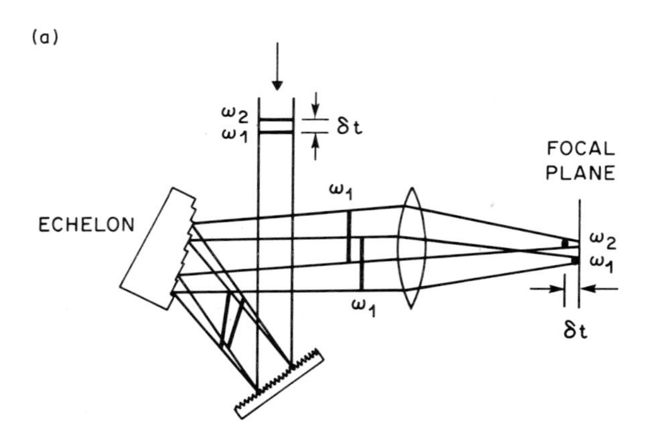

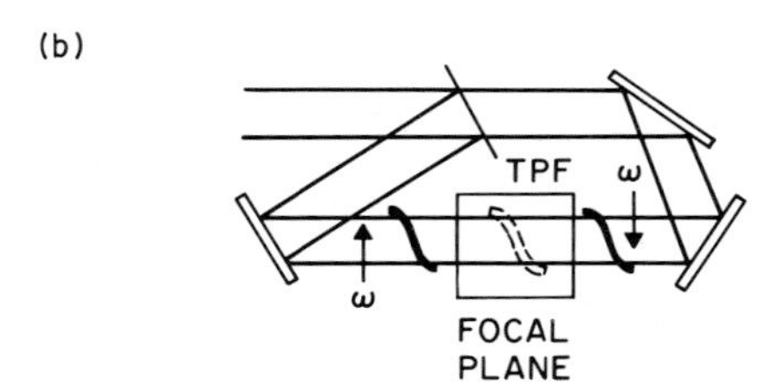

Fig. 3.11a and b. Treacy's scheme for dynamic spectroscopy with a single pulse. (a) The echelon compensated grating produces a proper spectrogram in the focal plane. (b) A TPF configuration with image inversion is used for recording in the focal plane

TPF image is a direct measure of chirp, which in Treacy's experiments [3.93, 94] was found to be nonlinear with the slowest sweep in the center of the pulses. Some information is lost due to the antisymmetrical nature of this experiment. A complete dynamic spectrogram would require using a shorter, transform-limited pulse as one of the TPF beams. The ultimate two-dimensional resolution in any dynamic spectrogram is given by the uncertainty principle $\Delta\omega\Delta t \leq 2\pi$.

Another clever scheme for obtaining dynamic spectrograms was devised by *Auston* [3.95]. He used the angular dependence of phase matching for four-photon parametric mixing to provide the frequency dispersion and multiple probe beams (all derived from the same pulse) to obtain temporal sampling. His results verified the positive nature of the chirp in Nd:glass laser pulses and revealed the considerable variability in shape and chirp rate of such pulses.

Complete dynamic spectrograms with subpicosecond resolution have been obtained of pulses from the modelocked cw dye laser [3.52]. The major advantages of working with this system are the high repetition rate of pulse output

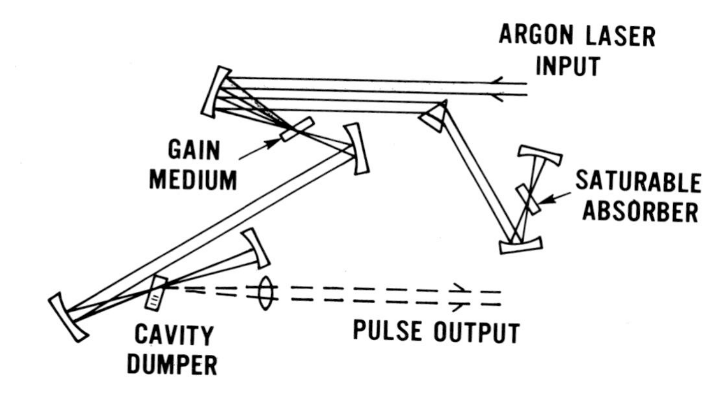

Fig. 3.12. Passively modelocked cw dye laser. The acousto-optic dumper provides an output of single, subpicosecond pulses at rates greater than 10^5 pps

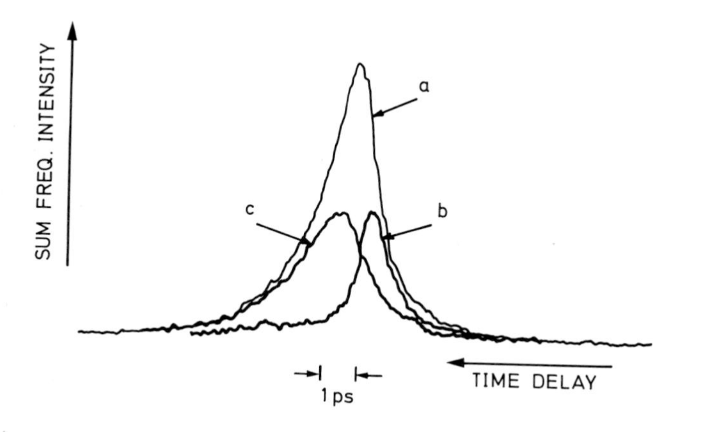

Fig. 3.13a-c. Dynamic spectrogram of pulses from the modelocked cw dye laser showing (a) the asymmetric envelope of the pulse, (b) the temporal behavior of the longer wavelength components and (c) the temporal behavior of the shorter wavelength components. The frequency sweep is positive and nonlinear

and excellent reproducibility of pulse characteristics. A schematic of the laser configuration used in these experiments is shown in Fig. 3.12. With acousto-optic cavity dumping, single pulses with peak powers of several kilowatts are available at repetition rates up to 10^6 pps. The output is divided into two beams in an arrangement like that shown in Fig. 3.4. One of the beams is compressed and filtered to produce short, transform-limited probe pulses (see Fig. 3.10) before being cross correlated by SHG with the other, uncompressed beam. In successive correlation traces, detection was limited by a spectrometer to particular sum-frequency components. Three traces are shown in Fig. 3.13: the overall pulse profile and the profiles of the highest and lowest frequency components. The rate of positive chirp is seen to slow down in the longer trailing edge of the pulse. This result clearly explains how additional pulse shortening is obtained in Fig. 3.10 by filtering out the highest frequency components.

3.1.6 Higher Order Nonlinear Methods

The techniques described above rely on second-order nonlinearities for pulse correlation. In most cases they provide a measure of $G^2(\tau)$, but we have seen that higher order correlation is also possible. TPF and SHG experiments that utilize two second-harmonic pulses can yield the fourth-order correlation $G^4(0, \tau, \tau)$ of the fundamental pulse. If the fundamental is correlated with its second harmonic, the result is $G^3(0, \tau)$. The optical Kerr shutter can be used to obtain second-, third-, and fourth-order correlations. All of these schemes suffer the limitation that only one delay is involved. A technique for obtaining two-dimensional information about $G^3(\tau_1, \tau_2)$ has been suggested by *Weber* and *Dändliker* [3.9, 13]. The first delay would be provided by generating a background-free second-harmonic pulse with two fundamental pulses delayed with respect to each other. This second-harmonic pulse could then be correlated with a third fundamental pulse using one of the standard techniques.

Third-order optical nonlinearities can be used to perform more direct measurements of $G^3(\tau_1, \tau_2)$. The use of third-harmonic generation (THG) for pulse measurement was first proposed by *Bey* et al. [3.96] and demonstrated by *Eckhardt* and *Lee* [3.97]. This particular scheme provides high contrast, but results in a measurement of $G^3(0, \tau)+G^3(\tau, 0)$ which is inherently symmetric. A THG method that removes this symmetry restriction has been demonstrated by *Shelton* and *Shen* [3.91]. In general, THG suffers the disadvantage that point by point plotting of the correlation is required, and its low efficiency with respect to SHG makes it impractical to use with low power cw lasers.

For single-shot measurements *Rentzepis* et al. [3.98] first demonstrated the use of three-photon fluorescence (3PF). In their experiments a three-photon absorbing dye was used in the usual TPF arrangement. This scheme again provides high contrast (10:4:1) but measures $G^3(0, \tau)+G^3(\tau, 0)$ and is insensitive to pulse asymmetry. Rentzepis et al. [3.98] pointed out that a complete measurement of $G^3(\tau_1, \tau_2)$ could be obtained in a series of shots in which two pulses with relative delay entered the 3PF medium from one end and a third pulse entered from the other. A more elegant scheme [3.99, 100] is one in which three pulses enter the 3PF medium from three different directions. If the beams are spatially uniform, $G^3(\tau_1, \tau_2)$ can be obtained directly by measuring the fluorescence intensity along an appropriate contour determined by the angles between the three beams. The three-photon response of the bi-alkali cathode on an image orthicon has also been used to obtain single-shot images of $G^3(0, \tau)+G^3(\tau, 0)$ [3.101], but the restricted dynamic range of the cathode was apparently responsible for the low contrast ratios observed.

The highest order correlation experiment performed to date is that of *Auston* [3.26]. He detected the four-photon parametric emission generated by mixing a pulse $I(t)$ and its second-harmonic $I^2(t)$ in methanol. With relative delay τ between the two, the fluorescence intensity in this case is proportional to $G^5(0, 0, 0, \tau)$. Such an experiment can be very sensitive to pulse asymmetry and to low intensity levels in the wings of a pulse. In an extension of this method, *Auston* [3.95] exploited the variation in the phase-matching angle for parametric mixing to display the dynamic spectral behavior as well. Although Auston's work was performed with entire pulse trains from a Nd:glass laser, it confirmed the presence of positive frequency chirp and provided early evidence for slight asymmetry in the pulse wings and for the high contrast between pulse and background.

A different technique which relies on the generation of short probe pulses has been used by *von der Linde* et al. [3.92, 102] to accurately characterize pulses from a Nd:glass laser. A second-harmonic pulse is used to generate transient stimulated Raman scattering. The short phonon pulse produced in the medium provides a means for sampling a second optical pulse introduced at the proper angle for anti-Stokes Raman scattering. Alternatively, the Stokes light pulse produced can be correlated with the laser pulse in a two-frequency TPF arrangement. *Von der Linde* et al. [3.92, 102] found that the leading edges of their glass laser pulses approximated a gaussian shape while the trailing edges exhibited exponential behavior. *Von der Linde's* careful series of experiments [3.28] have demonstrated the importance of selecting single pulses from the output train of the Nd:glass laser.

3.1.7 Summary

If used properly, a combination of nonlinear correlation techniques can provide reliable information about the duration, shape, and dynamic spectral behavior of picosecond laser pulses. Linear measurements with streak cameras [3.25] and quantitative studies of nonlinear conversion efficiencies [3.28] have provided additional calibration of these methods and have been especially helpful in verifying contrasts between pulse energy and background. Accurate characterization of picosecond pulses has been necessary for the development of understanding of the pulse generation process itself and is of utmost importance in the use of such pulses to study other ultrafast processes. Although the results of previous pulse measurements provide valuable guidelines for picosecond experiments, they should not be relied upon when precise knowledge of pulse characteristics is needed.

Pulses from flashlamp-pumped lasers have been notoriously unpredictable. Their characteristics vary with changes in laser configuration, from shot to shot in a particular system, and also within a single pulse train. Because of this variability we have not made an attempt to tabulate pulse measurements made under a variety of circumstances. The weight of evidence indicates that most pulses from the Nd:glass laser have durations in the range 4 to 10 ps. With a carefully designed system and the selection of single pulses from the beginning of the train, *von der Linde* [3.28] has achieved a degree of reliability with an average pulsewidth of 5 ps and an rms deviation of 0.5 ps. Modelocked ruby lasers are more likely to produce pulses in the range 10 to 20 ps. Frequency-tunable pulsed dye lasers offer pulsewidths of 2 to 5 ps at wavelengths between 5800 Å and 7000 Å [3.159]. When any of these lasers are used in experiments requiring accurate time resolution, single pulses should be selected and independent measurements made of each pulse used to accumulate data.

Development of the modelocked cw dye laser [3.52, 59, 105, 123, 161] has led to the generation of pulses under more reproducible steady-state conditions. Pulses as short as 0.3 ps can be produced at high repetition rate. This facilitates the use of extensive signal averaging in conjunction with time-resolved measurements, and permits the study of fast processes under conditions of low as well as high intensity excitation.

3.2 Techniques for Measuring Picosecond Events

3.2.1 Pump and Probe Techniques

Early in 1967 *Shelton* and *Armstrong* [3.106] reported the first application of picosecond pulses to measurement. A train of intense picosecond pulses was used to saturate the absorption of a Q-switching dye. An attenuated image of the saturating pulse train was used to monitor the absorption recovery as a function of relative time delay between the pumping and probing trains of pulses. Since the time of that experiment a number of techniques have been developed to excite, probe, delay, gate and synchronize picosecond pulses. In this section we will review, compare and examine the pitfalls of various picosecond measurement techniques.

Two basic philosophies have emerged for making measurements with pulses from modelocked lasers. In the first, a single pulse is selected from a modelocked laser train and the entire measurement is obtained in a single event. In the second approach, the data from numerous events is collected and averaged. The first approach is more likely applicable with laser systems which produce pulses at a low repetition rate (i.e., Nd:glass, ruby), whereas the second approach is quite useful for a highly repetitive stable system such as the modelocked cw dye laser.

For a burst modelocked system such as Nd:glass, there is considerable variation from burst to burst as well as from pulse to pulse within the modelocked pulse train. It is often advantageous to select a single pulse from the modelocked pulse to get a pulse with the most reliable pulsewidth and amplitude. Techniques for achieving pulse selection are described in Chapter 2.

In making measurements with picosecond pulses it is necessary to provide accurate relative timing and delay between different pulses. In the case of *Shelton* and *Armstrong* [3.106] a beam splitter was used to pick off a small portion of the pumping beam and a delay path was provided mechanically by a mirror corner cube arrangement on a translation stage. The probe beam transmission through the sample was measured at different relative delay settings to monitor the absorption recovery. A number of shots at each delay setting was required to average out fluctuations. Clearly, in order to take such data with a single laser shot a mechanical delay is inadequate.

Malley and *Rentzepis* [3.107] demonstrated a "crossed-beam" technique in which the relaxation of a dye from an excited state can be measured with a single pulse. The experimental diagram is shown in Fig. 3.14. The technique uses two pulses which pass through the sample cell at right angles. As an intense beam is passed through a dye-containing sample cell, a second interrogating pulse, travelling at right angles to the first, uniformly illuminates the cell. The short duration of the pulses actually "freezes" the bleaching pulse within the cell. From the transparency of the dye measured along the path of the bleaching, a complete history of the event is obtained including before, during, and after passage of the intense pulse. The technique is applicable to both fluorescing and nonfluorescing species, with the only requirement being that the absorption band is saturable with the picosecond pulse. The extinction coefficient of the dye must be large enough so that enough photons are absorbed to make an appreciable change in the ground-state population.

The result obtained on the photographic plate is the convolution of the two picosecond pulses and the saturation recovery time of the dye. The thick-

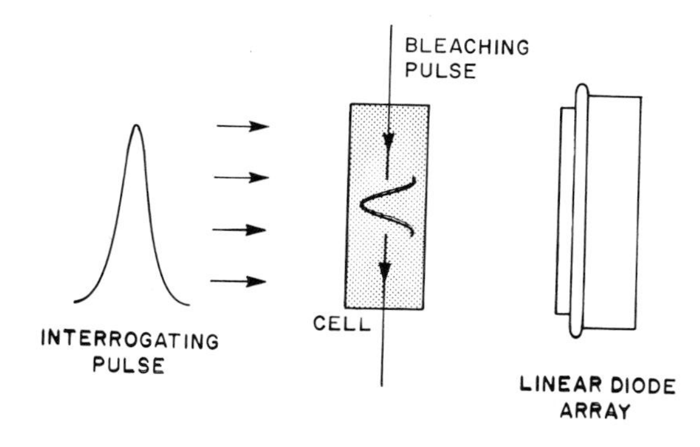

Fig. 3.14. The crossed-beam technique diagrammed above enables the complete time history of the bleaching and relaxation process to be taken in a single shot. The short duration of the bleaching and interrogating pulse "freezes" the bleaching pulse within the cell. The light entering the camera is a record of the density of the dye as a function of time

ness of the dye cell provides a time resolution limitation owing to the fact that the bleaching pulse moves while it is being interrogated. Thinner cells can minimize this effect but there must be a sufficient optical density change to allow a detectable contrast ratio between bleached and unbleached portions of the recording.

An improved technique which overcomes the cell thickness time-resolution problem in the technique just described utilizes an innovation introduced by *Topp* et al. [3.108]. The technique allows the strong bleaching beam to travel almost parallel to the interrogating beam which is divided up into parallel sections with a step progression of time delays. As can be seen in Fig. 3.15 the interrogating pulse is split up into a number of delayed pulses by means of an echelon. The echelon is formed by a stack of glass plates arranged in a step sequence. A plate of glass introduces approximately 1.7 ps of delay relative to air per millimeter of glass thickness. It is assumed in this technique that the beams are very nearly spatially uniform, a condition that can be strictly met only if the laser operates in the lowest-order mode.

The echelon technique is useful for making single pulse recordings but care must be taken to account for group dispersion effects when several frequencies are used in the experiment. *Topp* and *Orner* [3.109] have shown that group dispersion effects can be very important for interpreting time delay in a trans-

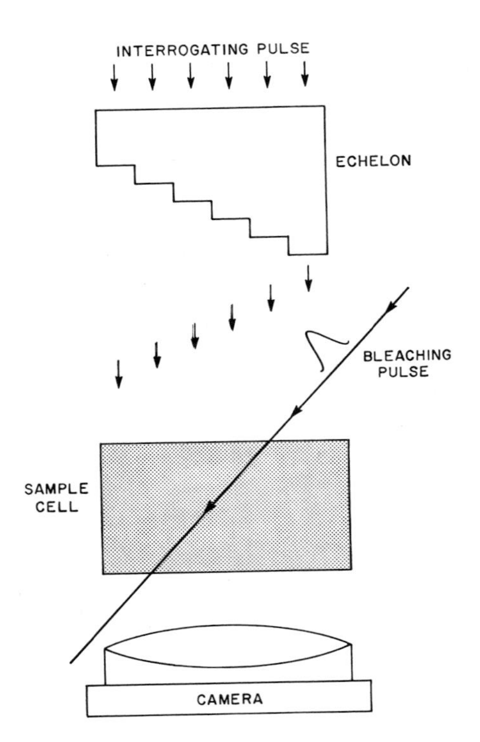

Fig. 3.15. Echelon technique adjacent segments of the interrogating beam are delayed by various thicknesses of glass. This allows transmission of the sample cell to be sampled at various times relative to the bleaching pulse

mission etalon. Group dispersion effects may be minimized by using a reflecting echelon.

Recently *Topp* [3.110] has outlined a technique which actually utilizes group dispersion to provide a means of producing a continuous time dispersion across a beam profile similar to the step progression of time delays produced by an echelon. By passing a beam through an equilateral prism a time dispersion Δt can be produced across the beam profile given by

$$\Delta t = -\frac{\lambda d \sec\theta}{c}\frac{dn}{d\lambda} \quad (3.17)$$

where d is the beam diameter, θ is the angle of incidence and n the index of refraction. The relative dispersion at different frequencies $\Delta\tau_\lambda$ is approximately

$$\Delta\tau_\lambda \approx \frac{2Bd \sec\theta}{c\lambda^2} \quad (3.18)$$

where B is almost constant.

To measure an entire absorption spectrum in a single shot a picosecond interrogating pulse with a broad bandwidth is required. *Alfano* and *Shapiro* [3.111] first suggested that a picosecond continuum, generated when an intense picosecond laser pulse is collimated into a nonlinear medium, could be used for monitoring absorption spectra of transient species on a picosecond time scale. Although various interpretations have been proposed the mechanism of picosecond continuum generation is not yet fully understood. See Chapter 4.

Numerous substances have been used to generate continua. *Alfano* and *Shapiro* [3.111] initially demonstrated the effect in liquid argon and krypton, sodium chloride, glasses, quartz and calcite crystals. H_2O and D_2O have been found useful in a number of experiments in which dynamic processes have been investigated [3.112, 113]. Recently it has been possible to stabilize variations in intensity over the continuum bandwidth by using CCl_4 [3.114]. *Nakashima* and *Mataga* [3.115] have used polyphosphoric acid and have found it to produce a broad stable spectrum. Care should be taken to measure the duration of the continuum pulses because quite often the pulses are appreciably broadened by material dispersion.

A typical configuration to measure time and wavelength resolved absorption spectra of excited states as described by *Malley* [3.116] is shown in Fig. 3.16. In this technique a modelocked laser pulse of a given frequency is used to "pump" the molecules of interest into the excited state and a series of picosecond continuum pulses is used to interrogate the excited-state absorption. A spectrograph is then used for recording the absorption spectrum. The image on the focal plane of the spectrograph is a three-dimensional recording of intensity, time and wavelength. Light intensity versus wavelength is plotted perpendicular to the slit, and intensity versus time is recorded parallel

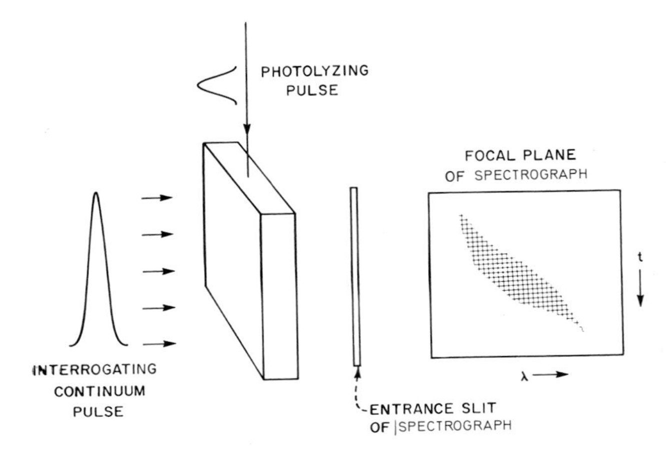

Fig. 3.16. A broadband picosecond continuum pulse interrogates the photolysis cell and enters the spectrograph. The spectral dependence is displayed in a direction perpendicular to the entrance slit and the time dependence in a direction parallel to the entrance slit

to the entrance slit. This information may be conveniently recorded on film, on an image intensifier or an array of detectors which can be read out electronically. A larger dynamic range can be obtained with an electronic read out.

Picosecond spectroscopy with single optical pulses has a number of attractive features. Samples that recover slowly or are irreversibly damaged can be moved or replaced between excitations. Each pulse used in an experiment can be characterized separately and results obtained with poor pulses can be rejected. This is especially important in studies of highly nonlinear processes which are sensitive to even small fluctuations in intensity or beam profile. Averaging over a train of nonidentical pulses can lead to incorrect results. The high powers available from single pulse systems also permit the efficient use of nonlinear processes to generated pump and probe pulses at a variety of different wavelengths.

On the negative side, single pulse measurements must induce relatively large perturbations to the system being measured in order to make the measurement. Often this results in artifacts such as self-focusing, stimulated emission, damage, etc., dominating the measurement of fundamental properties. An interesting case in point is the measurement of absorption recovery in DODCI [3.117–121] where several groups reported lifetimes from 10 ps to 1.2 ns. In this case stimulated emission was responsible for reducing the recovery time from its 1.2 ns value to shorter times depending on pump intensity.

It is also extremely important to have a uniform beam profile, particularly with the case of the echelon technique where different spatial regions of the beam sample different points in time.

In some cases effects being studied are so small that averaging of several laser shots is necessary to make the measurement. One experiment which illustrates this point is the investigation of transient electron hole plasmas in germanium [3.160]. An optical pulse incident on a Ge surface produces a small negative index change as the result of the formation of an electron hole plasma. By monitoring the refractive index change as a function of time, the diffusion of electrons can be measured under conditions of high density.

An ellipsometer is used to measure the induced index change by monitoring the change in ellipticity of light reflected from the Ge surface. The experimental diagram for making this measurement is shown in Fig. 3.17. A single pulse is selected from a modelocked Nd:glass laser pulse train and split into two parts. An intense part which impinges on the crystal surface and a weak probing

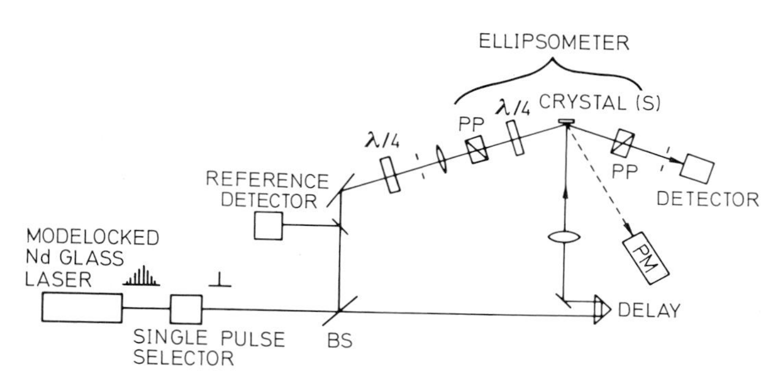

Fig. 3.17. Picosecond ellipsometer used to measure the time evolution of optically generated electron-hole plasmas in *Ge*

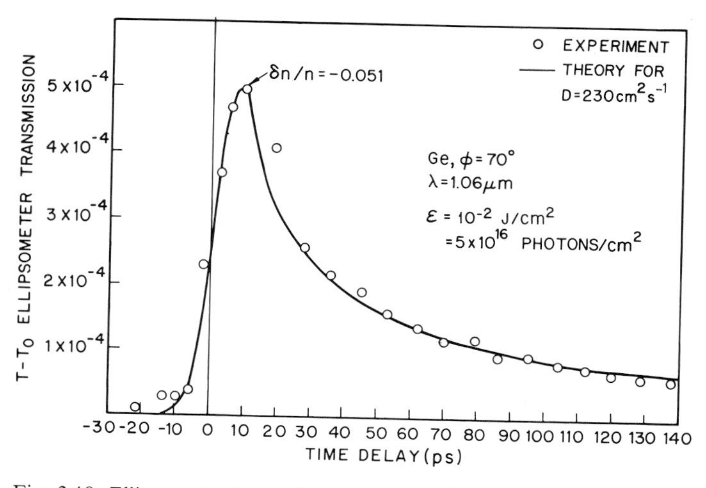

Fig. 3.18. Ellipsometer transmission $\Delta T \propto |\delta n/n|^2$ versus time delay between plasma generation and probing. Each point is a simple (unweighted) average of ten laser shots. The dispersion of the data was typically 11% per shot and 4% for the mean. The solid line is a theoretical fit for $D=230$ $cm^2\ s^{-1}$

beam which passes through the ellipsometer. The ellipsometer is set for extinction with the pump pulse blocked. The transmission through the ellipsometer is

$$\Delta T \propto \left|\frac{\delta n}{n}\right|^2 \tag{3.19}$$

where n is the index of refraction and δn is the induced change in index. The experimental results are given in Fig. 3.18. Each point on the curve is the average of ten laser shots. Note that a reference detector was used to monitor the intensity of each shot so that the transmission could be calculated. The solid curve is the result of diffusion theory. From the curve fit a $D = 230\ \mathrm{cm^2\ s^{-1}}$ is obtained for a plasma density of $1.7 \times 10^{20}\ \mathrm{cm^{-3}}$. The low density diffusivity is $65\ \mathrm{cm^2\ s^{-1}}$. The larger measured diffusion constant can be accounted for with a degenerate electron hole plasma model.

Taking data point by point as in the previous experiment and averaging a large number of shots is often a time-consuming and tedious task. By using stepping motors and electronics to automatically increment the time delay, fire the laser, and accumulate and average data, experiments of this kind are much less arduous.

3.2.2 Time-Resolved Measurements with Continuously Operated Systems

The passively modelocked cw dye laser [3.59, 105, 123] is a unique source of optical pulses for investigating ultrafast processes on a picosecond or even a subpicosecond time scale. Optical pulses have been generated as short as a few tenths of a picosecond at high repetition rates [3.52, 161]. The short duration of these pulses has provided a means of performing time-resolved spectroscopy with almost an order of magnitude higher resolution than is possible with other systems.

One advantage of a high repetition rate system is that powerful signal averaging detection methods can be applied to detect very small effects. Thus optical power levels can be kept low enough to eliminate measurement artifacts that often arise at high optical intensities.

A typical experimental setup for time-resolved spectroscopy measurements is shown in Fig. 3.19. [3.162] The incident pulses are acousto-optically dumped from the dye laser cavity at a repetition rate variable up to 10^6 pps. The pulse energy is typically 5×10^{-9} J. The pulses are generated in the $\mathrm{TEM_{00}}$ spatial mode and are easily focused near the diffraction limit.

The pulse train is divided in a modified Michelson interferometer into two beams, with one beam (the probe) about 10 times weaker than the pump beam. The pump beam alone passes through the mechanical chopper. Relative delay between the two beams is provided by varying one of the path lengths with a stepping motor which also indexes the multichannel analyzer in which data is stored and averaged. The pump and probe beams, parallel but not collinear, are focused by a simple lens to the same point in a 0.1 mm thick cell containing the sample. After transmission through the cell the pump beam is blocked and the probe beam passes to a photomultiplier. Modulation of the probe by the chopped pump is detected with a phase-lock amplifier whose output is recorded as a function of time delay in the multichannel analyzer. Averaging is achieved by repetitive scanning of the desired delay interval to average out slow drifts. A single measurement made over a period of several minutes utilizes more than 10^7 pulses.

An experimental trace of the absorption recovery of malachite green in methanol is shown in Fig. 3.20. The absorption recovery is characterized by a simple exponential behavior with a time constant of 2.1 ps. This to our knowledge is the shortest absorption recovery time ever directly measured in an organic molecule.

At this point it is probably instructive to consider potential artifacts in measurements made when both the pump beam and the probe beam are derived from the same pulse. A careful examination of the experimental trace in Fig. 3.20 reveals an unexpected peaking at zero time delay. This is due to an ad-

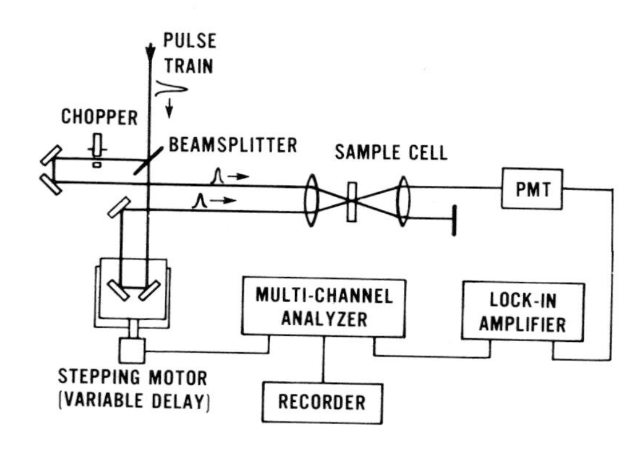

Fig. 3.19. Experimental arrangement for time-resolved spectroscopy measurements

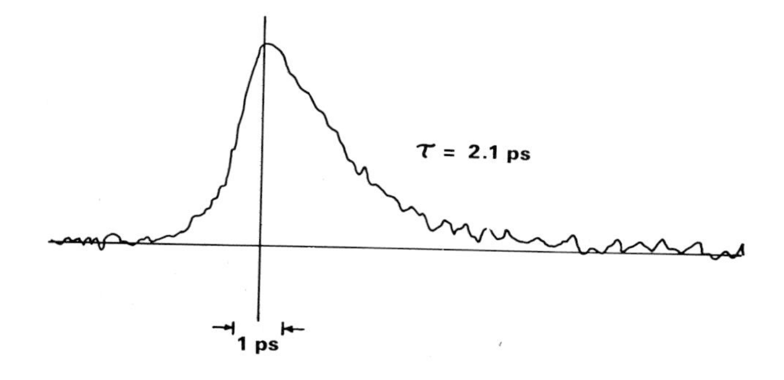

Fig. 3.20. Absorption recovery of malachite green in methanol

ditional coupling between the pump and probe beams that is unavoidable over the region near zero time delay where the pump and probe beams interact coherently.

The absorption change induced by a pulse of intensity $I(t)$ is given by

$$\Delta\alpha(t)=\int_{-\infty}^{\infty} I(t')A(t-t')dt' \tag{3.20}$$

where $A(t)$ is the impulse response of the molecular absorption. In this analysis we assume that the induced absorption change is a small perturbation to the system.

The optical intensity in the sample is given by

$$I(t)=|E_{\mathrm{p}}(t)+E_{\mathrm{m}}(t)|^2 \tag{3.21}$$

where $E_{\mathrm{p}}(t)$ and $E_{\mathrm{m}}(t)$ are the field amplitudes of the pump and measuring beams, respectively. We shall assume that the experimental geometry is non-collinear so that the two fields will have different propagation vectors. The fields will overlap in the sample but be separable thereafter. Temporally, $E_{\mathrm{m}}(t)$ is a weaker, delayed image of $E_{\mathrm{p}}(t)$ so that $E_{\mathrm{m}}(t)=aE_{\mathrm{p}}(t+\tau)$ with a $\ll 1$. Along the measuring beam axis in the sample we have

$$\frac{\partial E_{\mathrm{m}}}{\partial z}=-[\alpha_0+\Delta\alpha(t)]E_{\mathrm{m}} \tag{3.22}$$

where α_0 is the unperturbed absorption constant. Now substituting (3.20) and (3.21) into (3.22) we solve for the perturbation of the transmitted measuring beam as a function of delay τ. The resulting intensity modulation, $M(\tau)$, impressed on the transmitted measuring beam is given by a constant times the sum of two terms

$$M(\tau)=-b[\gamma(\tau)+\beta(\tau)] \tag{3.23}$$

where

$$\gamma(\tau)=\int_{-\infty}^{\infty} A(\tau-\tau')\int_{-\infty}^{\infty} I_{\mathrm{m}}(t+\tau')I_{\mathrm{m}}(t)dtd\tau' \tag{3.24}$$

and

$$\beta(\tau)=\int_{-\infty}^{\infty} E_{\mathrm{m}}(t+\tau)E_{\mathrm{m}}^*(t)\int_{-\infty}^{\infty} E_{\mathrm{m}}(t'+\tau)E_{\mathrm{m}}^*(t')A(t-t')dt'dt. \tag{3.25}$$

The first term $\gamma(\tau)$ is just the expected impulse response convolved with a pulse of finite width and the second term $\beta(\tau)$ is the artifact coherent coupling term. Note that the impulse response is convolved with the electric field auto-correlation function thus giving a contribution only for delays where the pump and probe are coherent. This term provides no information about the time behavior of the system. To illustrate this point consider a system where the impulse response is a step function (e.g., molecules removed from the ground state with a long recovery time). In such a system $A(t)$ is a constant for $t>0$. The measured response of the system is then the sum of two terms shown in Fig. 3.21.

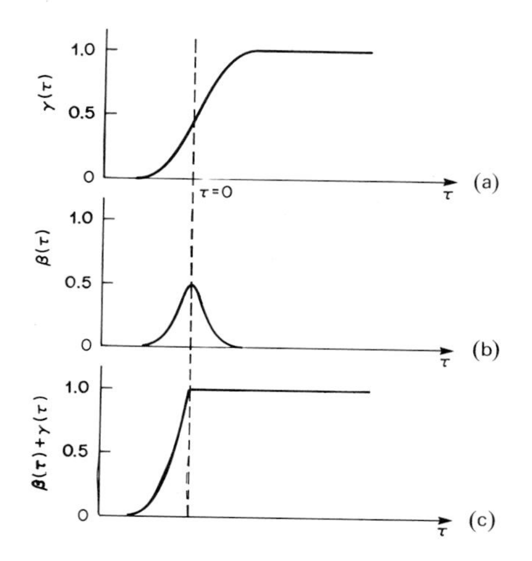

Fig. 3.21 a-c. The above curves illustrate the role of coherent coupling artifacts near $\tau=0$. For purposes of illustration the response has been normalized. Curve (a) is the expected response $\gamma(\tau)$. Curve (b) is the coupling artifact $\beta(\tau)$. Curve (c) is $-M(\tau)/b$ the measured response which is the sum of $\gamma(\tau)$ and $\beta(\tau)$

The first term is just the integral of the pulse autocorrelation function which is the actual response of the system. The second term is a sharp spike whose width is determined by the coherence of the pulse. Note that at $\tau=0$, $\beta(0)=\gamma(0)$. The measured function is the sum of these two terms and is peaked near $\tau=0$.

In a similar manner we could substitute an exponential $A(t)$ function into (3.24) and (3.25) to obtain the measured response shown in Fig. 3.20.

Coherent beam coupling artifacts are not usually observed in measurements made with nontransform-limited pulses from the Nd:glass laser. Usually the region of pulse coherence is so narrow compared to the pulsewidth that coherent effects are overlooked. However, there have been exceptions where pulse coherence has played a key role in interpreting a picosecond measurement [3.122].

The coherence of optical pulses can be used to advantage for making picosecond pulse measurements using a novel technique described by *Phillion* et al. [3.124]. The technique described as the induced transient-grating method is illustrated in Fig. 3.22. In this technique the excitation beam is split into two beams which intersect at an angle in the sample to be measured. The path lengths in the two arms are adjusted so that the excitation pulses arrive simultaneously at the sample. Interference between the two coherently related exciting beams then creates an interference pattern or grating in the sample.

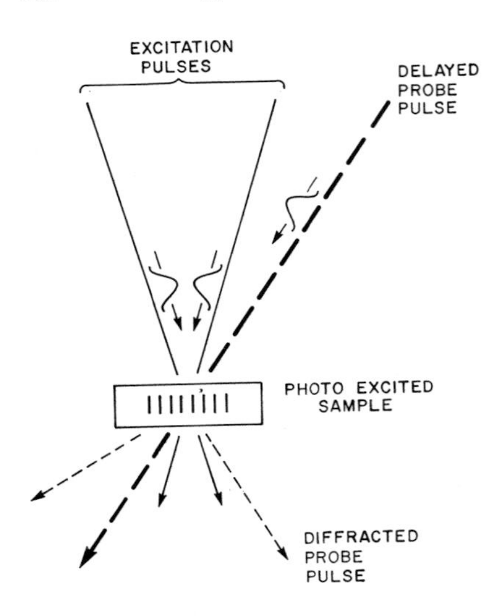

Fig. 3.22. The induced transient-grating measurement scheme

The grating spacing of this interference pattern is determined by the angle between the beams, pumping wavelength, and sample index of refraction.

The reference pattern will cause photoexcitation within the sample to occur in the grating pattern. The grating will decay away in time as the photoexcited states either relax or diffuse. A delayed probe pulse with variable delay to the exciting pulse is then passed through the transient grating at some angle. This probe pulse will be diffracted by the transient grating into one or more diffraction orders with a diffraction intensity appropriate to the residual strength of the transient grating phenomena. By measuring the scattered intensity in one of the side diffraction orders of the probe beam, one can measure the transient buildup or decay of the sample response. This technique was used by *Phillion* et al. [3.124] to study the rotational relaxation of rhodamine 6G molecules in solution. It has the advantage that the measured signal is background free when the detector looks only at the diffracted beam. Also, it provides a means of studying spatial diffusion effects by varying grating spacing.

The orientational relaxation time of organic molecules has been studied by inducing a dichroism by saturating molecules aligned parallel to the pump polarization [3.125, 126]. A weaker probe pulse is then used to measure and time resolve this induced dichroism as it decays away during subsequent randomization of molecular orientations. The anisotropic behavior is separated from the isotropic saturation by measuring the rotation of the polarization of the probe beam. Experimentally, the transmission of the probe beam is measured through crossed polarizers. The temporal decay of the induced transmission is a direct measure of the orientational relaxation time, providing that the orientational relaxation is much faster than decay to the ground state. If the decay rates of both processes are comparable, the measured rate is the sum of the two.

The transmission through the crossed polarizers for a small induced dichroism is given by

$$T=\tfrac{1}{8}\,\sigma^2(N_{\|}-N_{\perp})^2\,e^{-\alpha l} \tag{3.26}$$

where α is the average absorption constant, σ the absorption cross section, and $N_{\|}-N_{\perp}$ the anisotropic population difference between molecules oriented parallel and perpendicular to the pumping polarization. It is apparent from (3.26) that the measured transmission varies as the square of the induced population difference so that the measured exponential decay constant must be multiplied by a factor of 2. In general, to obtain the orientational relaxation time, we use the relation

$$\frac{1}{\tau_{OR}}=\frac{1}{2\tau_m}-\frac{1}{\tau} \tag{3.27}$$

where τ_m is the measured exponential decay constant and τ is the excited singlet lifetime.

In Fig. 3.23 we show an example of an orientation lifetime measurement for DODCI in methanol [3.126] obtained using the cw modelocked dye laser system. The measured decay constant is 44 ps. Using the measured value for $\tau=1.2$ ns an orientational lifetime of 95 ps is obtained.

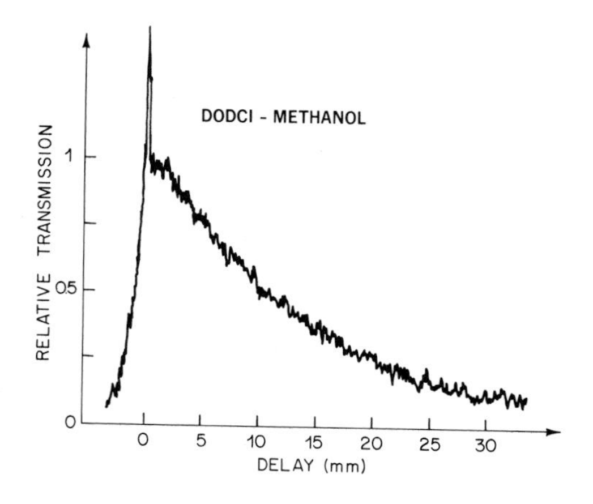

Fig. 3.23. Experimental result for orientational relaxation of DODCI in methanol. The exponential decay constant is 44 ps

3.2.3 Light-Gating Techniques for Time-Resolved Emission Studies

In Subsection 3.1.4 we analyzed the operation of the Kerr cell light gate. In this subsection we will discuss the use of light gating techniques for time-resolved emission spectroscopy.

In Fig. 3.24 we have diagrammed a typical experimental geometry for obtaining time-resolved luminescence spectra. Such an arrangement was first used by *Duguay* and *Hansen* [3.127] to observe the picosecond fluorescence lifetime of the dye DDI. The fluorescence to be measured is passed through polarizers and CS_2 cell with the polarizers set for extinction. The fluorescence is gated into the detector by passing an intense pulse through the CS_2 sample. The time resolution of the gate is limited by a number of factors. First, the pulsewidth of the gating pulse provides an obvious limitation. As the pulsewidth gets comparable to the Kerr lifetime the Kerr response can set a lower bound for the gate open time. Using subpicosecond optical pulses from a modelocked cw dye laser *Ippen* and *Shank* [3.71] have succeeded in operating a CS_2 gate at the Kerr resolution limit and in the process actually made direct measurement of the Kerr lifetime. The measured gate response is shown in Fig. 3.25. The measured Kerr response lifetime was determined to be $\tau = 2.0 \pm 0.3$ ps, one of the shortest lifetimes to be measured directly by optical techniques.

Several factors limit the time response of the gate in addition to pulsewidth and the Kerr lifetime. One is the group velocity mismatch between the gating and the gated beams. In CS_2 a green light pulse will be delayed by 2 ps with

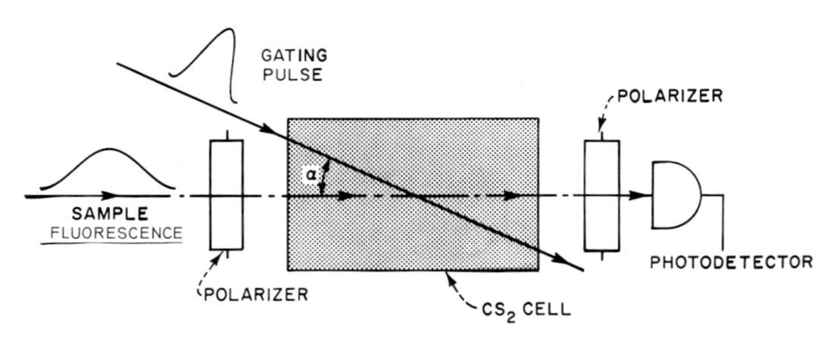

Fig. 3.24. Typical geometry for measuring time-resolved fluorescence with an optical Kerr gate

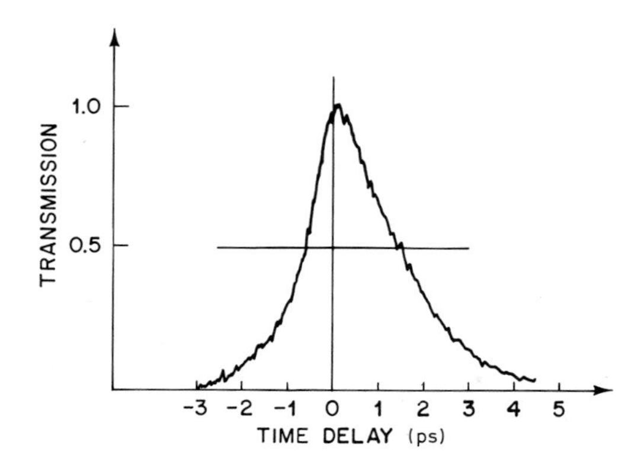

Fig. 3.25. Measured transmission response of a CS_2 optical Kerr gate

respect to an infrared pulse after 1 cm of travel. Another factor limiting the time response for the gate shown in Fig. 3.24 is the spread in arrival times at the gate for different parts of the infrared gating beam owing to a finite angle α between the gated fluorescence and the gating beams. This time spread is given by $d \tan \alpha/c$ where d is the gating beam diameter. In the experiment of *Duguay* and *Hansen* [3.66] $d=0.5$ cm, $\alpha=0.08$ so that $d \tan \alpha/c$ was 1.4 ps.

This geometrical time-smearing effect can be eliminated by using a collinear geometry for the gated and gating beams. Geometrical time smearing was eliminated by *Malley* and *Mourou* [3.128] by letting the gating beam go through the first polarizer of type Polaroid HN22 which is almost completely transparent to 1.06 μm radiation.

The above gating techniques have been adapted for single shot recording using crossed-beam or echelon techniques. A travelling-wave Kerr cell with crossed-beam geometry has been used to make single shot recordings with a pulse from a Nd : glass laser [3.103]. In Fig. 3.26 we show a crossed-beam light

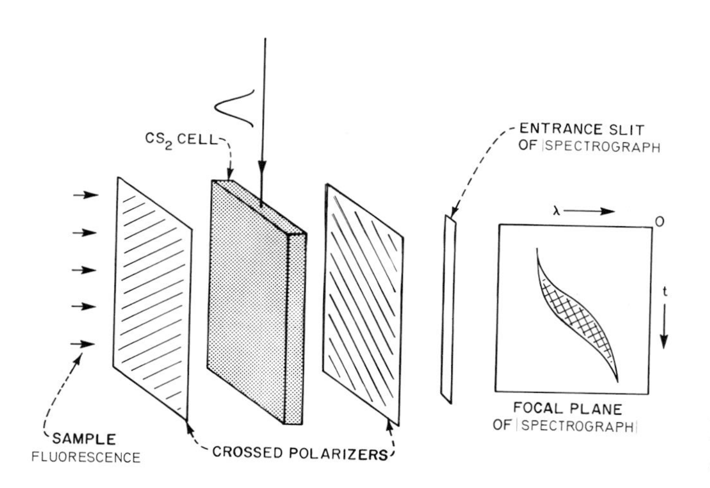

Fig. 3.26. A travelling-wave Kerr cell with crossed-beam geometry combined with a spectrograph to obtain time and spectrum resolved recordings

shutter combined with a spectrograph to obtain time- and spectrum-resolved recordings. The light intensity versus time is measured along the length of the entrance slit and the wavelength perpendicular to the slit. This technique has a timeresolution limitation owing to the passage of light through a cell of finite thickness. A 1 mm thick CS_2 light gate would have a 5.3 ps time-resolution limit.

Topp et al. [3.129] have demonstrated a nearly collinear geometry with an echelon which allows the recording of time-resolved emission in a single shot without the sacrifice of time resolution. The technique is illustrated in Fig. 3.27. The sample beam is reflected from an echelon stepped delay and travels nearly parallel to the gate pulses.

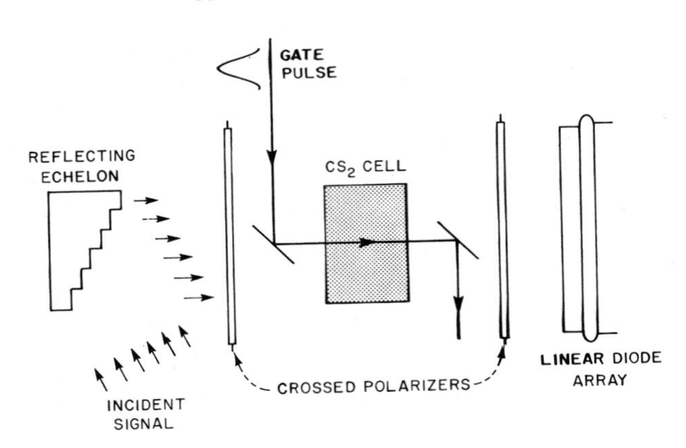

Fig. 3.27. Collinear geometry for CS_2 light gate for obtaining time-resolved emission

The optical Kerr gate has been used in a novel configuration to perform gated picture ranging [3.70]. The technique of gated picture ranging has been used in the nanosecond range to improve the visibility of targets such as ships and airplanes which have been obscured by fog. A powerful nanosecond light pulse is sent out from the observation point towards the target. At a preselected time, determined by the distance from the object, an electronic image-converter tube is gated on for a few nanoseconds. Earlier or later echoes from the fog corresponding to different ranges are not recorded because the picture tube is gated off at those times.

Picosecond pulses of course provide the possibility of ranging on a centimeter or even millimeter scale using an ultrafast Kerr shutter. The experimental set up of *Duguay* and *Mattick* [3.70] is shown in Fig. 3.28 (a). A second-harmonic green picosecond pulse was sent through a piece of thin paper tissue towards the target carrying the stylized drawing of a bell (shown unobscured in the upper left corner of Fig. 3.28 (b)). When photographed under room light illumination, the target is completely obscured by the tissue. When the green pulse is sent through the tissue and the ultrafast shutter is properly timed by the gating pulse, only the echo from the target is recorded by the camera. The result is shown in the lower part of Fig. 3.28 (b).

The gating technique provides the possibility of actually ranging through biological tissue. With a 2 ps Kerr gating time a spatial resolution of better than 1 mm should be possible. The technique is being used to range through primate eyes to determine the location of index discontinuities and to measure scattering particle sizes in microwave-induced cataracts [3.130]. Such a technique could also provide a means of detecting tumors in the human breast [3.131].

The optical Kerr shutter along with optical fiber delay lines has been used to make an optical sampling oscilloscope. With the configuration described by *Duguay* and *Savage* [3.132, 133] a picosecond event was displayed on a nanosecond response oscilloscope.

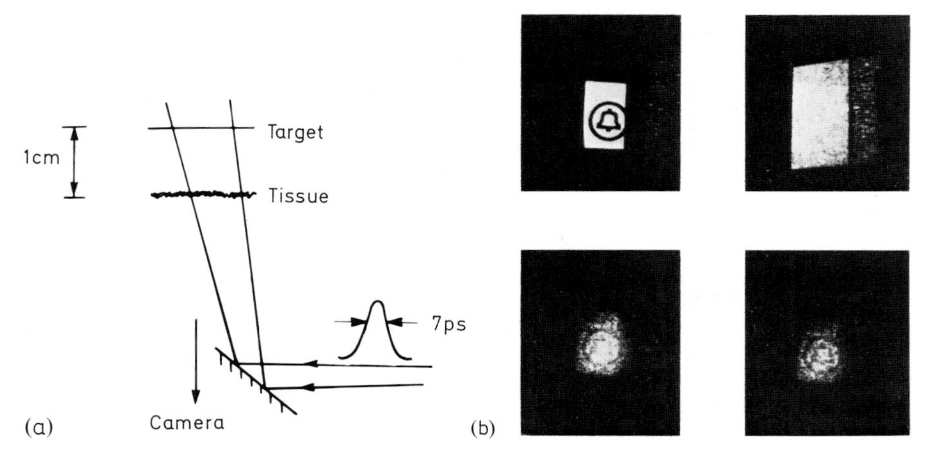

Fig. 3.28a and b. Experimental demonstration of picosecond optical gating. (a) Kerr-shutter camera views pulse reflections from a bell target hidden behind tissue paper; (b) Upper left: a direct photograph of the target, upper right: view from camera under room illumination, bottom: two photographs with camera shutter synchronized with the pulse reflection from the target

Mahr and *Hirsch* [3.104] have reported the use of nonlinear optical mixing techniques to make a light gate with picosecond response. The technique is based upon mixing the optical signal to be sampled with a short optical pulse in a crystal of ADP phase-matched for sum generation [3.134]. The opening time and resolution of the gate are determined by the optical pulse width. This technique allows the measurement of low level light signals with good wavelength discrimination.

The up-conversion experiments described by *Mahr* and *Hirsch* were performed with 20-picosecond pulses from a tunable, continuously modelocked dye laser. The luminescence measured was mixed with a modelocked dye laser pulse in a phase-matched crystal of ADP. The up-converting crystal in the presence of short optical pulses acted as an optical light gate which is "open" only when the pump pulse is present. Because the up-converted power is proportional to the instantaneous signal power, the temporal shape of the signal can be mapped out by delaying the dye laser pump with respect to the signal and measuring the up-converted power as a function of delay.

With the angle-tuned ADP crystal an efficiency of 0.2% conversion was obtained with 100 W pump power. However with a more optimum crystal and higher pump powers conversion efficiencies on the order of 10% should be possible.

One advantage of this technique is that the up-converted power is in the uv and can be spectrally isolated. This aids in detection. It has the disadvantage that the signal, the pump and the phase-match angle must be accurately aligned to upconvert the desired band of frequencies. Up-conversion promises to be a useful technique for time-resolved luminescence studies.

3.2.4 Streak Camera Techniques for Time-Resolved Emission Studies

Ultrafast, picosecond resolution streak cameras have proven useful in the study of ultrashort pulse emission from modelocked lasers [3.25, 29, 135–152]. The popularity of the streak camera as a diagnostic tool originates from its ability to measure the intensity of a pulse as a function of time, $I(t)$, directly and unambiguously with high temporal resolution. Resolutions of less than 10 ps were first obtained with a mesh design [3.141]. The use of the streak camera for pulse shape measurements as well as recent advances in streak tube design is discussed in detail in Chapter 2. The application of streak cameras for studying rapid temporal events other than laser pulse shapes is briefly discussed here.

Fluorescence decay times have now been measured by a number of investigators by using a streak camera technique [3.118, 153, 121] (see also Chapter 7). For measuring these lifetimes a streak camera technique has several special advantages. Experimental arrangements are simple, usually consisting of nothing more than the laser pulse, the sample, a collection lens to project the fluorescence onto the slit of the streak camera, and appropriate filters. The lifetimes can be measured on a single shot.

The streak camera may also be used to advantage for resolving spatial structure as a function of time. This is accomplished by imaging different spatial positions on different portions of the entrance slit of the camera. In this way events such as the rapid expansion of a plasma [3.154] or the contraction or expansion of a laser beam due to nonlinear optical effects [3.155, 156] can be effectively time resolved.

Often processes are encountered that are either intensity dependent or whose initiation is of a statistical nature, and these processes are not exactly reproducible from shot to shot. Examples of such events where the streak camera has been used to advantage include transient stimulated Raman scattering [3.157], dielectric breakdown accompanying plasma formation [3.155], and also exciton annihilation. The streak camera has also been applied in the X-ray region (see Chapt. 2).

The streak camera may be easily calibrated by passing a laser pulse through an etalon consisting of two mirrors of reflectivity R_1 and R_2 separated by a distance L. A series of pulses emerges, each spaced in time by $2L/c$ giving a calibration of the sweep speed. Each successive pulse is attenuated by the product of the mirror reflectivities, $R_1 R_2$, yielding an intensity calibration. When used in conjunction with an optical multichannel analyzer display system, the streak camera recording technique is especially simple. At present, the primary disadvantage of streak cameras is relatively large initial cost.

3.2.5 Conclusion

In the latter part of this chapter we have endeavored to describe the great number of experimental techniques which have been developed in the past decade to study phenomena on a picosecond or even a subpicosecond time scale. Which of these techniques will be developed into a routine laboratory tool with the passage of time remains to be seen.

References

3.1 A. J. DeMaria, D. A. Stetser, H. Heynau: Appl. Phys. Lett. **8**, 174 (1966)
3.2 H. P. Weber: J. Appl. Phys. **38**, 2231 (1967)
3.3 M. Maier, W. Kaiser, J. A. Giordmaine: Phys. Rev. Lett. **17**, 1275 (1966)
3.4 J. A. Armstrong: Appl. Phys. Lett. **10**, 16 (1967)
3.5 J. A. Giordmaine, P. M. Rentzepis, S. L. Shapiro, K. W. Wecht: Appl. Phys. Lett. **11**, 216 (1967)
3.6 D. J. Bradley, W. Sibbett: Appl. Phys. Lett. **27**, 382 (1975)
3.7 H. P. Weber: Phys. Lett. **27A**, 321 (1968)
3.8 J. R. Klauder, M. A. Duguay, J. A. Giordmaine, S. L. Shapiro: Appl. Phys. Lett. **13**, 174 (1968)
3.9 H. P. Weber, R. Dändliker: IEEE J. QE-**4**, 1009 (1968)
3.10 R. J. Harrach: Phys. Lett. **28A**, 393 (1968)
3.11 R. J. Harrach: Appl. Phys. Lett. **14**, 148 (1969)
3.12 E. I. Blount, J. R. Klauder: J. Appl. Phys. **40**, 2874 (1969)
3.13 A. A. Grutter, H. P. Weber, R. Dändliker: Phys. Rev. **188**, 300 (1969)
3.14 T. I. Kuznetsova: Soviet Phys. JETP **28**, 1303 (1969)
3.15 K. H. Drexhage: Appl. Phys. Lett. **14**, 318 (1969)
3.16 R. H. Picard, P. Schweitzer: Phys. Lett. **29A**, 415 (1969)
3.17 R. H. Picard, P. Schweitzer: Phys. Rev. A **1**, 1803 (1970)
3.18 H. E. Rowe, T. Li: IEEE J. QE-**6**, 49 (1970)
3.19 H. A. Pike, M. Hercher: J. Appl. Phys. **41**, 4562 (1970)
3.20 W. H. Glenn: IEEE J. QE-**6**, 510 (1970)
3.21 R. Dändliker, A. A. Grutter, H. P. Weber: IEEE J. QE-**6**, 687 (1970)
3.22 M. A. Duguay, J. W. Hansen, S. L. Shapiro: IEEE J. QE-**6**, 725 (1970)
3.23 D. H. Auston: IEEE J. QE-**7**, 465 (1971)
3.24 D. M. Kim, P. L. Shah, T. A. Rabson: Phys. Lett. **35A**, 260 (1971)
3.25 D. J. Bradley, G. H. C. New: Proc. IEEE **62**, 313 (1974)
3.26 D. H. Auston: Appl. Phys. Lett. **18**, 249 (1971)
3.27 H. A. Haus: private communication (1975)
3.28 D. von der Linde: IEEE J. QE-**8**, 328 (1972)
3.29 D. J. Bradley, B. Liddy, A. G. Roddie, W. Sibbett, W. E. Sleat: Opt. Commun. **3**, 426 (1971)
3.30 M. A. Duguay, S. L. Shapiro, P. M. Rentzepis: Phys. Rev. Lett. **19**, 1014 (1967)
3.31 P. C. Magnante: J. Appl. Phys. **40**, 4437 (1969)
3.32 S. L. Shapiro, M. A. Duguay, L. B. Kreuzer: Appl. Phys. Lett. **12**, 36 (1968)
3.33 A. J. DeMaria, W. H. Glenn, M. J. Brienza, M. E. Mack: Proc. IEEE **57**, 2 (1969)
3.34 G. Kachen, L. Steinmetz, J. Kysilka: Appl. Phys. Lett. **13**, 229 (1968)
3.35 P. M. Rentzepis, M. A. Duguay: Appl. Phys. Lett. **11**, 218 (1970)
3.36 S. K. Kurtz, S. L. Shapiro: Phys. Lett. **28A**, 17 (1968)
3.37 M. Bass, D. Woodward: Appl. Phys. Lett. **12**, 275 (1968)
3.38 D. J. Bradley, G. H. C. New, B. Sutherland, S. J. Caughey: Phys. Lett. **28A**, 532 (1969)
3.39 H. Statz, M. Bass: J. Appl. Phys. **40**, 377 (1969)
3.40 N. G. Basov, P. G. Kryukov, V. S. Letokhov, Ya. V. Senatskii: IEEE J. QE-**4**, 606 (1968)
3.41 V. S. Letokhov: Sov. Phys. JETP **28**, 1026 (1969)
3.42 S. L. Shapiro, M. A. Duguay: Phys. Lett. **28A**, 698 (1969)
3.43 A. R. Clobes, M. J. Brienza: Appl. Phys. Lett. **14**, 287 (1969)
3.44 D. von der Linde, O. Bernecker, W. Kaiser: Opt. Commun. **2**, 149 (1970)

3.45 H.P. Weber, H.G. Danielmeyer: Phys. Rev. A **2**, 2074 (1970)
3.46 M.W. McGeoch: Opt. Commun. **7**, 116 (1973)
3.47 J.H. Bechtel, W.L. Smith: J. Appl. Phys. **46**, 5055 (1975)
3.48 D.J. Bradley, T. Morrow, M.S. Petty: Opt. Commun. **2**, 1 (1970)
3.49 D.J. Bradley, M.H.R. Hutchinson, H. Koetser, T. Morrow, G.H.C. New, M.S. Petty: Proc. R. Soc. Lond. A **328**, 97 (1972)
3.50 H.P. Weber: J. Appl. Phys. **39**, 6041 (1968)
3.51 H.A. Haus, C.V. Shank, E.P. Ippen: Opt. Commun. **15**, 29 (1975)
3.52 E.P. Ippen, C.V. Shank: Appl. Phys. Lett. **27**, 488 (1975)
3.53 W.H. Glenn, M.J. Brienza: Appl. Phys. Lett. **10**, 221 (1967)
3.54 E. Mathieu, Hj. Keller: J. Appl. Phys. **41**, 1560 (1970)
3.55 D. Gloge, R. Roldan: Appl. Phys. Lett. **14**, 3 (1969)
3.56 D. Gloge, T.P. Lee: J. Appl. Phys. **42**, 307 (1971)
3.57 D. Gloge, T.P. Lee: IEEE J. QE-**7**, 43 (1971)
3.58 A. Dienes, E.P. Ippen, C.V. Shank: Appl. Phys. Lett. **19**, 258 (1971)
3.59 E.P. Ippen, C.V. Shank, A. Dienes: Appl. Phys. Lett. **21**, 348 (1972)
3.60 J. Comly, E. Garmire: Appl. Phys. Lett. **12**, 7 (1968)
3.61 R.C. Miller: Phys. Lett. **26 A**, 177 (1968)
3.62 S.L. Shapiro: Appl. Phys. Lett. **13**, 19 (1968)
3.63 W.H. Glenn: IEEE J. QE-**5**, 284 (1969)
3.64 G. Mayer, F. Gires: Compt. Rend. Acad. Sci. (Paris) **258**, 2039 (1964)
3.65 F. Shimizu, B.P. Stoicheff: IEEE J. QE-**5**, 544 (1969)
3.66 M.A. Duguay, J.W. Hansen: Appl. Phys. Lett. **15**, 192 (1969)
3.67 L. Dahlström: Opt. Commun. **3**, 399 (1971)
3.68 M.A. Duguay, J.W. Hansen: IEEE J. QE-**7**, 37 (1971)
3.69 M.A. Duguay: Am. Scient. **59**, 550 (1971)
3.70 M.A. Duguay, A.T. Mattick: Appl. Opt. **10**, 2162 (1971)
3.71 E.P. Ippen, C.V. Shank: Appl. Phys. Lett. **26**, 92 (1975)
3.72 M. Paillette: Compt. Rend. Acad. Sci. (Paris) **262**, 264 (1966)
3.73 A. Owyoung, R.W. Hellwarth, N. George: Phys. Rev. B **5**, 628 (1972)
3.74 S.L. Shapiro, H.P. Broida: Phys. Rev. **154**, 129 (1967)
3.75 M.A. Duguay, J.W. Hansen: NBS Spec. Pub. No. 341 (Gov. Printing Office, Wash. D.C. 1970), p. 45–49
3.76 J.P. Herman, D. Ricard, J. Ducuing: Appl. Phys. Lett. **23**, 178 (1973)
3.77 J.P. Herman, J. Ducuing: J. Appl. Phys. **45**, 5100 (1974)
3.78 R.R. Cubeddu, O. Svelto: IEEE J. QE-**5**, 495 (1969)
3.79 T.K. Gustafson, J.P. Taran, H.A. Haus, J.R. Lifsitz, P.L. Kelley: Phys. Rev. **177**, 1196 (1969)
3.80 E.B. Treacy: Phys. Letters **28 A**, 34 (1968)
3.81 E.B. Treacy: Appl. Phys. Lett. **14**, 112 (1969)
3.82 E.B. Treacy: IEEE J. QE-**5**, 454 (1969)
3.83 D.J. Bradley, G.H.C. New, S.J. Caughey: Phys. Lett. **32 A**, 313 (1970)
3.84 R.A. Fisher, J.A. Fleck: Appl. Phys. Lett. **15**, 287 (1969)
3.85 J.A. Giordmaine, M.A. Duguay, J.W. Hansen: IEEE J. QE-**4**, 252 (1968)
3.86 M.A. Duguay, J.W. Hansen: Appl. Phys. Lett. **14**, 14 (1969)
3.87 F. Gires, P. Tournois: Compt. Rend. Acad. Sci. **258**, 6112 (1964)
3.88 R.A. Fisher, P.L. Kelley, T.K. Gustafson: Appl. Phys. Lett. **14**, 140 (1969)
3.89 A. Laubereau: Phys. Lett. **29 A**, 539 (1969)
3.90 R.H. Lehmberg, J.M. McMahon: Appl. Phys. Lett. **28**, 204 (1976)
3.91 J.W. Shelton, Y.R. Shen: Phys. Rev. Lett. **26**, 538 (1971)
3.92 D. von der Linde, A. Laubereau: Opt. Commun. **3**, 279 (1971)
3.93 E.B. Treacy: J. Appl. Phys. **42**, 3848 (1971)
3.94 E.B. Treacy: Appl. Phys. Lett. **17**, 14 (1970)
3.95 D.H. Auston: Opt. Commun. **3**, 272 (1971)
3.96 P.P. Bey, J.F. Giuliani, H. Rabin: Phys. Lett. **26 A**, 128 (1968)
3.97 R.C. Eckhardt, C.H. Lee: Appl. Phys. Lett. **15**, 425 (1969)

3.98 P.M. Rentzepis, C.J. Mitschele, A.C. Saxman: Appl. Phys. Lett. **17**, 122 (1970)
3.99 P.M. Rentzepis: In *Proc. 3rd Intern. Symp. Solvated Electron*, Hannita, Israel (1972)
3.100 K. Hamal, T. Daricek, V. Kubecek, A. Novatony, M. Vrobova: IEEE J. QE-**8**, 600 (1972)
3.101 D.C. Burnham: Appl. Phys. Lett. **17**, 45 (1970)
3.102 D. von der Linde, A. Laubereau, W. Kaiser: Phys. Rev. Lett. **26**, 954 (1971)
3.103 M.M. Malley, P.M. Rentzepis: Chem. Phys. Lett. **7**, 57 (1970)
3.104 H. Mahr, M.D. Hirsch: Opt. Commun. **13**, 96 (1975)
3.105 C.V. Shank, E.P. Ippen: Appl. Phys. Lett. **24**, 373 (1974)
3.106 J.W. Shelton, J.A. Armstrong: IEEE J. QE-**3**, 302 (1967)
3.107 M.M. Malley, P.M. Rentzepis: Chem. Phys. Lett. **3**, 534 (1969)
3.108 M.R. Topp, P.M. Rentzepis, R.P. Jones: Chem. Phys. Lett. **9**, 1 (1971)
3.109 M.R. Topp, G. Orner: Chem. Phys. Lett. **3**, 407 (1975)
3.110 M.R. Topp: Opt. Commun. **14**, 126 (1975)
3.111 R.R. Alfano, S.L. Shapiro: Chem. Phys. Lett. **8**, 631 (1971)
3.112 G.B. Busch, R.P. Jones, P.M. Rentzepis: Chem. Phys. Lett. **18**, 178 (1973)
3.113 H. Tashiro, T. Yajima: Chem. Phys. Lett. **25**, 582 (1974)
3.114 D. Magde, B.A. Bushaw, M.W. Windsor: Chem. Phys. Lett. **28**, 263 (1974)
3.115 N. Nakashima, N. Mataga: Chem. Phys. Lett. **35**, 350 (1975)
3.116 M. Malley: Picosecond Laser Techniques. In *Creation and Detection of the Excited State*, Vol. 2, ed. by William R. Ware, (Marcel Dekker 1974)
3.117 G.E. Busch, R.P. Jones, P.M. Rentzepis: Chem. Phys. Lett. **18**, 1978 (1973)
3.118 E.G. Arthurs, D.J. Bradley, A.G. Roddie: Chem. Phys. Lett. **22**, 230 (1973)
3.119 D. Magde, M.W. Windsor: Chem. Phys. Lett. **27**, 31 (1974)
3.120 C.V. Shank, E.P. Ippen: Appl. Phys. Lett. **26**, 62 (1975)
3.121 J.C. Mialocq, A.W. Boyd, J. Jaraudias, J. Sutton: Chem. Phys. Lett. **37**, 236 (1976)
3.122 C.V. Shank, D.H. Auston: Phys. Rev. Lett. **34**, 479 (1975)
3.123 F. O'Neill: Opt. Commun. **6**, 360 (1972)
3.124 D.W. Phillion, D.J. Kuizenga, A.E. Siegman: Appl. Phys. Lett. **27**, 85 (1975)
3.125 K.B. Eisenthal, K.H. Drexhage: J. Chem. Phys. **51**, 5702 (1969)
3.126 C.V. Shank, E.P. Ippen: Appl. Phys. Lett. **26**, 62 (1975)
3.127 M.A. Duguay, J.W. Hansen: Opt. Commun. **1**, 254 (1969)
3.128 G. Mourou, M.M. Malley: Opt. Commun. **10**, 323 (1974)
3.129 M.R. Topp, P.M. Rentzepis, R.P. Jones: Chem. Phys. Lett. **9**, 1 (1971)
3.130 A.P. Bruckner: to be published
3.131 R. Kompfner: private communication
3.132 M.A. Duguay, A. Savage: Opt. Commun. **9**, 212 (1973)
3.133 G.C. Vogel, A. Savage, M.A. Duguay: IEEE J. QE-**10**, 642 (1974)
3.134 M.A. Duguay, J.W. Hansen: Appl. Phys. Lett. **13**, 178 (1968)
3.135 V.V. Korobkin, M.Ya. Schelev: In *High-Speed Photography*, Proc. 8th Intern. Congr. High-Speed Photography, ed. by N.R. Nilsson, L. Högberg (John Wiley & Sons, New York 1968) pp. 36–40
3.136 A.A. Malyutin, M.Ya. Schelev: JETP Lett. **9**, 266 (1969)
3.137 V.V. Korobkin, A.A. Malyutin, M.Ya. Schelev: J. Photogr. Sci. **17**, 179 (1969)
3.138 V.V. Korobkin, A.A. Malyutin, M.Ya. Schelev: Sov. Phys.-Tech. Phys. **16**, 165 (1971)
3.139 A.J. Alcock, M.C. Richardson, M.Ya. Schelev: In *Proc. 9th Intern. Congr. High-Speed Photography*, ed. by W.G. Hyzer, W.G. Chace (Soc. Mot. Pict. and Tel. Eng., New York 1970) pp. 191–197
3.140 N. Ahmed, B.C. Gale, M.H. Key: Advan. Electron. Phys. **28**, 999 (1970)
3.141 D.J. Bradley, B. Liddy, W.E. Sleat: Opt. Commun. **2**, 391 (1971)
3.142 M.Ya. Schelev, M.C. Richardson, A.J. Alcock: Appl. Phys. Lett. **18**, 354 (1971)
3.143 M.Ya. Schelev, M.C. Richardson, A.J. Alcock: Rev. Sci. Instr. **43**, 1819 (1972)
3.144 D.J. Bradley, B. Liddy, W. Sibbett, W.E. Sleat: Appl. Phys. Lett. **20**, 219 (1972)
3.145 D.J. Bradley, B. Liddy, A.G. Roddie, W. Sibbett, W.E. Sleat: Advan. Electron. Phys. **33B**, 1145 (1972)
3.146 S.W. Thomas, L.W. Coleman: Appl. Phys. Lett. **20**, 83 (1972)

3.147 E.G. Arthurs, D.J. Bradley, B. Liddy, F. O'Neill, A.G. Roddie, W. Sibbett, W.E. Sleat: In *Proc. 10th Intern. Conf. High-Speed Photography*, Nice, France (1972) pp. 117–122
3.148 V.V. Korobkin, A.A. Malyutin, A.M. Prokhorov, R.V. Serov, M.Ya. Schelev: In *Proc. 10th Intern. Conf. High-Speed Photography*, Nice, France (1972) pp. 122–127
3.149 N.G. Basov, M.M. Butslov, P.G. Kriukov, Yu.A. Matveets, E.A. Smirnova, S.D. Fanchenko, S.V. Chekalin, R.V. Chirkin: Preprint 82, Lebedev Phys. Inst., Moscow, USSR (1972) pp. 1–25
3.150 S.D. Fanchenko, B.A. Frolov: JETP Lett. **16**, 101 (1972)
3.151 D.J. Bradley, W. Sibbett: Opt. Commun. **9**, 17 (1973)
3.152 M.C. Richardson: IEEE J. QE-**9**, 768 (1973)
3.153 S.L. Shapiro, R.C. Hyer, A.J. Campillo: Phys. Rev. Lett. **33**, 513 (1974)
3.154 M.C. Richardson, K. Sala: Appl. Phys. Lett. **23**, 420 (1973)
3.155 A.J. Campillo, R.A. Fisher, R.C. Hyer, S.L. Shapiro: Appl. Phys. Lett. **25**, 408 (1974)
3.156 V.V. Korobkin, A.M. Prokhorov, R.V. Serov, M.Ya. Schelev: JETP Lett. **11**, 94 (1970)
3.157 W.H. Lowdermilk, G.I. Kachen: Appl. Phys. Lett. **27**, 133 (1975)
3.158 P.W. Smith, M.A. Duguay, E.P. Ippen: Mode-locking of lasers. In *Progress in Quantum Electronics*, Vol. 3, Part 2 (Pergamon Press 1974)
3.159 E. G. Arthurs, D. J. Bradley, A. G. Roddie: Appl. Phys. Lett. **20**, 125 (1972)
3.160 D. H. Auston, C. V. Shank: Phys. Rev. Lett. **32**, 1120 (1974)
3.161 I. S. Ruddock, D. J. Bradley: Appl. Phys. Lett. **29**, 296 (1976)
3.162 E. P. Ippen, C. V. Shank, A. Bergman: Chem. Phys. Lett. **38**, 611 (1976)

4. Picosecond Nonlinear Optics

D. H. Auston

With 24 Figures

The extremely active interest in recent years in the study with picosecond pulses of nonlinear optical effects can be accounted for by a number of reasons. First, the peak power available from modelocked laser and laser amplifier systems greatly exceeds that obtainable from Q-switched and other longer pulsed lasers. Also, the threshold for optical damage in most materials is considerably higher with picosecond pulses. This has enabled the observation of many higher order nonlinear effects which would have been difficult, if not impossible, by other means. Even when optical damage does not impose a major limitation, ultrashort pulses can be used to discriminate against undesirable effects having long time constants such as sample heating. Probably the most important use of picosecond pulses is for the study of transient nonlinear optical interactions. In these instances the material response has a finite memory, whereby the nonlinear polarization is determined by the previous history of the applied optical field. Examples are transient stimulated Raman scattering and transient self-focusing. Typical time constants for these nonlinearities are in the range of 10^{-13} to 10^{-10} s, and are thus amenable to study with picosecond pulses. The measurement of these effects provides important information about the dynamics of propagation of ultrashort pulses in nonlinear media, and about the physical mechanisms responsible for their nonlinear response.

The intent of this chapter is to review the topic of nonlinear optics with particular emphasis on the utilization of picosecond pulses. Where possible, the picosecond work has been placed in context with related nonlinear optics work on larger time scales. For more complete background information, the reader is referred to the books by *Bloembergen* [4.1], *Arecchi* and *Schulz-Dubois* [4.2], *Rabin* and *Tang* [4.3] and other volumes in the "Topics in Applied Physics" series.

4.1 Nonlinear Optical Effects

Although the material in this chapter is clearly delineated by sections dealing with specific physical effects, many of the experiments described span more than one section. This arises from the fact that nonlinear optical effects are often not isolated events but can frequently occur in combinations. This is especially true with picosecond pulses. For example, stimulated Raman scattering is often accompanied by self-focusing and sometimes by self-phase modulation, which can complicate an experiment, making its interpretation more difficult. A carefully designed experiment will attempt to isolate the desired effect, or at least control other competing effects in a predictable way.

A popular representation of the nonlinear properties of a material is to express the polarization as a perturbation expansion in successively higher orders of the optical electric field E:

$$
\begin{aligned}
P = {} & \chi^{(1)} \cdot E && \left\{ \text{complex index of refraction} \right. \\
& + \chi^{(2)} : EE && \left\{ \begin{array}{l} \text{second harmonic generation} \\ \text{sum, difference frequency generation} \\ \text{parametric fluorescence} \\ \text{optical rectification} \end{array} \right. \\
& + \chi^{(3)} \vdots EEE && \left\{ \begin{array}{l} \text{third harmonic generation} \\ \text{four photon parametric mixing} \\ \text{four photon parametric fluorescence} \\ \text{nonlinear index of refraction} \\ \text{Raman scattering} \\ \text{two photon absorption} \end{array} \right. \\
& + \ldots
\end{aligned}
\tag{4.1}
$$

Some nonlinear effects associated with each order of nonlinear susceptibility $\chi^{(n)}$ are shown as examples. Quantum mechanical derivations of the $\chi^{(n)}$'s for various effects have been derived by *Armstrong* et al. [4.4]. A recent comprehensive review of various theoretical models has been made by *Flytzanis* [4.5]. When the nonlinear response has a transient character, the nonlinear susceptibilities are complex and strongly frequency dependent. In many cases it is more illuminating to represent the nonlinearity by an explicit dynamical equation describing the excitation directly in the time domain, rather than in the form of the expansion (4.1).

The specific topics that have been selected for this review do not comprise a complete account of all the work on this subject, but are merely representative of the more important developments. I regret that limitations imposed on the size and scope of this article have resulted in some omissions. For related aspects of picosecond nonlinear optics dealing with applications to specific areas such as generation and measurement techniques, physics, chemistry, and biology, the reader is referred to the accompanying chapters in this volume.

4.2 Optical Harmonic Generation and Mixing

4.2.1 Second Harmonic Generation

Second harmonic generation (SHG) was the first nonlinear optical effect to be observed with lasers (*Franken* et al. [4.6]). Since that time a large amount of work has been devoted to this topic which has resulted in a detailed understanding of the physical basis of SHG and has enabled the development of materials and experimental techniques for the efficient generation of harmonics. Reviews have been written by *Kleinman* [4.7], *Kurtz* [4.8] and *Flytzanis* [4.5], and tables of second harmonic coefficients of materials have been published by *Bechman* and *Kurtz* [4.9] and *Singh* [4.10].

Second-harmonic generation occurs in materials without inversion symmetry and arises from a polarization which is second order in the optical electrical field having the form

$$P_{\mathrm{i}}(2\omega)=\sum_{j,k} d_{\mathrm{ijk}} E_{\mathrm{j}}(\omega) E_{\mathrm{k}}(\omega). \tag{4.2}$$

By popular convention, the magnitude of the nonlinearity is usually denoted by the "d" coefficient rather than $\chi^{(2)}$ (for a complete discussion see *Kurtz* [4.8]). In general, d is a third rank tensor having nine components corresponding to the different possible polarizations of P and E. Usually these nine components are not independent, but are related to each other by the symmetry properties of the crystal. For the ideal case of plane monochromatic waves, the second harmonic intensity, neglecting pump depletion, is given by the expression

$$I(2\omega)=\left(\frac{2^7\pi^3\omega^2 d_{\mathrm{eff}}^2 l^2}{n^3 c^3}\right) I^2(\omega)\left[\frac{\sin\left(\frac{\Delta k l}{2}\right)}{\left(\frac{\Delta k l}{2}\right)}\right]^2 \tag{4.3}$$

where l is the length of the crystal and $\Delta k = k_2 - 2k_1$ is the phase mismatch. As shown by *Giordmaine* [4.11] and *Maker* et al. [4.12], phase matching ($\Delta k = 0$) is often possible in birefringent materials by the suitable choice of the direction of propagation and polarization of the fundamental wave. In a uniaxial crystal with ordinary and extraordinary indices n_0 and n_{e}, an extraordinary wave propagating at an angle θ between its wave vector and the crystal axis has an index of refraction given by the equation:

$$\left[\frac{1}{n_{\mathrm{e}}(\theta)}\right]^2=\frac{\cos^2\theta}{n_0^2}+\frac{\sin^2\theta}{n_{\mathrm{e}}^2}. \tag{4.4}$$

Two types of phase matching are generally possible, one for which two fundamental photons of the same polarization combine to make a second harmonic photon, and the other for which two orthogonally polarized photons combine:

$$\text{type I:}\begin{cases} n_e^{2\omega}(\theta_m)=n_0^{\omega}; & \text{if} \quad n_e<n_0 \\ n_0^{2\omega}=n_e^{\omega}(\theta_m); & \text{if} \quad n_e>n_0 \end{cases} \tag{4.5}$$

$$\text{type II:}\begin{cases} n_e^{2\omega}(\theta_m)=\frac{1}{2}[n_e^{\omega}(\theta_m)+n_0^{\omega}]; & \text{if} \quad n_e<n_0 \\ n_0^{2\omega}=\frac{1}{2}[n_e^{\omega}(\theta_m)+n_0^{\omega}]; & \text{if} \quad n_e>n_0\,. \end{cases} \tag{4.6}$$

Phase matching can also be accomplished in some crystals by temperature tuning the birefringence (*Miller* et al. [4.13]).

When the intensity of the fundamental wave is very large, pump depletion occurs and the theoretical conversion efficiency approaches 100% (if $\Delta k=0$) (*Armstrong* et al. [4.4]). Naively, one might expect it would be straightforward to obtain conversion efficiencies approaching 100% with high intensity picosecond pulses. In practice, however, conversion efficiencies for SHG with picosecond pulses are often much less than expected from expression (4.3).

Miller [4.14] and *Comly* and *Garmire* [4.15] have shown that the short duration and finite bandwidth of picosecond pulses can significantly affect second-harmonic generation. For type I phase matching in a negative uniaxial crystal, *Miller* [4.14] has derived the following expression for the phase matching bandwidth:

$$\delta\lambda_1=\pm 1.39\,\lambda_1/2\pi l\left[\left(\frac{\partial n_0^{\omega}}{\partial\lambda_1}-\frac{1}{2}\,\frac{\partial n_e^{2\omega}}{\partial\lambda_2}\right)\right]\,. \tag{4.7}$$

For efficient SHG, the phase matching bandwidth must be at least as great as the laser bandwidth. For example, in KDP at $\lambda_1=1.06\ \mu m$, the phase matching bandwidth for a 2 cm crystal is ~135 Å, which can readily accommodate the laser bandwidth. In $LiNbO_3$, however, for the same crystal length the phase matching bandwidth is only 3.5 Å. Hence, in spite of its smaller nonlinear coefficient, KDP is a better material than $LiNbO_3$ for SHG of picosecond pulses from modelocked Nd:glass lasers.

For transform-limited pulses, *Comly* and *Garmire* [4.15] and *Glenn* [4.16] have shown that the duration of the second harmonic pulse can be stretched if the laser bandwidth exceeds the phase matching bandwidth. Also, the peak amplitude of the harmonic saturates and does not continue to grow with increasing crystal length, but only becomes broader (Fig. 4.1). This is due to the group velocity mismatch between the fundamental and second harmonic pulses. To avoid this effect, the group delay τ_D should be much less than the pulse duration, where (*Glenn* [4.16])

$$\tau_D=\frac{\lambda_1 l}{c}\left(\frac{\partial n^{\omega}}{\partial\lambda_1}-2\,\frac{\partial n^{2\omega}}{\partial\lambda_2}\right) \tag{4.8}$$

which is 0.07 ps per cm in KDP and 6.1 ps per cm in $LiNbO_3$ for a fundamental wavelength of 1.06 μm. When the group delay is much greater than the fundamental pulse duration, *Comly* and *Garmire* have shown that the temporal shape of the harmonic pulse will be rectangular, and its width will equal the group delay.

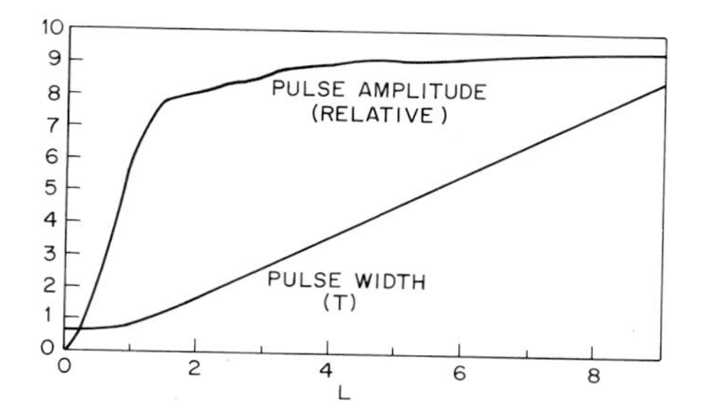

Fig. 4.1. Calculated second harmonic peak power and pulse width as a function of crystal length L normalized with respect to the distance over which the harmonic and fundamental pulses separate by one pulsewidth due to group velocity dispersion (from *Comly* and *Garmire* [4.15])

Second harmonic generation of picosecond optical pulses has been studied by *Shapiro* [4.17] who compared the harmonic pulsewidths from KDP and $LiNbO_3$ crystals by two photon fluorescence. The $LiNbO_3$ crystal of 9.2 mm consistently produced SHG pulses which were longer than those from a 1 cm KDP crystal (typically 6 and 3 ps, respectively). As expected, shorter harmonic pulses were produced in shorter $LiNbO_3$ crystals. Considerable spectral narrowing of the harmonic emission was observed in the longer $LiNbO_3$ crystals, consistent with estimates based on Miller's expression (4.7). A mode-locked Nd:glass laser was used as a source of fundamental pulses.

Glenn [4.16] has made calculations and measurements of second harmonic generation including the effects of a linear frequency modulation (chirp) on the fundamental. When the phase modulated spectral width of the laser $\Delta\omega_L$ exceeds the phase matching bandwidth of the SHG crystal he finds the conversion efficiency is reduced and varies inversely with $\Delta\omega_L$. The harmonic spectrum is narrowed and has a periodic structure. Similar calculations have been made by *Akhmanov* et al. [4.18], *Orlov* et al. [4.19] and *Karamzin* and *Sukhorukov* [4.20].

Deviations from a plane wave can also affect the conversion efficiency by introducing a spread in k vectors which alters the phase matching. For multimode beams such as those produced by the modelocked Nd:glass laser, the phase matching acceptance angle $\delta\theta$ can be estimated directly from (4.4). For type I phase matching in a negative uniaxial crystal, the acceptance angle of the fundamental (in the plane containing the c axis) is approximately (*Hagen* and *Magnante* [4.21]):

$$\delta\theta \simeq \frac{0.44\lambda_1 n_0}{l|n_e - n_0|\sin^2\theta_m}. \tag{4.9}$$

For KDP, $\delta\theta$ is approximately $1.6 \times 10^{-3}/l$ rad cm at $\lambda_1 = 1.06$ µm. For a non-critically phase matched geometry ($\theta = 90°$) the acceptance angle is usually much greater and varies as $l^{-\frac{1}{2}}$.

Boyd and *Kleinman* [4.22] have calculated the second harmonic power for monochromatic focused gaussian beams in uniaxial crystals accounting for both finite beam divergence and "walkoff" between the second harmonic and fundamental due to double refraction. For a fundamental mode with beam waist w_0, and power $P(\omega)$, they find the following approximate expressions for the SHG power:

$$P(2\omega) = \frac{32\pi^2\omega^2 P^2(\omega) d^2}{n^3 c^3 w_0^2} \begin{cases} l^2 & ; l_a, l_f \gg l \\ l l_a & ; l_f \gg l \gg l_a \\ l_f l_a & ; l \gg l_f \gg l_a \\ 4 l_f^2 & ; l \gg l_a \gg l_f \\ 4.75 l_f^2 & : l_a \gg l \gg l_f \end{cases} \tag{4.10}$$

where l_a and l_f are the aperture length and effective focal length, defined as

$$l_a = \frac{\sqrt{\pi} w_0}{\rho}; \quad \text{where} \quad \tan\rho = \frac{1}{2} n_e^2(\theta_m) \left(\frac{1}{n_e^2} - \frac{1}{n_0^2} \right) \sin 2\theta_m \tag{4.11}$$

$$l_f = \frac{\pi^2 w_0^2}{\lambda_1} \tag{4.12}$$

where ρ is the birefringence or double refraction angle. Optimum focusing occurs when the effective focal length l_f is comparable to the crystal length (see *Boyd* and *Kleinman* [4.22] for details).

Orlov et al. [4.19] have examined the second harmonic generation of picosecond pulses in focused gaussian beams. They find the conditions for optimum focusing are essentially the same as those *Boyd* and *Kleinman* derived for monochromatic beams. The conversion efficiency, however, is a complicated function of pulse duration, and spectral width, as well as the focusing geometry, and crystal properties. For the same spectral widths, they find that chirped pulses have smaller energy conversion to SHG than do transform limited pulses, if the spectral width exceeds the phase matching bandwidth.

Rabson et al. [4.23] have produced efficient SHG in a CsH_2AsO_4 (CDA) crystal located inside a folded modelocked Nd:glass laser. Using a crystal 1.3 cm long in a 90° phase matched geometry ($T = 45$ °C) they obtained second harmonic output intensities of as much as 10^9 W/cm^2 for fundamental intensity of 4×10^9 W/cm^2.

With modelocked Nd:YAG lasers, relatively high conversion efficiencies can be readily obtained for SHG in KDP and ADP. For example, *Kung* et al. [4.24] have doubled 50 ps pulses from 1.06 µm to 0.53 µm with an energy conversion up to 80%. The green pulses were then subsequently mixed with the fundamental in a second KDP crystal to produce 3547 Å pulses with an overall energy conversion of 10%.

Recently, *Ippen* and *Shank* [4.25] have frequency doubled the picosecond pulse from a modelocked cw dye laser in a phase-matched lithium iodate crystal with an efficiency of ~10%. The short duration (<1 ps) and lower peak power (~10 kW) of the laser pulses necessitated tight focusing in a thin crystal.

4.2.2 Third and Higher Order Optical Harmonic Generation and Mixing

Third harmonic generation (THG) is a third order nonlinear effect which can occur in both acentric and centric materials, including liquids and gases (*Maker* and *Terhune* [4.26]). As with second harmonic generation, phase matching can be accomplished in certain birefringent crystals. However, for the larger class of materials where this is not possible, other methods of phase matching can sometimes be used such as mixing normally dispersive and anomalously dispersive materials (*Bey* et al. [4.27], *Harris* and *Miles* [4.28]); using periodic or layered materials (*Freund* [4.29], *Bloembergen* and *Sievers* [4.30], *Shelton* and *Shen* [4.31]); acoustic modulation (*Boyd* et al. [4.32]); and waveguide dispersion (*Anderson* and *Boyd* [4.33]).

Eckardt and *Lee* [4.34] have used phase matched THG to measure the third order intensity correlation of picosecond pulses. They used a mixture of 37.5 gm/liter of the dye fuchsin red in hexafluoroisopropanol to obtain phase matching in a 0.1 mm cell after *Bey* et al. [4.27]. The anomalous dispersion of the strong visible absorption band of the dye enabled them to compensate the normal dispersion of the solvent between the third harmonic (0.353 μm) and the fundamental (1.06 μm). Conservation of angular momentum prohibits the third harmonic generation of circularly polarized light in isotropic materials. They measured the third harmonic produced by right- and left-handed circularly polarized pulses which produced linearly polarized light when they overlapped, enabling a background-free result (for more details see the chapter on measurement techniques).

Wang and *Baardsen* [4.35] have observed optical third harmonic generation in ADP crystals both in reflection from the crystal surface and in the bulk along phase-matched directions. They compared conversion efficiencies for both non-modelocked *Q*-switched operation and modelocked operation of a Nd:glass laser. The relatively inefficient THG for the modelocked case led them to suggest that most of the energy was not contained in the picosecond pulses but was distributed between the pulses, a conclusion that was later shown to be incorrect (see chapters on generation and measurement techniques).

Shelton and *Shen* [4.31, 36] have demonstrated phase-matched THG of picosecond pulses in cholesteric liquid crystals. These materials can be visualized as consisting of many thin layers, in which the molecular orientation between successive layers is rotated by an angle φ, producing a helical structure. This helical structure introduces an additional dispersion for circularly polarized waves which can be used to phase match the THG by compensating for the normal color dispersion. With a mixture of 1.75 parts cholesteryl chloride and 1 part cholesteryl myristrate the authors were able to adjust the pitch angle of the helix over a wide range by varying the temperature. As shown in Fig. 4.2 sharply peaked THG was produced at temperatures of 49.4 and 54.2 °C corresponding to pitches $p = -17.3$ and $+17.4$ μm for left-handed and right-handed circularly polarized waves, respectively. The results were in good agreement with theoretical calculations based on *de Vries* [4.37] model of cholesteric liquid crystals.

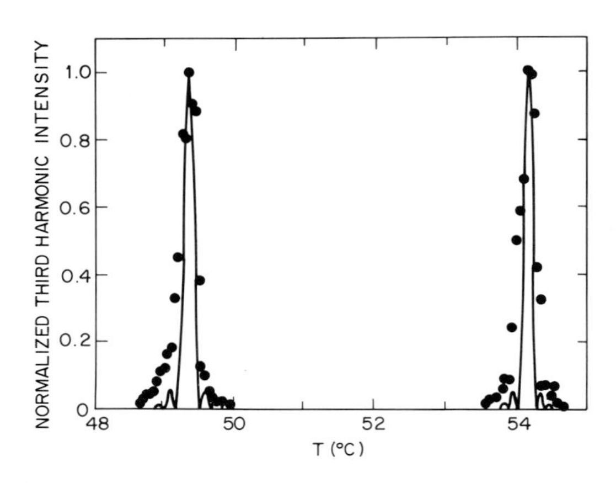

Fig. 4.2. Normalized third harmonic intensity versus temperature near the phase-matching temperatures for a mixture of 1.75 cholesteryl chloride and 1 cholesteryl myristate, in a cell 30 μm thick. The peak at the lower temperature (corresponding to left helical structure) is generated by left-circularly polarized fundamental waves and the one at the higher temperature by right-circularly polarized fundamental waves. The solid line is the theoretical phase-matching curve and the dots are experimental data points (from *Shelton* and *Shen* [4.36])

Shelton and *Shen* [4.31] have also demonstrated higher order phase matching in cholesteric liquid crystals in which the harmonic and one or more of the fundamental waves are travelling in opposite directions. In this case the phase matching is attained by a type of nonlinear Bragg reflection in which the crystal momentum arising from the periodic nature of the helical structure is used to compensate the momentum mismatch between the fundamental and harmonic waves. By analogy with the scattering of electrons in crystals, the authors have designated this process "umklapp optical third harmonic generation". They have applied it to the measurement of third order intensity autocorrelations of picosecond pulses. Using oppositely directed circularly polarized 1.06 μm fundamental waves, they were able to eliminate the background and determine shape and asymmetry information about the pulses.

Third harmonic generation of picosecond pulses by reflection from metal and semiconductor surfaces has been measured by *Bloembergen* et al. [4.38]. Reflected third harmonic from silicon, germanium, silver and gold was measured and compared to a reference sample of lithium fluoride. The angular and polarization dependence agreed well with theory. The magnitudes of the nonlinear susceptibilities varied from $\sim 4 \times 10^3$ to 4×10^4 times larger than LiF ($\chi^{(3)} = 1.17 \times 10^{-15}$ esu (*Maker* and *Terhune* [4.26])), with silicon being the largest. The use of picosecond pulses from a modelocked Nd:glass laser enabled the generation of readily measurable signals without damage to the surface of the materials.

An extensive body of work has been directed at the use of metal vapors and inert gases for phase matched third and higher order harmonic generation and mixing of picosecond pulses. An important goal of this work has been the generation of coherent vacuum ultraviolet and soft x-ray radiation by up-conversion of visible laser sources. *Ward* and *New* [4.39] first observed third harmonic generation in gases and made detailed measurements of the unphase matched THG of a focused *Q*-switched ruby laser in a variety of inert gases. *Harris* and *Miles* [4.28] proposed using metal vapors for efficient phase matched THG. They showed how the strong visible absorption resonances of the alkali metal vapors result in an anomalous dispersion between the uv and infrared. The addition of normally dispersive inert buffer gas of the correct amount can compensate the dispersion and produce phase matching. The strong visible absorption lines of the alkali metal vapors also provide a strong enhancement of the nonlinearity by as much as 10^6 relative to the inert gases. The nonlinear susceptibility for THG can be expressed in the form of products of dipole matrix elements μ_{jk} between various virtual intermediate states (*Armstrong* et al. [4.4]):

$$\chi^{(3)}(3\omega)=\frac{N}{\hbar^2}\sum_{ijk}\mu_{gi}\mu_{ij}\mu_{jk}\mu_{kg}A_{ijk} \tag{4.13}$$

where the A_{ijk} are energy denominators of the form

$$A_{ijk}=\frac{1}{(E_{ig}\pm 3\hbar\omega)\,(E_{jg}\pm 2\hbar\omega)\,(E_{kg}\pm\hbar\omega)} \tag{4.14}$$

and the E_{kg}'s are the energies of the intermediate states relative to the ground state and N is the density of atoms. *Harris* and *Miles* [4.28] point out that the group velocity dispersion for THG in metal vapors is relatively small, enabling efficient conversion with picosecond pulses. Calculations of the nonlinear susceptibilities for a variety of alkali metal vapors have been made by *Miles* and *Harris* [4.40]. *Bjorklund* [4.41] has made a detailed theoretical study of phase matched third order nonlinear processes in isotropic media.

Phase matched THG in metal vapors was first observed by *Young* et al. [4.42]. They used a *Q*-switched Nd:YAG laser to generate 0.3547 μm THG in Rb vapor. Phase matching was accomplished by the addition of Xe gas in the ratio of 412:1. The nonlinear susceptibility of the rubidium atom was determined to be 1.4×10^{-32} esu. With improved heat pipes an efficiency of 10% was later obtained using picosecond pulses (*Bloom* et al. [4.43]).

In a subsequent experiment, *Kung* et al. [4.24] generated radiation at 1773, 1520 and 1182 Å by frequency tripling and summing of picosecond pulses in a phase matched mixture of cadmium and argon. A modelocked Nd:YAG oscillator and amplifier system which produced pulses of 50 ps duration and 10 mJ of energy at 1.06 μm was frequency doubled and tripled to 5320 Å and 3547 Å in two KDP crystals (see Fig. 4.3). These pulses were then further tripled and mixed in the Cd:Ar cell to form the sixth, seventh and ninth harmonics of the laser frequency ω by the interactions:

(1) $2\omega+2\omega+2\omega\rightarrow 6\omega$ (1773 Å)
(2) $3\omega+3\omega+\ \omega\rightarrow 7\omega$ (1520 Å)
(3) $3\omega+3\omega+3\omega\rightarrow 9\omega$ (1182 Å)

Phase matched third harmonic generation of picosecond pulses from 3547 to 1182 Å has also been produced in mixtures of inert gases by *Kung* et al. [4.44]. In this case, anomalous dispersion arises from strong transitions in the uv between the fundamental (3547 Å) and third harmonic (1182 Å). Using mixtures of argon and xenon, strong phase matched THG emission was produced with efficiencies as high as 2.8% for an input peak power of 13 MW. A tightly focused geometry was used having a confocal parameter of 0.25 cm in a 9.5 cm gas cell, producing peak intensities of $\sim 6\times 10^{12}$ W/cm^2.

In general, the maximum conversion efficiencies are limited by optical breakdown due to multiphoton and avalanche ionization. Offsetting the phase matching by altering the atomic level populations due to trace absorption is a less serious complication (*Bloom* et al. [4.43]). *Harris* [4.45] has calculated optimum conversion efficiencies for harmonic and sum frequency generation when the applied optical field is limited only by multiphoton absorption or ionization. He estimated conversion efficiencies for a number of high order nonlinear processes capable of producing vacuum uv and soft x-ray radiation, including the 15th harmonic generation of 177 Å from 2660 Å. The shortest

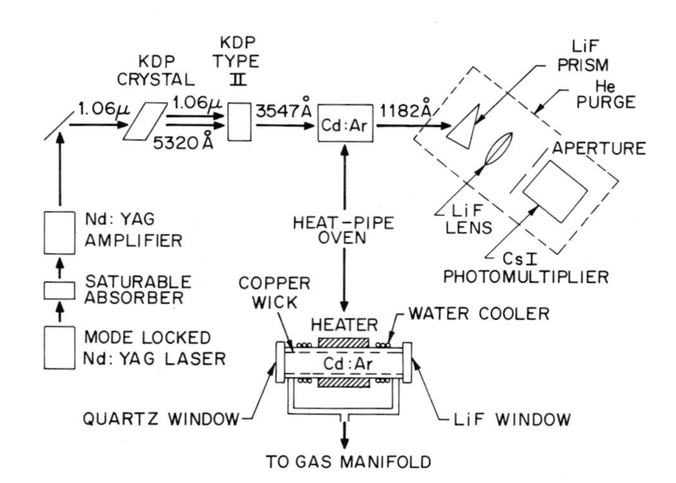

Fig. 4.3. Schematic of experimental apparatus for generation of 1182 Å radiation by phase-matched third harmonic generation in Cd vapor (from *Kung* et al. [4.24])

wavelength that has been produced thus far by these techniques is 570 Å obtained by tripling a 1710 Å pulse from a Xe_2 laser in an argon cell (*Hutchinson* et al. [4.46]).*

One of the practical limitations with the metal vapor/buffer gas nonlinear conversion schemes is the difficulty in obtaining long coherence lengths due to inhomogeneities of the mixture. This problem has been alleviated somewhat by the use of mixed metal vapors without buffer gases in a two component heat pipe (*Bloom* et al. [4.47]).

Phase matched fifth harmonic generation of 1.06 μm picosecond pulses in calcite has been observed by *Akhmanov* et al. [4.48]. The large birefringence enabled phase matching either by direct fifth harmonic generation or by cascade processes involving a combination of third harmonic generation and four photon frequency mixing, i.e., $5\omega = 3\omega + \omega + \omega$. The magnitude of the nonlinear susceptibility $\chi^{(5)}$ was estimated to be $\sim 10^{-27}$ esu. Despite the phase matching, the detected signal at 2120 Å was very small, typically only a few photoelectrons per shot.

Four photon parametric mixing of the type $2\omega_2 - \omega_1 \rightarrow \omega_3$ has been used by *Auston* [4.49, 50] to measure the shape and frequency chirp of picosecond pulses using higher order intensity correlation functions. By mixing the second harmonic (ω_2) and the fundamental (ω_1) of 1.06 μm pulses in a phase matched noncollinear geometry, it was possible to produce efficient generation at $\omega_3 = 3\omega_1$ in a variety of low dispersion liquids. The intensity of the emission was proportional to the correlation of the fourth power of the fundamental $[I(\omega_2) \propto I^2(\omega_1)]$ with itself, and was considerably more sensitive to pulse shape than the usual second order intensity correlation (see the chapter on measurement techniques). The variation of phase matching angle with wavelength permitted a measurement of the pulse chirp by dynamic nonlinear spectroscopy (*Auston* [4.50]).

Sorokin and *Lankard* [4.51] have observed higher order nonlinear mixing in Cs vapor cells pumped by a train of second harmonic pulses at 3470 Å from a modelocked ruby laser. These interactions were "primed" by a strong stimulated electronic Raman scattering to the 10s state which due to its near resonance with the pump produced a Stokes wave in the far infrared at ~ 20 μm. The Stokes wave ω_s then mixed with the pump to produce emission at 3600 Å by the four wave interaction $\omega_L + \omega_s + \omega_s \rightarrow \omega$. The strong visible absorption lines of the Cs vapor resulted in a resonant enhancement of the nonlinearity and a long coherence length of ~ 16 cm. A less strong six photon mixing produced radiation at ~ 3900 Å.

* Recently, the shortest coherent wavelength to date of 38 nm, was obtained by seventh-harmonic conversion of modelocked laser pulses by *J. Reintjes, C. Y. She, R. C. Eckardt, N. E. Karangelen, R. A. Andrews, R. C. Elton:* to be published.

4.3 Parametric Emission

4.3.1 Three Photon Parametric Fluorescence and Amplification

The basic physical process for three photon parametric fluorescence is the spontaneous decay of one photon ω_p, usually designated the pump, into two new photons, ω_s, and ω_i, the signal and idler, i.e.,

$$\omega_p \rightarrow \omega_s + \omega_i . \tag{4.15}$$

Except for dispersion, the nonlinear susceptibility responsible for three photon parametric fluorescence is the same as the nonlinear susceptibility for second harmonic generation and optical mixing. If the incident pump photon flux is sufficiently intense, stimulated parametric emission can occur, whereby signal photons can mix with the pump to produce new idler photons by difference frequency mixing. Similarly, idler photons can mix with the pump to produce more signal. This produces a gain at signal and idler frequencies, resulting in amplification and oscillation if appropriate feedback is provided. The specific idler and signal frequencies are usually determined by phase matching.

Theoretical discussions of optical parametric emission were first made by *Kingston* [4.52], *Kroll* [4.53], *Akhmanov* and *Khokhlov* [4.54] and *Armstrong* et al. [4.4]. The first experimental demonstration of optical parametric oscillation was made by *Giordmaine* and *Miller* [4.327]. Recent reviews of this topic have been made by *Smith* [4.55], *Byer* [4.56] and *Tang* [4.57].

The use of picosecond pulses for tunable parametric generation has been proposed by *Glenn* [4.58], *Akhmanov* et al. [4.59], *Akhmanov* et al. [4.60] and *Akhmanov* et al. [4.61]. They have pointed out the importance of group velocity matching and possible pulse steepening when the pump is a picosecond pulse. Experimental observation of stimulated parametric fluorescence by picosecond pump pulses was made by *Rabson* et al. [4.62]. They used a train of picosecond pulses from a doubled modelocked Nd:glass laser to pump a 5 mm crystal of barium sodium niobate ($Ba_2NaNb_5O_{15}$) in a single pass geometry with a 100 cm focusing lens. Emission from 0.96 to 1.16 μm was produced by temperature tuning the phase matching. Optical damage in the crystal limited the conversion efficiency to $\sim 10^{-4}$. A similar experiment was performed by *Burneika* et al. [4.63]. They used instead a 6 cm KDP crystal, however, whose higher damage threshold and longer length permitted energy conversion efficiencies as large as $\sim 1\%$ on a single pass. Tuning from 0.87 to 1.2 μm was obtained. At wavelengths shorter than 0.87 μm, the spectral emission was obscured by self-phase modulation of the pump. Most of the parametric emission seemed to emanate from self-focused filaments.

Laubereau et al. [4.64] have demonstrated efficient parametric generation in $LiNbO_3$ pumped by a single picosecond pulse at 1.06 μm from a modelocked Nd:glass laser-amplifier system. The use of a single pulse instead of a complete train enabled the use of much higher pump intensities (up to 10 GW/cm^2)

without damaging the crystal. Energy conversion as high as 3% was measured in a crystal 2 cm long (Fig. 4.4). Angular tuning of the crystal produced emission over the range of 700 to 2500 cm^{-1} enabling efficient generation of tunable picosecond pulses in the infrared. Measured spectral widths of the signal and idler near degeneracy were typically 100 cm^{-1}, in agreement with calculations based on estimates of the phase mismatch due to finite pump divergence ($\sim 10^{-3}$ rad) and signal divergence ($\sim 2\times10^{-2}$ rad).

Dychyus et al. [4.65] have observed single pass parametric generation in a 4 cm α-HIO_3 crystal pumped with the second harmonic train of a modelocked Nd:glass laser. Efficiencies of $\sim 0.5\%$ and angle tuning from 0.71 to 1.115 μm were obtained. *Kushida* et al. [4.66] have observed parametric generation in a 2 cm $LiNbO_3$ crystal pumped by a train of second harmonic pulses from a modelocked Nd:glass laser. Temperature tuning of the signal from 0.64 to 0.75 μm was achieved, but the conversion efficiency was appararently too low to make a practical source.

Kung [4.67] has obtained efficient tunable picosecond pulses in the visible by parametric generation in ADP. The pump was a quadrupled (2660 Å) passively modelocked Nd:YAG laser amplifier system. Two ADP crystals each 5 cm long were used in temperature controlled ovens. The first crystal acted as a source whose emission was spatially filtered and then amplified in the second

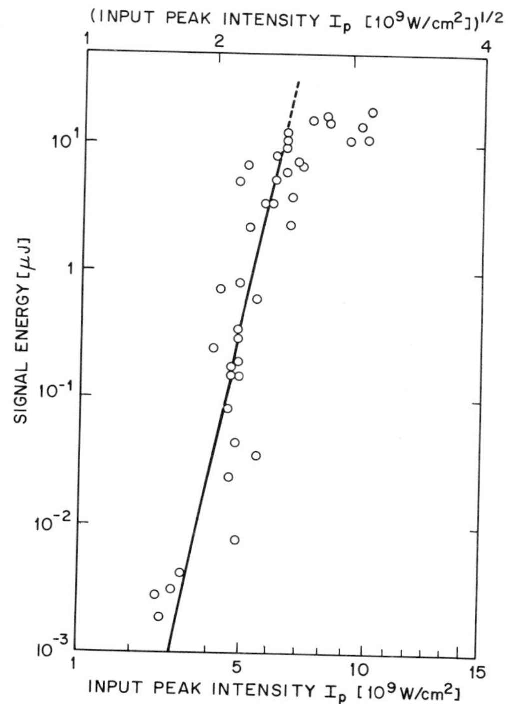

Fig. 4.4. Measured energy of signal pulses at 6500 cm^{-1} from a travelling-wave parametric oscillator pumped by picosecond pulses at 9450 cm^{-1} from a modelocked Nd:glass laser. The straight line is a calculation using experimental parameters (from *Laubereau* et al. [4.64])

crystal, both in a single pass, collinear geometry. By varying the temperature of both crystals from 50 to 105 °C, a tuning range of 4200 to 7200 Å was obtained. Peak conversion efficiencies were greater than 10% (>100 μJ tunable output for 1 mJ uv input).

Massey et al. [4.68] have demonstrated an ADP parametric amplifier pumped by the fourth harmonic of a modelocked Nd:glass laser. The amplifier was primed with a low power cw HeNe laser of 25 mW at 632.8 nm. Output pulses of ~10 ps duration and peak powers of 67 kW and 49 kW were produced at 458.9 and 632.8 nm, repectively.

4.3.2 Four Photon Parametric Interactions

Four photon parametric interactions are governed by $\chi^{(3)}$ and consequently can occur in a wide variety of materials including isotropic solids, liquids and gases as well as acentric crystals. The most commonly ocurring process is one in which two laser photons are converted to a signal and idler photon, i.e.,

$$2\omega_L \rightarrow \omega_s + \omega_i \tag{4.16}$$

where by definition $\omega_s > \omega_L$ and $\omega_i < \omega_L$. If two high intensity laser pump waves are present, a more general interaction can occur of the form

$$\omega_1 + \omega_2 \rightarrow \omega_3 + \omega_4 . \tag{4.17}$$

A single laser photon can also decay into three photons, but the conversion efficiency for this process is much smaller. When three incident photons combine, a fourth photon can be emitted by parametric frequency conversion, i.e.,

$$\omega_1 \pm \omega_2 \pm \omega_3 \rightarrow \omega_4 . \tag{4.18}$$

This is a four wave optical mixing and the intensity of the generated wave is simply proportional to the product of the intensities of the incident waves. The dynamics of the parametric emission processes (4.16) and (4.17) are more complicated, however, since they start from quantum noise and are amplified by the pump waves.

Stimulated four photon parametric fluorescence and amplification were first discussed in detail by *Chiao* et al. [4.69]. They calculated the gain of a weak wave due to its interaction with a strong wave in a Kerr liquid, for the nearly degenerate case $\omega_s \simeq \omega_i \simeq \omega_L$ (where $2\omega_L = \omega_s + \omega_i$). Both signal and idler waves were shown to have significant gains in directions slightly off the laser axis. They pointed out that self-focusing could be viewed as a degenerate four photon parametric emission process whereby on-axis spatial frequency components of the beam are converted to off-axis components, thereby increasing the spatial frequency spectrum of the beam which is equivalent to decreasing the beam diameter.

Degenerate four photon parametric interactions were first observed by *Carman* et al. [4.70], who used a Q-switched ruby laser to pump a 3 mm cell of nitrobenzene. Weak beams at the same frequency were found to have significant gains at phase matched angles of ± 8 mrad, in good agreement with the theory of *Chiao* et al. [4.69].

Nondegenerate four-photon parametric emission was first observed by *Alfano* and *Shapiro* [4.197]. They used a train of picosecond pulses from a frequency-doubled, modelocked Nd:glass laser as a pump in borosilicate BK-7 glass. Upshifted (signal) and downshifted (idler) spectra were recorded showing

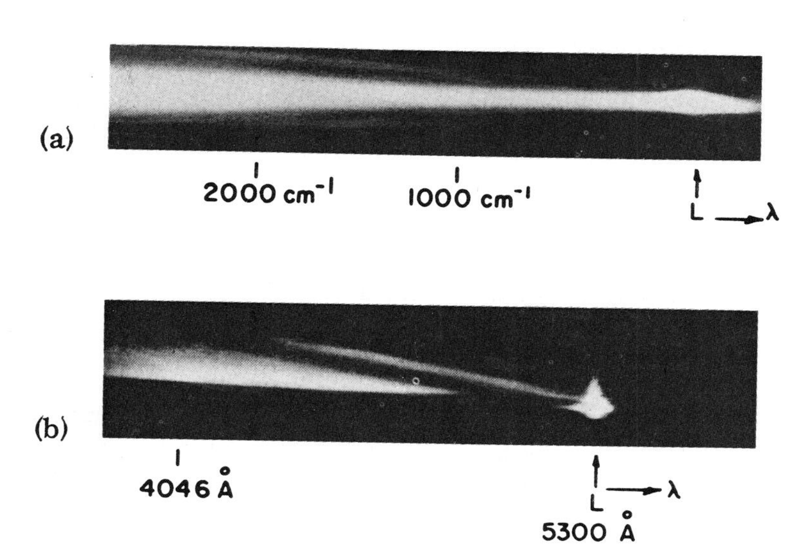

Fig. 4.5 a and b. Emission spectra due to four-photon parametric generation in borosilicate glass. Pump is a train of frequency doubled picosecond pulses from a modelocked Nd:glass laser. (a) Anti-Stokes emission with pump beam at slit center. Light in center is self-phase modulation while outer curves are due to four-photon emission. (b) Entire angular anti-Stokes emission curve from 4000 Å to 5300 Å (from *Alfano* and *Shapiro* [4.197])

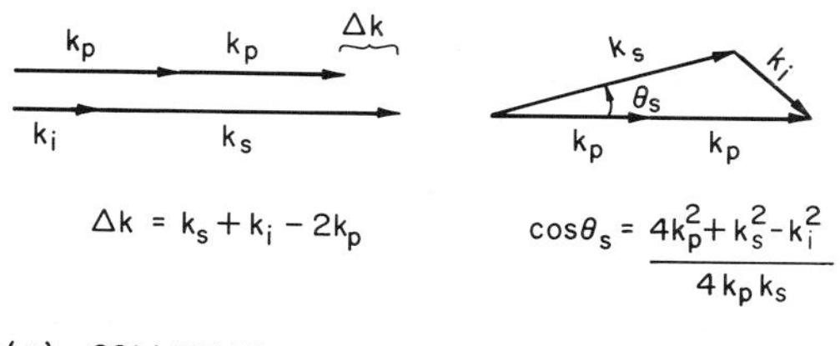

Fig. 4.6 a and b. Four photon parametric generation, $2\omega_p \to \omega_s + \omega_i$. In normally dispersive media, phase matching is often possible in a noncollinear geometry

emission on axis due to self-phase modulation, and off axis due to phase matched parametric emission of the type $2\omega_L \rightarrow \omega_s + \omega_i$ (Fig. 4.5). In a normally dispersive medium, it is generally not possible to phase match a collinear four photon parametric emission process. As indicated in Fig. 4.6, however, a noncollinear geometry will often permit phase matching. From the dispersion of the linear index of refraction in the glass, *Alfano* and *Shapiro* could calculate the signal and idler emission angles as a function of frequency. Good agreement with experiment was obtained if corrections for self-focusing of the pump and the nonlinear index of refraction were made. Self-focusing causes an angular spread in wavevectors of the laser pump, and the nonlinear index of refraction increases the magnitude of k_p. As shown by *Chiao* et al. [4.69] the weaker signal and idler waves experience twice as large a nonlinear index change as the laser pump, so that a net change of phase matching angles results.

Penzkofer et al. [4.71] have attributed the extreme spectral broadening of intense 1.06 μm pulses in water to four photon parametric interactions. They point out that the nonlinear index of refraction is too small to account for the broadening which can span the entire visible spectrum from the infrared to the uv. Figure 4.7 shows the spectra produced by single 1.06 μm pulses in a 2 cm cell of water for incident intensities from 2×10^{10} to 10^{11} W/cm^2. The magnitude of the broadened spectrum increases dramatically for a relatively small increase of incident intensity. At low intensity a strong structure is evident which becomes smooth at higher intensity. The emission angle was ~40 mrad

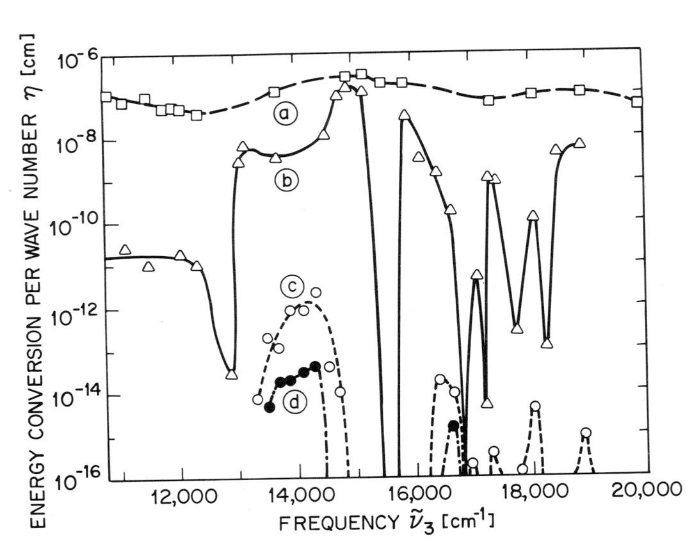

Fig. 4.7. Four photon parametric generation from a 2 cm water cell pumped by an intense single picosecond pulse at 9450 cm^{-1}. The conversion efficiency η is shown for four pump intensities: (a), 10^{11}; (b), 5×10^{10}; (c), 3×10^{10}; (d), 2×10^{10} W/cm^2 (from *Penzkofer* et al. [4.71])

and no evidence of self-focusing was observed. Penzkofer et al. have attributed the structure in the spectra to resonances in the nonlinear susceptibility, $\chi^{(3)}$, responsible for the four photon parametric generation. For an interaction of the form $\omega_1+\omega_2\rightarrow\omega_3+\omega_4$, they have derived the following expression for the emission energy at ω_3 (or ω_4)

$$W(\omega_3, z)=W(\omega_4, o)|\sinh \gamma z|^2 \frac{G|\chi|^2\omega_3}{4|\gamma|^2\omega_4} \cdot \exp\left[-(\alpha_3+\alpha_4)/2\right] \tag{4.19}$$

where

$$G=\frac{1024\pi^4\omega_3\omega_4 I_{01}(z) I_{02}(z)}{n_1 n_2 n_3 n_4 c^4}$$

$$\gamma=\tfrac{1}{4}\{(\alpha_4-\alpha_3)^2+4[G\chi^2-(\Delta k)^2+i\Delta k(\alpha_4-\alpha_3)]\}^{\frac{1}{2}}$$

$$\chi=\chi^{(3)}(-\omega_3, \omega_1, \omega_2, -\omega_4)=\chi'+i\chi''$$

$$\Delta k=k_1+k_2-k_3-k_4 .$$

Penzkofer et al. have considered two types of parametric generation; first a stimulated parametric conversion of the laser photons, $2\omega_L\rightarrow\omega_s+\omega_i$ (i.e., $\omega_1=\omega_2=\omega_L$ and $\omega_3=\omega_s$, $\omega_4=\omega_i$) and then a second order process in which signal photons combine with pump photons to produce new frequencies $\omega_s+\omega_L\rightarrow\omega_s'+\omega_i'$. The primary parametric generation process accounts for the emission at the lower pump intensities. Although the absorption at the infrared idler frequency and the collinear phase mismatch are both large, calculations of the gain indicate that it is sufficient to account for the observed emission. The nonlinear susceptibility $\chi^{(3)}$ has contributions from both single-frequency and difference frequency (Raman) resonances which account for the strong structure observed (Fig. 4.7). At higher laser intensities, the spectrum becomes smoother due to saturation of the primary emission process and the onset of the secondary emission process which tends to fill in the gaps.

Recently, *Penzkofer* et al. [4.72] have made further measurements of spectral broadening in H_2O, D_2O, Infrasil and NaCl. These materials have relatively small nonlinear refractive indices, low dispersion and small stimulated Raman gains, and all produced strong spectral broadening of intense 1.06 µm picosecond pulses. Conversion efficiencies in H_2O were typically $\sim 10^{-6}$ per cm^{-1} for a total conversion of $\sim 10\%$. *Penzkofer* [4.73] has considered a possible contribution to the spectral broadening in H_2O and NaCl due to electron avalanche (*Bloembergen* [4.74] and concludes that it is not important at intensities up to 10^{12} W/cm^2 in NaCl and up to 10^{13} W/cm^2 in H_2O, i.e., up to approximately a factor of two below the threshold for optical breakdown. Similar calculations by *Braunlich* and *Kelley* [4.75] support this conclusion.

In other materials such as carbon disulphide and nitrobenzene which have a large nonlinear index of refraction, the spectral broadening is usually accompanied by self-focusing and is more likely due to self-phase modulation and possible avalanche ionization rather than four photon parametric generation (see Subsec. 4.5.5).

4.4 Stimulated Scattering

4.4.1 Transient Stimulated Raman Scattering

Stimulated Raman scattering with picosecond pulses has been a very active field of study. Interest in this topic has generally fallen into one of two categories: 1) the nonlinear optics of stimulated Raman scattering (SRS) with picosecond pulses, or 2) the application of SRS with picosecond pulses to the measurement of the physical properties of materials such as vibrational dephasing times. The nonlinear optics of SRS with picosecond pulses is now reasonably well understood and is the central topic of this review. The applications to the measurement of material properties are discussed in the following chapter.

Stimulated Raman scattering was first observed in the nanosecond regime (*Woodburg* and *Ng* [4.76], *Eckhardt* et al. [4.77]). Since that time considerable work has been done, leading to a fairly complete understanding of SRS with nanosecond pulses. For reviews of this subject, the reader is referred to the articles by *Bloembergen* [4.78], *Kaiser* and *Maier* [4.79] and *Wang* [4.80]. Our aim in this review is to focus on picosecond SRS and in particular to delineate those aspects of SRS which are unique to picosecond pulses.

Of particular importance is the transient nature of SRS which occurs when the laser pulse duration is comparable to the dephasing relaxation time of the particular material excitation responsible for the scattering. In this situation, which frequently occurs with picosecond excitation, the dynamics of the growth of the Stokes wave are quite different. As we shall see, inertial delays, pulse compression, and other unique features often occur. Before dealing with the details of this topic, however, we first summarize the basic properties of SRS to enable a clear distinction between the steady-state and transient regimes.

Raman scattering can be broadly defined as an interaction of a light wave with a material in which incident light at frequency ω_L is converted to light at a new frequency ω_S by virtue of a material excitation with a characteristic frequency ω_V. The particular material excitation can be either a molecular vibration or rotation, an electronic or spin excitation, an acoustic, thermal or entropy wave, or even combinations of these excitations. The vast majority of cases, however, involves molecular vibrations and for this reason we will refer to ω_V as the vibrational frequency. Conservation of energy of the scattered photons determines the frequency ω_S, denoted the Stokes frequency:

$$\omega_S = \omega_L - \omega_V, \tag{4.20}$$

which is less than the laser frequency. Conservation of linear momentum imposes a similar constraint on the wave vectors:

$$\boldsymbol{k}_{\mathrm{S}}=\boldsymbol{k}_{\mathrm{L}}-\boldsymbol{k}_{\mathrm{V}}. \tag{4.21}$$

The nature of the coupling between the Stokes and laser waves can best be illustrated by introducing the notion of *Placzek* [4.81] wherein the optical polarizability α is assumed to be a linear function of the vibrational coordinate Q, i.e.,

$$\alpha=\alpha_0+\left(\frac{\partial\alpha}{\partial Q}\right)Q \tag{4.22}$$

where $\left(\frac{\partial\alpha}{\partial Q}\right)$ is a constant. Equation (4.22) simply states that the electronic and vibrational (i.e., nuclear) systems are not independent but are coupled so that an electronic distortion due to an optical polarization will induce a nuclear motion, and conversely, a nuclear vibration will affect the electronic polarizability. An excitation of the vibrational coordinate produces a polarization

$$\boldsymbol{P}=N\left(\frac{\partial\alpha}{\partial Q}\right)Q\boldsymbol{E} \tag{4.23}$$

which is proportional to the product of Q and the electric field of the laser. N is the total density of molecules. If Q is oscillating at frequency ω_{V} it will produce a polarization at a frequency $\omega_{\mathrm{L}}-\omega_{\mathrm{V}}$ which radiates to generate the Stokes wave. Conversely, the nonlinear mixing of the Stokes and laser waves drives the amplitude of the molecular vibration. These processes are more completely described by the following three coupled partial differential equations for the total electric field E (Stokes plus laser), the amplitude Q of the vibrational wave, and the excitation density n_{V} of the molecular system (*Maier* et al. [4.82], *Akhmanov* et al. [4.60], *Kaiser* [4.83])

$$\frac{\partial^2 E}{\partial z^2}-\frac{1}{c^2}\frac{\partial^2 \varepsilon E}{\partial t^2}=\frac{4\pi}{c^2}N\frac{\partial\alpha}{\partial Q}\frac{\partial^2}{\partial t^2}(QE) \tag{4.24}$$

$$\frac{\partial^2 Q}{\partial t^2}+\frac{1}{\tau}\frac{\partial Q}{\partial t}+\omega_{\mathrm{V}}^2 Q=\frac{1}{2m}\frac{\partial\alpha}{\partial Q}n_{\mathrm{V}}E^2 \tag{4.25}$$

$$\frac{\partial n_{\mathrm{V}}}{\partial t}+\frac{n_{\mathrm{V}}-1}{\tau'}=\frac{1}{2\hbar\omega_{\mathrm{V}}}\frac{\partial\alpha}{\partial Q}E^2\frac{\partial Q}{\partial t}. \tag{4.26}$$

Here n_{V} is the population difference between the ground state $|g\rangle$ and the first vibrational state $|v\rangle$, normalized relative to its equilibrium value. m is an effective mass of the molecular vibration. Generally $n_V \approx 1$. It is important to distinguish between the excitation level n_V and the vibrational amplitude Q, which is a coherent wave having a well defined frequency and phase representing the polarization of the molecular system in a superposition of states $|v\rangle$ and $|g\rangle$ due to the driving optical fields. Q is derived from the "off-diagonal" components of the density matrix of the $|v\rangle|g\rangle$ system, whereas n_V is determined by the diagonal components (*Shen* and *Bloembergen* [4.84]).

An equally important distinction should be made between the relaxation times τ and τ' appearing in these equations. The relaxation time τ is known as the dephasing time and is a measure of the rate of loss of phase coherence of the vibrational amplitude Q due to collisions and other damping mechanisms. The relaxation time τ' is a measure of the deexcitation rate of the molecular system due to real transitions from state $|v\rangle$ to ground state $|g\rangle$. Consequently τ' determines the rate of energy loss in the molecular system, whereas τ determines the loss of phase. Generally $\tau > \tau'$. *Fischer* and *Laubereau* [4.85] have developed a model of the dephasing of vibrationally excited molecules based on semiclassical collision theory.

In the spectral domain, the dephasing time τ is inversely proportional to the width of the spontaneous Stokes spectrum when the material is illuminated by a quasi-monochromatic light source. Typical dephasing times range from 10^{-1} to 10^{+1} ps for molecular vibrations in liquids and optical phonons in solids, a range that clearly overlaps the durations of available picosecond pulses.

When the duration of the laser pulse is very long relative to the dephasing time τ, a quasi-steady state exists wherein the molecular system is heavily damped, and the approximation $\partial Q/\partial t \ll (1/\tau)\, Q$ can be made. In this situation, which is typical of Q-switched pulses, the damping of the vibrational wave is very nearly balanced by the driving term due to the mixing of Stokes and laser waves. With shorter duration pulses of high intensity, however, the rate of growth of the driving term can greatly exceed the decay due to damping. In this transient regime, the dynamics of the stimulated Raman scattering will clearly differ significantly from the corresponding quasi-steady state situation. The use of picosecond pulses for stimulated Raman scattering has been aptly described as "impact excitation" by *Carman* et al. [4.98].

4.4.2 Stimulated Raman Scattering of Picosecond Pulses: Experiments

Before dealing with the specific details of the theory of transient stimulated Raman scattering, we will first review the important experimental observations.

From an experimental point of view, it is important to keep in mind that stimulated Raman scattering with picosecond pulses is very rarely observed as an isolated nonlinear effect. It is usually accompanied by strong self-focusing which may enhance the conversion efficiency but tends to break up the beam into numerous filaments, making accurate control of the geometry of the interaction difficult. Self-phase modulation can also occur, producing considerable spectral broadening, which in extreme cases can quench the stimulated Raman scattering. Four photon parametric scattering and two photon absorption can also be important. One of the few simplifying features of the use of picosecond pulses is the absence of competition from Brillouin scattering which, due to the much longer dephasing times of acoustic excitations (typically a few nanoseconds), is highly transient and has a very small gain relative to the faster vibrational excitations.

The first observation of transient stimulated Raman scattering in the picosecond regime actually preceded the use of passively modelocked lasers. Using a pump pulse of 12 ns duration from a Q-switched ruby laser, *Maier* et al. [4.87] observed intense backward stimulated Raman scattering in carbon disulphide having an intensity approximately nine times the incident pulse. Using second harmonic generation, they made an autocorrelation measurement of the duration of the Stokes pulse and found it was only 30 ps. They attributed the growth of the intense backward Stokes wave to the fact that the short backward propagating pulse sees a fresh supply of incoming pump wave enabling it to grow to large intensities without saturating due to pump depletion. The experimental observations were adequately accounted for by a theory (*Maier* et al. [4.82]) which allowed for the transient buildup of the Stokes pulse. Similar measurements were also made in hydrogen by *Culver* et al. [4.88]. *Loy* and *Shen* [4.89] have recently pointed out that self-focusing may contribute to this effect by producing backward moving foci.

The first observation of stimulated Raman scattering with picosecond optical pulses was made by *Shapiro* et al. [4.90]. They used a doubled mode-locked Nd:glass laser to produce a train of pulses at 530 nm which was focused into a 10 cm liquid cell with a 30 cm lens. Stimulated scattering in the forward direction was observed for a number of liquids such as carbon disulfide, benzene, toluene, chlorobenzene, bromobenzene, and nitrobenzene, and various mixtures of these. The conversion efficiencies varied widely for different liquids and did not seem to correlate with the conversion efficiencies for the comparable situation with nanosecond excitation. Overall, the magnitude of the scattering was lower than anticipated by the authors. The scattering was predominantly in the forward direction and no stimulated Brillouin scattering was observed. Near field photographs of the end of the cell showed evidence of strong self-focusing with the Raman emission emanating entirely from filaments. The authors attributed the lower scattering efficiencies in part to reduced self-focusing arising from a transient ac Kerr effect in those materials having an orientational relaxation time longer than the pulse duration. They also pointed out the importance of group velocity mismatch between the Stokes and laser pulses and calculated relative group delays of approximately 1 ps/cm for typical liquids.

Bret and *Weber* [4.91] made measurements of stimulated Raman scattering with picosecond pulses under conditions of low excitation for which self-focusing was reduced. They correlated Raman gain with the linewidths for spontaneous Raman scattering in different liquids. For liquids having spontaneous linewidths less than the laser linewidth they found the observed gain to be considerably less than the expected gain for steady-state stimulated scattering, and attributed this to a transient feature of the stimulated scattering. The implication was that in those cases where the laser spectrum exceeded the spontaneous Raman linewidth, the laser pulse duration was probably shorter than the dephasing relaxation time and consequently a steady state would not be reached so the gain would be smaller. This argument was primarily based on earlier work of *Hagenlocker* et al. [4.92] who made detailed studies of stimulated Raman and Brillouin scattering with nanosecond pulses in gases for which they varied the relaxation time over a wide range by adjusting the gas pressure. As we shall see when we come to a detailed discussion of the theory however, the importance of the laser spectrum in the case of picosecond pulses is considerably more complicated.

Bol'shov et al. [4.93] also made measurements of stimulated Raman scattering with picosecond pulses and recorded the Stokes emission spectra from benzene, carbon disulfide and water. Previously, stimulated Raman scattering had not been observed from water due to the strong damping of the OH vibrational modes and consequent small gain. The greater intensity of picosecond pulses, however, was sufficient to produce a sizeable gain. In CS_2 they observed a normal Stokes spectrum at low and moderate pump levels, but at high pump intensities the Stokes gain was reduced and a broad asymmetric emission was observed extending from the laser pump out to 1000 cm^{-1} on the Stokes side. They attributed this spectral broadening to a self-phase modulation by the nonlinear index of refraction in CS_2, and suggested that the gain was reduced by a "phase mismatching" of the molecular vibrations when driven by a broad spectrum, an idea due to *Akhmanov* [4.94]. More detailed measurements and calculations by *Akhmanov* et al. [4.95] confirmed this view.

At about the same time *Colles* [4.96] also observed SRS with picosecond pulses in a wide variety of liquids including water, the alcohols, and in a solid, calcite. He also observed a quenching of the scattering in CS_2. In addition, he made time-resolved measurements of the scattering due to each pulse in the modelocked train, and noticed a strong tendency for the Stokes emission to be greatest for the earlier pulses in the train and much smaller for later pulses. The envelope of the Stokes pulse train was noticeably different than the laser train. He attributed the reduced gain to the larger spectral and temporal widths of the pulses in the latter part of the train arising from a combination of self-phase modulation and dispersion in the glass rod of the laser. An alternative explanation due to thermal self-focusing in the SRS liquid cell was suggested by *Alfano* and *Shapiro* [4.97]. Colles showed that the energy conversion efficiencies, when measured at the peak of the Stokes train, were very large, up to 50% or more, especially for the liquids with low dispersion such as ethyl alcohol and acetone.

Carman et al. [4.98] measured the width of the Stokes pulses by two photon fluorescence. They chose CCl_4 for their measurements because of its relatively low dispersion and small nonlinear index of refraction. A modelocked ruby laser was used. Comparing TPF tracks for Stokes and laser pulses, they found that the Stokes pulse was on the average about 0.75 times the duration of the laser pulse. This result is expected from a theoretical treatment of transient SRS (see next section). Although some self-focusing existed, most of the Stokes emission appeared to emanate from regions outside the self-focused filaments. They also reported observation of weaker anti-Stokes emission at $\omega_L + \omega_V$ both in the forward direction and in a ring pattern. They also made measurements in methyl alcohol and photographed the Stokes emission spectrum and compared it to the spontaneous scattering (Fig. 4.8). In both spectra two lines are evident, one having Stokes shifts of 2837 cm^{-1} and 2942 cm^{-1}. The line at 2942 cm^{-1} has a broader width and consequently is not observed in steady-state SRS with longer pulses. In the transient regime with picosecond pulses, however, the gain depends only on the total Raman-scattering cross section, while the steady-state gain is inversely proportional to linewidth.

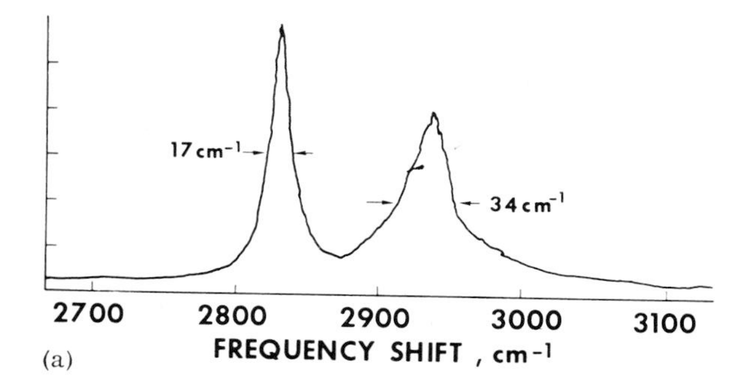

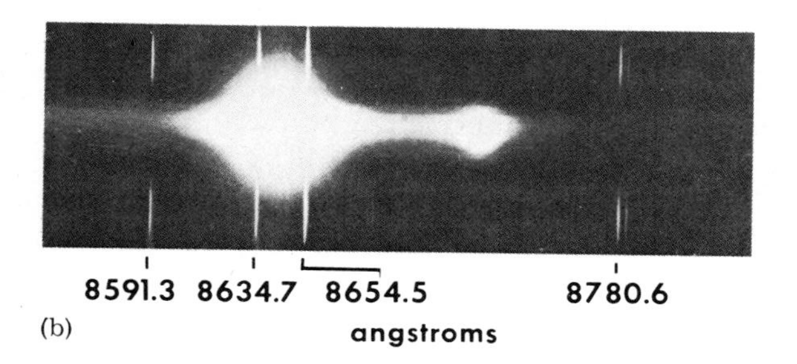

Fig. 4.8 a and b. (a) Spontaneous Raman spectrum of methanol. (b) Picosecond-pulse Raman emission of methanol. The stimulated line at 8726 Å is not seen with longer pulses (from *Carman* et al. [4.98])

The tendency for Stokes pulse shortening in transient SRS was used by *Colles* [4.99] to make a synchronously pumped Raman oscillator which produced Stokes pulse which were as short as one-tenth the laser pulses. He used the frequency doubled pulse train from a modelocked Nd:glass laser to pump a Raman oscillator consisting of a benzene cell and two mirrors forming an optical resonator with a spacing almost equal to the laser. Fine adjustment of the mirror spacing permitted him to compensate the differential group delay between the laser and Stokes pulses. Pulsewidth measurements were made by TPF.

Carman and *Mack* [4.100] have made detailed measurements of transient SRS with modelocked ruby pulses in 18 atmospheres of SF_6, a gas with very low dispersion. Particular care was taken to avoid self-focusing. The pump intensity was also carefully adjusted to avoid pump depletion. Using TPF correlations of pump and Stokes, they observed both a Stokes pulse shortening of a factor of 0.6 and a transient delay between the Stokes and the laser pulses of 6 ps for a 15 ps pump. This delay is a unique property of the transient scattering and was not caused by dispersion. From a measurement of the width of the spontaneous Raman spectrum of the 775 cm^{-1} vibrational line of SF_6 at 18 atm., they deduced a dephasing time of 7 ps. Using this parameter, they found good agreement with theory (*Carman* et al. [4.101]) for the Stokes pulsewidth and delay.

More recently *Adrain* et al. [4.102] have observed shortened Stokes pulses with the improved detection capability of high speed streak cameras. They used a flashlamp pumped passively modelocked rhodamine 6G dye laser and amplifier to pump a 7.5 cm cell of ethyl alcohol. Under conditions of high pump intensity ($\sim 5 \times 10^{11}$ W/cm^2) Stokes pulsewidths of 6 to 9 ps were observed compared to the laser pulsewidth of 12 ps. Using near threshold pump levels, however, even shorter pulses of ~ 3 ps were observed. Broader anti-Stokes pulses of ~ 16 ps were also observed at high pump levels. They estimated the time resolution of their camera to be 570 fs.

In other measurements, *Mack* et al. [4.103] have observed transient SRS from rotational excitations of gases. Using a modelocked ruby laser, they observed both rotational and vibrational SRS in a variety of high pressure gases where previously no stimulated scattering had been reported. Pulse energy conversion efficiencies as high as 70% were obtained. In many instances, the rotational Stokes spectra were broadened and skewed from line center or even split into several components, effects the authors attributed to the optical Stark effect. Transient SRS with picosecond pulses has also been observed in high pressure N_2 by *Chatelet* and *Oksengorn* [4.104].

Transient SRS with picosecond pulses has been used to make direct measurements of dephasing times of molecular vibrations (*Alfano* and *Shapiro* [4.105], *von der Linde* et al. [4.106]). *Von der Linde* et al. used a fast optical Kerr shutter to select a single pulse from a modelocked train. A probe pulse with a variable delay was scattered from the coherent vibrational wave producing anti-Stokes emission enabling a direct measurement of the dephasing time. In liquids with heavily damped vibrations, the technique permitted an accurate measurement of the laser pulse shape. Concurrently, *Alfano* and *Shapiro* [4.105] made a similar measurement of the optical phonon dephasing time in calcite. For details of these measurements and other applications of SRS to the measurement of relaxation times, the reader is referred to related chapters in this volume.

Examples of SRS with picosecond pulses involving excitations other than vibrational modes are *Sorokin* and *Lankard* [4.51] who generated pulses at 500 cm^{-1} by electronic Raman scattering in Cs, and *Laubereau* et al. [4.107] who observed SRS by polaritons in GaP.

4.4.3 Stimulated Raman Scattering of Picosecond Pulses: Theory

The theoretical understanding of transient SRS as applied to picosecond pulses is principally due to the work of *Akhmanov* et al. [4.108] and *Carman* et al. [4.101], and this discussion will be based primarily on their results. The reader should not, however, overlook the important influence of earlier work, some of which dealt with transient SRS in other contexts. To mention a few, we refer to *Kroll* [4.109], *Shen* and *Bloembergen* [4.84], *Maier* et al. [4.82], and *Wang* [4.110].

Returning to the general equations (4.24) to (4.26) of SRS introduced earlier in this section, the fields are expressed in terms of plane waves propagating in the positive z direction with slowly varying complex amplitudes:

$$E_{\mathrm{L}}=\tfrac{1}{2}\mathscr{E}_{\mathrm{L}}(z,t)\,\mathrm{e}^{ik_{\mathrm{L}}z-i\omega_{\mathrm{L}}t}+\mathrm{c.c.}, \tag{4.27}$$

$$E_{\mathrm{s}}=\tfrac{1}{2}\mathscr{E}_{\mathrm{s}}(z,t)\,\mathrm{e}^{ik_{\mathrm{s}}z-i\omega_{\mathrm{s}}t}+\mathrm{c.c.}, \tag{4.28}$$

$$Q=\tfrac{1}{2}Q_{\lambda}(z,t)\,\mathrm{e}^{ik_{\mathrm{v}}z-i\omega_{\mathrm{v}}t}+\mathrm{c.c}, \tag{4.29}$$

where the wave vectors and frequencies satisfy (4.20) and (4.21) for conservation of linear momentum and energy. Neglecting scattering in the backward z direction, and assuming a weak excitation of the molecular system ($n_{\mathrm{v}}\simeq 1$), the coupled equations for the Stokes and vibrational waves transform to the simplified form:

$$\frac{1}{v_{\mathrm{s}}}\frac{\partial\mathscr{E}_{\mathrm{s}}}{\partial t}+\frac{\partial\mathscr{E}_{\mathrm{s}}}{\partial z}=-iK_2Q^*\mathscr{E}_1 \tag{4.30}$$

$$\frac{\partial Q_{\mathrm{v}}}{\partial t}+v_{\mathrm{v}}\frac{\partial Q_{\mathrm{v}}}{\partial z}+\frac{1}{\tau}Q_{\mathrm{v}}=iK_1\mathscr{E}_{\mathrm{s}}\mathscr{E}_1^* \tag{4.31}$$

where v_{s} and v_{v} are the group velocities of the Stokes and vibrational waves and the coupling constants K_1 and K_2 are

$$K_1=\frac{N}{2\omega_{\mathrm{v}}}\frac{\partial\alpha}{\partial Q};\quad K_2=\frac{2\pi N\omega_{\mathrm{s}}^2}{c^2k_{\mathrm{s}}}\frac{\partial\alpha}{\partial Q}. \tag{4.32}$$

The group velocity of the vibrational wave is usually very small and consequently the second term in (4.31) can be ignored. In the quasi-steady-state situation, all time derivatives are assumed negligible and the solution of (4.30) and (4.31) is readily found to be:

$$\mathscr{E}_s(z,t)=\mathscr{E}_s(o,t)\,e^{\frac{1}{2}g_{ss}z} \tag{4.33}$$

where the steady-state gain per unit length g_{ss} at line center is

$$g_{ss}=2K_1K_2\tau|\mathscr{E}_L|^2=\frac{2\pi N^2\omega_s^2\tau}{c^2\omega_v k_s}\left(\frac{\partial\alpha}{\partial Q}\right)^2|\mathscr{E}_L|^2. \tag{4.34}$$

The initial Stokes field $\mathscr{E}_s(0,t)$ is determined by the spontaneous Raman scattering signal.

The solution for the transient case, including all time derivatives, can be obtained in the dispersionless case where the group velocities of the Stokes and laser waves are equal so that

$$\mathscr{E}_L(z,t)=\mathscr{E}_L(t-z/v_s)=\mathscr{E}_L(t'). \tag{4.35}$$

For the special case of a rectangular laser pulse of duration τ_p, the approximate solution for the Stokes wave in the transient regime with large gain is (*Akhmanov* et al. [4.108]):

$$|\mathscr{E}_s(z,t)|^2\sim\frac{1}{\sqrt{\dfrac{2g_{ss}(t-z/v_s)}{\tau}}}\exp\left\{2\sqrt{\frac{2g_{ss}(t-z/v_s)}{\tau}}\right\};\quad 0<t<\tau_p. \tag{4.36}$$

As indicated in Fig. 4.9 the Stokes wave lags the laser, rising from zero at the beginning of the laser pulse ($t=0$) to a maximum at the trailing edge of the laser pulse ($t=\tau_p$) where it abruptly falls to zero. The Stokes power gain at its peak ($t=\tau_p$) is

$$G_T=\sqrt{4G_{ss}\tau_p/\tau}\,, \tag{4.37}$$

where $G_{ss}=g_{ss}\cdot l$, with $l=$ the cell length. Notice that the transient gain depends on the square root of both the laser intensity and interaction length, whereas in the steady-state case the gain depends linearly on these two parameters. In the high gain limit, the transition between the steady-state and transient regimes is found to be

$$\left.\begin{array}{ll}\text{steady-state:} & \tau_p>G_{ss}\tau\\ \text{transient:} & \tau_p<G_{ss}\tau\end{array}\right\}G_{ss}\gg 1. \tag{4.38}$$

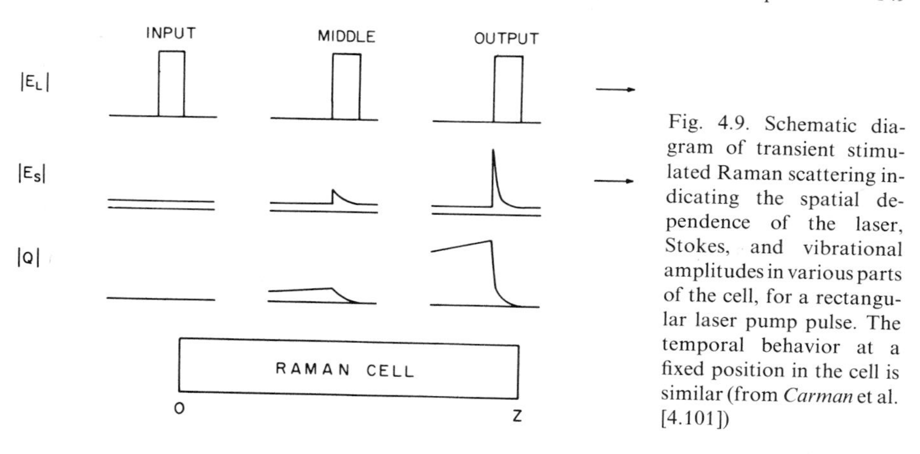

Fig. 4.9. Schematic diagram of transient stimulated Raman scattering indicating the spatial dependence of the laser, Stokes, and vibrational amplitudes in various parts of the cell, for a rectangular laser pump pulse. The temporal behavior at a fixed position in the cell is similar (from *Carman* et al. [4.101])

Since G_{ss} can easily be 30 or more in an experiment, the transient regime can extend to situations where the laser pulsewidth is even much greater than the molecular dephasing time. For arbitrary laser pulse shapes the gain is given by the expression:

$$G_{\mathrm{T}} = \log|(\mathscr{E}_{\mathrm{s}})_{\max}/\mathscr{E}_{\mathrm{s}}(0)|^2 \approx \sqrt{\frac{4G_{\mathrm{ss}}}{\tau}\int_{-\infty}^{t}\left|\frac{\mathscr{E}_{\mathrm{L}}(t)}{\mathscr{E}_{\mathrm{L}}(0)}\right|^2 dt}. \tag{4.39}$$

Notice that G_{T} is independent of the dephasing time unlike G_{ss} which is linearly proportional to τ (and hence cancels τ in the denominator of (4.39)). Also, since G_{T} depends on the time integral of $|\mathscr{E}_{\mathrm{L}}|^2$ it is not strongly sensitive to the particular shape of the laser pulse, but is primarily determined by the energy in the laser pulse. Fig. 4.10 shows numerical calculations of G_{T} for gaussian-

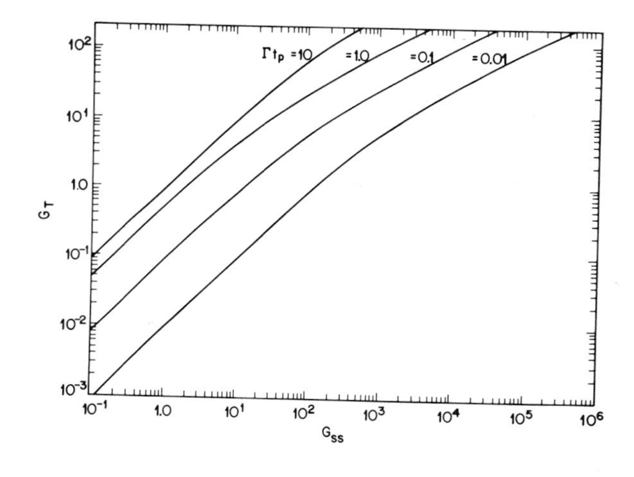

Fig. 4.10. Calculated transient Raman gain coefficient for gaussian laser input pulses with the same total energy, but different pulse widths. The steady-state gain coefficient G_{ss} corresponds to a constant-intensity laser output equal to the maximum laser pulse intensity (from *Carman* et al. [4.101])

shaped laser pulses for different values of the ratio of pulsewidth to dephasing time plotted vs G_{ss}.

The shape of the Stokes pulse, however, is very sensitive to the detailed shape of the laser pulse. In some cases the Stokes pulse can be considerably shorter than the laser pulse. This arises primarily from the fact that the inertia of the molecular system causes the buildup of the vibrational wave to lag the peak of the laser pulse so that the Stokes wave is delayed. The Stokes wave however is driven by the product of the vibrational amplitude *and* the instantaneous laser field (4.24) and consequently is abruptly terminated by the trailing edge of the laser pulse (see Fig. 4.9). The vibrational wave on the other hand persists for the dephasing time τ, which may be longer than τ_p. Fig. 4.11 shows numerical calculations of the Stokes pulse width and delay for gaussian-shaped laser pulses plotted for different values of τ_p/τ and G_T. Clearly at large transient gains the Stokes pulse can be appreciably shortened and delayed relative to the laser. Recent streak-camera measurements by *Lowdermilk* and *Kachen* [4.111] of transient SRS in H_2 and SF_6 provide convincing evidence for these effects (Fig. 4.12).

The dynamics of the vibrational amplitude have been calculated by *von der Linde* et al. [4.106] for gaussian and hyperbolic secant shaped laser pulses. In the transient regime, the buildup of $|Q|^2$ shows a delay comparable to the Stokes pulse delay, and then decays asymptotically as $\exp(-t/\tau)$. It was found that $|Q|^2$ approached the asymptotic exponential behavior more rapidly for gaussian pulses than for hyperbolic secant pulses. These results were used in the interpretation of the experimental determination of vibrational dephasing times.

In the high gain regime, G_T is found to be independent of the width of the laser spectrum if there is no dispersion between the laser and Stokes frequencies. This result applies even in the case where self-phase modulation may cause the laser bandwidth to greatly exceed the spontaneous Raman linewidth. This unique property follows from the fact that the Stokes wave is free to adjust its phase to follow any rapid phase variations in the laser so that the product $\mathscr{E}_L^* \mathscr{E}_s$ appearing in (4.31) is slowly varying and can efficiently drive the vibrational wave within its linewidth of $2\tau^{-1}$. In a dispersive medium, however, the Stokes and laser pulses lose synchronism due to unequal group velocities and the phases no longer track but tend to randomize as they propagate through the medium. The vibrational wave is driven off resonance and the gain is reduced. For a laser with a spectral width of $\Delta\omega_L$, the phases will be scrambled in a distance

$$z_0 = \frac{\pi}{\left(\frac{1}{v_L} - \frac{1}{v_s}\right)\Delta\omega_L}. \tag{4.40}$$

Although the phase scrambling is compensated to a certain extent by the natural tendency of the Raman process to restore the phases, in general there

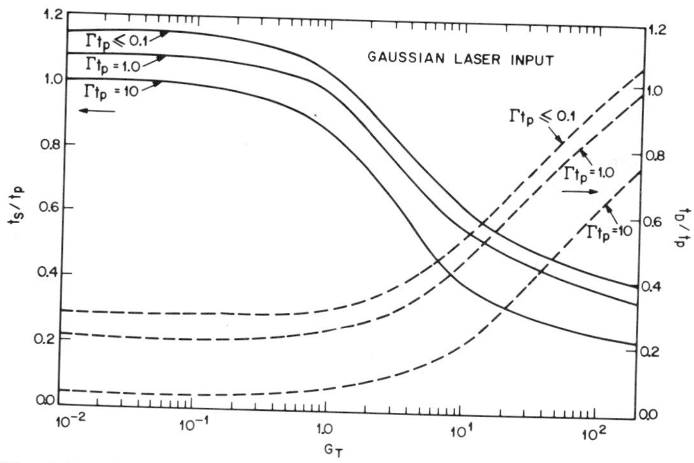

Fig. 4.11. The variation of Stokes pulse width t_S and delay t_D with transient gain coefficient for gaussian laser input pulses of various widths t_p given in terms of the vibrational dephasing time $\Gamma^{-1} = \tau_v$ (from *Carman* et al. [4.101])

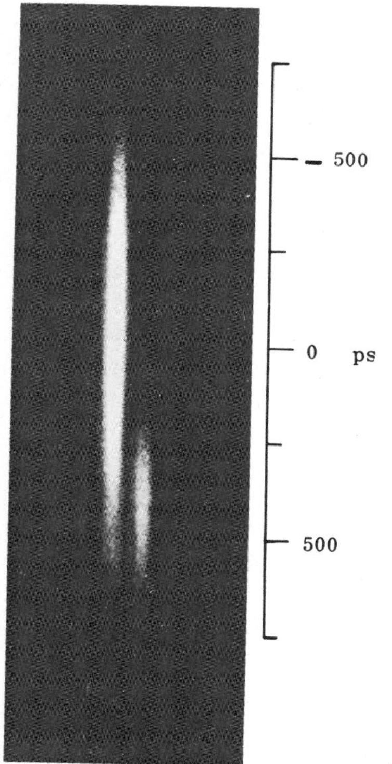

Fig. 4.12. Streak camera photograph of the laser pulse (L) and Stokes pulse (S) due to transient stimulated Raman scattering in H_2 at 10 atm. The delay and shortening of the Stokes pulse are clearly evident (from *Lowdermilk* and *Kachen* [4.111])

will be a reduction in interaction length which in severe cases such as CS_2 is sufficient to quench the scattering (*Akhmanov* et al. [4.95]). Except in the limiting case of a transform limited laser pulse (i.e., $\Delta\omega_L = 2/\tau_p$) the effects of phase scrambling are usually more important than the loss of overlap of the Stokes and laser pulse amplitudes, the latter being a problem only in cases of large dispersion and/or long interaction lengths.
The presence of strong phase modulation on the pump can have an additional effect unrelated to dispersion whereby the amplitude of the Stokes wave can be reduced due to a decrease in the initial spontaneous scattering. For a phase modulated laser pulse with a bandwith $\Delta\omega_L$, reduction in Stokes intensity (not gain) in the transient case is approximately equal to $1/\tau\Delta\omega_L$ (*Akhmanov* et al. [4.108], *Bloembergen* et al. [4.112]). Even in the limit of a transform limited pulse ($\Delta\omega = 2/\tau_p$), a reduction of τ_p/τ is to be expected relative to the steady-state case.

The spectrum of the Stokes emission in the transient case depends strongly on the pump excitation level. At very low pump levels, the spontaneous scattering is observed which has a spectrum equal to the convolution of the laser spectrum and the steady-state spontaneous spectrum, i.e., $\Delta\omega_s \simeq \Delta\omega_L + 1/\tau$. At moderate pump levels, in the absence of phase modulation, the Stokes spectrum tends to narrow, approaching the laser spectrum similar to the steady-state case. With strong pump levels, however, the temporal narrowing of the Stokes pulse can cause the spectrum to increase again. If strong self-phase modulation exists on the pump, so that $\Delta\omega_L \gg \tau^{-1}$, τ_p^{-1}, the Stokes spectrum tends to approximate the laser spectrum at all pump levels.

All of the theoretical results discussed in this section have assumed the Stokes intensity was not so large as to deplete the laser pump. *Daree* and *Kaiser* [4.113] and *Neef* [4.114] have recently considered the problem of estimating the Stokes pulse properties in the saturation regime. *Daree* and *Kaiser* [4.113] found by numerical calculations that the transient gain begins to saturate at considerably higher pump intensities than does the steady-state gain.

Thus far, we have assumed the population difference n_v between the ground state $|g\rangle$ and first excited vibrational state $|v\rangle$ was not appreciably different from its equilibrium value. In the extreme transient case where $\tau_p \ll \tau$ one might expect to observe coherent population fluctuations of the molecular system at sufficiently high intensities analogous to the resonant one photon coherent pulse propagation effects (*McCall* and *Hahn* [4.115], *Slusher* [4.116]). Numerous authors have addressed themselves to this question of coherent Raman pulse propagation (*Akhmanov* et al. [4.108], *Medvedev* [4.117], *Courtens* [4.118], *Shimoda* [4.119], *Medvedev* et al. [4.120], and *Tan-no* et al. [4.121]). The analysis, which requires the complete solution of all three coupled equations (4.24–26) predicts coherent population fluctuations with n_v going both negative and positive, pulse breakup, and spectral modulation. Since a population inversion can occur, forward anti-Stokes emission is possible (*Akhmanov* et al. [4.108], *Medvedev* [4.117] with the direct generation of a wave at $\omega_A = \omega_L + \omega_v$ along the laser axis. The requirements on laser pulse intensity and duration are rather demanding for these effects, however, and consequently positive experimental observation have not been made.

Although anti-Stokes emission is frequently observed in SRS with picosecond pulses (*Carman* et al. [4.98], *Griffiths* et al. [4.122]) it is more likely attributed to a third order nonlinear mixing of the Stokes and laser waves (*Laubereau* [4.123]). This process can be represented by a nonlinear susceptibility of the form (*Bloembergen* [4.78]):

$$P(\omega_A)=\chi(\omega_A=2\omega_L-\omega_s)\mathcal{E}^2(\omega_L)\mathcal{E}_s^*(\omega_s). \tag{4.41}$$

The physical basis for this process is the scattering of the laser wave off the coherent vibrational wave produced by the SRS (*Giordmaine* and *Kaiser* [4.124]). Since the propagation vector of the vibrational wave is predetermined by the dispersion between the Stokes and laser waves, i.e., $\boldsymbol{k}_v=\boldsymbol{k}_L-\boldsymbol{k}_s$, the direction of emission of the anti-Stokes wave must satisfy the condition $\boldsymbol{k}_A=\boldsymbol{k}_v+\boldsymbol{k}_L=2\boldsymbol{k}_L-\boldsymbol{k}_s$. In a normally dispersive medium the anti-Stokes emission is generally in the form of cones about the laser axis (*Garmire* [4.125], *Bloembergen* [4.78]). When strong self-focusing is present, however, the emission angles of the anti-Stokes may be altered significantly (*Shimoda* [4.126]). *Herman* [4.127] has recently considered the specific case of the theory of anti-Stokes emission with picosecond pulses.

4.4.4 Other Light Scattering

Except for Raman scattering, there are relatively few other stimulated light scattering phenomena which have been extensively studied with picosecond pulses. This situation contrasts sharply with work in the nanosecond domain which covers a wider range of topics such as stimulated Brillouin, Rayleigh wing, thermal, and concentration scattering (for a review see *Kaiser* and *Maier* [4.79]). The absence of observation of these effects with picosecond pulses is in some cases caused by smaller gains due to the inertia of the material excitation, e.g., Brillouin scattering, concentration scattering. In other cases, the frequency shifts are too small to resolve them relative to the laser pulse spectra. Although stimulated Rayleigh wing scattering has not been directly observed with picosecond pulses, it contributes to self-phase modulation (*Gustafson* et al. [4.128]).

One topic which has been the subject of both experimental and theoretical interest in the picosecond domain is stimulated thermal scattering. This phenomenon was originally proposed by *Herman* and *Gray* [4.129] who made calculations of stimulated gain due to thermally induced refractive index changes in absorbing media. Experiments with nanosecond pulses by *Rank* et al. [4.130] confirmed their predictions. *Mack* [4.131] has observed stimulated thermal scattering with picosecond pulses. He used a train of pulses from a modelocked ruby laser and a cell containing a liquid with a nonsaturable absorption, such as quinoline. Two beams, one strong, and the other weak, were sent into the cell, crossing at a small angle θ. On transmission through the cell, the weak beam was observed to experience a gain. In addition, higher order beams were generated at angles $\pm n\theta$. Mack interpreted these results as being caused by an index grating formed by the instantaneous thermal response of the liquid to the crossed laser beams. He argued that an instantaneous response arising from entropy waves must be responsible for the scattering on a picosecond time scale rather than the slower thermally induced density waves originally suggested by *Herman* and *Gray* [4.129]. However, calculations by *Pohl* [4.132] and *Starunov* [4.133] showed that the gain for pure thermal scattering (at constant density) was too small to explain his results. *Pohl* [4.132] suggested instead that the gain was due to the slow buildup of density waves from successive pulses in the complete modelocked train. Time-resolved experiments by *Scarlet* [4.134] confirmed this interpretation. More complete calculations by *Rangnekar* and *Enns* [4.135] agree with Pohl's results. Consequently, the stimulated thermal scattering observed by *Mack* was not a unique feature of picosecond pulses, and probably would not occur with a single pulse. A review of stimulated thermal scattering has been written by *Battra* et al. [4.136].

Yu and *Alfano* [4.137] have observed multiple photon light scattering in diamond due to hyper-Raman (*Terhune* et al. [4.138] and hyper-Rayleigh (*Kielich* et al. [4.139]) effects. Using a train of 1.06 μm pulses from a mode-locked Nd:glass laser to excite a 3 mm long type II-b diamond sample, they were able to produce scattered light at 90° to the pump axis at frequencies equal to $3\omega_L - \omega_{ph}$ and $2\omega_L - \omega_{ph}$, where ω_{ph} is the 1332 cm^{-1} optic phonon mode of the diamond lattice. The scattered signals were very weak and varied with the cube and square of the pump intensity, respectively. Scattered signals at 2ω and 3ω at 90° to the laser axis were also observed and attributed to hyper-Rayleigh scattering. Theoretical treatments of these higher order nonlinear scattering phenomena have been given by *Cyvin* et al. [4.140] and *Kielich* [4.141].

Stimulated scattering of picosecond pulses has also been observed from high density gaseous plasmas [4.333–335, 340, 341]. Forward and backward stimulated scattering has been observed at multiples of one-half the laser frequency [4.333–335]. Theoretical calculations [4.336–339] have suggested that this SRS emission is due to a nonlinear steepening of resonant electron-plasma oscillations when the plasma frequency equals one-half the laser frequency. Stimulated Brillouin scattering from low frequency in waves has also been observed [4.340]. For review of laser plasma studies see *Bobin* et al. [4.341].

4.5 Self-Focusing, Self-Phase Modulation and Self-Steepening

The three effects of self-focusing, self-phase modulation and self-steepening are caused by the same physical mechanism: the nonlinear index of refraction.

Experimentally, they often occur together for the same reason. The term "self-action" is used to describe effects such as these whereby the spatial, spectral, and temporal properties of an intense optical pulse are modified by propagation through a nonlinear medium. These topics have fascinated many workers and have resulted in extensive experimental and theoretical studies both in the nanosecond and picosecond domains. Reviews have been written by *Akhmanov* et al. [4.142], *Svelto* [4.143], *Shen* [4.144], and *Marburger* [4.145].

Although self-focusing has a definite intrinsic appeal as a unique nonlinear wave propagation effect, from the experimentalist's point of view it is often detrimental. By competing with other nonlinear effects, it can readily complicate an experiment, altering the interaction in a way that is difficult to control. Indeed, it is probable that some of our current understanding of this topic is due to the efforts of conscientious experimentalists who shifted their attention to self-focusing as a result of the frustation of trying to study other effects at the same time. In extreme cases, self-focusing leads to optical damage, a topic which will be discussed in the next section. Self-phase modulation is usually undesirable also, producing unwanted spectral broadening. This broadening, however, can also be useful, providing a "white light" source of short optical pulses which can be used as a probe to measure material properties such as absorption spectra with picosecond time resolution.

Self-focusing and self-phase modulation are generally more prevalent with picosecond pulses than with nanosecond pulses. This is due to both their shorter duration and higher available intensities. In addition, transient effects can also be important in those cases where the characteristic relaxation time of the nonlinear refractive index is comparable to or longer than the pulse duration. Consequently, the dynamics of self-focusing and the detailed nature of self-phase modulated spectra can exhibit features which are unique to picosecond pulses. The same is true for self-steepening.

4.5.1 Nonlinear Index of Refraction

At very high optical intensities, the index of refraction of materials can be a weak function of the electric field. Neglecting transient effects for the moment, this property can be expressed as

$$n = n_0 + n_2 \langle E^2 \rangle \ldots \tag{4.42}$$

where $\langle E^2 \rangle$ is the time average value of the square of the instantaneous electric field. If $E = \frac{1}{2}\mathscr{E}\exp(i\omega t) + \text{c.c.}$, then $\langle E^2 \rangle = \frac{1}{2}|E|^2$. (See *Marburger* [4.145] for a discussion of the notations used by different authors.) Odd powers of E cannot contribute a dc (or slowly varying) component. If $n_2 > 0$, an intense optical beam with a tapered cross section will produce an incremental index change $\delta n(\boldsymbol{r}) = n_2 E^2(\boldsymbol{r})$ which is a maximum on axis and tapers to zero off axis. This produces a lens-like effect which leads to self-focusing of the beam. In the spectral domain, a time varying phase shift $\delta\varphi = \delta k z = \delta n \omega z / c$ occurs which can lead to a frequency modulation

$$\delta\omega = -\frac{\partial}{\partial t}(\delta\varphi) = -\frac{\omega z}{c} n_2 \frac{\partial}{\partial t}(E^2). \tag{4.43}$$

For $n_2 > 0$ the frequency shift will be negative on the leading edge of the pulse and positive on the trailing edge.

The Orientational Kerr Effect

The effect which is generally the strongest and has consequently received the most attention is the molecular orientational Kerr effect (*Mayer* and *Gires* [4.146], *Bloembergen* and *Lallemand* [4.147]). This effect occurs in liquids and gases composed of anisotropic molecules whose optical polarizabilities in different coordinate directions are unequal. The simplest case is that of the symmetric top molecule whose polarizability along its axis α_{zz} is not equal to its polarizability perpendicular to its axis α_{xx}. In the absence of an electric field, the molecules are oriented in random directions so that the index of refraction is isotropic. The application of an optical field can induce a dipole moment in the direction of the field. The field can then further interact with this induced moment to exert a torque on the molecule proportional to $(\alpha_{zz}-\alpha_{xx})E^2$. Since the induced moment and the field have the same frequency and phase this torque will have a dc component which will tend to rotate the molecules into alignment with the field. The incremental rotation of a molecule produces the following change in index of refraction parallel to the electric field (*Close* et al. [4.148]):

$$\delta n = \tfrac{3}{2}\delta n_s[\cos^2\theta - \tfrac{1}{3}], \tag{4.44}$$

where

$$\delta n_s = \frac{4\pi}{3}\frac{(n_0^2+2)^2}{9n_0} N(\alpha_{zz}-\alpha_{xx}), \tag{4.45}$$

N is the density of molecules, and θ is the angle between the molecule and the electric field. The average is over all orientations with a weighting factor of a Boltzmann distribution which reflects the interaction energy: $(\alpha_{zz}-\alpha_{xx})E^2 \cos^2\theta$. For complete alignment ($E\to\infty$), $\langle\cos^2\theta\rangle = 1$ and $\delta n = \delta n_s$. For a small fractional alignment, *Close* et al. have derived the following approximate expression for the nonlinear index

$$n_2 = \frac{1}{15n_0}\left(\frac{n_0^2+2}{3}\right)^4\left(\frac{4\pi}{3}N\right)\frac{(\alpha_{zz}-\alpha_{xx})^2}{kT}. \tag{4.46}$$

As an example, in the case of CS_2, $\delta n_s = 0.58$ and $n_2 \simeq 1.7 \times 10^{-11}$ esu. The index of refraction perpendicular to the beam is reduced by an amount equal to one-half the value that the index is changed in the parallel direction. Consequently, the birefringence is:

$$\delta n_{\parallel} - \delta n_{\perp} = \tfrac{1}{2} n_2 \langle E \rangle^2 . \tag{4.47}$$

If circularly polarized instead of linear light is used, the change in index of refraction is only one-fourth as large.

If the optical field is suddenly turned off, the incremental alignment of the molecules will tend to randomize by collisions with neighboring molecules. The same relaxation occurs for the orientation of polar molecules in a dc electric field (the dc Kerr effect). *Debye* [4.149] has shown that the relaxation time of the dc dielectric constant is related to the viscosity η by the expression

$$\tau_D = \frac{4\pi\eta a^3}{kT} \tag{4.48}$$

where a is the radius of the molecular volume. If the randomization of the alignment proceeds by a diffusive process of a sequence of small changes, the relaxation time τ of the ac Kerr effect is equal to one-third of the dielectric relaxation time (Debye relaxation time), i.e.,

$$\tau = \tfrac{1}{3} \tau_D . \tag{4.49}$$

In some liquids, the alignment randomization may result from a single collision, in which case $\tau = \tau_D = \tau_c$, where τ_c is the mean collision time (*Ivanov* [4.150], *Pinnow* et al. [4.151]). More typically, an intermediate situation applies (*Ho* et al. [4.152]).

The relaxation of the nonlinear index can be accounted for by modifying (4.42) to (*Debye* [4.149]):

$$\tau \frac{\partial n}{\partial t} + n = n_0 + n_2 \langle E \rangle^2 . \tag{4.50}$$

Nonlinear indices of Kerr liquids have been calculated by *Kasprowicz* and *Kielich* [4.153] and *Kielich* [4.154]. Reviews have been written by *Pailette* [4.155], *Kielich* [4.156], and *Sala* and *Richardson* [4.157]. The latter authors have made calculations of the induced index changes for picosecond pulses of specific temporal and spatial characteristics.

Electronic Hyperpolarizability

In fluids composed of isotropic molecules and in solids, an important contribution to the nonlinear index is due to nonlinear electronic polarizability

(*Maker* et al. [4.158]). This effect, which arises from an optically induced distortion of the electronic charge distribution, is essentially instantaneous ($\tau \sim 10^{-15}$ s) and in general is considerably smaller in magnitude than the molecular orientational Kerr effect. Typically n_2(electronic)$\simeq 10^{-13}$ esu. In this case the difference between the induced index changes parallel and perpendicular to the electric field (birefringence) is determined by the symmetry properties of the particular material. For an isotropic material, $\delta n_\perp = (1/3)\delta n_\parallel$ (*Maker* and *Terhune* [4.26]). Examples of materials in which the electronic nonlinear index is important are liquid argon (*McTague* et al. [4.159], *Alfano* and *Shapiro* [4.160], CCl_4 (*Hellwarth* et al. [4.161], glass (*Owyoung* et al. [4.162]), and β-carotene (*Herman* et al. [4.163]).

The electronic contribution to the nonlinear index of refraction is present in all materials. Except for a correction due to dispersion, it can be directly related to the third order nonlinear susceptibility $\chi^{(3)}$ which accounts for third harmonic generation, four photon parametric mixing, etc., (*Wang* [4.164]). *Owyoung* [4.165] has recently determined that the electronic nonlinearity accounts for 13% of the total n_2 in CS_2 and 20% in benzene.

Electrostriction

Electrostriction can also make a contribution to the nonlinear index of refraction (*Shen* [4.166]). The internal energy density $\varepsilon|E|^2$ of an intense electromagnetic wave can cause a stress, resulting in a density change which alters the index of refraction. The magnitude of the steady-state nonlinear index is (*Shen* [4.166]):

$$n_2(\text{electrostriction}) = \frac{n_0 \beta}{2\pi} \left(\rho \frac{\partial n}{\partial \rho}\right)^2 \tag{4.51}$$

where ρ is the density and β is the bulk compressibility. The magnitude of n_2 can be quite large, typically 10^{-11} to 10^{-12} esu. The speed of response, however, is very slow, since the growth of the density change is governed by acoustic propagation (*Kerr* [4.167]). The response time is on the order of the beam cross section divided by the sound velocity, which typically might be 10^{-8} to 10^{-9} s. Consequently, electrostriction is not usually important for picosecond pulses except possibly when integrated over an entire train of pulses (*Kerr* [4.168]).

Collective Mechanisms

In gases composed of spherically symmetric molecules such as argon, krypton, etc., a contribution to the nonlinear index can arise from molecular collisions (*McTague* and *Birnbaum* [4.169]). During a collision between two molecules, a distortion in their electronic distributions can occur so that the polarizability of the pair may have a small anisotropic component. A nonlinear index can then result analogous to the molecular orientational Kerr effect. For binary collisions, the magnitude of the effect is proportional to the square of the gas pressure. The equivalent relaxation time for this effect is approximately equal to the mean collision time which at room temperature can typically be 10^{-13} s.

Hellwarth [4.170–172] has proposed an additional physical mechanism for the nonlinear refractive index of fluids composed of spherical molecules or atoms based on a molecular redistribution or clustering model. In this mechanism, a pair of atoms or molecules attracts each other through their optically induced dipole moments; they are then redistributed in space so as to produce a macroscopic change in index of refraction. *McTague* et al. [4.159] estimate that in liquid argon the electronic and clustering mechanisms are comparable and approximately equal to 1.8×10^{-14} esu. The relaxation time of this mechanism is determined by the random translational motion of the molecules (Brownian motion) and is comparable to reorientational relaxation time (typically 10^{-12} to 10^{-10} s).

Thermal Index Changes

Thermal effects can also contribute to a nonlinear index of refraction (*Gordon* et al. [4.173], *Litvak* [4.174], and *Akhmanov* and *Sukhorukov* [4.175]). Usually the thermal n_2 is negative, and produces a self-defocusing effect. In general, there are two components to the thermally induced index change, one arising from a purely thermal effect at constant volume and the other due to thermal expansion, i.e.,

$$\frac{dn}{dT} = \left(\frac{\partial n}{\partial T}\right)_{\mathrm{V}} + \left(\frac{\partial n}{\partial \rho}\right)_{\mathrm{T}} \frac{\partial \rho}{\partial T}. \tag{4.52}$$

The first term is due to a change in entropy, and can be quite rapid, although its magnitude is usually not large. The contribution from thermal expansion will have a slow response, however, due to the required growth of a density wave similar to electrostriction. The relaxation of both contributions will be very slow, being primarily determined by diffusion of the heat away from the optical beam (typically 10^{-3} to 1 s). Thermally induced index changes are usually not important for picosecond pulses, except in those cases where the effects may be integrated over an entire pulse train (*Pohl* [4.132], *Alfano* and *Shapiro* [4.160]). For a complete review of thermal self-focusing and defocusing see *Akhmanov* et al. [4.142].

Other Mechanisms

In pyroelectric materials a nonlinear index can arise due to the generation of a pyroelectric polarization which produces a δn by the electro-optic effect (*Glass* and *Auston* [4.176]). An additional contribution can also arise from an optically induced polarization in absorbing polar materials in which the dipole moment of an excited state differs from the ground state (*Glass* and *Auston* [4.176], *Auston* et al. [4.177]).

Tzoar and *Gersten* [4.178] have shown that in semiconductors with small band gaps, the nonparabolicity of the bands can produce a large nonlinear index of refraction. For InSb, their calculations suggest an n_2 even larger than in CS_2. As yet, this effect has not been observed.

Table 4.1. Nonlinear index of refraction

A. Liquids

Liquid	n_2 ($\times 10^{13}$ esu)	τ (ps)	
		Dynamic	Spectroscopic
Carbon disulphide	200[a,b]; 170[c]; 110[d]	2.1[k]	2.0[j]; 1.4[t]
Benzene	20[a]; 25[c]		3.3[l,m]
Nitrobenzene	140[a]; 160[b]	32[p,r]	36[n]; 39[o]
Toluene	45[a]		5.3[m]
m-Nitrotoluene		50[r]	55[u]
Chloroform	8.1[a]		5.9[q]
α-Chloronapthalene	130[a]		
Carbon tetrachloride	2.5[a]		
Acetone	6.8[a]		
Water	1.4[a]		
Chlorobenzene	160[b]		8.5[s]; 6[t]

B. Solids

Solid	n_2 ($\times 10^{13}$ esu)	Solid	n_2 ($\times 10^{13}$ esu)
KDP	3.6[h]	Lucite	2.7[d]
Fused quartz	1.2[e]; 1.4[h]	YAG(111)	3.5[d]; 3.2[f]
BK-7 glass	1.7[e]	Al_2O_3	1.3[g]
SF-7 glass	6.9[e]	Lithium fluoride	2.4[h]
ED-2 glass	1.5[d,f]; 1.6[i]	Sodium chloride	6.5[h]
Ruby	1.5[d]	Sodium fluoride	0.95[h]
β-Carotene	100[w]	Potassium bromide	14.2[h]
Calcium fluoride	2.8[h]	Potassium chloride	3.3[h]

References:

[a] Paillette [4.155]
[b] Mayer and Gires [4.146]
[c] Owyoung [4.165]
[d] Moran et al. [4.313]
[e] Owyoung et al. [4.162]
[f] Bliss et al. [4.314]
[g] Smith and Bechtel [4.265]
[h] Smith et al. [4.263]
[i] Duguay et al. [4.217]
[j] Shapiro and Broida [4.315]
[k] Ippen and Shank [4.316]
[l] Gabelnick and Strauss [4.317]
[m] Fabelinski [4.318]
[n] Alms et al. [4.319]
[o] Stegeman and Stoicheff [4.320]
[p] Duguay and Hansen [4.208]
[q] Starunov [4.321]
[r] Ho et al. [4.152]
[s] Craddock et al. [4.322]
[t] Rouch et al. [4.323]
[u] Foltz et al. [4.324]
[w] Herman et al. [4.163]

Yablonovich and *Bloembergen* [4.179] have pointed out that an important nonlinear contribution to the index of refraction can result from plasma formation by avalanche ionization at field strengths near the breakdown field.

We conclude this discussion of physical mechanisms with Table 4.1 which is a compilation of the values of n_2 and relaxation times of some representative materials which exhibit a significant nonlinear index. The table is divided into two parts, the first dealing with liquids, and the second with solids. In both cases the values of n_2 quoted are measured values of the total nonlinear index. In the case of liquids with large n_2's, the dominant contribution is likely the orientational Kerr effect, and the electronic hyperpolarizability is likely the dominant mechanism in the other materials. The relaxation times indicated are measured values determined either directly with picosecond pulses (dynamic) or indirectly from Rayleigh scattering (spectroscopic). The relationship between the optical Kerr effect and molecular light scattering has been discussed by *Kielich* [4.154]. For details of the measurement techniques, the reader is referred to the specific references quoted in the table. Relaxation times for the solid materials are not given, but they are estimated to be $\sim 10^{-14}$ s.

4.5.2 Self-Focusing of Picosecond Pulses: Experiments

The possibility of self-focusing of an optical beam was first suggested by *Askaryan* [4.180] who pointed out the importance of the nonlinear index of refraction. Approximate solutions of the nonlinear wave equation were made by *Talanov* [4.181] and *Chiao* et al. [4.182]. The latter authors suggested a self-trapping model whereby the beam diameter would collapse to a thin filament which propagated in a dielectric waveguide channel. The first convincing experimental demonstration of self-focusing was made with nanosecond pulses by *Garmire* et al. [4.183]. They observed the gradual reduction of the beam diameter while propagating through a long CS_2 cell. High resolution photographs (*Chiao* et al. [4.184]) showed the formation of filaments ~ 5 µm in diameter. The distance required to reduce the beam to this diameter seemed to agree well with the calculations by *Talanov* [4.185] and *Kelley* [4.186], who predicted a self-focusing distance:

$$z_{\mathrm{f}} = \tfrac{1}{4}\, r_0^2 n_0 \sqrt{\frac{c}{n_2}}\, \frac{1}{\sqrt{P} - \sqrt{P_{\mathrm{cr}}}} \tag{4.53}$$

where r_0 is the initial beam radius at the cell entrance (unfocused), c is the velocity of light, P is the optical power in the beam, and P_{cr} is the critical power for self-focusing given by (*Chiao* et al. [4.182]):

$$P_{\mathrm{cr}} = \frac{c\lambda^2}{32\pi^2 n_2} \ \text{esu.} \tag{4.54}$$

The notion of the self-trapped filament, however, led to considerable controversy and is now generally believed to have been incorrect as applied to nanosecond pulses. Although the theoretical calculations accurately predicted the reduction in beam diameter, they all diverged at the focus and were not able to account for a stable steady-state mode of propagation in a dielectric waveguide filament. This difficulty has to a certain extent been solved by the model of moving foci (see *Shen* [4.144] for a complete discussion). *Lugovoi* and *Prokhorov* [4.187] proposed a nonstationary model of self-focusing whereby different time segments of the incoming pulse focus at different distances in the cell due to their unequal powers, and these focal spots move with the rising and falling power of the incoming pulse. This model can be better understood with the aid of Fig. 4.13 which shows the trajectories in space and time of the foci due to different portions of the incoming pulse. When integrated over the duration of the pulse the motion of these foci will give the appearance of filaments, or narrow streaks of light. Experimental confirmation of this model has been made by *Loy* and *Shen* [4.188–190], by *Zverev* et al. [4.191], and by *Korobkin* et al. [4.192]. The length of the focal spot is very short and it can propagate either in the backward direction or in the forward direction with a velocity that can exceed the velocity of light. As the moving focus passes through a point on the beam axis, the optical intensity at that particular point will rapidly increase to a large value and then fall within a time which is much less than the laser pulse duration. The effective duration of the focal spot has been estimated to range from a few picoseconds to 100 ps in typical cases (*Gustafson* et al. [4.128]; *Brewer* et al. [4.193]). Consequently, even with nanosecond excitation, self-focusing has important features which extend into the picosecond time domain. The influence of self-focusing on stimulated Raman scattering has been discussed by *Shen* and *Shaham* [4.194], and

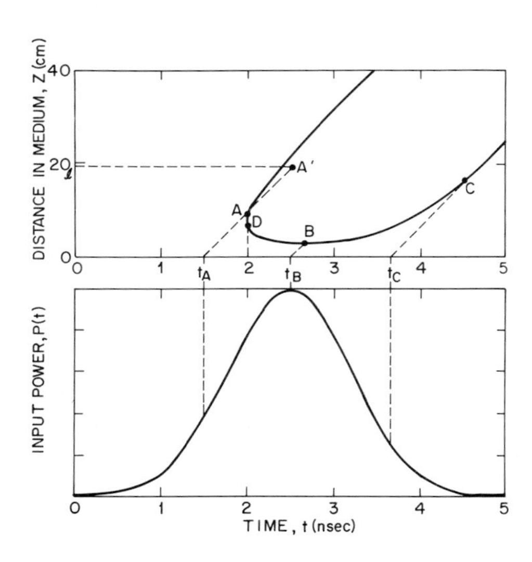

Fig. 4.13. Moving focus model of self-focusing. Lower trace describes input power $P(t)$ as a function of time. Upper trace is a calculation which describes the position of the focal spot as a function of time. The dotted lines, with the slope equal to the light velocity, indicate how light propagates in the medium along the z-axis at various times. The first focal spot appears at D and then splits into two: one which first moves backward, and then forward, the other keeps on moving forward with a velocity faster than light. Calculations are based on CS_2 with a peak laser power of 8 kW, and input beam diameter of 400 μm (from *Loy* and *Shen* [4.89])

Lallemand and *Bloembergen* [4.195]. For complete details of self-focusing of ns pulses, the reader is referred to the review articles mentioned previously.

When the excitation pulse is of picosecond duration, the situation is complicated by two factors. First, the pulse duration, or more important the spatial length of the pulse $c\tau_p/n_0$ is usually considerably shorter than the interaction length, whereas with nanosecond pulses the opposite is true. Secondly, the transient character of the nonlinear index of refraction means that in many cases δn will not follow the rise and fall of the pulse but can integrate so that the trailing edge of the pulse may see a greater index of refraction than the leading edge.

The first detailed measurements of self-focusing of picosecond pulses were made by *Brewer* and *Lee* [4.196]. The propagation of picosecond pulses from a modelocked Nd:glass laser through liquid cells caused the beam to self-focus into multiple spots each having diameters of a few microns similar to what is observed with nanosecond pulses. From careful photographs of the optical field at the exit face of the cell, they were able to measure the diameters and energy of each spot. Fig. 4.14 shows a microdensitometer trace from a photograph of a self-focused filament in nitrobenzene. The energy per pulse in this filament was estimated to be 10^{-7} J. Usually a number of such filaments were produced due to spatial breakup of the multimode output of the Nd:glass laser. Also, the entire train of pulses was used for their measurements so the results represent an average over many pulses, each not necessarily having the same characteristics. The authors also made measurements in CS_2, CCl_4 and a mixture of ethanol, ether, and toluene which could be continuously adjusted from a low to a high viscosity liquid and to a glassy solid ($\eta \sim \infty$) by varying the temperature from *RT* to -165 °C. The surprising feature of their results

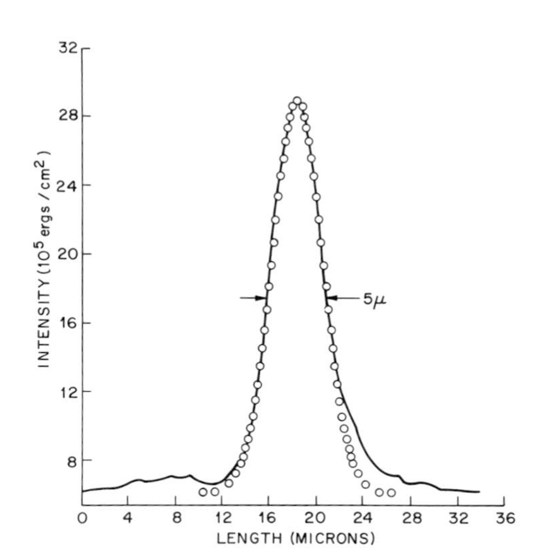

Fig. 4.14. The radial intensity profile of a self-focused picosecond pulse in nitrobenzene at 23 °C. The profile was obtained from a microdensitometer trace, and the instrument's resolution of 1.2 μm or less did not alter the shape. The solid curve is experimental, while the circles are a gaussian fit (from *Brewer* and *Lee* [4.196])

was the similarity for such a wide range of materials. Filament diameters varied only a factor of 3 and pulse energies varied a factor of 4, except for the glassy solid which had 10 times the energy of CS_2. This occurred in spite of a variation in rotational relaxation time from a few picoseconds to longer than one microsecond, and consequently led the authors to conclude that the molecular reorientation and clustering effects were not dominant, but rather that the main contribution was electronic. Also, the results for CCl_4, a symmetric molecule with no orientational Kerr effect, were almost identical to CS_2 and nitrobenzene, both of which are known to have a large orientational Kerr effect. Later work, however, has since shown that the orientational Kerr effect is the dominant mechanism in CS_2, nitrobenzene and other Kerr liquids.

Alfano and *Shapiro* [4.197] and *McTague* et al. [4.159] have observed self-focusing of picosecond pulses in liquid argon. They both emphasized the importance of the electronic nonlinearity, although McTague et al. have estimated that molecular clustering could be equally important. The latter conclusion was based on an estimate of n_2 from spontaneous scattering measurements (*McTague* et al. [4.198]). *Alfano* and *Shapiro* [4.197], however, observed a strong similarity between liquid and solid krypton and concluded that molecular clustering was not as important as the electronic nonlinearity.

Extensive measurements of self-focusing of trains of TEM_{00} mode pulses from a modelocked ruby laser of special design have been made by *Svelto* and co-workers [4.199–203]. Their observations of self-focusing and self-phase modulation in anisotropic molecular liquids led them to suggest that a molecular "rocking" or librational motion was responsible for the nonlinear index in addition to the orientational Kerr effect. The magnitude of the librational effects, however, was estimated to contribute no more than 20% (*Svelto* [4.143]).

The importance of the orientational Kerr effect for self-focusing of picosecond pulses in anisotropic molecular liquids has been firmly established by a sequence of experiments in which direct time resolved measurements of the refractive index changes have been made. *Shimizu* and *Stoicheff* [4.204] used a train of pulses from a modelocked Nd:glass laser to produce self-focusing in CS_2. The second harmonic was used as a probe to illuminate the cell transversely through crossed polarizers so that the induced birefringence could be photographed with picosecond time resolution. The duration of the birefringence was observed to be less than 10 ps, and was probably limited by the pulse duration. Its magnitude was estimated to be $\sim 2\times 10^{-3}$. Similar, but more detailed measurements were later done by *Reintjes* and *Carman* [4.205] and *Reintjes* et al. [4.206]. They used a single picosecond pulse and compressed the SHG probe pulse (*Treacy* [4.207]) for improved time resolution to observe birefringence tracks in CS_2, nitrobenzene, toluene, and benzene (Fig. 4.15). As indicated in Fig. 4.16 they observed birefringence tracks which persisted following the main pulse for a duration comparable to the orientational relaxation time. Their results were in good agreement with earlier measurements of orientational relaxation times made by *Duguay* and *Hansen* [4.208] by optical gating in a geometry without self-focusing. More recently, *Ippen* and *Shank* [4.209] have made direct time resolved measurements of the orientational relaxation in CS_2 and find $\tau = 2.1$ ps.

Reintjes et al. [4.206] also followed the evolution of the self-focusing in CS_2 from the front to the rear of the cell. They observed both gross beam self-focusing and breakup into multiple filaments in the first 4 cm of a 20 cm cell. In the central region of the cell quasi-steady-state filaments were observed which seemed to propagate for distances as much as 3 cm without change in diameter or spatial distribution. Following this, filaments are observed to expand and disappear with few reaching the end of the cell.

Topp and *Rentzepis* [4.210] have also made time resolved measurements of the birefringence induced by self-focusing in Kerr liquids. They used a longitudinal geometry with multiple delayed weak probe pulses following a single intense self-focusing pulse. This permitted direct measurements of the relaxation of the nonlinearity. In most cases their results supported the orientational Kerr model with viscous damping, although some evidence suggested the possibility of additional faster contributions. They also employed a novel method of viewing self-focused filaments by adding a small amount of an organic scintillator to the liquid cell to produce fluorescence by three photon absorption. Being proportional to the sixth power of filament diameter, this provided an accurate method of determining the point of maximum focusing.

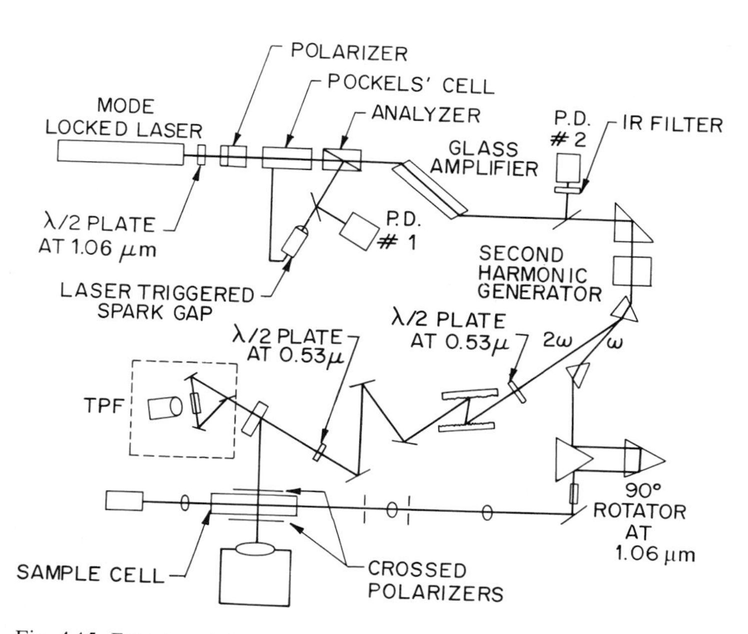

Fig. 4.15. Experimental apparatus used to study the self-focusing of picosecond pulses in Kerr liquids by observing the optically induced birefringence with a transverse probing pulse (from *Reintjes* and *Carman* [4.205])

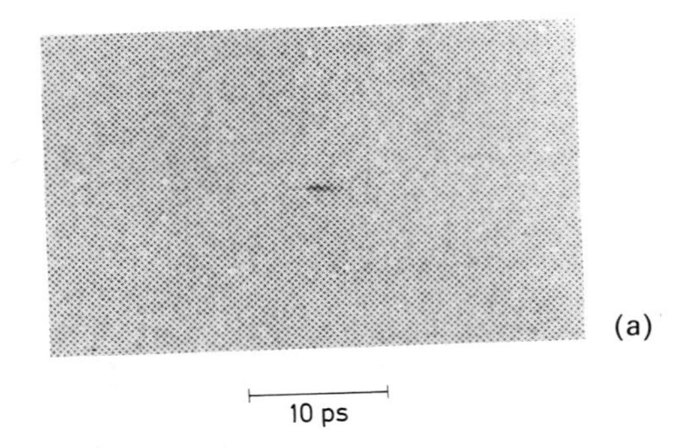

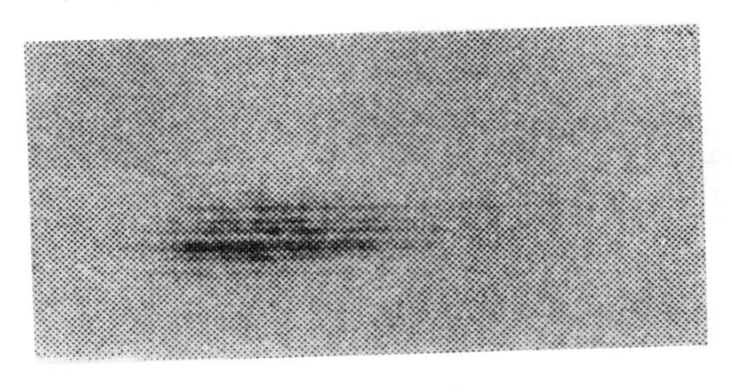

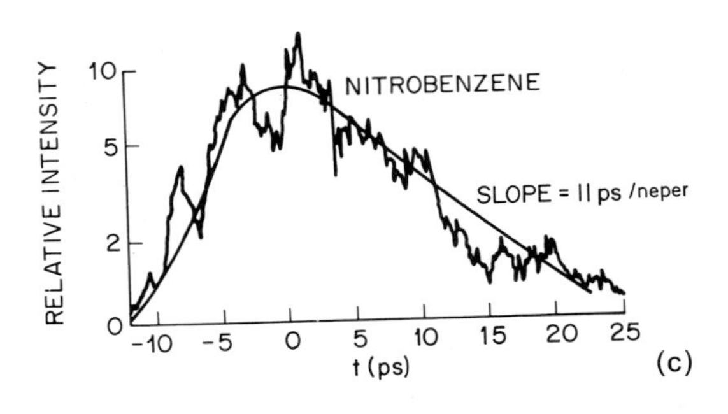

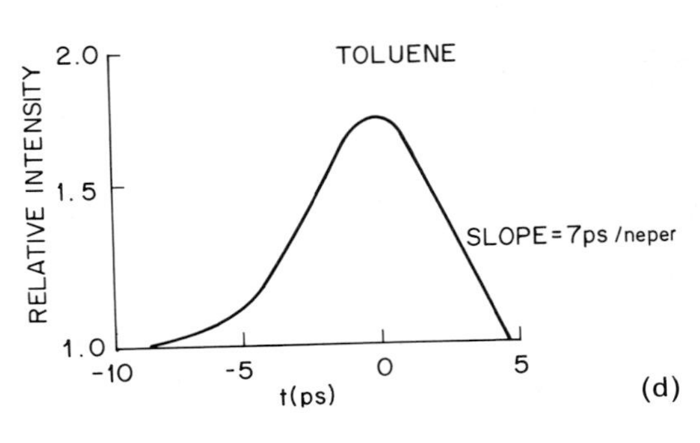

Fig. 4.16 a-d. Experimental observations of self-focusing in Kerr liquids using the apparatus in Fig. 4.16. (a) Birefringence in CS_2. (b) Birefringence in nitrobenzene. (c) Microdensitometer trace of a birefringence streak in nitrobenzene showing decay of the probe signal after the peak of the laser pulse with a time constant of 11 ps. (d) Smoothed densitometer trace of birefringence in toluene showing decay of ~7 ps. (from *Reintjes* and *Carman* [4.205])

Alfano and *Shapiro* [4.211] have observed self-focusing of picosecond pulses in crystalline quartz, calcite, sodium chloride, and glass. Although the nonlinear index of refraction in these materials is considerably smaller than the Kerr liquids, they were nevertheless able to produce strong self-focusing in

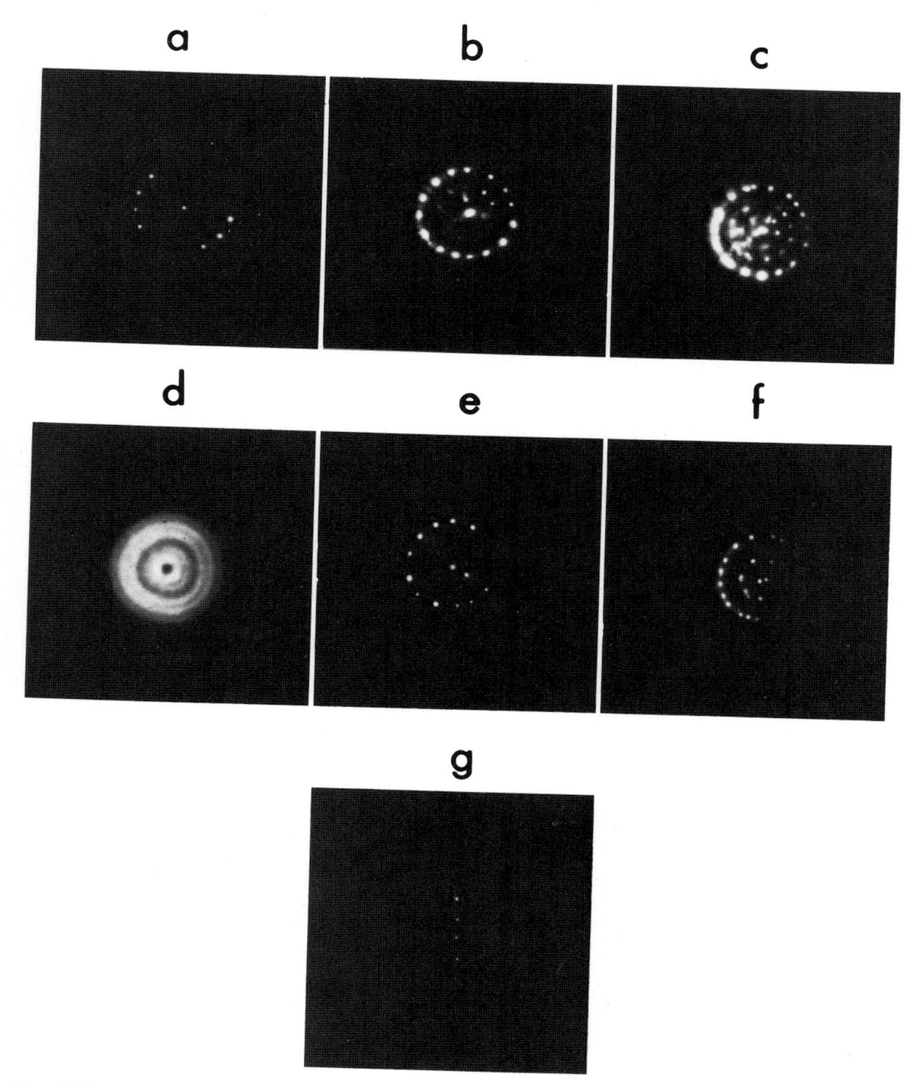

Fig. 4.17 a-g. Photographs showing the spatially periodic breakup of beam due to self-focusing. The patterns (a)–(c) show the effect of progressively higher laser powers for a Fresnel number of seven; (d) is the diffraction pattern at the end of the cell for $F=4$; while (e) and (f) are the corresponding focal spot patterns (from *Campillo* et al. [4.215])

multiple filaments. In some, but not all, cases the materials were damaged. Strong spectral broadening accompanied the self-focusing. They concluded the electronic nonlinearity was most likely responsible.

Self-focusing of picosecond pulses in gases has been observed by *Mack* et al. [4.103]. They used N_2O and CO_2 at ~ 50 atmos. with a modelocked ruby laser, and believe that the orientational Kerr effect is responsible.

Lehmberg et al. [4.212] have observed self-defocusing of picosecond pulses from a modelocked Nd:YAG laser in cesium vapor. A near resonance for a two-photon absorption is believed responsible for the relatively strong and negative n_2 which was comparable to that of glass at an atomic density of only 5×10^{16} cm^{-3}. Theoretical estimates of n_2 gave good agreement with experiments.

When the optical power greatly exceeds the critical power for self-focusing, a spatial instability develops which tends to make the beam break up into multiple beams (*Bespalov* and *Talanov* [4.213]). Since $P_{cr}\sim10^4$ W for CS_2, this is an effect of considerable importance for high power picosecond pulses, which generally tend to self-focus in multiple filaments rather than in single beams. An initial disturbance δE in the optical field, having a transverse spatial frequency k_T, will grow with a gain α given by (*Campillo* et al. [4.214]):

$$\alpha=\frac{c}{2\omega n_0}\,k_T(2\gamma|E_0|^2-k_T^2)^{\frac{1}{2}} \tag{4.55}$$

where $\gamma=3\omega^2 n_0 n_2/2c^2$. The gain is largest for $k_T=\sqrt{\gamma}|E_0|$. Regular spatial breakup of high power beams from a Q-switched laser have been observed by *Campillo* et al. [4.215]. The tendency is for the beam to break up into multiple spatial "cells", each containing a few critical powers, which then self-focus independently (Fig. 4.17). As a consequence, the self-focusing length is inversely proportional to the intensity for high-power beams, and the beam self-focuses in a much shorter distance than predicted by (4.53). Although *Campillo* et al. used nanosecond pulses for their measurements, the concept of spatial breakup is crucial for high-power picosecond pulses. It is very likely the reason for the strong random spatial structure in the output of the modelocked Nd:glass laser (*Korobkin* et al. [4.216]; *Duguay* et al. [4.217]). *Eckardt* [4.218] has made detailed numerical calculations of the self-focusing in this laser. *Fournier* and *Snitzer* [4.219] have calculated the electronic nonlinear index of refraction of glass.

4.5.3 Self-Focusing of Picosecond Pulses: Theory

The theoretical understanding of self-focusing with picosecond pulses is somewhat less satisfactory than the corresponding situation in the nanosecond domain. In the transient case ($\tau\geq\tau_p$), however, a theoretical model has been developed that seems to explain most of the experimental observations. The basic features of this model, which was first suggested by *Akhmanov* et al. [4.220], depend on the inertial property of the nonlinearity which causes the refractive index change to build up slowly and then to persist after the passage of the peak of the pulse. As a result, the leading portion of the pulse will not self-focus but will spread by diffraction. However, the central and trailing sections of the pulse will see a significant δn and will contract by self-focusing. As a net result, the pulse assumes the shape of a horn as indicated in Fig. 4.18. The neck of the horn, where the intensity is very high, is responsible for the observed filaments. If the relaxation time is considerably longer than the pulse duration, the index change will not reach steady state and its peak magnitude will be smaller, increasing the threshold for self-focusing as observed experimentally. The inertia of the nonlinearity also has a tendency to retard the self-focusing action, so that the horn-shaped pulse may propagate over a significant distance before expanding by diffraction. *Akhmanov* et al. found approximate analytic solutions of the scalar wave equation with a nonlinear index having a finite relaxation time, and determined that the profile of the horn has an exponential flare.

For axially symmetric geometry, the scalar wave equation can be reduced to the following nonlinear equations for the amplitude and phase factors (*Akhmanov* et al. [4.220]):

$$\frac{\partial A}{\partial z}+\left(\frac{\partial s}{\partial r}\right)\left(\frac{\partial A}{\partial r}\right)+\frac{A}{2}\left(\frac{\partial^2 s}{\partial r^2}+\frac{1}{r}\frac{\partial s}{\partial r}\right)=0 \tag{4.56}$$

$$2\left(\frac{\partial s}{\partial t}\right)+\left(\frac{\partial s}{\partial z}\right)^2=2\left(\frac{n-n_0}{n_0}\right)+\frac{1}{k^2 A}\left(\frac{\partial^2 A}{\partial r^2}+\frac{1}{r}\frac{\partial A}{\partial r}\right) \tag{4.57}$$

$$\frac{\tau\partial n}{\partial \xi}+n=n_0+\tfrac{1}{2}n_2A^2 \tag{4.58}$$

where

$$E=\tfrac{1}{2}A(r,z,\xi)\exp\left[ikz-i\omega t+iks(r,z,\xi)\right]+c.c \tag{4.59}$$

and

$$\xi=t-\frac{n_0 z}{c}. \tag{4.60}$$

Numerical solutions of these equations by *Fleck* and *Kelley* [4.221], *Fleck* and *Carman* [4.222], *Shimizu* and *Courtens* [4.223], and *Shimizu* [4.224] have elucidated the details of the relaxation model of self-focusing of short pulses. Their results show a quasi-stationary solution in which the effects of diffraction and self-focusing are approximately balanced permitting a horn-shaped pulse

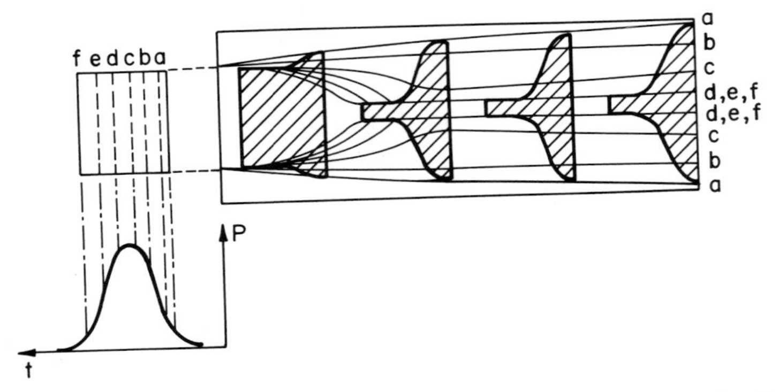

Fig. 4.18. "Horn" model of transient self-focusing of a picosecond pulse in a Kerr liquid. Different parts (*a*, *b*, *c*, etc.) of the pulse focus and defocus along different ray paths. The pulse first gets deformed into a horn shape and then propagates on without much further change (from *Shen* [4.144])

to propagate almost unchanged over extended distances. It is not a true steady-state trapped mode, however, and does eventually expand by diffraction. Nevertheless, the gross features of the model are in good agreement with experimental results, as shown by *Wong* and *Shen* [4.225] who made observations of self-focusing of nanosecond pulses in the nematic liquid crystal MBBA. This material has a large Kerr effect and its orientational relaxation time can be temperature tuned from 40 to 800 ns, permitting detailed time-resolved measurements.

The most serious failure of the theory is its inability to account for the limiting filament diameter. It predicts a filament diameter which approaches zero with increasing incident power. This contrasts sharply with the experimental situation where the diameter approaches a finite limit of a few microns and shows surprisingly little variation between materials. Other physical mechanisms are necessary to account for the limiting filament diameters.

In glasses, crystalline solids, non-Kerr fluids, and other materials where the electronic nonlinearity is believed to be dominant, the theoretical picture is less clear. In this case relaxation of the nonlinearity is extremely fast and the refractive index perturbation should follow the instantaneous optical intensity. The direct application of the moving focus model, however, is not straightforward and requires further study. As pointed out by *Loy* and *Shen* [4.89], the moving focus should yield a filament pulse with local transients of a few relaxation times, or $\sim 10^{-14}$ s. Although this transient might account for the extreme spectral broadening usually observed in these materials, further work is needed to clarify the situation. Possibly, when additional physical mechanisms are included in the theory to account for the limiting filament diameters, a more complete picture of the self-focusing dynamics will emerge.

4.5.4 Limiting Filament Diameters

Numerous mechanisms have been suggested to account for the limiting filament diameter. Saturation of the nonlinear index due to complete molecular alignment is a possibility (*Goldberg* et al. [4.226]). However, even with the inclusion of steric (*Gustafson* and *Townes* [4.227]) and other corrections (*Piekara* and *Gustafson* [4.228]), the saturated δn is too large to account for the observed δn of $\sim 10^{-3}$. Multiphoton absorption (*Goldberg* et al. [4.226], *Dyshko* et al. [4.86]) is another possibility, but in most materials the nonlinear absorption cross sections are too small to have an appreciable limiting effect on the pulse intensity. *Kelley* and *Gustafson* [4.325] have suggested that backward stimulated Raman scattering could have a limiting effect by depleting the laser pulse energy. However, measurements by *Alfano* and *Zawadzkas* [4.326] have shown that with picosecond pulses the conversion to backward Stokes is too small ($\sim 10^{-4}$) to have any effect. For nanosecond pulses, *Rahn* and *Maier* [4.229] have suggested that forward SRS can limit the filament diameter. At present, the most plausible explanation for the limiting filament diameters is pre-breakdown ionization (*Yablonovitch* and *Bloembergen* [4.179], *Bloembergen* [4.74]). The generation of free carriers by avalanche ionization can make a negative contribution to the refractive index which is an extremely sensitive function of electric field. Estimates of the size of the filament diameters seem to be in fair agreement with experimental observations. Ionization also results in some absorption which can also limit the duration of the filament. Finally, *Shen* [4.144] has cautioned that the limiting filament diameter may be considerably smaller than indicated by the experimental data due to the extreme difficulty of making accurate measurements of this type.

4.5.5 Self-Phase Modulation

Extreme spectral broadening of optical pulses in nonlinear media was first observed from the self-focused filaments of nanosecond pulses (*Bloembergen* and *Lallemand* [4.147], *Brewer* [4.230]). At first, this broadening which extended to as much as 1000 cm^{-1} on the Stokes side was puzzling and difficult to explain, without somehow accounting for an effective lifetime of the filament of only a few picoseconds. The moving focus model, however, provided a plausible explanation which gave good agreement between theory and experiment (*Lugovoi* and *Prokhorov* [4.187], *Loy* and *Shen* [4.189, 190]). For complete details the reader is referred to the recent reviews by *Shen* [4.144] and *Marburger* [4.145].

Shimizu [4.231] observed strong spectral broadening in Kerr liquids with a *Q*-switched ruby laser operating in a partially modelocked regime having amplitude fluctuations of an estimated mean duration of $\sim 10^{-11}$ s. Quasi-periodic spectra were obtained extending in both Stokes and anti-Stokes directions to many hundred wavenumbers. He interpreted his results as due to a rapid time varying phase shift arising from the nonlinear index of refraction.

The Stokes broadening was determined by the leading edge of the "pulse(s)" and the anti-Stokes by the trailing edge. The period of the fine structure within the spectra was related to the mean "pulse" duration (~ 6 ps).

Shimizu's ideas were further elaborated by *Gustafson* et al. [4.128] who made detailed numerical calculations of the spectra of self-phase modulated picosecond pulses, including the effects of dispersion and relaxation of the nonlinearity. The anti-Stokes component of the spectrum was found to be greatly reduced by a relaxation time $\tau \gtrsim \tau_p$. This property arises from the integrating feature of an inertial nonlinearity which produces an index change $\delta n(t)$ which decays more slowly than it builds up. Consequently the magnitude of the instantaneous frequency deviation $\delta\omega \approx -kz\delta n$ is smaller on the trailing portion of the pulse where it is positive than on the leading edge where it is negative. The index change δn is given by the solution of (4.58):

$$\delta n(z,\xi) = \frac{1}{\tau} \int_{-\infty}^{\xi} \frac{1}{2} n_2 A^2(z,\eta) \exp\left(-\frac{\xi-\eta}{\tau}\right) d\eta . \tag{4.61}$$

Dispersion was found to be less important, although in some cases, it can contribute to the formation of shock waves (see next section).

Svelto and coworkers have made extensive measurements of the phase modulated spectra of picosecond pulses in Kerr liquids ([4.199–201] and *Cubeddu* and *Zaraga* [4.203]). Using the complete train of pulses from a mode-locked ruby laser, they observed strong spectral broadening in CS_2, bromobenzene, and toluene similar to Shimizu's results. Although the Stokes component of the spectra was generally stronger, containing most of the energy, the extent of the frequency spread in the anti-Stokes direction was comparable to the Stokes spread and was larger than expected for a simple relaxation model of the orientational Kerr effect. This led them to postulate an alternative model for the nonlinearity with a much faster relaxation time due to molecular librational motion. Similar spectra showing symmetric Stokes and anti-Stokes broadening were also obtained by *Reintjes* et al. [4.206] using single picosecond pulses from a modelocked Nd:glass laser. They suggested that four-wave parametric mixing of Stokes and laser light was responsible for the anti-Stokes component rather than a librational motion. Other similar measurements in Kerr liquids have also been made by *Bol'shov* et al. [4.232], and *Il'ichev* et al. [4.233].

Loy and *Shen* [4.89] have shown that the symmetric feature of the self-phase modulated spectra can be adequately explained by the relaxation model of self-focusing, without the need for introducing additional physical mechanisms. The plane wave analysis of *Gustafson* et al. was found to be inadequate since it did not account for additional phase variations due to the propagation of the self-focused pulse in the neck of the dielectric channel (Fig. 4.18). In this region, on the trailing edge of the pulse, the axial component of the propagation vector is rapidly changing due to the purely geometric factor of propagation in a horn-shaped waveguide so that the phase shift can also vary rapidly even though δn is not. This accounts for the upshifted Stokes component. A numerical calculation of a spectrum based on this model is shown in Fig. 4.19. The general features are in good agreement with experimental results.

An alternative explanation for the anti-Stokes broadening has been made by *Bloembergen* [4.74] who suggested that avalanche ionization could produce a negative contribution to the refractive index, causing δn to decrease on the trailing edge of the pulse. Because the growth of the plasma density increases vary sharply, this effect could produce a rapid phase change and correspondingly large frequency upshift.

In glasses, crystalline solids, non-Kerr fluids, and other materials where the relaxation of the nonlinear index is very fast, the self-phase modulated spectra are generally quite similar to the spectra from Kerr liquids. The spectra usually show strong Stokes and anti-Stokes broadening but tend to be smoother without a well-defined periodic substructure. *Alfano* and *Shapiro* [4.211] observed phase modulated spectra extending to a few thousand wavenumbers in both Stokes and anti-Stokes directions in calcite, NaCl, and various glasses. They also obtained similar results in liquid argon and in solid and liquid krypton (*Alfano* and *Shapiro* [4.160]). In a later work (*Alfano* et al. [4.234]), they made detailed measurements of a broader range of materials, including CCl_4 and liquid nitrogen. The extent of the Stokes and anti-Stokes spectra were equal to within 20% and the energy in each component was also approximately equal. Most of their data was taken with a prism spectrometer

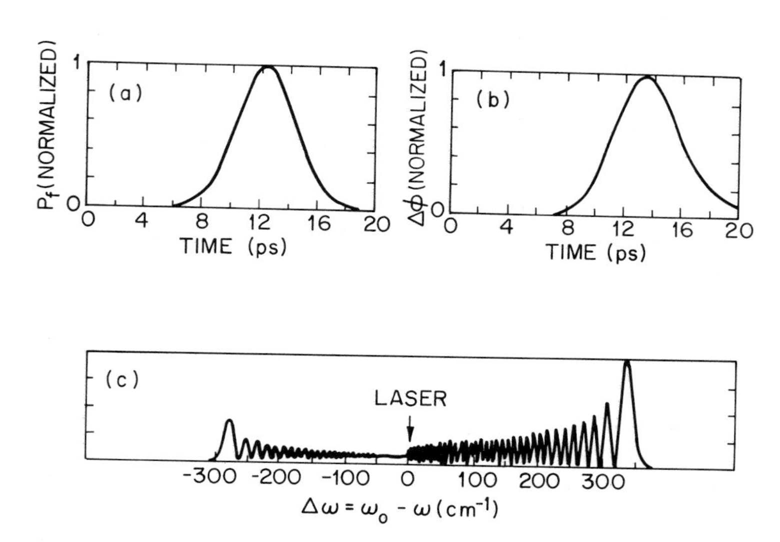

Fig. 4.19 a-c. Theoretical power spectrum due to self-phase modulation of a picosecond pulse in a Kerr liquid including the effects of transient self-focusing. (a) laser pulse shape (b) phase-shift (c) power spectrum (from *Loy* and *Shen* [4.89])

and a thin wire spatial filter to block the unmodulated laser signal, which was typically 10^2 to 10^3 times brighter than the broadened spectrum. All of the light in the phase modulated spectra emanated from self-focused filaments. The researchers concluded that the electronic nonlinearity was the dominant mechanism in all the materials which they studied. Calculations of the spectra based on this assumption were in reasonably good agreement with their results (*Alfano* et al. [4.234]). Other physical mechanisms however can also contribute. In glasses, a contribution from phase-matched four photon parametric mixing was observed when the emission angles of the anti-Stokes spectrum was resolved (*Alfano* and *Shapiro* [4.197] (see also Subsec. 4.3.2 for a complete discussion). *Shen* [4.144] has also pointed out that the extremely broad spectra could be the result of very short ($\sim 10^{-14}$ s) local transients associated with moving foci. *Bloembergen* [4.74] has suggested that the rapid phase change due to the formation of free electrons by avalanche ionization could also produce large anti-Stokes frequency broadening which is consistent with the results of *Alfano* and *Shapiro* [4.197]. Recent measurements of self-phase modulation in KBr by *Yu* et al. [4.235] also tend to support this model. They observed anti-Stokes broadenings of 4900 cm^{-1} and 3200 cm^{-1}, respectively. *Bloembergen* [4.74] has suggested that the maximum Stokes broadening is limited by the avalanche ionization mechanism which fixes the optical electric field at a maximum value comparable to the breakdown field. Since the plasma does not have an appreciable effect during the rise of the pulse, the maximum positive $\partial n/\partial t$ is limited only by the pulse rise time and the breakdown field. A similar suggestion has been made by *Marburger* [4.236]. It is becoming increasingly clear that the avalanche ionization mechanism of Yablonovich and Bloembergen plays a crucial role in both self-phase modulation and self-focusing as well as optical damage.

The effect of nearby electronic resonances on the electronic nonlinear index of refraction was studied by *Alfano* et al. [4.329]. They recorded self-phase modulated spectra of 0.53 μm picosecond pulses in PrF_3 which has electronic resonances at 0.45 μm and 0.59 μm. The spectra showed significant modifications due to the resonances, as expected from theoretical calculations.

Self-phase modulation in the Nd:glass laser was first observed by *Treacy* [4.207] in the form of a frequency chirp on the pulse. *Duguay* et al. [4.217] have shown that an electronic nonlinear index of refraction in the glass rod can account for the chirp, and can also explain the observed growth of the spectrum from pulse to pulse in the train. Direct observation of the time resolved spectra of successive pulses has been made with streak cameras by *Korobkin* et al. [4.216] and *Eckardt* et al. [4.237]. A review of self-phase modulation in the Nd:glass laser has been written by *Eckardt* et al. [4.238].

Ippen et al. [4.239] have observed self-phase modulation of picosecond pulses in a CS_2-filled glass fiber. Pulses from a modelocked cw dye laser were coupled into a fiber of 7 μm core diameter. Although long fiber lengths of one meter were used, the spectral broadening was not large due to the relatively low peak power of the pulses, and waveguide dispersion.

An important application of self-phase modulation of picosecond pulses is its use as a "white-light" probe for picosecond spectroscopy. The first measurement of this type was made by *Alfano* and *Shapiro* [4.240] who used the spectral broadening in glass to generate a probe to observe induced absorption in liquids and solids due to the inverse Raman effect. *Busch* et al. [4.241] used multiple probe beams, each spectrally broadened in a water cell and incrementally delayed to provide a capability for picosecond spectroscopy with a single laser shot. For the details of these and other applications of self-phase modulation, the reader is referred to the chapter on measurement techniques.

The extremely broad emission spectra produced by the propagation of intense picosecond pulses through water have received considerable attention. *Werneke* et al. [4.242] observed spectral broadening of as much as 14,000 cm^{-1} with picosecond pulses from a modelocked ruby laser. They attributed the broadening to self-phase modulation and self-steepening. They observed that in most other liquids for which the threshold for stimulated Raman scattering is lower, the spectral broadening was much reduced. *Busch* et al. [4.241] and *Varma* and *Rentzepis* [4.243, 244] have made mesurements of the spectral broadening in water by 0.53 μm picosecond pulses. Using an ultrafast light gate (*Duguay* and *Hansen* [4.208] and multibeam echelon (*Topp* and *Rentzepis* [4.245]), they measured the duration of the broadband emission and found it was comparable to the pump pulse duration. Single pulses were used by *Busch* et al. [4.241]. They attributed the spectral broadening to a combination of stimulated Raman and Rayleigh scattering, self-phase modulation and self-steepening. *Penzkofer* et al. [4.71] have pointed out that the magnitude of the nonlinear index of refraction in water is too small to account for the spectral broadening by self-phase modulation, and have proposed that it is due instead to four-photon parametric emission (see Subsec. 4.3.2 for details).

4.5.6 Self-Steepening of Picosecond Pulses

In a material with a nonlinear index of refraction, an intense optical pulse can sustain dramatic changes in its shape in the direction of propagation and can develop a shock wave. This phenomenon, known as self-steepening, was first discussed by *Landauer* [4.246] in regard to electrical waves on transmission lines, and was applied to optical waves by *Joenk* and *Landauer* [4.247] and by *DeMartini* et al. [4.248]. The formation of shock fronts arises from an intensity-dependent velocity of propagation which if $n_2 > 0$ causes the peak of a pulse to be retarded relative to the leading and trailing portions so that the pulse distorts to a triangular shape with a sharp trailing edge. The distance required to form the shock was estimated by *DeMartini* et al. to be approximately

$$z_s \simeq 0.19\ \tau_p v_0 \left(\frac{\delta n}{n_0}\right)^{-1} \tag{4.62}$$

for a gaussian input pulse, where τ_p is the pulse duration, $v_0 = c/n_0$, and $\delta n/n_0$ is the peak refractive index change. Estimates for typical picosecond pulses of duration 5 ps and intensities of 10 GW/cm^2 ($\delta n \approx 10^{-3}$) suggest shock distances of ~ 30 cm in CS_2. *Haus* and *Gustafson* [4.249] have obtained approximate analytic solutions for the propagation of steady-state shocks in ideal Kerr liquids with no dispersion and infinitely fast relaxation of the nonlinear index.

In dispersive media, *Ostrovskii* [4.250] has emphasized that a finite relaxation of the nonlinear index is necessary to account for the existence of shocks. In this circumstance, the formation of the shock is due to a distinctly different mechanism than originally proposed by *DeMartini* et al. The action of self-phase modulation causes different portions of the pulse waveform to propagate at slightly different velocities due to group dispersion. In some cases the result will be a concentration of energy at a particular point and a shock is formed. This concept was proposed by *Fisher* et al. [4.251] as a method of compressing optical pulses to shorter durations (see also Chap. 3 for a discussion of applications of pulse compression). Numerical calculations by *Shimizu* [4.252] including the effects of dispersion and finite relaxation show the growth of a shock on the leading edge of a one picosecond pulse in CS_2. *Fisher* and *Bischel* [4.253] have made similar, but more extensive, calculations of shock formation in CS_2 and glass. They have also derived an approximate formula for the shock distance:

$$z_s \approx c\tau_p \Big/ \sqrt{\lambda^2 \frac{\partial^2 n}{\partial \lambda^2} \delta n_{\max}} \,. \tag{4.63}$$

In most circumstances this shock distance is considerably shorter than the corresponding expression (4.62) derived by *DeMartini* et al. Both this expression and (4.62) are based on plane wave analysis and do not take into account the possible complications of self-focusing.

The existence of optical shock waves in media with a nonlinear refractive index has yet to be demonstrated experimentally. The conditions required to observe these effects are unfortunately difficult to attain. Usually shock distances exceed self-focusing distances, and although shocks might exist in the self-focused filaments, their direct measurement is a nontrivial matter, especially with picosecond pulses. The numerical calculations of pulse propagation with self-focusing discussed in the previous section do suggest the existence of pulse distortion and possible shock formation. The extent of the self-steepening, however, is small and is generally considered to be of lesser importance in self-focusing.

4.6 Optical Damage

An experimentalist working with high intensity picosecond pulses very quickly develops a keen appreciation for the limitations imposed by optical damage. Its avoidance is essential for the generation and manipulation of optical pulses and for the successful observation of weak nonlinear effects in materials. Considering the greater propensity for picosecond pulses to produce damage, it is unfortunate that more work has not been devoted to this topic. Undoubtedly, the reason for this situation is the difficulty of meeting the demanding requirements of accurate calibration of the spatial and temporal properties of picosecond laser pulses. Recently, however, some very well calibrated measurements have been made with modelocked Nd:YAG lasers.

Before discussing the specifics of damage with picosecond pulses, it is worthwhile to summarize the basic features of optical damage derived from studies with longer pulses. (For complete details and references the reader is referred to the review by *Bloembergen* [4.254], and to the set of articles in the April 1973 issue of Applied Optics and references therein.) There are four basic types of optical damage:

1) Simple thermal damage in strongly absorbing materials.
2) Damage arising from absorption at inclusions, or other impurities or bulk defects.
3) Surface damage due to dirt, cracks, scratches or other surface imperfections.
4) Intrinsic bulk damage in transparent materials due to avalanche or multiphoton ionization – usually accompanied by self-focusing.

The thresholds for simple thermal damage in uniformly absorbing materials are merely determined by the maximum permissible temperature rise for the associated strain to cause fracturing or to melt the material. Typical maximum energy densities are a few tens of joules per cm^3. Inclusions, such as submicron metallic particles and other impurities in solid materials, cause damage by local heating due to absorption by the impurities. Cleaning and polishing can greatly increase the threshold for surface damage and can in many cases approach the bulk threshold. Cracks and other surface irregularities are important since the electric field at sharp points can be greatly enhanced. The threshold for bulk damage in transparent materials which are free of inclusions and other defects results from the generation of a hot plasma by electron avalanche ionization. Generally, this is the most important mechanism for picosecond pulses.

The avalanche ionization model of optical damage (*Yablonovich* and *Bloembergen* [4.179]) is essentially an extrapolation to optical frequencies of the theory of dc dielectric breakdown (*von Hippel* [4.255, 256]). The basic physical mechanism is the heating of free electrons by the electric field to the point where they gain sufficient energy to ionize more free electrons which are similarly heated to produce more, etc. To gain energy from an optical field the electronic motion must be heavily damped so that an in-phase (absorbing) component of the electronic current results; otherwise the interaction is purely reactive and only alters the index of refraction. This means that the electronic momentum relaxation time τ must be less than one optical cycle (i.e., $\omega\tau \lesssim 1$). *Yablonovich* and *Bloembergen* [4.179] and *Bloembergen* [4.254] point out that τ can be as short as 10^{-15} s for energetic electrons in most solid materials (see also *Seitz* [4.257]). This suggests that the threshold fields for ac breakdown should be comparable to the dc breakdown fields all the way up to visible frequencies. In general the breakdown field at frequency ω, $E_b^{rms}(\omega)$ is expected to have the following relationship to the dc breakdown field:

$$E_b^{rms}(\omega) = E_b^{dc}\sqrt{1+\omega^2\tau^2}. \tag{4.64}$$

Measurements of the optical breakdown field in alkali-halide crystals at 10.6 μm, 1.06 μm and 0.69 μm with nanosecond pulses were found to be comparable to the dc breakdown field in each material (*Yablonovich* [4.258], *Fradin* and *Bass* [4.259]), suggesting that $\omega\tau \lesssim 1$ at least up to frequency of the ruby laser. Time resolved measurements of optical damage in NaCl with *Q*-switched ruby pulses showed an abrupt decrease in transmission at the instant of damage, suggesting an extremely rapid buildup of the electron avalanche (*Fradin* et al. [4.260]). The rate of increase of the free electron density N is determined by the equation

$$\frac{\partial N}{\partial t} = \alpha(E)N - \left(\frac{\partial N}{\partial t}\right)_{\text{losses}} \tag{4.65}$$

where $\alpha(E)$ is the avalanche ionization rate. The losses are due to recombination, trapping, and diffusion, and in most cases are relatively slow processes and can be neglected. The solution of (4.65) is then

$$N(t) = N_0 \exp\left\{\int_{-\infty}^{t} \alpha[E(t')]dt'\right\}. \tag{4.66}$$

The initial electron density N_0 is responsible for starting the avalanche process. *Bloembergen* [4.254] estimated that in most solids at room temperature N_0 is at least 10^8 electrons/cm^3 due to photoionization or thermal activation of shallow donor states. Even for this density, the number of free electrons in a small focal volume of a few microns in diameter could be only 1 to 10. With such small numbers, a statistical fluctuation in the occurrence of damage might be expected. *Bloembergen* [4.254] has suggested this might account for the probabilistic interpretation of optical damage of *Bass* and *Barrett* [4.261]. In those cases where the initial number of electrons is negligibly small, other mechanisms may be responsible for the initiation of the avalanche process such as multiphoton ionization, which in the high frequency limit is equivalent to Zener tunneling. *Bloembergen* [4.254] estimates that a final electron density of $\sim 10^{18}$ cm^{-3} is necessary to produce damage. At this density the hot plasma has acquired an energy sufficient to fracture the material.

It is clear from (4.66) that the breakdown threshold will depend on pulse duration. For a shorter pulse duration, a larger ionization rate α and hence a greater field E_{max} will be required to produce the same multiplication N/N_0 (typically $\sim 10^{10}$). The specific functional dependence of α on E will determine the sensitivity to pulse duration. *Fradin* et al. [4.262] and *Smith* et al. [4.263] have made measurements of the breakdown threshold field in alkali-halides at 1.06 μm with pulses of duration from 15 ps to 10 ns. The results for NaCl are shown in Fig. 4.20. The theoretical curve is based on an empirical form for $\alpha(E)$ derived from dc breakdown data. *Holway* and *Fradin* [4.264] have made detailed calculations of electron avalanche breakdown using classical models based on a diffusion approximation to the Boltzmann equation. Good agreement is found with experiments for NaCl.

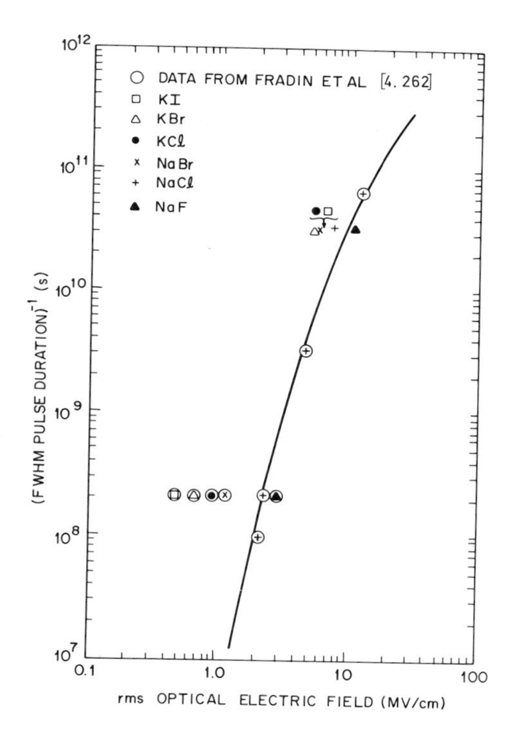

Fig. 4.20. Functional relationship between the breakdown threshold electric field and pulse duration. The solid curve is a semi-empirical prediction based on the avalanche ionization model of optical breakdown (from *Smith* et al. [4.263])

Smith et al. [4.263] and *Smith* and *Bechtel* [4.265] have also measured damage thresholds for picosecond pulses in a wide range of optical materials. They used a modelocked Nd:YAG laser with a single TEM_{00} mode having a FWHM pulse duration of 30 ps. Particular care was taken to accurately calibrate the laser and to account for self-focusing in the tested materials. Their results are shown in Table 4.2. Breakdown fields vary from 3.4 MV/cm for RbI to 22.28 MV/cm for KDP.

Table 4.2. Optical breakdown field strengths (from *Smith* et al. [4.263, 265])

Material	E_b (MV/cm)	Material	E_b (MV/cm)
NaF	10.77	ED-4	9.90
NaCl	7.34	YAG:Nd	9.82
NaBr	5.67	Fused SiO_2	11.68
KF	8.34	CaF_2	14.44
KCl	5.86	KDP	22.28
KBr	5.33	$La_2Be_2O_5$:Nd	15.6
KI	5.87	LHG-5:Nd	13.0
LiF	12.24	LHG-6:Nd	13.0
RbI	3.40	Al_2O_3	12.9

At shorter wavelengths, it is expected that multiphoton ionization is more important than avalanche ionization, but there are not yet sufficient experimental data to determine at what frequency the crossover between the two effects occurs (*Bloembergen* [4.254]).

Breakdown of gases by intense picosecond pulses has been studied extensively [4.342–345]. Measurements with nanosecond pulses suggest a cascade theory of ionization, with the breakdown threshold being strongly dependent on pressure. With picosecond pulses, the mechanism is less clear. Although theory [4.346] suggests a multiphoton ionization process with a weak pressure dependence, this model has not been confirmed. Experiments generally show a significantly longer threshold for decreasing both gas pressure and pulse duration.

4.7 Coherent Pulse Propagation

As a general rule, coherent pulse propagation effects occur only when the pulse duration is much shorter than the phase memory time (dephasing time) of the particular atomic or molecular system involved. On this basis, one might expect that picosecond pulses would be ideal for the observation of coherent propagation effects. However, as a careful reading of the reviews on this subject by *Slusher* [4.116] and *Courtens* [4.118] shows, experiments of this type require extremely well calibrated and "clean" optical pulses. Also the wavelength of the pulses should closely match the particular material resonance of interest (preferably a relatively strong, allowed, nondegenerate transition). Although Raman echoes are an exception to this requirement, they have yet to be observed even with nanosecond pulses. Considering the amplitude and pulsewidth fluctuations, the excess bandwidth and fixed wavelength of the modelocked Nd:glass and ruby lasers, it is clear that they are not ideal sources for this type of experiment. Modelocked dye lasers would appear to be better choices, although as yet they have not seen applications to this topic. Despite these exacting requirements, a number of experiments involving attempts to observe coherent pulse propagation effect have been made, some successfully, and others with rather dubious results.

4.7.1 Coherent Birefringence Echoes in Kerr Gases

In the opinion of this reviewer, the most interesting coherent effect observed to date with picosecond pulses is the experiment of *Heritage* et al. [4.266] in which the birefringence of CS_2 gas was found to exhibit delayed echoes following excitation with an intense picosecond pulse. This novel effect which had not been previously observed was first suggested in a theoretical proposal by *Lin* et al. [4.267]. The effect is based on the coherent rotational motion of anisotropic molecules in a gas. The dephasing time is determined by the collision rate and can be quite long, as much as a few hundred ps at moderate pressures. In the absence of an optical field, the random rotational motion of the individual molecules can be represented by the occupation probabilities of a set of stationary states with energies

$$\hbar\omega_J = \frac{\hbar^2}{2I} J(J+1), \tag{4.67}$$

where J is the rotational angular momentum quantum number, and I is the moment of inertia of the molecule (see for example *Townes* and *Schawlow* [4.268]). The energies of each rotational state relative to the ground state depend quadratically on J, and hence the energy differences between successive states are harmonically related, i.e.,

$$\Delta\omega_J = \omega_{J+1} - \omega_J = \frac{\hbar}{I}(J+1). \tag{4.68}$$

Superpositions of these states correspond in the semiclassical picture to molecules rotating at frequencies which are harmonics of some fundamental frequency determined by the zero-point rotational energy, $\Delta\omega_0 = \hbar/I$. For nonpolar linear molecules such as CS_2, the odd rotational states do not exist, since by symmetry a rotation of the molecule by 180° is indistinguishable from no rotation. In this instance the fundamental rotational frequency is $\Delta\omega_0 = 4\hbar/I$, which for CS_2 is equal to 2.4×10^{10} Hz equivalent to a rotational period of 38 ps (actually 1/2 a rotation). In equilibrium, then, in the semiclassical picture, the gas consists of an ensemble of molecules, each rotating at some harmonic of $\Delta\omega_0$. The application of an intense optical pulse induces an instantaneous incremental alignment of the anisotropic molecules and a birefringence results due to the Kerr effect. Following the application of the pulse, the molecules continue to rotate and the macroscopic incremental alignment disappears. A short time later, however, corresponding to the fundamental rotational period, the orientation of each molecule will return to their original positions and the macroscopic birefringence reappears as an echo, assuming no collisions have occurred. It will continue to reappear at equal time intervals until the collisions randomize the rotational motion.

In the experiment of *Heritage* et al. [4.266], birefringence echoes in CS_2 were observed at time delays of 38 and 76 ps following excitation with an intense picosecond pulse from a modelocked Nd:glass laser. They used the second harmonic pulses at 0.53 µm to probe the birefringence in a 1 m cell of CS_2 vapor at 325 K. Their experimental results are shown in Fig. 4.21 where the echoes are clearly visible. The diminished amplitude of the echoes is due to dephasing collisions and interference of overlapping vibrational levels. A complete quantum mechanical analysis of the birefringence echo effect has been made by *Lin* et al. [4.328].

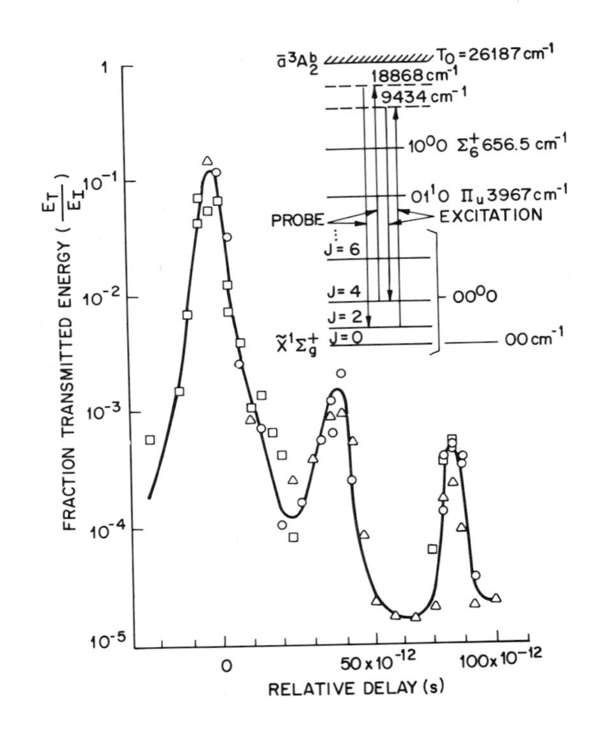

Fig. 4.21. Observation of coherent transient birefringence in CS_2 vapor. Fractional transmission of probe pulse is a measure of time evolution of birefringence following intense pump pulse at $t=0$. Inset shows rotational states of CS_2 molecule (from *Heritage* et al. [4.266])

4.7.2 Self-Induced Transparency

The phenomenon of self-induced transparency was first considered by *McCall* and *Hahn* [4.115], who showed that under special conditions an intense optical pulse could propagate through an absorbing material with negligible attenuation. A coherent interaction with a resonant atomic system can occur whereby energy is absorbed on the leading edge of the pulse and is then reradiated on the trailing edge. If the pulse duration is much shorter than the phase memory of the atomic system, the reradiated energy will be phase coherent with the original pulse. Since energy is stored momentarily in the atomic system, a delay occurs and the effective group velocity of propagation of the pulse can be significantly decreased. *McCall* and *Hahn* [4.115] have shown that a necessary condition for the observation of self-induced transparency is that the pulse "area"

$$\theta=\frac{\mu}{\hbar}\int_{-\infty}^{+\infty}Edt \tag{4.69}$$

be equal to a multiple of 2π, where μ is the dipole matrix element of the transition. Since *McCall* and *Hahn's* original demonstration of self-induced transparency, a considerable body of work on coherent effects has evolved, dealing almost exclusively with nanosecond and longer pulses. The interested reader is referred to the reviews of *Slusher* [4.116] and *Courtens* [4.118].

The original experiments of *McCall* and *Hahn* [4.115], were done in ruby at a temperature near 4 K with a Q-switched ruby laser as a source of pulses. Recently, *Leontovich* and *Mozharovskii* [4.269] have made observations of self-induced transparency in ruby at 105 K with picosecond pulses from a modelocked ruby laser operating at the same temperature. At this temperature, the dephasing time of the R_1 transition in ruby is estimated to be ~ 100 ps, considerably longer than their pulsewidth of 35 ps. In a ruby crystal measuring 11 cm, they observed an increase in transmission from 10^{-16} to 10^{-1} for various pulse areas. The fact that only 10% transmission was observed for a 2π pulse was attributed to scatter, a finite dephasing time, level degeneracy, and poor beam divergence. The authors also observed delays for a 2π pulse of typically 400 ps, although fluctuations made precise measurements difficult.

Self-induced transparency in semiconductors was first proposed by *Poluektov* and *Popov* [4.270] who made calculations of pulse velocities for direct gap semiconductors. *Bruckner* et al. [4.271, 331, 332] have performed experiments in CdS_xSe_{1-x} crystals with the second harmonic pulses from a modelocked Nd:glass laser. Two different compositions of crystals were used, one for direct absorption just above the band gap at 130 K, and the other for resonant absorption with free A excitons. In both cases, increased transmission and pulse delays were observed at high excitation levels. When a weak probe pulse was used to examine the time evolution of the transmission following a strong excitation pulse, the transmission was observed to decay with characteristic time of approximately 50 ps and 150 ps for the band-to-band and exciton cases, respectively. In both cases the authors attribute their results to coherent self-induced transparency. They assumed the carrier dephasing times were larger than the optical pulse duration, a condition that is usually difficult to satisfy in semiconductors. Although the use of resonant excitation helps to reduce the influence of the electron-phonon interaction, impurity and carrier-

carrier scattering are probably more important in the experimental conditions used by *Bruckner* et al. and could readily produce dephasing times of one picosecond or less. A possible alternative explanation for the observed self-induced transparency is an incoherent saturation effect. More extensive measurements in the future will be likely to clarify these effects.

A similar attempt to observe self-induced transparency in semiconductors has also been made by *Gvardzhaladze* et al. [4.272, 273]. In this case, however, 1.06 μm ($\hbar\omega = 1.17$ eV) pulses were used in GaAs ($E_g = 1.4$ eV) so that a one photon interaction was not possible and the intended coherent effect was by two photon transitions (*Belenov* and *Poluektov* [4.274]). Although pulse delays and anomalous two photon absorption at both *RT* and 77 K were apparently observed by the authors, the interpretation of this work is even more difficult than the previous one photon experiment. In this case the optically generated carriers are extremely hot since $\hbar\omega - Eg \sim 0.23$ eV and the momentum relaxation times for electrons and holes are certainly much less than one picosecond (see for example *Seeger* [4.275]). Even if the electron-phonon interaction is suppressed by a strong interaction with the optical field, as suggested by *Aleksandrov* et al. [4.276], it seems unlikely that the carrier dephasing times would be sufficiently long to make plausible an interpretation of these experiments as coherent effects.

4.8 Device Applications

In an early review of picosecond pulses, *DeMaria* [4.277] emphasized the importance of potential device applications. He mentioned a number of possibilities such as optical communications, high-speed electronics, millimeter wave generation, and ultrasonic generation. Since then experimental demonstrations of these and other applications have occurred. However, it is fair to say that as yet, picosecond pulses have not been utilized on a significant scale for industrial or commercial purposes. An important reason for this is the impracticality and low repetition rates of most methods of generating picosecond pulses. However, with the recent technological advances in pulse generation such as the modelocked cw dye laser (see Chap. 2), a renewed interest in this topic will likely develop, and some practical device applications may yet evolve. In that sense, the material covered in this section may provide a background for further developments.

4.8.1 Optical Rectification

Optical rectification is the inverse of the electro-optic effect and produces a dc polarization by virtue of a second order mixing of an intense light beam with itself in acentric crystals. This was predicted theoretically by *Armstrong* et al. [4.4] and was first observed experimentally by *Bass* et al. [4.278]. *Bass* et al.

used crystals of KDP and KD*P which were electroded on the faces perpendicular to their optic axes. A Q-switched ruby laser pulse propagating through the transparent crystals induced a dc voltage across the electrodes of a few hundred microvolts. The dc polarization can be represented by a nonlinear susceptibility of the form

$$P_i(0)=\chi_{ijk}(0=\omega-\omega)\mathscr{E}_j(\omega)\mathscr{E}_k^*(\omega) \tag{4.70}$$

where

$$\chi_{ijk}=-\tfrac{1}{2}n_j^2n_k^2r_{kji} \quad \text{(mks units)} \tag{4.71}$$

and r_{kji} is the electro-optic tensor (see *Ward* [4.279] and *Kaminow* [4.280] for a complete discussion). The intrinsic speed of response is very fast ($\sim 10^{-13}$ s) being limited only by the ionic resonances of the material, and consequently optical rectification has a potential for high speed optical detection and the generation of short electrical impulses. In practice the speed of response of the measurement is usually limited by the RC time constant of the crystal capacitance and load circuit.

Optical rectification of picosecond pulses was first observed by *Brienza* et al. [4.281]. With a spectrum analyzer, they measured rectified signal at frequencies equal to the harmonics of the repetition rate (275 MHz) of a train of 1.06 μm pulses from a modelocked Nd:glass laser. Signals up to ~9 GHz were observed in $LiNbO_3$ and KDP. *Morris* and *Shen* [4.282] proposed that millimeter and far-infrared waves could be produced by optical rectification of picosecond pulses, and pointed out how phase matching could be used to intensify a particular frequency component. *Yang* et al. [4.283] have produced phase matched radiation up to 400 GHz by this method.

A detailed travelling-wave analysis of the generation of the rectified wave due to picosecond pulses has been made by *Gustafson* et al. [4.330]. They included the effects of forward and backward generated waves and boundary reflections. The actual shape of the propagating rectified field is proportional to the derivative of the optical intensity, having both a positive and negative lobe. Usually this effect is not observed since in most measurements the induced charge on the electrodes is measured, which is directly proportional to optical intensity. However, if the current response ($J\equiv\partial P/\partial t$) is measured in a low capacitance geometry a bipolar signal is observed (*Auston* et al. [4.177]). In this case, signals up to 5 volts in amplitude have been produced with a time resolution less than 300 ps. *Gustafson* et al. have also pointed out how the rectified wave can induce self-phase modulation on the optical pulse by the electro-optic effect.

Auston et al. [4.177] have demonstrated the use of two additional physical mechanisms for optical rectification of picosecond pulses. These are the pyroelectric effect and the excited state dipole effect (*Glass* and *Auston* [4.176]) and occur in polar (i.e., pyroelectric) crystals which are either naturally absorbing or have been made absorbing by the introduction of impurities. The rectified polarization produced by these effects is determined by the solution of the equation

$$\frac{\partial P}{\partial t}+\frac{1}{\tau}P=\frac{\alpha\Delta\mu}{\hbar\omega}I(t)+\frac{\alpha}{C_{\mathrm{v}}}\left(\frac{\partial P_3}{\partial T}\right)\frac{(1-\eta)}{\tau}\int_{-\infty}^{t}I(t)dt' \quad (4.72)$$

where τ is the relaxation of the upper absorbing state, α is the linear absorption constant, C_{v} is the specific heat, $\partial P_3/\partial T$ is the pyroelectric coefficient, $I(t)$ is the optical intensity, $\Delta\mu$ is the difference between the dipole moments of the excited and ground states, and η is the radiative quantum efficiency. The first term on the right of (4.72) accounts for the excited state dipole effect, and the second for the pyroelectric effect. Measurements of the rectification of picosecond pulses by these effects in copper-doped $LiTaO_3$ showed them to be considerably stronger than the inverse electro-optic effect in the same material. The speed of response of these effects was determined by using the rectified signal to switch a Pockel's cell (*Auston* and *Glass* [4.284]). Direct observation of the rectified signals showed their duration was pulsewidth limited ($\sim$8 ps) and their amplitude was as much as 300 V.

Roundy and *Byer* [4.285] have used the pyroelectric effect for a fast optical detector of picosecond pulses. They used a black absorbing surface rather than bulk absorption, and consequently the speed of response was slower ($\sim$500 ps) due to the time delay required to transmit the temperature rise to the crystal. The black surface, however, has the advantage of being absorbing over a wide spectral range.

4.8.2 Picosecond Electronics

Elementary electronic operations such as switching, generation of impulses, and linear gating are limited by the response times of conventional semiconductor devices of $\sim 10^{-10}$ to 10^{-9} s. Although special devices such as sampling gates have speeds as fast as 25 ps, their usage is restricted to special circumstances. Picosecond optical pulses offer a unique potential for extending electronics measurement capabilities to the picosecond domain. Recently, a relatively simple technique was demonstrated which permits picosecond optical pulses to control basic electronic functions in high speed semiconductor circuits with a speed which is comparable to the duration of the optical pulses (*Auston* [4.286]). As illustrated in Fig. 4.22, the basic device consists of a thin wafer of high resistivity ($\sim 10^4$ Ωcm) silicon on which a high speed microstrip transmission line has been fabricated by evaporation of aluminum electrodes. A gap in the top electrode prevents the passage of an electrical signal from the input (left) to output (right). The absorption of a picosecond optical pulse at 0.53 µm from a doubled modelocked Nd:glass laser creates a high conductivity at the gap by electron-hole pair generation, enabling the electrical signal to pass. Recombination of the electron-hole plasma is generally too slow and an alternative method must be used to turn off the device. To accomplish this a second picosecond optical pulse at 1.06 μm is absorbed at the gap. For this pulse, however, the absorption depth is much greater (~1 mm) and a region of high conductivity is produced which penetrates through the wafer to the ground plane. This causes a short circuit which terminates the transmission of the device by reflecting all subsequent incoming signals. As illustrated in Fig. 4.22, a time delay between the two controlling optical pulses enables adjustment of the duration of the transmission aperture. When operated with a dc input on the left, the device acts as a pulse generator of variable duration. Alternatively it can be used as a linear gate for arbitrary incoming electrical signals.

The speed of response of the device was measured by correlating the responses of two linear gates in tandem, each having an aperture of ~15 ps, as illustrated in Fig. 4.23. In this instance, the first gate on the left was biased with a dc voltage and acted as a pulse generator, while the second gate was used to sample the incoming signal. Variation of the timing between the generating gate and measuring gate enabled a measurement of the overall response. After accounting for a slight broadening due to dispersion of the

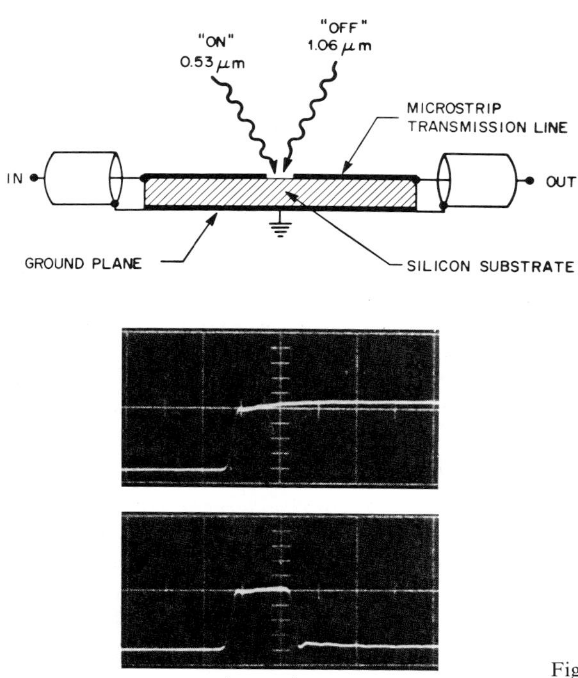

Fig. 4.22. A picosecond electrical switch which is activated by optical pulses (from *Auston* [4.286])

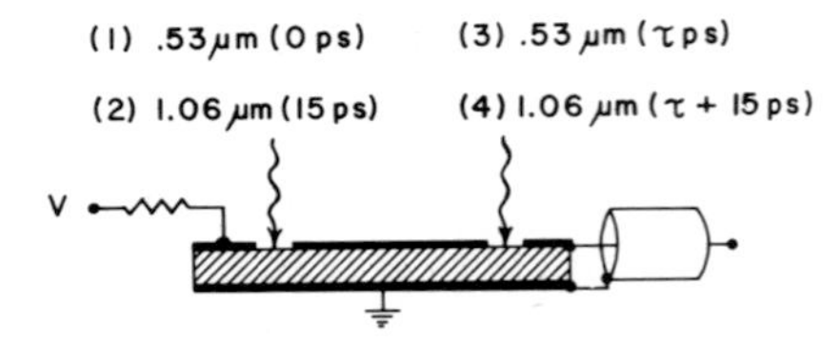

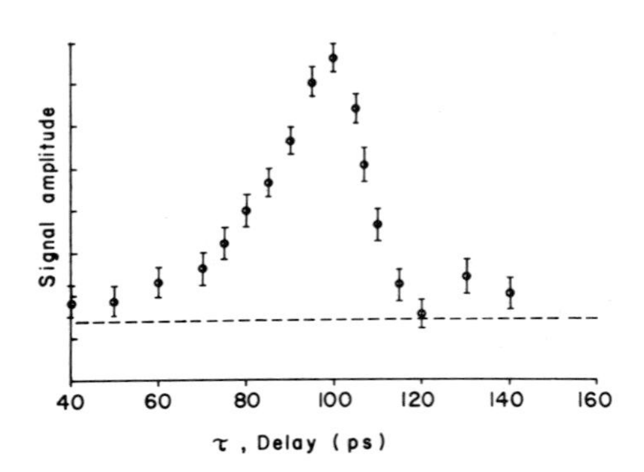

Fig. 4.23. Measurement of speed of switch shown in Fig. 4.22. Two gates in tandem are correlated, the first acting as a pulse generator, and the second as a sampling gate (from *Auston* [4.286])

electrical pulse while propagating between the gaps, it was determined that the risetime of each gap was ~10 ps.

To ensure efficient switching, the net resistances produced in the gap by the optical pulses must be much less than the characteristic impedance of the electrical transmission structure. For typical gap dimensions of 0.2 mm × 0.2 mm, one microjoule of optical energy is adequate to produce resistances of ~1 Ω in a structure with a characteristic impedance of 50 Ω. These devices clearly differ from fast photoconductive detectors which usually operate in a linear, rather than a saturated regime, and rely on carrier sweepout and recombination for speed of response. An example of the latter type of device has been demonstrated by *Lawton* and *Scavannec* [4.287] who observed photoconductive response times ~90 ps in a GaAs detector illuminated by pulses from a modelocked cw dye laser.

Since the device illustrated in Fig. 4.22 must necessarily be very broadband to respond to the extremely fast switching transients (5 ps ≡ 100 GHz), it follows that microwave and millimeter wave signals can also be switched and gated. This has been done with microwave signals of 1 GHz and 10 GHz, for which gated signals as short as a single rf cycle were demonstrated (*Johnson* and *Auston* [4.288]).

With a continuous bias, the maximum operating voltages are limited by thermal dissipation to ~100 V in typical geometries. By using a pulsed input in the nanosecond range, however, they can be used to switch kilovolt pulses.

In a recent experiment by *LeFur* and *Auston* [4.289], voltages up to 1.5 kV were switched by picosecond optical pulses and used to drive a fast Pockels cell. Measurement of the response time of the Pockels cell with a second picosecond optical probe pulse determined a risetime of ~25 ps. The combination of a fast electrical switch and Pockels cell makes an efficient light triggered optical switch. The optical trigger pulse can even be of lower power than the gated optical pulse. It was found that ~5 μJ was adequate to switch 1.5 kV. At these high electric fields the switch operates in a current-limited mode, since the electron and hole drift velocities are saturated. Air breakdown across the gap, rather than intrinsic bulk breakdown in the semiconductor, appears to limit the maximum permissible signal amplitudes. Although avalanche ionization was not evident in the experiments, it could be used to improve the switching efficiency and also possibly to enhance the switching speed.

Current work on this topic has been concerned with the use of smaller geometries with gaps ~15 μm to enable the use of a modelocked cw dye laser for picosecond electronics at high repetition rates (10^5 Hz) and low pulse energies (~10^{-9} J) (*Auston* et al. [4.290]). In addition, the shorter pulses of this laser offer a potential for even faster devices.

The maximum attainable switching speed of these devices is determined by the following four considerations:

1) The duration of the optical pulses.
2) The dielectric relaxation time of the electron-hole plasma.
3) The energy and momentum relaxation times of the carriers.
4) Geometric factors such as the gap size, and dispersion in the electrical transmission structures.

Currently available optical pulses have durations as short as 0.3 ps (see Chap. 2). The dielectric relaxation time of the electron-hole plasma is simply the time required for the plasma to polarize sufficiently to result in the collapse of the electric field in the gap. It is given by the formula $\tau_0 = \varepsilon/\sigma$, where ε is the dielectric constant of the semiconductor substrate, and σ is the conductivity of the electron-hole plasma. Even at the lowest anticipated carrier densities of 10^{17} to 10^{18} cm^{-3}, the dielectric relaxation time is less than 10^{-13} s in silicon. Dielectric relaxation is a macroscopic effect, whereas energy and momentum relaxation are microscopic effects related to the dynamics of individual carriers. Under the influence of an electric field, charged carriers respond in a time equal to the momentum relaxation time, which is determined by electron-phonon coupling and is ~10^{-13} s in silicon at room temperature for carriers in thermal equilibrium with the lattice (see for example *Seeger* [4.275]). Initially, however, the carriers are "hot" due to the excess energy imparted to them by the photons (i.e., $\hbar\omega - \mathscr{E}_g > 0$), and the momentum relaxation times are even faster. Also, the mobility of the carriers is initially small, and then increases as the carriers thermalize. The carrier thermalization time (energy relaxation time) is typically ~5 ps at room temperature in nonpolar semiconductors. This transient is not a serious limitation, however, since the initial mobility, although reduced, can still be adequate for fast switching if the excitation is increased. An interesting application of picosecond electronics would be to measure these times, especially at low temperature where they should be much longer.

The limitations due to geometric factors can in principle be overcome if the devices are made sufficiently small. An approximate rule of thumb is that the substrate thickness and gap dimensions should be much less than the distance an electrical wave propagates during the desired risetime of the device. In silicon this is $\sim 10^{-2}$ cm/ps. Dispersion and radiation losses can also be minimized by the use of smaller geometries. Frequency-dependent absorption due to skin effect, however, becomes worse as the dimensions are reduced. With all the forgoing factors considered, a practical speed limit of ~ 1 ps would seem to be attainable.

The maximum repetition rate of these devices is determined by the decay time of the electron-hole plasma due to sweepout or recombination. In indirect gap materials, such as silicon, low density recombination times are typically 10^{-3} to 10^{-6} s. However, in direct gap materials, such as GaAs, repetition rates up to 1 GHz should be possible, although the average power dissipation of the optical pulses may present a problem. In small gaps, sweepout times can be typically 1 ns for 100 μm.

The need for a second optical pulse with an absorption depth comparable to the substrate thickness is a restriction which severely limits the use of different lasers and materials. One way to overcome this constraint is to use two-photon absorption. Another, simpler, technique is to use only a single wavelength and shallow surface absorption in coupled microstrip structures in which the ground terminal is located on the top surface.

Wiczer and *Merkelo* [4.291] have proposed and demonstrated a picosecond optical detector which uses ultrashort electrical pulses to sample the photocurrent produced by an optical pulse in a vacuum photodiode. The photodiode consists of a photoemissive surface and a special anode which together comprise a microstrip transmission structure enabling a sequence of fast electrical sampling pulses to collect the photocurrent. The spacing is chosen to be very small (e.g., 10 to 50 μm) so that drift times are 10 ps or less with sampling voltages of ~ 10 V. A preliminary device designed for 80 ps resolution was constructed and used with modelocked laser pulses.

4.8.3 Holography

Holographic information storage is a promising technique for high density optical memories (for a recent review see *Glass* [4.292]). One of the most attractive systems is pyroelectric crystals in which optically induced refractive index changes (the photorefractive effect) are utilized to write and read reversible phase holograms (*Chen* [4.293]). The physical basis for the photorefractive effect is the displacement of photoexcited free electrons along the polar axis of the crystal to metastable trapping sites. The resulting space-charge field produces an index change by the electro-optic effect which is proportional to optical intensity. One of the difficulties of the use of the photorefractive effect for volume holographic storage, however, is that recording and reading must be done with the same wavelength. Consequently the stored information tends to be erased during readout.

An approach which solves the erasure problem has been demonstrated by *von der Linde* et al. [4.294]. They used picosecond optical pulses to write holograms by two photon absorption (Fig. 4.24). The use of relatively intense pulses from a modelocked Nd:glass laser enabled them to write holograms with excellent efficiency, while permitting nonerasable readout with lower intensity beams. Both the second harmonic ($2\omega + 2\omega$) alone, and the second harmonic plus the fundamental ($\omega + 2\omega$) were used for two photon absorption to the conduction band of $LiNbO_3$. Low intensity beams at either ω or 2ω had negiligible absorption and could be used for readout. Diffraction efficiencies of 25% were obtained with ~ 0.4 J/cm^2 exposure. Erasure could be accomplished by exposure with a high intensity spatially uniform second harmonic beam.

In another experiment, they demonstrated a greatly enhanced writing efficiency using two photon absorption of picosecond pulses in a potassium tantalate niobate (KTN) crystal. The application of a dc electric field of ~ 5 kV/cm resulted in recording sensitivities of 100 μJ/cm^2, for a few percent reconstruction efficiency, which is comparable to typical holographic silver halide emulsions. The enhanced writing efficiency was attributed to the greater drift distances of the photoexcited free electrons in KTN under the influence of a high electric field. *Von der Linde* et al. [4.295] have also reported the use of two-step excitation via long-lived real intermediate states. This approach has the same advantages of two-photon absorption, but does not require extremely intense pulses for writing.

Holographic recording on thin bismuth films with picosecond pulses has been demonstrated by *Fourney* and *Barker* [4.296] and *Olsen* [4.297]. Fourney and Barker used modelocked Nd : glass. The extremely short coherence

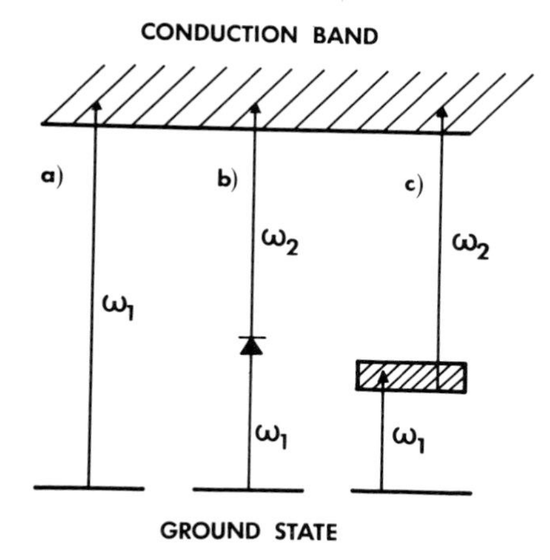

Fig. 4.24. Comparison of different excitation mechanisms for photorefractive recording. (a) Single photon absorption: readin, readout and erasure with the same frequency ω_1. (b) Two-photon absorption via a virtual intermediate state: readin with frequencies ω_1 and ω_2, readout at ω_1 (or ω_2). (c) Two-step absorption via a real intermediate level: readin and erasure with ω_1 and ω_2, readout at ω_2. Note: when using two photon or two step absorption with $\omega_1 \neq \omega_2$, three beams must be used for readin, with two at the same frequency to provide interference (from *von der Linde* et al. [4.295])

length of their laser (~32 μm) prevented the interference of reference and object beams over large areas. Illumination of particular regions of the exposed film was found to produce partial reconstruction of specific points in the object. *Olsen* [4.297] was able to make large area holograms by using transform limited pulses from a modelocked Nd:glass laser of duration ~20 ps. Holographic gratings were recorded with spatial frequencies of 1730 lines/mm having a diffraction efficiency of 5% for an exposure of 50 mJ/cm^2. The physical mechanism of recording is the simple evaporation of the bismuth film due to heating. The highly nonlinear response of this process produced gratings with square wave profiles which had as many as 15 diffraction orders.

4.8.4 Optical Communications

Binary On-Off pulse code modulation (PCM) is generally regarded as one of the best methods of transmitting information in optical communication systems (for a review, see *Hoverstein* [4.298]). On this basis one might expect picosecond pulses would be ideal for use in high information rate systems. In principle, a train of picosecond pulses of ~5 ps duration and ~10 ps spacing could be gated On (binary: *1*) and Off (binary: *0*) to provide an information rate of ~10^{11} bits per second, capable of transmitting ~10^8 simultaneous telephone conversations. Unfortunately, there are a number of problems with this approach. A basic difficulty is the requirement that a practical optical communication system must interface with compatible electronic systems capable of handling the same information rate at both the transmitting and receiving ends. High speed transducers such as electro-optic modulators and optical detectors are necessary to interface optics and electronics. Although experimental electro-optic modulators have been demonstrated with speeds as high as ~10 ps (*Auston* and *Glass* [4.284], *LeFur* and *Auston* [4.289]), practical modulation systems are generally limited to speeds of a few tenths of a nanosecond (for a review, see *Kaminow* [4.280]). The problem with electro-optic modulators is not an intrinsic speed limitation, but rather it is their relatively weak interaction with the optical field which necessitates the use of large drive signals, which are difficult to generate at high speeds and produce a high average power dissipation. For demodulation, fast optical detectors are required (for a review, see *Melchior* et al. [4.299]). The best risetimes of silicon *p-i-n* diodes are ~50 ps. Although improved silicon diodes and point contact metal-insulator metal diodes (see for example *Faris* et al. [4.300]) might have speeds of ~10 ps, practical devices with measured risetimes in this range have yet to be demonstrated. Even with fast modulators and detectors, it is not clear that the electronic circuitry on each end could handle such rapidly varying signals.

An approach which relaxes some of these difficulties is time division multiplexing (*Kinsel* [4.301]). The essence of this technique is to use a number of separate pulse trains, each having identical low repetition rates. Modulation of each channel (train) can then be accomplished with relatively low speed electro-optic modulators. The trains are then optically mixed into a single beam, each train being incrementally delayed in time to avoid overlap. The number of separate trains of pulses of duration τ and separation T that can be multiplexed in this manner is approximately $N = T/2\tau$. Before detection, the multiplexed optical beam is first demultiplexed into separate beams by a sequence of electro-optical modulators. These modulators separate the beams by polarization switching. They are driven synchronously and are timed to operate at multiples of the multiplexed pulse period T/N. First, every second pulse is switched out to generate two beams, then every second pulse in these beams is switched out to generate four, etc. Following the demultiplexing operation, each beam is demodulated in a separate low speed detector.

Kinsel and *Chen* [4.302] have demonstrated a time division multiplexing system which utilizes a modelocked Nd:YAG laser for a source of optical pulses. A $LiNbO_3$ internal phase modulator produced pulses of duration 34 ps, spaced 3.57 ns. The system was designed to multiplex a total of 24 PCM channels for an information rate of 6.7 Gbit/s. In the multiplexed train the pulse spacing was 150 ps. Although the complete system of 24 channels was never completed, a system employing the first two channels clearly demonstrated the feasibility of this approach. For a discussion of similar approaches to optical communications systems and applications to space communications, the reader is referred to the review by *Ross* [4.303].

Current approaches to optical communications systems emphasize the use of optical fibers as the transmission medium (for reviews, see *Miller* et al. [4.304] and *Li* [4.305]). The small size and low cost of glass fibers enables the use of fiber bundles for the transmission of high information rates, without the necessity for high speed modulation and demodulation. In addition, dispersion in optical fibers limits the use of picosecond pulses (*Gloge* et al. [4.306]). Although graded index fibers show considerably reduced dispersion, pulse spreading rates of ~ 150 ps/km remain (*Cohen* et al. [4.307]). Except possibly for short links, it is not considered likely that picosecond pulses will see widespread use in optical fiber communication systems.

4.8.5 Other Applications

Brienza and *DeMaria* [4.308] have demonstrated the generation of ultrashort acoustic pulses by picosecond optical pulses. They directed a train of picosecond optical pulses from a modelocked Nd:glass laser at a thin metal film on a crystalline bar. Partial absorption of the laser pulses in the film resulted in a fast rising thermal stress which produced an acoustic wave which propagated through the bar. At the other end a piezoelectric transducer detected the acoustic pulses. They estimated the risetime of the acoustic wave to be significantly less than 0.5 ns. Similar results were obtained by *Peercy* et al. [4.309] and *Jones* [4.310]. They used single, amplified pulses of ~ 20 J. Targets were X-cut quartz transducers with one of the metal film electrodes acting as the target material. The peak acoustic pressures were ~ 5 kbar. The generation technique was destructive, however, and required a fresh metal film for each shot.

Basov et al. [4.311] have developed a system of optical logic devices based on transient instabilities in semiconductor lasers. Their basic logic element is an injection diode laser which is split into two or more components. One of these is biased with a current which is adequate to meet the threshold for lasing; the other is either not biased at all or has a low bias and acts as a saturable absorber. This results in a bistable mode, either On or Off, which can be switched rapidly by the injection of a fast light or current pulse. Potential switching speeds are estimated to be as fast as 10 ps. Basic logic devices performing the functions of AND, OR, NOT, and others have been demonstrated. For a review, see *Basov* et al. [4.312].

References

4.1 N. Bloembergen: *Nonlinear Optics* (Benjamin, New York 1965)
4.2 F. T. Arecchi, E. O. Schulz-Dubois: *Laser Handbook* (North Holland, Amsterdam 1972)
4.3 H. Rabin, C. L. Tang: *Quantum Electronics* (Academic Press, New York 1975)
4.4 J. A. Armstrong, N. Bloembergen, J. Ducuing, P. S. Pershan: Phys. Rev. **127**, 1918 (1962)
4.5 C. Flytzanis: In *Quantum Electronics*, Vol. I, part A, chap. 2, ed. by H. Rabin, C. L. Tang (Academic Press, New York 1975)
4.6 P. Franken, A. Hill, C. Peters, G. Weinreich: Phys. Rev. Letters **7**, 118 (1961)
4.7 D. A. Kleinman: In *Laser Handbook*, Vol. 2, Chap. E4, ed. by F. T. Arecchi, E. O. Schulz-Dubois (North Holland, Amsterdam 1972)
4.8 S. K. Kurtz: In *Quantum Electronics*, Vol. 1, part A, Chap. 3, ed. by H. Rabin, C. L. Tang (Academic Press, New York 1975)
4.9 R. Bechman, S. K. Kurtz: In *Landolt-Börnstein, New Series*, Group III: Crystal and Solid State Physics, Vol. 2, Chap. 5 (Springer-Verlag, Berlin, New York 1969)
4.10 S. Singh: *Handbook of Lasers*, Chap. 18, ed. by R. J. Pressley (Chemical Rubber Company, Cleveland 1971)
4.11 J. A. Giordmaine: Phys. Rev. Lett. **8**, 19 (1962)
4.12 P. D. Maker, R. W. Terhune, M. Nisenoff, C. M. Savage: Phys. Rev. Lett. **8**, 21 (1962)
4.13 R. C. Miller, G. D. Boyd, A. Savage: Appl. Phys. Lett. **6**, 77 (1965)
4.14 R. C. Miller: Phys. Lett. **26 A**, 177 (1968)
4.15 J. Comly, E. Garmire: Appl. Phys. Lett. **12**, 7 (1968)
4.16 W. H. Glenn: IEEE J. QE-**5**, 284 (1969)
4.17 S. L. Shapiro: Appl. Phys. Lett. **13**, 19 (1968)
4.18 S. A. Akhmanov, A. P. Sukhorukov, A. S. Chirkin: Sov. Phys. JETP **28**, 748 (1969)
4.19 R. U. Orlov, T. Usmanov, A. S. Chirkin: Sov. Phys. JETP **30**, 584; Rws: **57**, 1069 (1970)
4.20 Yu. N. Karamzin, A. P. Sukhorukov: Sov. J. Quant. Electron. **5**, 496 (1975)
4.21 W. F. Hagen, P. C. Magnante: J. Appl. Phys. **40**, 219 (1969)
4.22 G. D. Boyd, D. A. Kleinman: J. Appl. Phys. **39**, 3597 (1968)
4.23 T. A. Rabson, H. J. Ruiz, P. L. Shah, F. K. Tittel: Appl. Phys. Lett. **20**, 282 (1972)
4.24 A. H. Kung, J. F. Young, G. C. Bjorklund, S. E. Harris: Phys. Rev. Lett. **29**, 985 (1972)
4.25 E. P. Ippen, C. V. Shank: *Digest of IX International Quantum Electronics Conf.*, Amsterdam (1976)
4.26 P. D. Maker, R. W. Terhune: Phys. Rev. **137**, 801 (1965)
4.27 P. P. Bey, J. F. Giuliani, H. Rabin: Phys. Rev. Lett. **26 A**, 128 (1968)
4.28 S. E. Harris, R. B. Miles: Appl. Phys. Lett. **19**, 385 (1971)

4.29 I. Freund: Phys. Rev. Lett. **21**, 1404 (1968)
4.30 N. Bloembergen, A. J. Sievers: Appl. Phys. Lett. **17**, 483 (1970)
4.31 J. W. Shelton, Y. R. Shen: Phys. Rev. Lett. **26**, 538 (1971)
4.32 G. D. Boyd, F. R. Nash, D. F. Nelson: Phys. Rev. Lett. **24**, 1298 (1970)
4.33 D. B. Anderson, J. T. Boyd: Appl. Phys. Lett. **19**, 266 (1971)
4.34 R. C. Eckardt, C. H. Lee: Appl. Phys. Lett. **15**, 425 (1969)
4.35 C. C. Wang, E. L. Baardsen: Appl. Phys. Lett. **15**, 396 (1969)
4.36 J. W. Shelton, Y. R. Shen: Phys. Rev. Lett. **25**, 23 (1970)
4.37 H. N. de Vries: Acta Crystallogr. **4**, 219 (1951)
4.38 N. Bloembergen, W. K. Burns, M. Matsuoka: Opt. Commun. **1**, 195 (1969)
4.39 J. F. Ward, G. H. C. New: Phys. Rev. **185**, 57 (1969)
4.40 R. B. Miles, S. E. Harris: IEEE J. QE-**9**, 470 (1973)
4.41 G. C. Bjorklund: IEEE J. QE-**11**, 287 (1975)
4.42 J. F. Young, G. C. Bjorklund, A. H. Kung, R. B. Miles, S. E. Harris: Phys. Rev. Lett. **27**, 1551 (1971)
4.43 D. M. Bloom, G. W. Bekkers, J. F. Young, S. E. Harris: Appl. Phys. Lett. **26**, 687 (1975)
4.44 A. H. Kung, J. F. Young, S. E. Harris: Appl. Phys. Lett. **22**, 301 (1973)
4.45 S. E. Harris: Phys. Rev. Lett. **31**, 341 (1973)
4.46 M. H. R. Hutchinson, C. C. Ling, D. J. Bradley: *IX International Quantum Electronics Conference*, Amsterdam, June 1976. Post-deadline paper #S-1
4.47 D. M. Bloom, J. F. Young, S. E. Harris: Appl. Phys. Lett. **27**, 390 (1975)
4.48 S. A. Akhmanov, V. A. Martynov, S. M. Saltiel, V. G. Tunin: JETP Lett. **22**, 65 (1975)
4.49 D. H. Auston: Appl. Phys. Lett. **18**, 249 (1971)
4.50 D. H. Auston: Opt. Commun. **3**, 272 (1971)
4.51 P. P. Sorokin, J. R. Lankard: IEEE J. QE-**9**, 227 (1973)
4.52 R. H. Kingston: Phys. Rev. **127**, 1207 (1962)
4.53 N. M. Kroll: Phys. Rev. **127**, 1207 (1962)
4.54 S. A. Akhmanov, R. V. Khokhlov: Sov. Phys. JETP **16**, 252 (1962)
4.55 R. G. Smith: In *Laser Handbook*, Vol. 1, Chap. C8, ed. by F. T. Arecchi, E. O. Schulz-Dubois (North Holland, Amsterdam 1972)
4.56 R. L. Byer: In *Quantum Electronics*, Vol. I, part B, Chap. 9, ed. by R. Rabin, C. L. Tang (Academic Press New York 1975)
4.57 C. L. Tang: In *Quantum Electronics*, Vol. I, part B, Chap. 6, ed. by R. Rabin, C. L. Tang (Academic Press, New York 1975)
4.58 W. H. Glenn: Appl. Phys. Lett. **11**, 333 (1967)
4.59 S. A. Akhmanov, R. V. Khokhlov, A. I. Kovrigin, V. I. Piskarskas, A. P. Sukhorukov: IEEE J. QE-**4**, 829 (1968)
4.60 S. A. Akhmanov, A. S. Chirkin, K. N. Drabovich, A. I. Kovrigin, R. V. Khokhlov, A. P. Sukhorukov: IEEE J. QE-**4**, 598 (1968)
4.61 S. A. Akhmanov, A. I. Kovrigin, A. P. Sukhorukov, R. V. Khokhlov, A. A. Chirkin: JETP Lett. **7**, 182 (1968)
4.62 T. A. Rabson, H. J. Ruiz, P. L. Shah, F. K. Tittel: Appl. Phys. Lett. **21**, 129 (1972)
4.63 K. P. Burneika, M. V. Ignatavichus, V. I. Kabelka, A. S. Piskarskas, Yu. A. Stabnis: JETP Lett. **16**, 257 (1972)
4.64 A. Laubereau, L. Greiter, W. Kaiser: Appl. Phys. Lett. **25**, 87 (1974)
4.65 G. A. Dychyus, V. I. Kabelka, A. S. Piskarskas, Yu. A. Stabnis: Sov. J. Quant. Electron. **4**, 1402 (1975)
4.66 T. Kushida, Y. Tanaka, M. Ojima, Y. Nakazaki: Japan. J. Appl. Phys. **14**, 1097 (1975)
4.67 A. H. Kung: Appl. Phys. Lett. **25**, 653 (1974)
4.68 G. A. Massey, J. C. Johnson, R. A. Elliot: IEEE J. QE-**12**, 143 (1976)
4.69 R. Y. Chiao, P. L. Kelley, E. Garmire: Phys. Rev. Lett. **17**, 1158 (1966)
4.70 R. L. Carman, R. Y. Chiao, P. L. Kelley: Phys. Rev. Lett. **17**, 1281 (1966)
4.71 A. Penzkofer, A. Laubereau, W. Kaiser: Phys. Rev. Lett. **14**, 863 (1973)
4.72 A. Penzkofer, A. Seilmeier, W. Kaiser: Opt. Commun. **14**, 363 (1973)
4.73 A. Penzkofer: Opt. Commun. **11**, 265 (1974)

4.74 N. Bloembergen: Opt. Commun. **8**, 285 (1973)
4.75 P. Braunlich, P. Kelley: J. Appl. Phys. **46**, 5205 (1975)
4.76 E. J. Woodbury, W. K. Ng: Proc. IRE **50**, 2367 (1962)
4.77 G. Eckhardt, R. W. Hellwarth, F. J. McClung, S. E. Schwarz, D. Weiner, E. J. Woodbury: Phys. Rev. Lett. **9**, 455 (1962)
4.78 N. Bloembergen: Am. J. Phys. **35**, 989 (1967)
4.79 W. Kaiser, M. Maier: In *Laser Handbook*, Vol. 2, Chap. E2, ed. by F. T. Arecchi, E. O. Schulz-Dubois (North Holland, Amsterdam 1972)
4.80 C. S. Wang: In *Quantum Electronics*, Vol. I, part A, Chap. 7, ed. by R. Rabin, C. L. Tang (Academic Press, New York 1975)
4.81 G. Placzek: In *Handbuch der Radiologie*, ed. by E. Marx, part II (Akademische Verlagsgesellschaft, Leipzig 1934) p. 209
4.82 M. Maier, W. Kaiser, J. A. Giordmaine: Phys. Rev. **177**, 580 (1969)
4.83 W. Kaiser: Sov. J. Quant. Electron. **4**, 1131 (1974)
4.84 Y. R. Shen, N. Bloembergen: Phys. Rev. **137**, A1786 (1965)
4.85 S. F. Fischer, A. Laubereau: Chem. Phys. Lett. **35**, 6 (1975)
4.86 A. L. Dyshko, V. N. Lugovoi, A. M. Prokhorov: JETP Lett. **6**, 146 (1967)
4.87 M. Maier, W. Kaiser, J. A. Giordmaine: Phys. Rev. Lett. **17**, 125 (1966)
4.88 W. H. Culver, J. T. A. Vanderslice, V. W. Townsend: Appl. Phys. Lett. **12**, 189 (1968)
4.89 M. M. T. Loy, Y. R. Shen: IEEE J. QE-**9**, 409 (1973)
4.90 S. L. Shapiro, J. A. Giordmaine, K. W. Wecht: Phys. Rev. Lett. **19**, 1093 (1967)
4.91 C. G. Bret, H. P. Weber: IEEE J. QE-**4**, 807 (1968)
4.92 E. E. Hagenlocker, R. W. Minck, W. G. Rado: Phys. Rev. **154**, 226 (1967)
4.93 M. A. Bol'shov, Y. I. Golayaev, V. S. Dneprovskii, I. I. Nurminskii: Sov. Phys. JETP **30**, 190 (1970)
4.94 S. A. Akhmanov: Mater. Res. Bull. **4**, 455 (1969)
4.95 S. A. Akhmanov, M. A. Bol'shov, K. N. Drabovich, A. P. Sukhorukov: JETP Lett. **12**, 388 (1970)
4.96 M. J. Colles: Opt. Commun. **1**, 169 (1969)
4.97 R. R. Alfano, S. L. Shapiro: Phys. Rev. A **2**, 2376 (1970)
4.98 R. L. Carman, M. E. Mack, F. Shimizu, N. Bloembergen: Phys. Rev. Lett. **23**, 1327 (1969)
4.99 M. J. Colles: Appl. Phys. Lett. **19**, 23 (1971)
4.100 R. L. Carman, M. E. Mack: Phys. Rev. A **5**, 341 (1972)
4.101 R. L. Carman, F. Shimizu, C. S. Wang, N. Bloembergen: Phys. Rev. A **2**, 60 (1970)
4.102 R. S. Adrain, E. G. Arthurs, W. Sibbett: Opt. Commun. **15**, 290 (1975)
4.103 M. E. Mack, R. L. Carman, J. Reintjes, N. Bloembergen: Appl. Phys. Lett. **16**, 209 (1970)
4.104 M. Chatelet, B. Oksengorn: Chem. Phys. Lett. **36**, 73 (1975)
4.105 R. R. Alfano, S. L. Shapiro: Phys. Rev. Lett. **26**, 1247 (1971)
4.106 D. von der Linde, A. Laubereau, W. Kaiser: Phys. Rev. Lett. **26**, 954 (1971)
4.107 A. Laubereau, D. von der Linde, W. Kaiser: Opt. Commun. **1**, 173 (1973)
4.108 S. A. Akhmanov, K. N. Drabovich, A. P. Sukhorukov, A. S. Chirkin: Sov. Phys. JETP **32**, 266 (1971)
4.109 N. M. Kroll: J. Appl. Phys. **36**, 34 (1965)
4.110 C. S. Wang: Phys. Rev. **182**, 482 (1969)
4.111 W. H. Lowdermilk, G. I. Kachen: Appl. Phys. Lett. **27**, 133 (1975)
4.112 N. Bloembergen, M. J. Colles, J. Reintjes, C. S. Wang: Ind. J. Pure, Appl. Phys. **9**, 874 (1971)
4.113 K. Daree, W. Kaiser: Opt. Comm. **10**, 63 (1974)
4.114 V. E. Neef: Ann. Phys. **32**, 191 (1975)
4.115 S. L. McCall, E. L. Hahn: Phys. Rev. Lett. **18**, 908 (1967)
4.116 R. E. Slusher: Progress in Optics, Vol. XII, ed. by E. Wolf (North Holland, Amsterdam 1974) p. 55
4.117 B. A. Medvedev: Sov. Phys. JETP **33**, 19 (1971)
4.118 E. Courtens: In *Laser Handbook*, Vol. 2, Chap. E5, ed. by F. T. Arecchi, E. O. Schulz-Dubois (North Holland, Amsterdam 1972)
4.119 K. Shimoda: Z. Phys. **234**, 293 (1970)

4.120 B.A. Medvedev, O.M. Parshkov, V.A. Gorshenin, A.E. Dmitriev: Sov. Phys. JETP **40**, 36 (1974)
4.121 N. Tan-no, T. Shirahata, K. Yokoto: Phys. Rev. A **12**, 159 (1975)
4.122 J.E. Griffiths, M. Clerc, P.M. Rentzepis: J. Chem. Phys. **60**, 3824 (1973)
4.123 A. Laubereau: J. Chem. Phys. **63**, 2260 (1975)
4.124 J.A. Giordmaine, W. Kaiser: Phys. Rev. **144**, 676 (1966)
4.125 E. Garmire: Phys. Lett. **17**, 251 (1965)
4.126 K. Shimoda: Japan. J. Appl. Phys. **8**, 1499 (1969)
4.127 J. Herman: Sov. J. Quant. Electron. **5**, 207 (1974)
4.128 T.K. Gustafson, J.-P. Taran, H.A. Haus, J.R. Lifsitz, P.L. Kelley: Phys. Rev. **177**, 306 (1969)
4.129 R.M. Herman, M.A. Gray: Phys. Rev. Lett. **19**, 824 (1967)
4.130 D.H. Rank, C.W. Cho, N.D. Foltz, T.A. Wiggins: Phys. Rev. Lett. **19**, 828 (1967)
4.131 M.E. Mack: Phys. Rev. Lett. **22**, 13 (1969)
4.132 D. Pohl: Phys. Rev. Lett. **23**, 711 (1969)
4.133 V.S. Starunov: Sov. Phys. JETP **30**, 553 (1970)
4.134 R.I. Scarlet: Phys. Rev. Lett. **26**, 364 (1971)
4.135 S.S. Rangnekar, R.H. Enns: Phys. Rev. A **6**, 1199 (1972)
4.136 I.P. Battra, R.H. Enns, D. Pohl: Phys. Stat. Sol. B. **48**, 11 (1971)
4.137 W. Yu, R.R. Alfano: Phys. Rev. A **11**, 188 (1975)
4.138 R.W. Terhune, P.D. Maker, C.M. Savage: Phys. Rev. Lett. **17**, 681 (1965)
4.139 S. Kielich, J.R. Lalanne, F.B. Martin: Phys. Rev. Lett. **26**, 1295 (1971)
4.140 S.J. Cyvin, J.E. Rauch, J.C. Decius: J. Chem. Phys. **43**, 4083 (1965)
4.141 S. Kielich: Bull. Acad. Pol. Sci. **312**, 53 (1964)
4.142 S.A. Akhmanov, R.V. Khokhlov, A.P. Sukhorukov: In *Laser Handbook*, Vol. 2, Chap. E3, ed. by F.T. Arecchi, E.O. Schulz-Dubois (North Holland, Amsterdam 1972)
4.143 O. Svelto: Progress in Optics, Vol. XII, ed by *E.* Wolf (North Holland, Amsterdam 1974)
4.144 Y.R. Shen: Progr. Quant. Electron. **4**, 1 (1975)
4.145 J. Marburger: Progr. Quant. Electron. **4**, 35 (1975)
4.146 G. Mayer, F. Gires: Compt. Rend. **258**, 2039 (1964)
4.147 N. Bloembergen, P. Lallemand: Phys. Rev. Lett. **16**, 81 (1966)
4.148 D.H. Close, C.R. Giuliano, R.W. Hellwarth, L.D. Hess, F.J. McLung, W.G. Wagner: IEEE J. QE-**2**, 593 (1966)
4.149 P. Debye: *Polar Molecules* (Chemical Catalogue Co., New York 1929), Chap. 5
4.150 E.N. Ivanov: Sov. Phys. JETP **18**, 1041 (1964)
4.151 D.A. Pinnow, S.J. Candau, T.A. Litovitz: J. Chem. Phys. **49**, 347 (1968)
4.152 P.P. Ho, W. Yu, R.R. Alfano: Chem. Phys. Lett. **37**, 91 (1976)
4.153 B. Kasprowicz, S. Kielich: Acta Phys. Pol. **33**, 495 (1968)
4.154 S. Kielich: Chem. Phys. Lett. **2**, 112 (1968)
4.155 M. Paillette: Ann. Phys. (Paris) **4**, 671 (1969)
4.156 S. Kielich: In *Dielectric and Related Molecular Processes*, ed. by M. Davies (The Chemical Society, London 1972) Chap. 5
4.157 K. Sala, M.C. Richardson: Phys. Rev. A **12**, 1030 (1975)
4.158 P.D. Maker, R.W. Terhune, C.M. Savage: Phys. Rev. Lett. **12**, 507 (1964)
4.159 J.P. McTague, C.H. Lin, T.K. Gustafson, R.Y. Chiao: Phys. Lett. **33A**, 82 (1970)
4.160 R.R. Alfano, S.L. Shapiro: Phys. Rev. Lett. **24**, 1217 (1970)
4.161 R.W. Hellwarth, A. Owyoung, N. George: Phys. Rev. A **4**, 2342 (1971)
4.162 A. Owyoung, R.W. Hellwarth, N. George: Phys. Rev. B **5**, 628 (1972)
4.163 J.P. Herman, D. Ricard, J. Ducuing: Appl. Phys. Lett. **23**, 178 (1973)
4.164 C.C. Wang: Phys. Rev. **152**, 149 (1966)
4.165 A. Owyoung: Opt. Commun. **16**, 266 (1976)
4.166 Y.R. Shen: Phys. Rev. Lett. **20**, 378 (1966)
4.167 E.L. Kerr: IEEE J. QE-**6**, 616 (1970)
4.168 E.L. Kerr: IEEE J. QE-**7**, 532 (1971)
4.169 J.P. McTague, G. Birnbaum: Phys. Rev. A **3**, 1376 (1971)

4.170 R.W.Hellwarth: Phys. Rev. **152**, 156 (1966)
4.171 R.W.Hellwarth: Phys. Rev. **163**, 205 (1967)
4.172 R.W.Hellwarth: J. Chem. Phys. **52**, 2128 (1970)
4.173 J.P.Gordon, R.C.C.Leite, R.S.Moore, G.P.S.Porto, J.R.Whinnery: J. Appl. Phys. **36**, 3 (1965)
4.174 A.G.Litvak: JETP Lett. **4**, 230 (1966)
4.175 S.A.Akhmanov, A.P.Sukhorukov: JETP Lett. **5**, 87 (1967)
4.176 A.M.Glass, D.H.Auston: Opt. Commun. **5**, 45 (1972)
4.177 D.H.Auston, A.M.Glass, A.A.Ballman: Phys. Rev. Lett. **28**, 897 (1972)
4.178 N.Tzoar, J.L.Gersten: Phys. Rev. Lett. **26**, 1634 (1971)
4.179 E.Yablonovich, N.Bloembergen: Phys. Rev. Lett. **29**, 907 (1972)
4.180 G.A.Askaryan: Sov. Phys. JETP **15**, 1088 (1962)
4.181 V.I.Talanov: Radiophysics **7**, 254 (1964)
4.182 R.Y.Chiao, E.Garmire, C.H.Townes: Phys. Rev. Lett. **11**, 1281 (1964)
4.183 E.Garmire, R.Y.Chiao, C.H.Townes: Phys. Rev. Lett. **16**, 347 (1966)
4.184 R.Y.Chiao, M.A.Johnson, S.Krinsky, H.A.Smith, C.H.Townes, E.Garmire: IEEE J. QE-**2**, 467 (1966)
4.185 V.I.Talanov: JETP Lett. **2**, 138 (1965)
4.186 P.L.Kelley: Phys. Rev. Lett. **15**, 1005 (1965)
4.187 V.N.Lugovoi, A.M.Prokhorov: JETP Lett. **7**, 117 (1968)
4.188 M.M.T.Loy, Y.R.Shen: Phys. Rev. Lett. **22**, 994 (1969)
4.189 M.M.T.Loy, Y.R.Shen: Phys. Rev. Lett. **25**, 1333 (1970)
4.190 M.M.M.T.Loy, Y.R.Shen: Appl. Phys. Lett. **19**, 285 (1971)
4.191 G.M.Zverev, E.K.Maldutis, V.A.Pashkov: JETP Lett. **9**, 61 (1969)
4.192 V.A.Korobkin, A.M.Prokhorov, R.V.Serov, M.Ja.Shehelev: JETP Lett. **11**, 94 (1970)
4.193 R.G.Brewer, J.R.Lifsitz, E.Garmire, R.Y.Chiao, C.H.Townes: Phys. Rev. **166**, 326 (1968)
4.194 Y.R.Shen, Y.J.Shaham: Phys. Rev. Lett. **15**, 1008 (1967)
4.195 P.Lallemand, N.Bloembergen: Phys. Rev. Lett. **15**, 1010 (1965)
4.196 R.G.Brewer, C.H.Lee: Phys. Rev. Lett. **21**, 267 (1969)
4.197 R.R.Alfano, S.L.Shapiro: Phys. Rev. Lett. **24**, 584 (1970)
4.198 J.P.McTague, P.A.Fleury, D.B.duPre: Phys. Rev. **188**, 303 (1969)
4.199 R.Polloni, C.A.Sacchi, O.Svelto: Phys. Rev. Lett. **23**, 690 (1969)
4.200 R.Cubeddu, R.Polloni, C.A.Sacchi, O.Svelto: Phys. Rev. A **2**, 1955 (1970)
4.201 R.Cubeddu, R.Polloni, C.A.Sacchi, O.Svelto, F.Zaraga: Phys. Rev. Lett. **26**, 1009 (1971)
4.202 R.Cubeddu, R.Polloni, C.A.Sacchi, O.Svelto, F.Zaraga: Opt. Commun. **3**, 9 (1971)
4.203 R.Cubeddu, F.Zaraga: Opt. Commun. **3**, 310 (1971)
4.204 F.Shimizu, B.Stoicheff: IEEE J. QE-**5**, 544 (1969)
4.205 J.F.Reintjes, R.L.Carman: Phys. Rev. Lett. **28**, 1697 (1972)
4.206 J.F.Reintjes, R.L.Carman, F.Shimizu: Phys. Rev. A **8**, 1486 (1973)
4.207 E.B.Treacy: Phys. Rev. Lett. A **28**, 34 (1968)
4.208 M.A.Duguay, J.W.Hansen: Appl. Phys. Lett. **15**, 19 (1969)
4.209 E.P.Ippen, C.V.Shank: Appl. Phys. Lett. **26**, 92 (1975)
4.210 M.R.Topp, P.M.Rentzepis: J. Chem. Phys. **56**, 1066 (1972)
4.211 R.R.Alfano, S.L.Shapiro: Phys. Rev. Lett. **24**, 592 (1970)
4.212 R.H.Lehmberg, J.Reintjes, R.C.Eckardt: Appl. Phys. Lett. **25**, 374 (1974)
4.213 V.I.Bespalov, V.I.Talanov: JETP Lett. **3**, 307 (1966)
4.214 A.J.Campillo, S.L.Shapiro, B.R.Suydam: Appl. Phys. Lett. **24**, 178 (1974)
4.215 A.J.Campillo, S.L.Shapiro, R.R.Suydam: Appl. Phys. Lett. **23**, 628 (1973)
4.216 V.A.Korobkin, A.A.Malyutin, A.M.Prokhorov: JETP Lett. **12**, 150 (1970)
4.217 M.A.Duguay, J.W.Hansen, S.L.Shapiro: IEEE J. QE-**6**, 725 (1970)
4.218 R.C.Eckardt: IEEE J. QE-**10**, 48 (1974)
4.219 J.T.Fournier, E.Snitzer: IEEE J. QE-**10**, 473 (1974)
4.220 S.A.Akhmanov, A.P.Sukhorukov, R.V.Khokhlov: Sov. Phys. JETP **24**, 198 (1967)
4.221 J.A.Fleck, P.L.Kelley: Appl. Phys. Lett. **15**, 313 (1969)
4.222 J.A.Fleck, R.L.Carman: Appl. Phys. Lett. **20**, 290 (1970)

4.223 F. Shimizu, E. Courtens: *Proceedings of the Esfahan Symposium* (Wiley, New York 1971) p. 67
4.224 F. Shimizu: IBM J. Res. **17**, 287 (1973)
4.225 G. K. L. Wong, Y. R. Shen: Appl. Phys. Lett. **21**, 163 (1972)
4.226 A. J. Goldberg, V. I. Talanov, R. Z. Irm: Radiophysika **10**, 674 (1967)
4.227 T. K. Gustafson, C. H. Townes: Phys. Rev. A **6**, 1659 (1972)
4.228 A. H. Piekara, T. K. Gustafson: Opt. Commun. **7**, 197 (1973)
4.229 O. Rahn, M. Maier: Phys. Rev. A **9**, 1427 (1974)
4.230 R. G. Brewer: Phys. Rev. Lett. **19**, 8 (1967)
4.231 F. Shimizu: Phys. Rev. Lett. **19**, 1097 (1967)
4.232 M. A. Bol'shov, G. V. Venkin, S. A. Zhilkin, I. I. Nurminskii: Sov. Phys. JETP **31**, 1 (1970)
4.233 N. N. Il'ichev, V. V. Korobkin, V. A. Korshunov, A. A. Malyutin, T. G. Okroashvili, P. P. Pashinin: JETP Lett. **15**, 133 (1972)
4.234 R. R. Alfano, L. L. Hope, S. L. Shapiro: Phys. Rev. A **6**, 433 (1972)
4.235 W. Yu, R. R. Alfano, C. L. Sam, R. J. Seymour: Opt. Commun. **14**, 344 (1975)
4.236 J. Marburger: Opt. Commun. **14**, 92 (1975)
4.237 R. C. Eckardt, C. H. Lee, J. N. Bradford: Appl. Phys. Lett. **19**, 420 (1971)
4.238 R. C. Eckardt, C. H. Lee, J. N. Bradford: J. Opto-Electron. **6**, 67 (1974)
4.239 E. P. Ippen, C. V. Shank, T. K. Gustafson: Appl. Phys. Lett. **24**, 190 (1974)
4.240 R. R. Alfano, S. L. Shapiro: Chem. Phys. Lett. **8**, 631 (1971)
4.241 G. E. Busch, R. P. Jones, P. M. Rentzepis: Chem. Phys. Lett. **18**, 178 (1973)
4.242 W. Werncke, A. Lau, M. Pfeifer, K. Lenz, J. H. Weigmann, C. D. Thag: Opt. Commun. **4**, 413 (1972)
4.243 C. A. G. O. Varma, P. M. Rentzepis: Chem. Phys. Lett. **19**, 162 (1973)
4.244 C. A. G. O. Varma, P. M. Rentzepis: J. Chem. Phys. **58**, 5237 (1973)
4.245 M. R. Topp, P. M. Rentzepis: J. Appl. Phys. **42**, 3451 (1970)
4.246 R. Landauer: J. Appl. Phys. **31**, 479 (1960)
4.247 R. J. Joenk, R. Landauer: Phys. Lett. **A 24**, 228 (1967)
4.248 F. DeMartini, C. H. Townes, T. K. Gustafson, P. L. Kelley: Phys. Rev. **164**, 312 (1967)
4.249 H. A. Haus, T. K. Gustafson: IEEE J. QE-**4**, 519 (1968)
4.250 L. A. Ostrovskii: Sov. Phys. JETP **27**, 660 (1968)
4.251 R. A. Fisher, P. L. Kelley, T. K. Gustafson: Appl. Phys. Lett. **14**, 140 (1969)
4.252 F. Shimizu: IEEE J. QE-**8**, 851 (1972)
4.253 R. A. Fisher, W. K. Bischel: J. Appl. Phys. **46**, 4921 (1975)
4.254 N. Bloembergen: IEEE J. QE-**10**, 375 (1974)
4.255 A. von Hippel: J. Appl. Phys. **8**, 815 (1937)
4.256 A. von Hippel: Phys. Rev. **54**, 1096 (1937)
4.257 F. Seitz: Phys. Rev. **76**, 1376 (1949)
4.258 E. Yablonovich: Appl. Phys. Lett. **19**, 495 (1971)
4.259 D. W. Fradin, M. Bass: Appl. Phys. Lett. **22**, 206 (1973)
4.260 D. W. Fradin, E. Yablonovich, M. Bass: Appl. Opt. **12**, 700 (1973)
4.261 M. Bass, H. H. Barrett: Appl. Opt. **12**, 690 (1973)
4.262 D. W. Fradin, N. Bloembergen, J. P. Letellier: Appl. Phys. Lett. **22**, 635 (1973)
4.263 W. L. Smith, J. H. Bechtel, N. Bloembergen: Phys. Rev. B **12**, 706 (1975)
4.264 L. H. Holway, D. W. Fradin: J. Appl. Phys. **46**, 279 (1975)
4.265 W. L. Smith, J. H. Bechtel: Appl. Phys. Lett. **28**, 606 (1976)
4.266 J. P. Heritage, T. K. Gustafson, C. H. Lin: Phys. Rev. Lett. **21**, 1299 (1975)
4.267 C. H. Lin, J. P. Heritage, T. K. Gustafson: Appl. Phys. Lett. **19**, 397 (1971)
4.268 C. H. Townes, A. L. Schawlow: *Microwave Spectroscopy* (McGraw Hill, New York 1966)
4.269 A. M. Leontovich, A. M. Mozharovskii: Sov. Phys. JETP **19**, 195 (1974)
4.270 I. A. Poluektov, Yu. M. Popov: JETP Lett. **9**, 330 (1969)
4.271 F. Brukner, V. S. Dneprovskii, D. G. Koshchug, V. U. Khattatov: JETP Lett. **18**, 14 (1973)
4.272 T. L. Gvardzhaladze, A. Z. Grasyuk, V. A. Kovalenko: Sov. Phys. JETP **37**, 227 (1973)
4.273 T. L. Gvardzhaladze, A. Z. Grasyuk, I. G. Zuborev, P. G. Kryukov, O. B. Shashberashvili: JETP Lett. **18**, 27 (1971)

4.274 E.M.Belenov, I.A.Poluektov: Sov. Phys. JETP **29**, 754 (1969)
4.275 K.Seeger: *Semiconductor Physics* (Springer, New York 1973), Chap. 6
4.276 A.S.Aleksandrov, V.F.Elesin, Yu.P.Lisovets, I.A.Polusktov, Yu.M.Popov, V.S.Roitberg: Sov. J. Quant. Electron. **5**, 188 (1975)
4.277 A.J.DeMaria: *Progress in Optics*, Vol. 9, ed. by E.Wolf (North Holland, Amsterdam 1971)
4.278 M.Bass, P.A.Franken, J.F.Ward, G.Weinreich: Phys. Rev. Lett. **9**, 446 (1962)
4.279 J.F.Ward: Phys. Rev. **143**, 569 (1966)
4.280 I. P. Kaminow: *An Introduction to Electrooptic Modulators* (Academic Press, New York 1974)
4.281 M. J. Brienza, A. J. DeMaria, W. H. Glenn: Phys. Lett. **26A**, 390 (1968)
4.282 J.R.Morris, Y.R.Shen: Opt. Commun. **3**, 81 (1971)
4.283 K.H.Yang, P.L.Richards, Y.R.Shen: Appl. Phys. Lett. **19**, 320 (1971)
4.284 D.H.Auston, A.M.Glass: Appl. Phys. Lett. **20**, 398 (1972)
4.285 C.B.Roundy, R.L.Byer: Appl. Phys. Lett. **21**, 512 (1972)
4.286 D.H.Auston: Appl. Phys. Lett. **26**, 101 (1975)
4.287 R.A.Lawton, A.Scavannec: Elect. Lett. **11**, 74 (1975)
4.288 A.M.Johnson, D.H.Auston: IEEE J. QE-**11**, 283 (1975)
4.289 P.LeFur, D.H.Auston: Appl. Phys. Lett. **28**, 21 (1976)
4.290 D.H.Auston, A.M.Johnson, P.LeFur, E.P.Ippen, C.V.Shank, O.Teschke: *Digest of IX International Quantum Electronics Conference*, Amsterdam (1976)
4.291 J.J.Wiczer, H.Merkelo: Appl. Phys. Lett. **27**, 397 (1975)
4.292 A.M.Glass: In *Photonics*, ed. by M.Balanski, P.Lallemand (Gauthier-Villars, Paris 1975)
4.293 F.S.Chen: J. Appl. Phys. **40**, 3389 (1969)
4.294 D.von der Linde, A.M.Glass, K.F.Rodgers: Appl. Phys. Lett. **25**, 155 (1974)
4.295 D.von der Linde, A.M.Glass, K.F.Rodgers: J. Appl. Phys. **47**, 217 (1976)
4.296 M.E.Fourney, D.B.Barker: Appl. Phys. Lett. **21**, 21 (1972)
4.297 J.N.Olsen: Appl. Phys. Lett. **24**, 220 (1974)
4.298 E.V.Hoverstein: In *Laser Handbook*, Vol. 2, Chap. F8, ed. by F.T.Arecchi, E.O.Schulz-Dubois (North Holland, Amsterdam 1972)
4.299 H.Melchior, M.B.Fisher, F.R.Arams: Proc. IEEE **58**, 1466 (1970)
4.300 S.M.Faris, T.K.Gustafson, J.C.Weisner: IEEE J. QE-**9**, 797 (1973)
4.301 T.S.Kinsel: Proc. IEEE **58**, 1666 (1970)
4.302 T.S.Kinsel, F.S.Chen: Appl. Opt. **11**, 1411 (1972)
4.303 M.Ross: *Laser Applications*, Vol. I (Academic Press, New York 1971), p. 239
4.304 S.E.Miller, E.A.J.Marcatili, T.Li: Proc. IEEE **61**, 1703 (1973)
4.305 T.Li: Bell Labs. Record **53**, 333 (1975)
4.306 D.Gloge, A.R.Tynes, M.A.Duguay, J.W.Hansen: *Digest of* 1971, CLEA, Washington, D.C. (1971)
4.307 L.G.Cohen, G.W.Tasker, W.G.French, J.R.Simpson: Appl. Phys. Lett. **28**, 391 (1976)
4.308 M.J.Brienza, A.J.DeMaria: Appl. Phys. Lett. **11**, 44 (1967)
4.309 P.S.Peercy, E.D.Jones, J.C.Bushnell, G.W.Gobeli: Appl. Phys. Lett. **16**, 120 (1970)
4.310 E.D.Jones: Appl. Phys. Lett. **18**, 33 (1971)
4.311 N.G.Basov, V.V.Nikitin, A.S.Semenov: Sov. Phys. Uspekhi **12**, 219 (1969)
4.312 N.G.Basov, W.H.Culver, B.Shah: In *Laser Handbook*, Vol. 2, Chap. F5, ed. by F.T. Arecchi, E.O.Schulz-Dubois (North Holland, Amsterdam 1972)
4.313 M.J.Moran, C.Y.She, R.L.Carman: IEEE J. QE-**11**, 259 (1975)
4.314 E.S.Bliss, D.R.Speck, W.W.Simmons: Appl. Phys. Lett. **25**, 728 (1974)
4.315 S.L.Shapiro, H.P.Broida: Phys. Rev. **154**, 129 (1967)
4.316 E.P.Ippen, C.V.Shank: Appl. Phys. Lett. **24**, 190 (1974)
4.317 H.S.Gabelnick, H.L.Strauss: J. Chem. Phys. **49**, 2334 (1968)
4.318 I.L.Fabelinski: Optik. Spectrosk. **2**, 510 (1957)
4.319 G.R.Alms, D.R.Bauer, J.I.Braumen, R.Pecora: J. Chem. Phys. **59**, 5310 (1973)
4.320 G.I.A.Stegeman, B.P.Stoicheff: Phys. Rev. A **7**, 1160 (1973)
4.321 V.S.Starunov: Sov. Phys. Dokl. **8**, 1205 (1964)
4.322 H.C.Craddock, D.A.Jackson, J.G.Powles: Molec. Phys. **14**, 373 (1968)

4.323 J. Rouch, R. Lochet, A. Sousset: Compt. Rend. **265**, 253 (1967)
4.324 N.D. Foltz, C.W. Cho, D.H. Rank, T.A. Wiggins: Phys. Rev. **165**, 396 (1968)
4.325 P.L. Kelley, T.K. Gustafson: Phys. Rev. A **8**, 315 (1973)
4.326 R.R. Alfano, G.A. Zawadzkas: Phys. Rev. A **9**, 822 (1974)
4.327 J.A. Giordmaine, R.C. Miller: Phys. Rev. Lett. **14**, 973 (1965)
4.328 C.H. Lin, J.P. Heritage, T.K. Gustafson, R.Y. Chiao, J.P. McTague: Phys. Rev. A **13**, 813 (1976)
4.329 R.R. Alfano, J.I. Gersten, G.A. Zawadzkas, N. Tzoar: Phys. Rev. A **10**, 698 (1974)
4.330 T.K. Gustafson, J.-P. Taran, P.L. Kelley, R.Y. Chiao: Opt. Commun. **2**, 17 (1970)
4.331 F. Bruckner, V.S. Dnestrovskii, D.G. Koshchug: JETP Lett. **20**, 4 (1974)
4.332 F. Bruckner, Ya.T. Vasilev, V.S. Dneprovskii, D.G. Koshchug, E.K. Silina, V.U. Khattatov: Sov. Phys. JETP **40**, 1101 (1975)
4.333 P. Lee, D.V. Giovanielli, R.P. Godwin, G.H. McCall: Appl. Phys. Lett. **24**, 406 (1974)
4.334 S. Jackel, J. Albritton, E. Goldman: Phys. Rev. Lett. **35**, 514 (1975)
4.335 S. Jackel, B. Perry, M. Lubin: Phys. Rev. Lett. **37**, 95 (1976)
4.336 K. Eidmann, R. Sigel: Phys. Rev. Lett. **34**, 799 (1975)
4.337 D. Biskamp, H. Welter: Phys. Rev. Lett. **34**, 312 (1975)
4.338 P. Koch, J. Albritton: Phys. Rev. Lett. **34**, 1616 (1975)
4.339 D.W. Forslund, J.M. Kindel, E.L. Lindman: Phys. Rev. Lett. **30**, 739 (1973)
4.340 B.H. Ripin, J.M. McMahon, E.A. McLean, W.M. Manheimer, J.A. Stamper: Phys. Rev. Lett. **33**, 634 (1974)
4.341 J.L. Bobin, M. Lubin, R. Sigel, J. Stamper, C. Yamamaka: In *Laser Interaction and Related Plasma Phenomena*, Vol. 3, ed. by H.J. Schwartz, H. Hora (Plenum Publishing Co., New York 1974)
4.342 A.J. Alcock, M.C. Richardson: Phys. Rev. Lett. **21**, 667 (1968)
4.343 R.J. Dewhurst, G.J. Pert, S.A. Ramsden: J. Phys. B: Atom. Molec. Phys. **7**, 2281 (1974)
4.344 C.L.M. Ireland, C. Grey Morgan: J. Phys. D: Appl. Phys. **7**, L87 (1974)
4.345 R.J. Dewhurst: J. Phys. D: Appl. Phys. **8**, L80 (1975)
4.346 F.V. Bankin, A.M. Prokhorov: Sov. Phys. JETP **25**, 1072 (1967)

5. Picosecond Interactions in Liquids and Solids*

D. von der Linde

With 31 Figures

The development of novel picosecond optical measuring techniques has greatly expanded our capability of studying very rapid events. Accurate and direct time measurements on a picosecond scale are now possible, offering many new opportunities in various scientific areas. In the past several years numerous experiments have been reported involving the use of picosecond optical pulses for measuring rapid relaxation processes. The present chapter reviews these developments emphasizing application concerning the physics of liquids and solids. Related work dealing with applications of picosecond laser pulses in chemistry and biology is reviewed in Chapter 6 and Chapter 7.

The experimental tools and techniques of picosecond spectroscopy have been discussed in detail in the preceding chapters of this book. The modelocked neodymium laser (Chap. 2) plays a major role as a source of ultrashort optical pulses in most of the work examined herein. Often frequency multiplication and frequency shifting of the fundamental light by nonlinear optical processes (Chap. 4) is used to generate picosecond pulses at new frequencies. An important measuring device is the excite-and-probe technique in which a strong pulse generates a phenomenon, and a different, time delayed pulse interrogates the state of the system at a known time interval after excitation. A detailed account of this method and of the various other detection and measuring techniques can be found in Chapter 3.

The present chapter is organized as follows. In Section 5.1 picosecond vibrational relaxation of molecules and crystals in the electronic ground state is discussed. Starting with an examination of the role of the Raman effect for the generation and detection of material vibrations, measurements of molecular vibrational relaxation times and of optical phonon lifetimes are described.

Picosecond spectroscopy is a new, rapidly expanding field, and there is a large number of experiments in which novel measuring techniques are developed and tested for the first time. To date many of these potentially fruitful applications have not yet been studied in great depth. Section 5.2 contains a discussion of a variety of experiments, most of which involve picosecond electronic excitations in solids.

In Section 5.3 the interaction of ultrashort laser pulses with excitons is discussed. Picosecond optical techniques have revealed a fascinating variety of rapid dynamic processes involving the creation and the decay of excitons

* Part of this work was performed when the author was with Bell Laboratories, Murray Hill, New Jersey, USA.

and exciton complexes in semiconductor crystals. The interpretation of some of these recent observations is still under discussion, but undoubtedly new insight may be gained on excitons and their interactions from picosecond time-resolved spectroscopy.

5.1 Vibrational Relaxation in the Electronic Ground State

Ultrashort light pulses permit novel and exciting investigations of vibrational relaxation, and this field is certainly among the most successful areas that utilizes modelocked laser pulses. Previously only indirect information on the relaxation of material vibrations in the condensed phase was available, very often inferred from infrared and Raman spectroscopic observations. For example, linewidth data suggested that the time constants of vibrational relaxation are of the order of 10^{-12} s, but very little was known about the details of the relaxation processes. Because many different physical effects can contribute to the experimentally observed line shape, the linewidth data offer only crude indirect, and integral information.

This section deals with vibrational relaxation in the electronic ground state. Vibrational relaxation in the excited state is discussed in Chapter 6. Most of that work is concerned with large organic molecules, in which the density of vibrational states is quite substantial. Electronic transitions generally couple to many different vibrational modes simultaneously, and usually the identity of the modes is not known. On the other hand, in the present discussion, the emphasis is on single, well-defined vibrational modes of relatively small molecules (liquids), and on well-defined lattice vibrations (solids).

Raman scattering has played an important role for vibrational relaxation studies. Very strong vibrational excitation is readily obtained by stimulated Raman scattering (SRS) of picosecond pulses. A review of transient SRS is given by *Auston* in Chapter 4 of this book. In the present section only a few aspects of SRS relevant to relaxation studies will be outlined. A discussion of coherent vibrational decay in liquids and solids follows. The section is concluded with an account of picosecond measurements of relaxation and transfer of vibrational energy in molecular liquids.

5.1.1 Excitation and Detection of Vibrational Waves by Means of the Raman Effekt

According to *Placzek* [5.1] the interaction between a given vibrational mode and the electromagnetic field can be described by the interaction Hamiltonian

$$H_I = -\tfrac{1}{2}\left(\frac{\partial\alpha}{\partial Q}\right) Q E^2 . \tag{5.1}$$

$\partial\alpha/\partial Q$ describes the change of electronic polarizability with the vibrational coordinate Q, and E is the electric field strength of the light wave.

Two conclusions follow immediately from (5.1). First, in the presence of an electromagnetic field, a force F is acting on the oscillators:

$$F = -\frac{\partial H_I}{\partial Q} = \tfrac{1}{2}\left(\frac{\partial \alpha}{\partial Q}\right) E^2. \tag{5.2}$$

Second, an electric polarization P exists when both a vibrational wave and an electromagnetic wave are present:

$$P = -\frac{\partial H_I}{\partial E} = N\left(\frac{\partial \alpha}{\partial Q}\right) QE \tag{5.3}$$

(N is the number density of oscillators).

Equation (5.2) implies excitation of vibrational modes by means of light, whereas (5.3) points to the possibility of detecting a given vibrational wave via the generated electric polarization. Let us briefly discuss these two aspects of the Raman interaction:

a) It follows that any desired material vibration having $\partial\alpha/\partial Q \neq 0$ (Raman active modes) can be excited by applying two light waves of frequency, say ω_L and ω_s in such a way that the difference $\omega_L - \omega_s$ equals the frequency ω_v of the vibrational mode of interest. The driving force (5.2) is proportional to the product $\mathscr{E}_L \mathscr{E}_s$ of the amplitudes of the electric fields at ω_L and ω_s (the notation of Chap. 4 is used). If the driving electric fields are two short synchronous pulses an impact excitation may be achieved. After the exciting pulses have passed, the vibrational wave oscillates freely, and the free relaxation of the wave may be studied.

b) It follows also that a vibrational wave can be detected by measuring the Raman light scattered from the excited vibrational wave. According to (5.3) Raman scattering is viewed as a nonlinear mixing of an electromagnetic wave with the material vibration. When a light wave of frequency ω_L and amplitude $\mathscr{E}_L$ interacts with a vibrational wave of frequency ω_v and amplitude Q_v, electric polarizations are created at two new frequencies $\omega_s = \omega_L - \omega_v$ and $\omega_{AS} = \omega_L + \omega_v$. These polarizations give rise to the scattered Stokes and anti-Stokes light. The field amplitude of the scattered light is proportional to the product $Q_v \mathscr{E}_L$. Thus, the intensity of the scattered light is a measure of the degree of vibrational excitation, and the Raman scattered light can be used to monitor the vibrational wave.

Before discussing in more detail the detection of material vibrations using the Raman effect a few remarks are made about vibrational excitation via SRS. When vibrational waves are driven by two light waves, as we have just discussed, the wave at ω_s increases at the expense of the wave at ω_L. In fact, speaking in terms of energy quanta, conservation of energy demands that for each phonon $\hbar\omega_v$ created, a photon $\hbar\omega_L$ is destroyed and a new Stokes photon $\hbar\omega_s = \hbar(\omega_L - \omega_v)$ is generated. Thus, there is light amplification for the Stokes wave due to stimulated Raman emission. The gain factor of the Stokes wave is proportional to the intensity of the laser wave at ω_L (for details see Chap. 4). When a strong laser pulse is sent through a slab of Raman active material, very large Stokes gains are readily obtained. Without any Stokes input the Stokes wave may grow from quantum noise to power levels approaching that of the original laser pulse while travelling through only a few centimeters of material. Thus, we actually require only one intense light pulse, and a strong Stokes wave builds up automatically at the proper frequency $\omega_s = \omega_L - \omega_v$ for driving the vibrational wave. Given sufficient Raman gain, the beating of the laser wave with the generated Stokes wave will give rise to intense material vibrations within a few centimeters of interaction length in the Raman medium. There is no need for a tunable source of short laser pulses, and the laser frequency can be chosen at one's convenience, the principal restriction being that the material be transparent at ω_L. These features have made stimulated Raman scattering an extremely useful tool for the generation of strong molecular and lattice vibrations, and this method has been used extensively for studying vibrational relaxation.

5.1.2 Dephasing Processes and Relaxation of the Vibrational Energy

For the subsequent discussion of vibrational relaxation we must clearly understand the nature of the vibrational excitation we are dealing with. Some of the necessary background information is given in this section. For more details see Chapter 4 and the review papers listed under [5.2].

In (5.1) we introduced the coordinate Q to describe the vibrational excitation. Actually, Q is to be regarded as an expectation value of the oscillation amplitude of all the contributing oscillators in a given element of volume:

$$Q = \langle q_j \rangle = \frac{1}{N} \sum_j^N q_j \,. \tag{5.4}$$

We are considering vibrational waves of frequency ω_v and wavevector $\boldsymbol{k}_v$. The dependence of ω_v on the wavevector implied by the phonon dispersion relations $\omega_v(\boldsymbol{k}_v)$ can be neglected, because for interaction with light $\boldsymbol{k}_v$ is always small in comparison with the reciprocal lattice constant, or with the reciprocal of the spacing of the molecules in a liquid. Then ω_v may be regarded as constant, independent of wavevector (zero group velocity).

As discussed in Chapter 4 the amplitude of the vibrational wave $Q(\boldsymbol{x}, t)$ obeys the following equation of motion:

$$\left(\frac{\partial^2}{\partial t^2} + \frac{1}{\tau}\frac{\partial}{\partial t} + \omega_v^2\right) Q(\boldsymbol{x}, t) = \frac{1}{m} F(\boldsymbol{x}, t) \mathrm{n}_v \,. \tag{5.5}$$

Equation (5.5) covers optical phonons at the center of the Brillouin zone and internal molecular vibrations (liquids), but sound waves are excluded from the discussion.

Let us ignore for the moment the factor n_v on the right-hand side of (5.5) i.e., $n_v = 1$. We recognize that the material vibrations are simply described by the equation of motion of a damped harmonic oscillator driven by a force $F(\boldsymbol{x}, t)$.

Two points should be noted here: The first remark concerns the driving term. Let us assume that the force is caused by two electric fields at frequencies ω_L and ω_s, as discussed in the preceding section. According to (5.2) the force will be proportional to the product of the fields, e.g.,

$$F(\boldsymbol{x}, t) \propto \mathscr{E}_L \mathscr{E}_s^* \exp[-i(\omega_L - \omega_s)t + i(\boldsymbol{k}_L - \boldsymbol{k}_s)\boldsymbol{x}], \tag{5.6}$$

where $\boldsymbol{k}_L$ and $\boldsymbol{k}_s$ are the wavevectors of the two driving light fields.

Efficient excitation of the oscillators obviously requires frequency resonance, i.e., $\omega_v = \omega_L - \omega_s$. Note, however, that the force field also imposes *spatial coherence* on the vibrational wave; the driving force (5.5) creates a coherent vibrational wave of the form

$$Q(\boldsymbol{x}, t) = \tfrac{1}{2} Q_v \exp[-i(\omega_v t - \boldsymbol{k}_v \boldsymbol{x})] + \text{c.c.} \tag{5.7}$$

where $\boldsymbol{k}_v = \boldsymbol{k}_L - \boldsymbol{k}_s$. The coherent character of the vibrational excitation resulting from Raman nonlinear mixing is of great importance in this discussion.

The second point is concerned with the phenomenological relaxation term in (5.4). The free vibrational wave will decay exponentially according to

$$Q(\boldsymbol{x}, t) \propto \exp(-t/2\tau). \tag{5.8}$$

The interesting question is: What are the physical processes leading to the decay of the coherent vibrational wave? We will consider two processes contributing to the damping of Q. First, let us assume that the vibrational energy of the individual oscillators relaxes. Then Q obviously decays because the magnitude $|q_j|$ of each of the contributing oscillators goes to zero. However, there is yet another, more interesting possibility for the decay of Q. Let the amplitudes $|q_j|$ stay constant, but let us introduce random phase shifts among the oscillators. For example, such random phase shifts may be due to elastic collisions of the oscillators. One can imagine a completely dephased state of the system with all the individual oscillators still vibrating but with such random phase relations that the expectation value Q equals zero:

$$Q = \frac{1}{N} \sum_j^N q_j = \frac{1}{N} \sum_j^N |q_j| e^{i\psi_j} = 0. \tag{5.9}$$

We are led to the conclusion that the amplitude Q of the coherent wave may decay at a different, faster rate than the vibrational energy proper.

The distinction between vibrational relaxation due to dephasing and relaxation due to dissipation of the vibrational energy is a key point of the present chapter.

Naturally, the question now arises how to describe the relaxation of the vibrational energy as distinct from dephasing. For dealing just with the energy aspect of the vibrational excitation it is useful to write the density of vibrational energy ε_v as

$$\varepsilon_v = \hbar\omega_v nN \tag{5.10}$$

where n is the occupation probability of the excited (vibrational) state, and N is the number density of oscillators. Note that n is expressed in terms of the population difference n_v used in Chapter 4 by

$$n = \tfrac{1}{2}(1 - n_v). \tag{5.11}$$

The change with time of n can be described by a power balance equation:

$$\left(\frac{\partial}{\partial t} + \frac{1}{\tau'}\right) n = \frac{1}{\hbar\omega_v} F(t) \frac{\partial Q}{\partial t}. \tag{5.12}$$

The term on the right represents the work done on the oscillator by the force $F(t)$ (divided by the phonon energy $\hbar\omega_v$). Note that the relaxation term of the excess population is written as n/τ', with $\tau' \neq \tau$ in accordance with our discussion above. The distinction between τ' and τ properly takes into account the different nature of energy relaxation and of the decay of the vibrational amplitude Q.

Because all energy relaxation mechanisms also contribute to the decay of Q, but not vice versa, τ and τ' obey the relaxation

$$\tau' \geq \tau. \tag{5.13}$$

The definition of τ' and τ parallels that of the spin-lattice relaxation time T_1 and of the spin-spin relaxation time T_2 in the Bloch equations of nuclear and electron spin resonance.

5.1.3 Coherent and Incoherent Raman Scattering

This section deals with the question how phase relaxation and energy relaxation can be distinguished experimentally. Let us first discuss the detection of a coherent vibrational wave by means of Raman scattering. In principle both Stokes and anti-Stokes scattering could be used, but we restrict the discussion to the anti-Stokes case.

Let us assume, then, that a coherent vibrational wave has been launched into a material and that an interrogating light wave is sent through the medium. The probe light beats with the vibrations according to (5.3) and the resulting polarization wave radiates anti-Stokes light at $\omega_{AS} = \omega_L + \omega_v$.

For the growth of the anti-Stokes light the phase relations between the waves are very important. The propagation of the polarization wave P (see (5.3)) is determined by the sum of the wavevector of the vibrational wave $\boldsymbol{k}_v$ and the probe light $\boldsymbol{k}$, because P is proportional to the product QE. The wavevector of the anti-Stokes light, on the other hand, is given by the usual dispersion law of light (color dispersion), i.e., $|\boldsymbol{k}_{AS}| = \omega_{AS} n_{AS}/c$, where n_{AS} is the refractive index at ω_{AS}.

The essential point is that the Stokes wave can grow efficiently only if it stays in phase with the driving polarization wave. Phase matching requires that $\boldsymbol{k}_{AS}$ equals the wavevector of the polarization wave, i.e.,

$$\boldsymbol{k}_{AS} = \boldsymbol{k} + \boldsymbol{k}_v . \tag{5.14}$$

The anti-Stokes light and the polarization wave propagate with the same phase velocity when (5.14) is satisfied and no dephasing can occur during propagation.

The wavevector phase-matching condition is well known [5.3] from nonlinear optical mixing processes. In our case it can be viewed as the conservation of momentum condition for a photon-phonon collision process in which the incident photon $\hbar\omega$ and the phonon $\hbar\omega_v$ are destroyed and a new photon $\hbar\omega_{AS}$ is generated.

Let us now consider the proper experimental arrangement that takes into account the phase-matching condition for the detection of a coherent vibrational wave by Raman scattering. Figure 5.1 shows a typical experimental

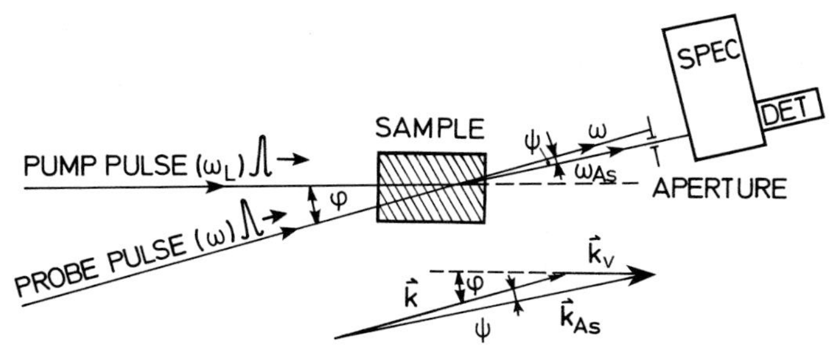

Fig. 5.1. Schematic of a coherent Raman scattering experiment for the measurement of dephasing times (τ). The wavevector diagram illustrates the interaction of the various waves. The pump process creates a coherent vibrational wave of wavevector $\boldsymbol{k}_v$. The probe light of wavevector $\boldsymbol{k}$ is incident at an angle φ, and the coherent anti-Stokes Raman signal is radiated at an angle ψ. These angles have to be adjusted such that the wavevector triangle is satisfied. The Raman light is measured with a spectrometer (SPEC) and a photodetector (DET). An aperture blocks the transmitted probe beam. The optical delay line for generating variable delays between the probe pulses and the pump pulses is not shown

setup. The wavevector matching requirement leads to the following important consequences. First, the probe light must be incident at a well-defined angle φ with respect to the direction of the vibrational wave to be detected. Second, the anti-Stokes light is emitted as a well-collimated beam at an angle ψ with respect to the incident probe light. The two angles φ and ψ are determined by the wavevector triangle.

Raman scattering from coherent material excitations subject to the phase-matching condition is called coherent Raman scattering. The directional properties of coherent Raman scattering differ sharply from ordinary spontaneous Raman scattering which arises from incoherent thermal vibrations. Incoherent spontaneous light is radiated into the full solid angle regardless of the direction of the incident light.

The relevance of coherent Raman scattering for vibrational relaxation studies is now obvious. The instantaneous amplitude of a coherent vibrational wave can be monitored by scattering a short, phase-matched probe pulse off the vibrational wave. By suitably delaying the probe pulse and by measuring the intensity of the scattered light as a function of delay time the decay of the wave can be measured directly. This is the basic principle for the picosecond measurements of dephasing times and phonon lifetimes to be discussed later.

Coherent Raman scattering was originally demonstrated by *Giordmaine* and *Kaiser* [5.4]. The same principle was used for the first time for picosecond measurements of vibrational relaxation in liquids by *von der Linde* et al. [5.5]. Before discussing in more detail these experiments let us consider the methods of measuring the relaxation of the vibrational energy (measurement of τ').

As outlined above, we expect different relaxation times for the vibrational energy and for the coherent vibrational amplitude; Q may have decayed due to the loss of phase coherence, but the individual oscillators may still be vibrating, i.e., the excited vibrational states are still populated. We are now interested only in the relaxation of the population of excited vibrational states.

We recall that in ordinary spontaneous Raman scattering the intensity of the anti-Stokes light is proportional to the number of excited oscillators. When all molecules are in the ground state, there is no anti-Stokes light. Some anti-Stokes light will be observed at temperatures $T>0$ due to the thermal population of the upper states. However, if we generate a large excess population, e.g., by stimulated Raman scattering, there will be a strongly enhanced anti-Stokes scattering. Note that the scattering from the dephased oscillators is not directional, nor does it depend on the direction of the incident light. We are dealing with spontaneous incoherent scattering, the only difference being that vibrational energy in excess of thermal equilibrium has been created by the pumping process (i.e., SRS).

The instantaneous population of the upper vibrational state can be measured by scattering a short probe pulse off the excited material. The energy relaxation time τ' is found by measuring the incoherent anti-Stokes light as a function of the delay time of the probe pulse.

A typical experimental arrangement for such a measurement is shown in Fig. 5.2. From our discussion it is clear that phase matching must be avoided, and that the scattered Raman light is emitted into the full solid angle. Just as in common spontaneous Raman spectroscopy a large aperture objective lens is used to collect as much of the scattered light as possible.

The incoherent anti-Stokes probe technique to measure excess population was first introduced by *DeMartini* and *Ducuing* [5.6] who measured the relatively slow relaxation of the vibrating hydrogen molecule. The method was applied for the first time for picosecond relaxation studies in liquids by *Laubereau* et al. [5.7]. In the following sections we will discuss how these new picosecond techniques made it possible to conduct novel studies of energy relaxation and energy transfer in vibrationally excited liquids.

In concluding this section we want to comment on the relation between the information obtained from the new direct picosecond techniques with that obtained from traditional linewidth data in the frequency domain.

First, let us consider a situation in which relaxation is dominated by energy dissipation of the individual oscillators. We assume that dephasing is unimportant. Then we have

$$\tau = \tau' = 1/\Delta\omega \qquad (5.15)$$

where $\Delta\omega$ is the full half width of the Lorentzian-shaped spontaneous line. The direct time domain techniques yield the same information as the spectroscopic data.

Second, we allow dephasing collisions to occur, i.e., we now have $\tau < \tau'$, but it is assumed that we are still dealing with an isolated resonance of a well-

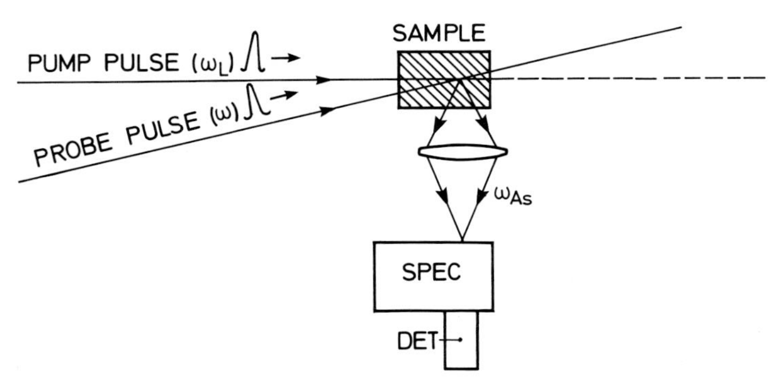

Fig. 5.2. Schematic of an incoherent Raman scattering experiment for the determination of the energy relaxation time (τ'). An intense pump pulse creates strong material vibrations. The incoherent Raman light scattered from the probe light is measured by a spectrometer (SPEC) and a photodetector (DET). The Raman light is radiated into the full solid angle, and a lens of large aperture collects the scattered light. The optical delay line for generating variable delays between probe pulses and pump pulses is not shown

defined transition. The spectral line then has a Lorentzian shape with a half width

$$1/\Delta\omega = \tau < \tau' . \tag{5.16}$$

The direct measurement of the decay time of the coherent material excitation gives the same information as the linewidth measurement. However, the energy relaxation is now obscured by the much faster dephasing processes, and τ' can no longer be inferred from the spectral data. Direct measurements in the time domain are needed to extract the energy relaxation time τ'.

Third, we consider a distribution of closely spaced resonance frequencies. For example, we allow additional degrees of freedom such as rotations and translations. A typical situation of this kind is a Doppler-broadened transition in a gas. In this case neither τ' nor τ is directly related to the experimentally observed line shape, which is determined by the distribution of resonance frequencies and not by the relaxation processes. However, as we shall see below, direct time-resolved measurements can yield the relaxation times τ' and τ even under these difficult conditions.

5.1.4 Measurement of Dephasing Times in Liquids

The first direct measurement of picosecond relaxation times of molecular vibrations was reported in [5.5]. Two organic liquids were investigated, carbon tetrachloride (CCl_4) and ethanol (C_2H_5OH). In these experiments the samples were contained in liquid cells, typically a few centimeters long. A single, intense picosecond pulse (duration $t_p \simeq 8$ ps) at a wavelength $\lambda = 0.53$ μm was used for exciting molecular vibrations via the stimulated Raman effect. Under these conditions the maximum material excitation is generated near the end of the sample cell, and the wave vector $\boldsymbol{k}_v$ of the vibrational wave approximately points into the forward direction. A delayed weak pulse (1% of the pump pulse) at the same wavelength served for probing the coherent material excitation. Anti-Stokes phase matching of the probe light and the vibrations was achieved at small values of the angles φ and ψ (see Fig. 5.1) so that the scattered probe light was radiated approximately into the forward direction.

Before discussing the experimental results it is instructive to consider calculations [5.5] of the time dependence of the vibrational amplitude $Q(t)$ generated by SRS of the pump pulse, and of the expected variation of the anti-Stokes light S scattered from those vibrations. The normalized vibrational excitation Q^2, say, at the end of the sample cell, is shown in Fig. 5.3a for various values of the ratio t_p/τ (t_p = pulse duration, τ = dephasing time), whereas Fig. 5.3b depicts the variation with delay time of the anti-Stokes intensity S generated by a weak probe pulse having the same shape as the pump pulse. $S(t_D)$ represents a convolution of the probe pulse with the vibrational excitation $Q^2(t)$. Comparing Fig. 5.3a and 5.3b it is seen that the decay time of the vibrational wave (dephasing time τ) can be directly inferred from the

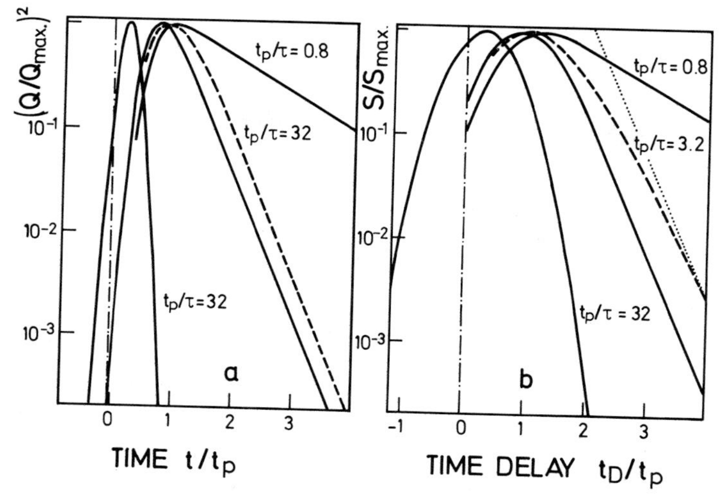

Fig. 5.3. (a) Calculated vibrational excitation $Q^2(t)$ generated by stimulated Raman scattering of a pump pulse of duration t_p with the maximum at $t=0$. τ is the desphasing time of the vibrational wave. Gaussian (full curves) and hyperbolic secant (dashed curves) pump pulses. (b) Coherent anti-Stokes scattering S from the probe pulse as a function of delay time t_D between probe and pump. S is the convolution of $Q^2(t)$ with the probe pulse, assumed to have the same shape as the pump pulse (after *von der Linde* et al. [5.5]

exponential decay of $S(t_D)$ if the ratio t_p/τ is not too large. For instance, with gaussian pump and probe pulses, $Q^2(t)$ is quite accurately approximated by $S(t_D)$ for values $t_p/\tau \lesssim 3$ (full curves). As indicated by the example of a hyperbolic second pulse (broken curve), the time resolution can be less for a different pulse shape. Knowledge of the pulse shape is therefore essential for determining the time resolution for a particular experimental situation.

Let us now turn to the experimental results of [5.5]. In carbon tetrachloride the totally symmetric A_1 mode at 459 cm^{-1} ($\nu_1(a_1)$ normal mode of the tetrahedral molecule), and in ethanol the CH bond stretching vibration at 2928 cm^{-1} were investigated. The experimental results are shown in Fig. 5.4. The measured anti-Stokes probe light is plotted versus the probe delay time on a semilogarithmic scale. Let us discuss the CCl_4 results first (full circles). It is seen from Fig. 5.4 that for delay times longer than 15 ps the variation of the anti-Stokes light can be represented by a straight line, which accurately accounts for the decay down to a factor of 10^3 below the maximum. From this exponential tail of the scattering curve a decay time of 4 ± 0.5 ps was obtained. This time constant was identified with the dephasing time of the excited vibrational wave involving the A_1 tetrahedral mode of carbon tetrachloride.

The directly measured value of the dephasing time was compared with spontaneous Raman linewidth data of the same mode. Carbon tetrachloride is an interesting system because the Raman band at 459 cm^{-1} consists of

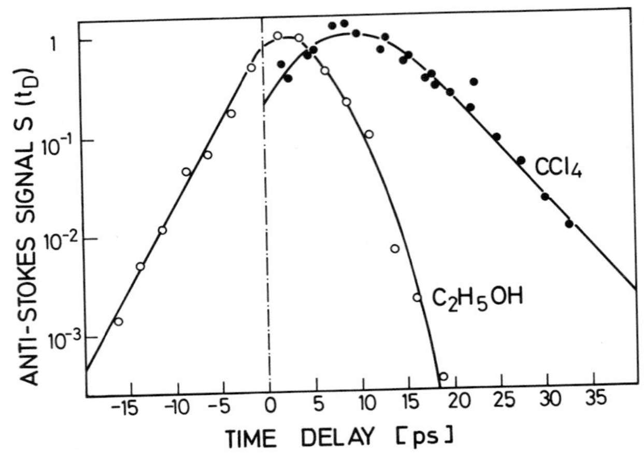

Fig. 5.4. The measured coherent anti-Stokes scattering $S(t_D)$ as a function of probe pulse delay time. In CCl_4 (full circles) the dephasing time τ is long, and $S(t_D)$ traces the exponential decay of the coherent vibrational wave generated by the pump process. In C_2H_5OH (open circles) τ is much shorter than the pulse duration. $S(t_D)$ approximately traces the shape of the probe pulse. $t_D>0$ corresponds to the leading edge of the probe pulse (after *von der Linde* et al. [5.5]

several different overlapping components due to the various combinations of carbon and chlorine isotopes that can form a CCl_4 molecule. This system is a typical case where only a crude estimate of the relaxation time can be obtained from the spectral line shape [5.8]. The total linewidth of the A_1 band of CCl_4 is about 7 cm^{-1}, suggesting a relaxation time as short as 0.75 ps. However, the isotopic structure is partially resolved, and one can guess the width of a single isotopic component to be roughly 1.5 cm^{-1}. The relaxation time measured in the picosecond experiment is in good agreement with this estimate.

Let us now discuss the data for ethanol (open circles in Fig. 5.4). A completely different variation of the anti-Stokes light is measured for this material. The scattered light decays much faster than in CCl_4. The data cannot be represented by an exponential decay; the curve drawn through the experimental points for $t_D > 5$ ps is gaussian. Apparently, the decay of the vibrational wave cannot be resolved for ethanol.

In fact, the CH bond stretching vibration of ethanol shows up in the spontaneous Raman spectrum as a broad band with a full width of approximately 20 cm^{-1}, suggesting a relaxation time of about 0.25 ps (if the line was homogeneously broadened). It was shown in [5.5] that the gaussian fall off of the anti-Stokes light in the case of ethanol represents the leading edge of the probe pulse and not the free relaxation of the molecular vibration.

The comparison of the results for carbon tetrachloride and ethanol sheds light on the question of time resolution and the reliability of the picosecond experiments involving stimulated Raman scattering. The steep decay of $S(t_D)$ in the ethanol measurement proves that the decay seen in CCl_4 is indeed due to molecular relaxation and not due to a feature of the light pulses (or the excitation mechanism). It must always be ascertained that the time resolution of the experimental system is adequate for a determination of τ from scattering curves such as Fig. 5.4. This problem may become particularly severe when the material excitation is generated via transient SRS with pulses short compared with the dephasing time of the relevant Raman transition. It is typical of SRS under highly transient conditions that parasitic intensity much less than the peak of the pump pulses, but of much longer duration than the nominal pulse duration, can build up vibrational excitation comparable to that generated by the proper pulse. In such a case the Raman probe technique would give incorrect results. This buildup of parasitic intensities may have been the reason for some discrepancies between the various experimental results reported in the literature.

Following the initial experiments of [5.5] almost a dozen different liquids have been investigated by the picosecond probe technique. Table 5.1 summarizes the experimental results by comparing the measured dephasing times with the available Raman linewidth data. The observed values of τ vary over a wide range, e.g., from 75 ps for a simple diatomic liquid such as liquid nitrogen down to a subpicosecond time constant for ethanol (not fully resolved).

Except for the tetrahalides there is generally good agreement between the directly measured values and the values estimated from spontaneous linewidth, suggesting that in these cases the linewidth is homogeneously broadened by dephasing processes. However, poor agreement is found for many tetrahalides. Due to isotope line splitting (a typical example is CCl_4), the Raman line in these compounds is composed of several overlapping components. For these inhomogeneously broadened lines the picosecond technique yielded accurate new data, previously not available.

The theory of dephasing processes is receiving increasing attention because precise data on dephasing times are now readily available. *Fischer* and *Laubereau* [5.11] have attempted to account for the observed relaxation times using a simple model of elastic collisions of vibrationally excited molecules.

Table 5.1

	$\bar{\nu}_0\,[\mathrm{cm}^{-1}]$	$\Delta\bar{\nu}_0\,[\mathrm{cm}^{-1}]$	$(2\pi c\Delta\bar{\nu}_0)^{-1}\,[\mathrm{ps}]$	$\tau\,[\mathrm{ps}]$	
CCl_4	459	~7	0.76	4.0 ± 0.5	[5.5]
				3.6 ± 0.4	[5.9]
$SiCl_4$	425	~8	0.66	3.0 ± 0.5	[5.9]
$SnCl_4$	368	~8	0.66	2.8 ± 0.3	[5.9]
$SnBr_4$	221	~3.2	1.7	3.0 ± 0.3	[5.9]
CH_3CCl_3	2939	~4.9	1.0	1.3 ± 0.7 [a]	[5.7]
C_2H_5OH	2928	~20	0.26	$0.1 - 0.4$ [a]	[5.7]
Liq. N_2	2326	0.067	79	75 ± 8	[5.10]

[a] Not fully resolved.

They considered pure dephasing processes arising from random adiabatic fluctuations of the vibrational frequency that result from pertubations of the molecular potential during an elastic collision with another molecule.

By assuming an exponential interaction potential of range L, they derived a simple expression for the dephasing time:

$$1/\tau = \frac{F(m_j)k_B T}{\omega_v^2 L^2 \tau_c}. \tag{5.17}$$

F is a function of the atomic masses of the colliding molecules, $1/\tau_c$ is the collision frequency, and the other symbols have their usual meaning.

Figure 5.5 shows a diagram from [5.11] with the measured dephasing times plotted versus the time constants calculated from (5.17). The collision frequency and the range of the interaction potential were estimated from viscosity and molecular density data.

Note that despite the widely differing chemical nature of the molecules, and despite a large variation of the measured relaxation times (by about two orders of magnitude) the model of [5.11] seems to be capable of predicting the dephasing time quite accurately for most of the considered molecules. Note also that the model takes into account pure dephasing processes only. Contributions to τ due to energy relaxation are neglected. The result of Fig. 5.5 confirms expectations that energy relaxation is generally slow compared with dephasing processes.

There is another interesting point. The calculations of [5.11] neglect contributions to the dephasing time due to resonant energy transfer between identical molecules. From the good agreement between the experimental data

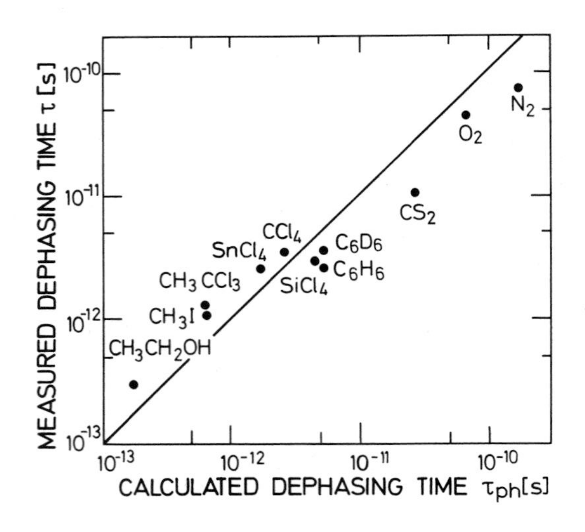

Fig. 5.5. Comparison of the measured and the calculated dephasing time for a variety of liquids (after *Fischer* and *Laubereau* [5.11]

and *Fischer* and *Laubereau's* model it appears that resonant energy transfer does not significantly affect the dephasing times in these molecular liquids.

5.1.5 Dephasing in Inhomogeneously Broadened Systems

A comparison of linewidth data with the relaxation times measured directly with picosecond pulses indicated substantial inhomogeneous broadening for the tetrahalides. The effect results from isotope line splitting in these compounds. In carbon tetrachloride the isotope structure is partially resolved, but in some other compounds the individual isotope bands overlap completely. In this section we will discuss the role of inhomogeneous broadening for the measurement of dephasing times, and the interesting interference phenomena that can arise from the line substructure. These effects can be used to obtain information on inhomogeneous frequency distributions. However, if not properly controlled, these interferences can readily obstruct dephasing time measurements.

Let us consider excitation of a distribution of N_j different oscillators with different, but closely spaced, resonance frequencies. When a short intense preparing pulse at ω_L interacts with this system via stimulated Raman scattering, a Stokes pulse is generated at a frequency [5.9]

$$\begin{aligned}\omega_s &= \omega_L - \textstyle\sum_j \omega_j N_j / \sum_j N_j \\ &= \omega_L - \omega_v .\end{aligned} \tag{5.18}$$

The essential point is that the oscillators are locked together in phase by the driving laser and Stokes pulses at a common vibrational frequency ω_v differing slightly from the individual resonance frequencies ω_j. However, the various vibrational components relax at their proper eigenfrequency after the exciting pulses have been turned off. Subsequently, as time progresses, the oscillators get out of phase in a way characteristic of the frequency distribution. For instance, if there is simply a constant frequency spacing $\Delta\omega$ between the resonances the total vibrational excitation will exhibit a typical beat note with a periodicity $2\pi/\Delta\omega$.

Let us now consider coherent Raman scattering from such a freely relaxing inhomogeneous system. There are two possibilities for the behavior of the scattered probe light, depending on the phase-matching condition of the scattering process.

First, assume that there is strict phase matching just for a single frequency component i out of total distribution, e.g.,

$$\Delta \boldsymbol{k}_i = \boldsymbol{k} + \boldsymbol{k}_i - \boldsymbol{k}_{AS} \ll 1/L \tag{5.19}$$

but

$$\Delta \boldsymbol{k}_j = \boldsymbol{k} + \boldsymbol{k}_j - \boldsymbol{k}_{AS} \gg 1/L \tag{5.20}$$

for $j \neq 1$. In these equations L is the interaction length of the probe light with the excited material and $\boldsymbol{k}_j$ is the wavevector of the vibrational component j of the distribution, while $\boldsymbol{k}$ and $\boldsymbol{k}_{AS}$ are the wavevectors of the probe pulse, and of the anti-Stokes scattered light, respectively.

Under these conditions we are clearly probing just the component i; the other oscillators $j \neq 1$ do not interact with the probe pulse because they are not phase matched. By varying the delay time of the probe pulse we are measuring the dephasing time of just that particular phase-matched component of the inhomogeneous distribution.

The measurement of the A_1 mode of carbon tetrachloride discussed in Subsection 5.1.5 is a good example of such a situation. In this experiment the scattering geometry was chosen so that only one isotope component was detected, and the τ value obtained from this experiment consequently represents the relaxation time of molecules of a specific isotopic composition ($CCl^{37}Cl_3^{35}$).

A very interesting situation arises when the phase-matching condition is satisfied for several different vibrational waves simultaneously. For example, if two components are phase matched the total material excitation seen by the probe pulse is modulated at the difference frequency $\Delta\omega$ of the two waves, and the scattered light generated by the probe pulse varies with delay time with a periodicity $2\pi/\Delta\omega$.

Beat notes due to interference of different isotopes have been observed for the first time in several tetrahalides [5.9, 12]. As a first example, let us discuss CCl_4 [5.12]. It is instructive to consider the geometry used in that experiment, because it stresses the relation between the measurement of relaxation times and the detection of the interference effects. The experimental setup of Fig. 5.6 is similar to the arrangement discussed earlier (see Fig. 5.1) except that now a variable aperture is used to select anti-Stokes light scattered into different directions.

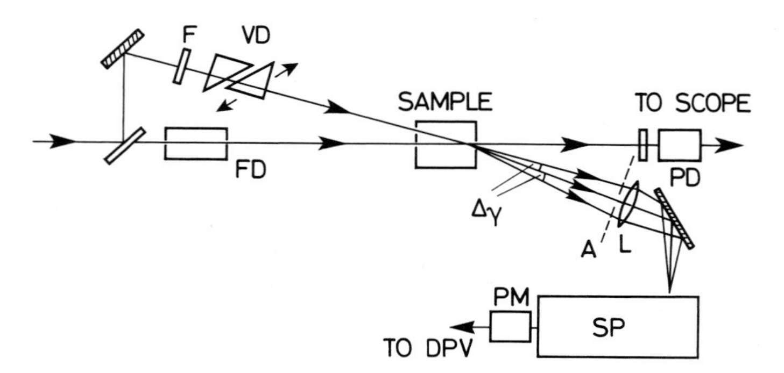

Fig. 5.6. Example of an experimental system for the observation of inhomogeneous dephasing. With the help of an aperture A scattering from a single vibrational component is selected (selective phase matching). With a larger opening scattering from several components is detected simultaneously (nonselective phase matching), and interference effects are observed. VD, FD, variable and fixed delay; PD, PM, photodetectors; SP, spectrometer; F, filter (after *Laubereau* et al. [5.12]

In the experiment of [5.12] the preparing pulse generated Stokes light at different, but small angles, and as a result the wavevectors of the material excitation were distributed over a certain solid angle close to the forward direction (see insert in Fig. 5.6). In [5.12] phase matching of the probe pulse was adjusted such that anti-Stokes scattering was generated from *two* different isotope components at the same time, in cones of about 2 mrad separated by 6 mrad. What is observed in the experiment depends on the setting of the aperture. With the aperture centered at one cone and stopped down to accept an angle of 2 mrad, the measured anti-Stokes light varied with delay time in the same way as observed in the earlier CCl_4 experiment [5.5], i.e., an exponential decay was observed with a time constant $\tau = 3.6 \pm 0.4$ ps. With this setup the relaxation time of a single isotopic component is measured.

However, a completely different time dependence was measured when the aperture was increased allowing anti-Stokes light from the two different cones to be detected simultaneously. The experimental result for this situation is shown in Fig. 5.7. The scattering curve now exhibits well-defined maxima and minima with an envelope given by the dephasing time $\tau \sim 4$ ps. The periodicity of the peaks corresponds to a frequency $\Delta\omega = 2.9\ \mathrm{cm}^{-1}$ which agrees very well with the separation of the two most abundant isotopic compositions $CCl^{37}Cl_3^{35}$ and CCl_4^{35} of approximately 3 cm^{-1}.

Figure 5.8 shows results for $SnBr_4$. This is a more interesting case because the isotopic line splitting is smaller and the components are not resolved. The decay curve of Fig. 5.8a was measured for selective phase matching of a single isotope component ($SnBr^{79}Br_3^{81}$). Under these conditions the scattered anti-Stokes light exhibits a strict exponential decay over 4 orders of magnitude, with a time constant of 3 ps. This value represents the dephasing relaxation time of the investigated isotope component.

On the other hand, when a nonselective phase-matching geometry was used permitting several isotopes to be probed simultaneously, the curve shown in Fig. 5.8b was measured. It is seen that now the anti-Stokes light decays much faster than expected for an exponential decay corresponding to a relaxation time of 3 ps. This observation is explained by the fact that we are now looking at inhomogeneous dephasing due to the interference of several isotope components. In $SnBr_4$ the frequency separation is much smaller than in CCl_4, and the beat period is too long to be seen over the range of delay times investigated in this experiment. The relaxation time of 3 ps inferred from the data of Fig. 5.8a corresponds to a homogeneous linewidth of 1.8 cm^{-1} which should be compared with the experimental linewidth of 3.5 cm^{-1} determined by the isotope effect.

A comparison of the results in Figs. 5.8a and 5.8b stresses again the importance of phase matching for the measurement of dephasing times. Quite obviously, very misleading results are obtained if measurements were made with a poorly defined scattering geometry. With proper choice of the phase matching, on the other hand, information on the frequency distribution

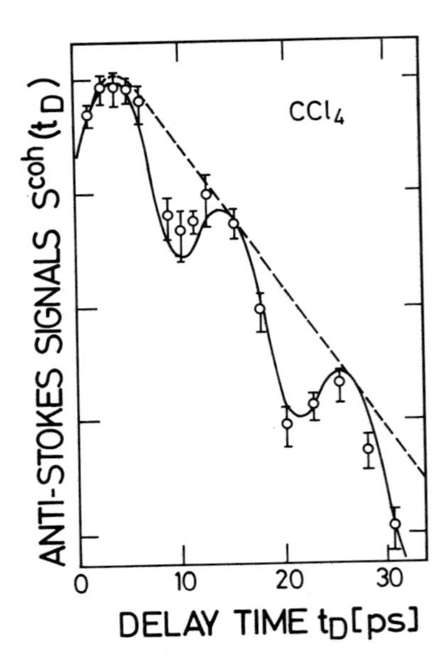

Fig. 5.7. Coherent anti-Stokes scattering versus delay time for nonselective phase matching. The experiment shows beating of different isotope components of CCl_4 (after *Laubereau* et al. [5.12])

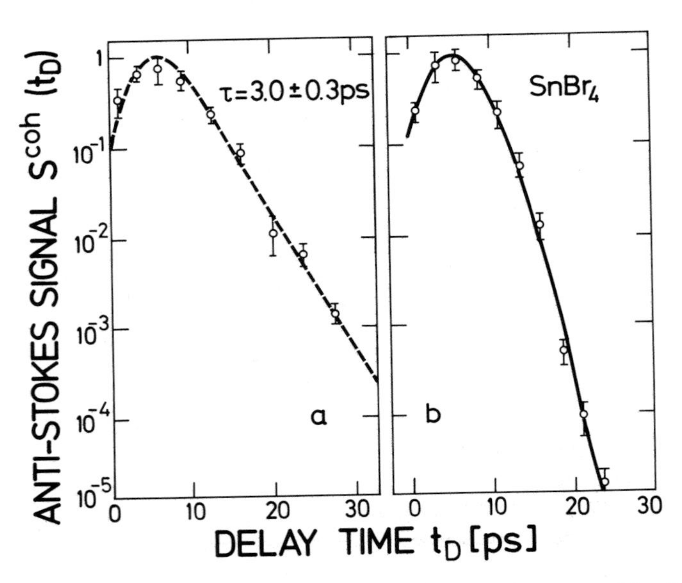

Fig. 5.8 a and b. Coherent anti-Stokes probe scattering versus delay time in $SnBr_4$ measured (a) with selective phase matching, and (b) with nonselective phase matching. Curve *a* represents the decay of the A_1 vibration of a single isotope component, $SnBr^{79}Br_3^{81}$. The non-exponential decay of curve *b* is due to the interference of the anti-Stokes scattered light from several different isotope components of $SnBr_4$ (after *Laubereau* et al. [5.12])

of an inhomogeneously broadened band can be obtained even in cases where the inhomogeneous structure is not resolved in the spectra.

Rapid dephasing among the components of an inhomogeneous frequency distribution also occurs in other inhomogeneously broadened systems. For example, effects similar to the dephasing of the isotopes of molecular liquids have been observed in calcium atomic vapor by *Matsuoka* et al. [5.13]. The electronic transitions of the atoms in the vapor are inhomogeneously broadened by the Doppler effect. Using a dye laser at $\omega_1 = 10925$ cm^{-1} synchronously pumped by a modelocked ruby laser ($\omega_2 = 14400$ cm^{-1}) *Matsuoka* et al. coherently excited the calcium atoms from the $^1S_0$ ground state to the lowest $^1D_2$ state via two-photon absorption. Following excitation the atoms oscillate at a frequency $\sim 2\omega_1$ corresponding to the $^1S_0 \to {}^1D_2$ plus an increment due to the individual Doppler shift. Dephasing occurs between atoms belonging to different velocity components of the Maxwell distribution, leading to a gaussian decay with a time constant $\tau_G = 2/\Delta\omega_G$, where $\Delta\omega_G$ is the frequency width of the Doppler-broadened transition. *Matsuoka* et al. monitored the Doppler dephasing by measuring the nonselectively phase-matched, anti-Stokes electronic Raman scattering from the excited Ca atoms produced by picosecond probe pulses at $\omega_2 = 14400$ cm^{-1} (ruby laser). The anti-Stokes light in the uv at $2\omega_1 + \omega_2 = 36250$ cm^{-1} (2758 Å) decayed with delay time according to a gaussian function, as expected, and the time constant was measured to be $\tau_G = 133$ ps ($\Delta\nu_G = 2.4$ GHz).

5.1.6 Measurement of Optical Phonon Lifetimes in Crystals

In this section we will discuss the application of picosecond Raman probe techniques for measuring relaxation of lattice vibrations. It should be recalled that in first order Raman scattering the light interacts with optical phonons near the center of the Brillouin zone. Phonons of larger wavevectors could be studied by higher order Raman scattering.

A few remarks may be required to explain the relation between the optical phonon lifetime and the time constants τ and τ' introduced in Subsection 5.1.3 for describing the different aspects of vibrational relaxation in molecular liquids.

1) Lattice vibrations are usually analyzed in terms of vibrational modes of well-defined frequency ω_v and wavevector $k_v(\omega_v)$. The coherent vibrational waves we were considering in the discussion of liquids are the exact analogue of these lattice modes. The dephasing time τ introduced for molecular vibrations is the equivalent of the lifetime of lattice modes, or optical phonon lifetime. Thus, the optical phonon lifetimes in solids can be measured by the same techniques as the dephasing times in liquids.

2) The energy relaxation time τ described the relaxation of a system of oscillators having lost all spatial correlation of the motion but still possessing excess vibrational energy (at the original frequency). The corresponding dephased state of a vibrating lattice can be described by a superposition of

phonon modes of the same vibrational frequency but with different wavevectors. A similar time constant τ' can be introduced for the solid describing relaxation of vibrational energy distributed over a number of modes in an extended region of the Brillouin zone of a certain phonon branch. The latter time constant τ' is to be distiguished from the phonon lifetime proper τ which determines the decay of the individual phonon modes.

By analogy with homogeneously broadened molecular transitions we expect the decay time of lattice modes in perfect crystals to be given by the reciprocal of the homogeneous linewidth. But picosecond techniques can now be used to measure phonon lifetimes directly, making an independent comparison with spectral data possible.

Alfano and *Shapiro* [5.14] were the first to attempt a direct measurement of optical phonon lifetimes by using the picosecond coherent Raman probe technique. They studied the A_{1g} mode at $\omega_v = 1086\ \text{cm}^{-1}$ of calcite (the same lattice mode of calcite was used for the first demonstration of coherent Raman scattering [5.4]). In the experiment of [5.14] the A_{1g} mode was excited by stimulated Raman scattering of picosecond pulses at 1.06 μm, and the phonon modes were interrogated by properly delayed probe pulses at the second harmonic (0.53 μm). Phase matching of the probe pulses was adjusted for coherent *Stokes* scattering off the excited A_{1g} phonons. By measuring the scattered Stokes light (at $\lambda = 5624$ Å) as a function of the delay time of the probe pulse the authors of [5.14] measured a decay of the Stokes intensity with a time constant of 8.5 and 19.1 ps at 100 K and 297 K, respectively. Spontaneous Raman linewidth data [5.15] of the same mode, on the other hand, suggested a phonon lifetime of 4.8 ± 0.6 ps ($\Delta\bar{\nu} = 1.1 \pm 0.12\ \text{cm}^{-1}$) at room temperature, and of 7.7 ± 1.4 ps ($\Delta\bar{\nu} = 0.69 \pm 0.12\ \text{cm}^{-1}$) at 100 K.

These discrepancies gave rise to speculations that the decay of highly excited coherent lattice vibrations differs from the relaxation of phonons close to thermal equilibrium. *Alfano* and *Shapiro* [5.14] estimated that in their experiment the number of phonons per mode exceeded the thermal equilibrium value by factors 10^6 and 10^9, respectively, for the two temperatures. They also suggested a possible wavevector dependence of the relaxation time by pointing out that phonons from different regions in k-space are involved for 90-degree detection of spontaneous Raman scattering (the usual geometry for measuring spontaneous Raman spectra), e.g., $k_v = 2.3 \times 10^5\ \text{cm}^{-1}$, and for stimulated Raman scattering close to the forward direction, $k_v = 1.1 \times 10^4\ \text{cm}^{-1}$.

A different investigation of optical phonon relaxation was published by *Laubereau* et al. [5.16]. In this experiment the lifetime of the transverse optical phonon mode at $1333\ \text{cm}^{-1}$ of diamond was measured. A careful theoretical and experimental analysis of the phonon population of the lattice mode at $1333\ \text{cm}^{-1}$ was made. It was established that the occupation number enormously exceeded thermal equilibrium value when the lattice mode was excited by stimulated Raman scattering of a single well-defined picosecond pulse. Factors of 10^{21} and 10^{13} were calculated for 77 K and 300 K, respectively.

The relaxation of these highly populated modes was measured by using anti-Stokes coherent Raman probing. The experimental result reproduced in Fig. 5.9 shows the observed variation with probe pulse delay time of the scattered anti-Stokes light. The decay of the lattice mode could be followed over almost four orders of magnitude to an occupation level still much higher than the thermal equilibrium value. A clean exponential decay was found for both temperatures. From these measurements the phonon lifetime was inferred to be 2.9 ps at 300 K, and 3.4 ps at 77 K. These data suggest Raman linewidth values of 1.83 cm^{-1} (300 K) and 1.56 cm^{-1} (77 K) in very good agreement with the observed spontaneous Raman linewidths of 1.48 cm^{-1} and 1.65 to 2.2 cm^{-1} [5.17] for 77 K and 300 K, respectively. Thus, there is no evidence for a different decay mechanism in the picosecond experiments. It was concluded in [5.16] that these extremely "hot" phonons generated by SRS decay with the same time constants as phonons at much lower densities in low power spontaneous Raman experiments.

In fact a careful remeasurement of the A_{1g} mode of calcite [5.18] ruled out the initial discrepancies [5.14] and gave very good agreement between the Raman linewidths and the directly measured phonon lifetimes.

One might intuitively be tempted to expect different relaxation times for those extremely hot phonons produced by SRS, but a closer analysis indicates why this is not the case. Although the phonon occupation of the modes greatly exceeded the thermal equilibrium by an enormous amount it turns out that the vibrational amplitude was still very small compared to the total thermal

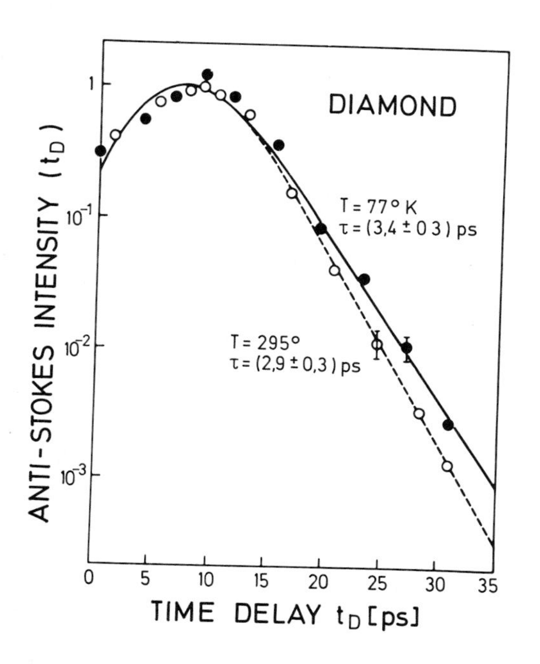

Fig. 5.9. Measurement of the TO phonon lifetime in diamond. Coherent anti-Stokes scattering versus delay time of the probe pulse. The maximum corresponds to approximately 10^{17} phonons/cm^3 generated by stimulated Raman scattering of the pump pulse (after *Laubereau* et al. [5.16])

amplitude. For diamond the vibrational amplitude was calculated to be only 10^{-4} Å for an occupation number of 10^{10} phonons per mode (at 1333 cm^{-1}) which is to be compared with the thermal rms amplitude of about 0.2 Å at 77 K (10^{-11} phonons per mode at 1333 cm^{-1}). The reason for this surprising result is the large density of vibrational states, $8.8 \times 10^{22}/cm^3$ in diamond. Because of the large number of modes, minute thermal excitations from all the individual modes add up to amplitude values by far in excess of the contributions to the vibrational amplitude from the few highly populated modes (in diamond, about 10^7 modes/cm^3 were populated by the SRS pumping process).

By looking at the degree of vibrational excitation from this point of view it is not suprising that the relaxation times of the hot phonons do not differ from those near thermal equilibrium. The phonon decay processes are due to the anharmonic terms of the lattice potential, and the degree of anharmonicity experienced by a lattice atom depends on its displacement, but the total displacement is still very small, even if a few modes are very highly populated.

The experiments in calcite and diamond were concerned with Raman active lattice modes coupling only weakly to the radiation field. A very interesting situation arises for infrared active phonons. It is well known that the strong coupling of these modes with the electromagnetic field gives rise to polariton states [5.19], i.e., to coupled, partially electromagnetic and partially vibrational modes with typical dispersion relations $\omega_v(k_v)$. Interestingly, the polariton shows strong dispersion near the center of the Brillouin zone for wavevectors comparable with the wavevector of light. The frequency of Raman phonons is constant in this region of k-space. In crystals lacking a center of inversion a lattice mode can simultaneously be infrared active *and* Raman active. Consequently, polaritons can be studied in noncentrosymmetric solids with the same Raman techniques that were used in diamond and calcite.

A direct picosecond measurement of the decay of a polariton mode involving the TO lattice phonon at 367 cm^{-1} in GaP was reported in [5.20]. This experiment differs from the calcite and diamond measurements in several interesting aspects. In GaP the polariton mode was excited by two intense synchronous picosecond pulses at frequencies $\omega_L = 9455$ cm^{-1} and $\omega_s = 9087$ cm^{-1}. The first pulse at 9455 cm^{-1} was directly obtained from a modelocked neodymium laser, while the second pulse at 9087 cm^{-1} was generated by stimulated Raman scattering of the first pulse in a separate liquid cell filled with $SnCl_4$ ($\omega_v = 368$ cm^{-1}). Thus the driving force (see (5.23)) oscillated at the difference frequency $\omega_L - \omega_s = 368$ cm^{-1}. Note that the relevant resonance frequency of the polariton follows from the dispersion curve with the polariton wavevector k_v determined by the wavevector difference of the two driving fields, i.e., $k_v = k_L - k_s$. The directions of the two preparing pulses were chosen such that k_v equaled 2770 cm^{-1}, somewhat below the TO-phonon frequency, i.e., the polariton was mostly phononlike. The light pulses were driving the polariton slightly *above* resonance (at 368 cm^{-1}).

In the experiment [5.20] the decay of the excited polariton mode was measured by coherent anti-Stokes Raman scattering, probing the material excitation with weak delayed pulses at 9455 cm^{-1} (the frequency of one of the pump pulses). From the exponential tail of the measured scattering curve the polariton relaxation time was found to be 5.5 ps at room temperature.

In GaP it is difficult to compare the polariton lifetime with the linewidth data because the TO phonon resonance shows up as an asymmetric band having a half width of about 4 cm^{-1} [5.21]. In principle it should be possible to predict the lifetime from the complex dielectric function $\varepsilon(\omega)$ [5.22]. However, a strong frequency dependence of the damping constant Γ is suggested by the infrared and Raman spectroscopic data. A completely satisfactory functional form for $\varepsilon(\omega)$ to account for the various experimental data has not yet been found [5.23].

Picosecond experiments of the type discussed here could yield accurate information on the frequency dependence of the damping function. Only one point along the dispersion curve has been measured so far, but with tunable synchronous picosecond pulses, measurements could cover the complete polariton branch.

5.1.7 Energy Relaxation and Energy Transfer

In Subsection 5.1.3 we mentioned that the decay of a vibrational wave may be caused by two different mechanisms, i.e., either by processes changing only the phase, or by relaxation due to dissipation of vibrational energy. The total effect of these two contributions determines the relaxation time τ, but there is no information on the energy relaxation time τ'. Previously, only crude estimates of the energy relaxation time of molecular liquids could be obtained, for example, from ultrasonic absorption and dispersion data, but accurate values of τ' for individual vibrational modes were not available. This situation has changed with the advent of picosecond measuring techniques. Today it is possible not only to measure the energy relaxation time, but also to study decay routes and to identify decay products of relaxing vibrational modes.

The principle for measuring the lifetime of excited vibrational levels was discussed in Subsection 5.1.3. The energy relaxation time τ' can be obtained from time-resolved observations of incoherent anti-Stokes Raman scattering. *Ducuing* and *DeMartini* [5.6] demonstrated the feasibility of this method by measuring the lifetime of the first excited state ($v=1$) of molecular hydrogen. For example, at a gas pressure of 50 at that lifetime was 21 μs. At this pressure the Raman linewidth is determined by molecular collision processes. Assuming homogeneous broadening a value of $\tau = 53$ ps is estimated from the linewidth, suggesting greatly different time constants for dephasing and energy relaxation.

The first measurement on a picosecond time scale of energy relaxation in liquids was reported by *Laubereau* et al. [5.7]. They investigated the CH bond stretching vibration of ethanol (C_2H_5OH) at 2928 cm^{-1}, and of 1,1,1 trichloroethane at 2939 cm^{-1}. In these experiments SRS of picosecond pulses at 1.06 μm were used to populate the vibrational levels of interest. It is instructive to consider the excess population created by the stimulated Raman pump process. The maximum number of excited molecules near the exit window of the liquid cell containing the sample can be estimated from the Stokes conversion efficiency, using the power balance equation (5.12). Typically, one works with a few percent conversion of laser energy to Stokes energy. For these conditions the occupation probability of the $v=1$ state is 10^{-4} to 10^{-3}. This value should be compared with the thermal equilibrium population. For a high frequency molecular vibration, e.g., the CH bond stretching modes, the probability of being thermally excited is very small: for ethanol at $\omega_v = 2928\ \mathrm{cm}^{-1}$ we have

$$n_{th} \simeq \exp(-\hbar\omega_v/k_B T) \simeq 5 \times 10^{-7} \tag{5.21}$$

at room temperature.

Thus, the population generated by SRS pumping exceeds thermal equilibrium by about three orders of magnitude. However, note that the anti-Stokes scattering off the excess population is still much weaker than the spontaneous Stokes scattering from the same vibrations. In the Stokes process the scattered intensity is proportional to the total number of molecules in the vibrational ground state. The anti-Stokes scattering, on the other hand, involves only the excited states, and it is therefore weaker than the Stokes scattering by a factor of typically $1/n \simeq 10^3$ to 10^4 even when the levels are strongly pumped by SRS [5.7].

The different intensity levels of the scattered light in the incoherent Raman process and in the coherent Raman process should also be realized. Phase-matched coherent scattering can readily amount to a few percent of the incident light. From the point of view of the available intensity of the scattered probe light the measurement of the energy relaxation time τ' by incoherent scattering is more difficult than the measurement of the dephasing time τ by phase-matched coherent scattering.

In the experiment of [5.7] the population of the vibrational levels was interrogated with a properly delayed probe pulse at the second harmonic of the pump pulse. The experimental geometry corresponded to the principal arrangement discussed in Fig. 5.2. The experimental results for ethanol and trichloroethane are presented in Figs. 5.10 and 5.11. Both the incoherent anti-Stokes scattering S^{inc} and the coherent scattering S^{coh} are shown for a ready comparison of the relevant relaxation processes. Normalized signals are plotted, but the different intensity scales for S^{inc} and S^{coh} should be kept in mind.

The very different variations with delay time of S^{inc} and S^{coh} are immediately apparent from the figures. The fast decay of the coherent anti-Stokes scattering in ethanol has already been discussed in Subsection 5.1.4. It was noted that the decay was not fully resolved, and a value of about 0.25 ps was estimated. On the other hand, it is seen that the incoherent scattering decreases at a much slower rate. The observed decrease of S^{inc} with probe delay time reflects the relaxation of the excess population at $2928\ \mathrm{cm}^{-1}$, and from the exponential

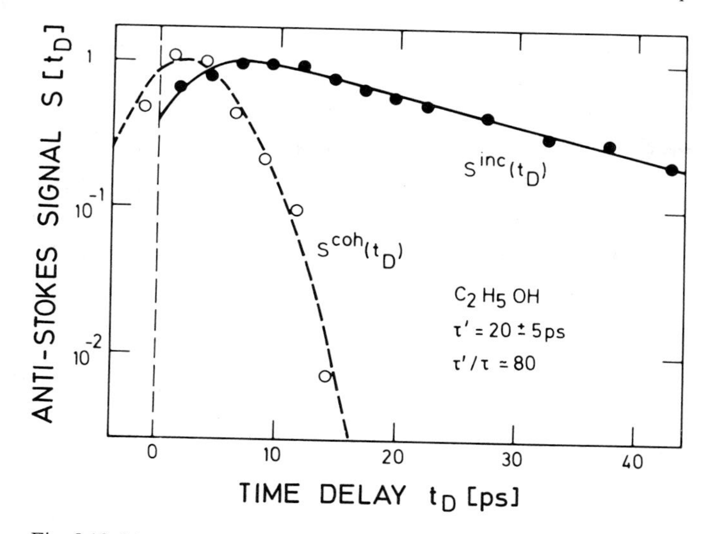

Fig. 5.10. Measurement of incoherent (full curve) and coherent (broken curve) anti-Stokes Raman scattering for the determination of the energy relaxation time τ' and the dephasing time τ of the CH bond stretching vibration of ehtanol at 2928 cm^{-1} (after *Laubereau* et al. [5.7])

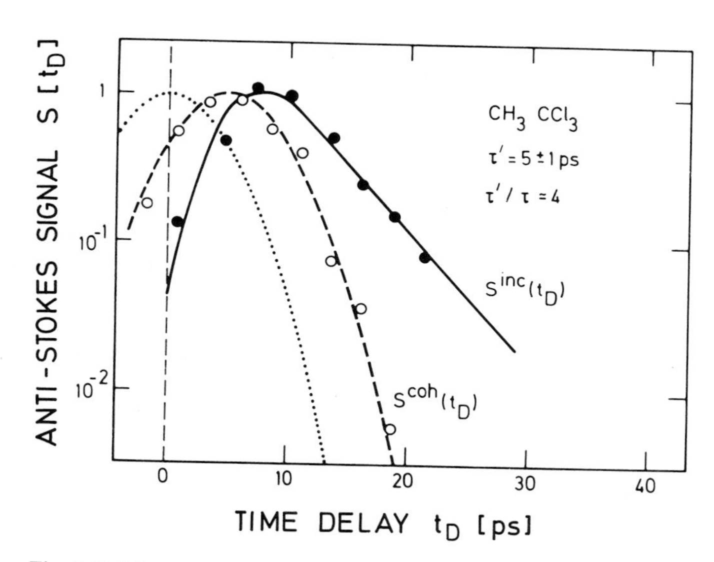

Fig. 5.11. Measurement of the energy relaxation time τ' and of the dephasing time τ of the CH bond stretching vibration at 2939 cm^{-1} of trichloroethane (after *Laubereau* et al. [5.7])

decay of S^{inc} the energy relaxation time of the CH bond stretching mode was inferred to be 20 ± 5 ps.

Similar results were obtained for trichloroethane. The coherent Raman signal also decays very fast; the experimental data could be fitted assuming that S^{coh} decays in 1.3 ps (dashed curve). On the other hand, the decay of S^{inc} is clearly resolved, and for the 2939 cm^{-1} vibration of CH_3CCl_3 τ' was 5 ± 1 ps.

These two experiments represent the first direct measurement of the energy relaxation of well-defined molecular vibrational modes in liquids, and the data firmly establish large differences between the time constants for dephasing processes and for energy relaxation, e.g., τ'/τ is ~ 80 for ethanol and ~ 4 for trichloroethane.

The results demonstrate the usefulness of incoherent Raman probing for studying picosecond energy relaxation in molecular liquids. These methods can be readily extended for investigations of other modes and other molecules, and also for studying the subsequent stages of energy relaxation, e.g., the decay routes by which thermal equilibrium is eventually established.

Before discussing these other applications of the Raman techniques we mention a different method for measuring vibrational lifetimes which utilizes excitation by absorption of infrared picosecond pulses rather than stimulated Raman scattering.

Laubereau et al. [5.24] have investigated vibrational relaxation in the electronic ground state of a dye molecule, coumarin 6. A picosecond infrared pulse at $\omega_1 = 2970$ cm^{-1} ($\lambda = 3.37$ µm) was used to excite molecular vibrations, and the number of excited molecules was probed by measuring the fluorescence induced by a interrogating pulse in the visible, at $\omega_2 = 18910$ cm^{-1} ($\lambda_2 = 0.53$ µm). The probe pulse produced transitions of the vibrationally excited molecules from the electronic ground state S_0 to the first excited singlet S_1. The coumarine molecules return to S_0 by emitting fluorescence at about 5000 Å. According to [5.24] the essential point is that the energy of the probe photons is not sufficient to excite the molecules to S_1 unless there is excess vibrational energy in the electronic ground state. Some background fluorescence was observed due to the thermal population of the vibrational states. However, when the molecules were excited by the infrared pulse the fluorescence was increased by a factor of about 30. By increasing the delay time between the visible probe pulse and the infrared pump pulse the enhanced fluorescence was observed to decay exponentially with a time constant of 1.3 ps at room temperature, and 1.7 ps at 253 K. According to *Laubereau* et al. the infrared pulse primarily populates a well-defined CH bond stretching mode of the ethyl groups of coumarin 6, and they believe that the observed decay of the induced fluorescence directly reflects the relaxation of those vibrational modes. It was not possible to decide whether the relaxation was due to decay to lower vibrational states or due to energy transfer to neighboring states of similar energy but smaller transition probabilities to S_1.

This question leads us to the discussion of the relaxation mechanisms responsible for picosecond vibrational decay. The first attempt to identify relaxation mechanisms with the help of picosecond optical techniques was made by *Alfano* and *Shapiro* [5.25], who investigated the same CH bond stretching vibration of ethanol discussed above [5.7]. Exciting the ethanol molecules via stimulated Raman scattering and probing the excess vibrational population by measuring the anti-Stokes Raman scattering of a probe pulse at the second harmonic they were able to detect the immediate decay product of the CH mode at 2928 cm^{-1}. Anti-Stokes probe scattering was found not only at a frequency corresponding to the mode excited directly by the pump pulses, but also at a new band centered around an anti-Stokes shift of about 1460 cm^{-1} approximately half the frequency of the original mode. This new scattering was interpreted as possible evidence that subsidiary daughter vibrations were populated by decay of the modes excited by the SRS pump process. However *Alfano* and *Shapiro* [5.25] also considered possible contributions to the anti-Stokes scattering at 1460 cm^{-1} due to combination Raman scattering between the excited vibration at 2928 cm^{-1} and the various possible modes of ethanol around 1460 cm^{-1}. They could not definitely rule out contributions of the observed probe scattering from this alternative mechanism.

A detailed study of the relaxation mechanism of the CH bond stretching mode of ethanol was reported by *Laubereau* et al. [5.26]. Their experimental result of the variation with probe delay time of the anti-Stokes light is shown in Fig. 5.12. The data for the scattering at 2920 cm^{-1} (full circles) agree with the earlier work [5.7], confirming the lifetime of about 20 ps. The open circles represent the rise and decay of the anti-Stokes scattering at the new frequency around 1450 cm^{-1}. Note that the secondary scattering decreases at a different, much slower rate. This observation rules out the possibility that the light at 1450 cm^{-1} is due to difference frequency Raman scattering.

A key point for the interpretation of the relaxation mechanism was the following additional observation. When the *spectrum* of the scattered probe light was measured at a fixed delay time of about 11 ps, anti-Stokes scattering was detected over a spectral region ranging from 2850 cm^{-1} up to about 3000 cm^{-1}, i.e., over the full width of the Raman bands in this frequency regime. This spectrum was much wider than one would expect if the scattering originated just from the pumped mode at 2928 cm^{-1} (linewidth 30 cm^{-1}). Apparently, a rapid energy transfer takes place from the excited mode to the neighboring vibrations. For an interpretation of these observations *Laubereau* et al. [5.26] used a model based upon the energy level scheme shown in Fig. 5.13. The Raman band around 2900 cm^{-1} is actually composed of a number of different overlapping components corresponding to the vibrational levels on the left of the upper rectangle in Fig. 5.13 ($v=1$ levels). There are additional energy levels around 2900 cm^{-1} due to the $v=2$ states of the CH bond bending modes (δ_H). The $v=1$ states of the same mode are represented by the levels in the lower rectangle. According to the model proposed in [5.26] the first step of the relaxation is a rapid thermalization of the vibrational energy among the levels of the upper rectangle with a time constant τ_c. Subsequently, the δ_H levels below are populated by decay from the equilibrated ensemble of states at 2900 cm^{-1}. The decay time is determined by a common relaxation time τ_1. Further, it is postulated that the lower δ_H levels are also in quasi-thermal equilibrium relaxing to the vibrational ground state with a different relaxation time τ_2.

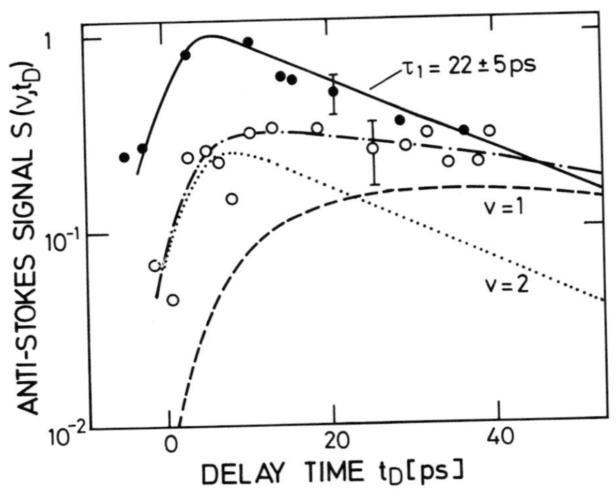

Fig. 5.12. Measurement of the energy relaxation time τ_1 of the group of vibrational levels around 2900 cm^{-1} of ethanol (full circles), and detection of the subsidiary modes around 1450 cm^{-1} (open circles) populated during relaxation of the levels at 2900 cm^{-1}. The dashed and the dotted curves are the calculated individual contributions from $v=1$ and $v=2$ levels; the dash-dotted curve is the total scattering due to both contributions (after *Laubereau* et al. [5.26])

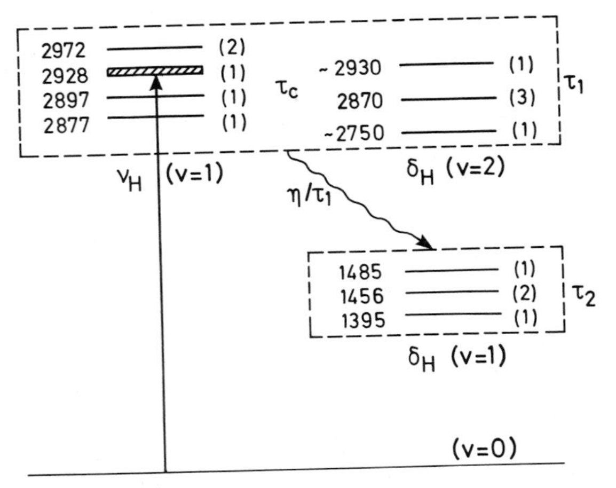

Fig. 5.13. Energy level scheme illustrating the interaction of the various vibrational modes of ethanol, according to *Laubereau* et al. [5.26]

There are two different contributions to anti-Stokes probe scattering around 1450 cm^{-1}. First, there is efficient scattering due to transitions from the $v=2$ to the $v=1$ states of the modes. The cross section for this process is twice as large as the cross section for the $v=1$ to $v=0$ transition of the same mode (i.e., much larger than that of the combination tone due to transitions from the $v=1$ state of the mode to the $v=1$ state of δ_H). The second contribution to the anti-Stokes light stems from transitions between the $v=1$ state of the δ_H modes and the vibrational ground state.

Based on this model *Laubereau* et al. [5.26] calculated the expected time dependence of the two different components. The dotted and the dashed curves of Fig. 5.12 represent the calculated individual $v=2$ and $v=1$ contributions, respectively, for values of $\tau_c=0.5$ ps and $\tau_2=40$ ps.

It is seen that neither of the two components accounts for the observed data, but that the sum of the contributions from $v=2$ and $v=1$ describes the observations very well. In particular, the decrease of the anti-Stokes scattering at 1450 cm^{-1} apparently is determined by the decay of the δ_H subsidiary levels, with a lifetime of approximately 40 ps.

Additional strong evidence for the significance of $v=2 \rightarrow v=1$ scattering was obtained from the observed frequency spectrum of the anti-Stokes light at 1450 cm^{-1}. The spectrum generated by the picosecond probe pulses differed from the ordinary spontaneous anti-Stokes scattering in thermal equilibrium by an additional peak shifted to lower energies. This structure could be accounted for by $v=2 \rightarrow v=1$ anti-Stokes scattering. The observed shift to lower energies of the $v=2 \rightarrow v=1$ peak with respect to the $v=1 \rightarrow v=0$ scattering is due to anharmonicity.

Let us now turn to yet another aspect of vibrational relaxation in liquids. Not only did picosecond investigations of the CH bond stretching mode of ethanol yield accurate values of the energy relaxation time τ', but these studies also disclosed a complex decay mechanism involving energy transfer between several different vibrational modes. However, we have not discussed whether these relaxation processes involve intramolecular or intermolecular energy transfer. Experimental evidence for the significance of intermolecular energy transfer was obtained from a picosecond investigation of binary mixtures of trichloroethane [5.27]. At the beginning of this section we descussed the measurement of the lifetime of the $v=1$ state of the ν_H bond stretching mode at 2939 cm^{-1} of pure trichloroethane. Using the same experimental techniques the energy relaxation time of the this vibration was measured in binary mixtures with carbon tetrachloride. It was found that τ decreased proportional to the square of the concentration of CH_3CCl_3 molecules, varying from 5 ps in the pure liquid to about 15 ps at a concentration of 0.6 [5.27]. This observation suggests a relaxation mechanism whereby an excited CH_3CCl_3 molecule loses its energy during a collision with two other molecules of the same species. Apparently the CCl_4 molecules do not accept vibrational energy; carbon tetrachloride just serves as a diluent.

A very different situation was encountered in binary mixtures of the same liquid with deuterated methanol. Some increase of τ' with dilution was also observed, but the effect was considerably less than in mixtures with CCl_4. It must be concluded that in mixtures with deuterated methanol the effect of dilutions has to compete with new decay channels opened up by the presence of the methanol molecules. In fact, the observed concentration dependence suggested that there is efficient energy transfer during a collision of an excited CH_3CCl_3 molecule with a ground state methanol molecule and another trichloroethane molecule ($v=0$).

The picosecond incoherent Raman probe technique permitted a direct experimental proof of these conclusions. For example, in CH_3CCl_3 anti-Stokes scattering was detected not only at a frequency corresponding to the original mode at 2939 cm^{-1}, but also at a new frequency of about 1450 cm^{-1} which corresponds to the δ_H bond bending modes of CH_3CCl_3. This result parallels the observations in ethanol where the decay products of the relaxing ν_H modes were also identified to be the δ_H modes at approximately half the frequency. But the experiments in binary mixtures with CCl_4 now provide experimental evidence that the secondary probe scattering at 1450 cm^{-1} arises from an intermolecular energy transfer process in which the vibrational energy is transferred from the excited molecule to two ground state molecules during a three-body collision.

A similar transfer mechanism was also found in the mixtures with deuterated methanol. Excess population was detected by incoherent Raman probing for the ν_D mode of CD_3OD at 2227 cm^{-1}. It is very likely that in this case the remaining vibrational energy is accepted by a ground state CH_3CCl_3 molecule. A very good energy resonance exists for a splitting of the vibrational quantum at 2939 cm^{-1} of the original ν_H mode into one quantum at 2227 cm^{-1} of CD_3OD and a ν_{Cl} quantum at 713 cm^{-1} of trichloroethane [5.27].

Laubereau et al. [5.27] concluded from the observations in binary mixtures that intermolecular energy transfer requiring collisions of the excited molecule with two reaction partners plays a major role for energy relaxation in these liquids. Resonance with suitable energy levels of the two collision partners is necessary for efficient transfer of vibrational energy. From this point of view it is not surprising that CD_3OD is an efficient acceptor of excess energy from modes of CH_3CCl_3 while CCl_4 is not. The highest vibrational levels of CCl_4 are around 800 cm^{-1}; there is no resonance for energy transfer to vibrational levels of CCl_4.

Concluding this section we mention the work of *Monson* et al. [5.28] who used the picosecond Raman excite-and-probe technique [5.7] to measure the energy relaxation times of CH-stretching modes in a number of hydrocarbons. They noted a correlation of the relaxation time with the ratio of the number of methyl groups to the total number of carbon atoms of the molecules. *Monson* et al. were led to believe that the methyl groups play a major role for the dissipation of the vibrational energy of the CH modes in hydrocarbons.

5.2 Picosecond Electronic Interactions

In this section we describe a number of experiments in which picosecond optical pulses are used for studying electronic excitations in solids. The first two subsections deal with nonlinearities of the optical absorption induced by intense ultrashort pulses, i.e., multiphoton absorption and optical saturation effects. We then describe investigations of very dense electron-hole plasmas generated by picosecond light pulses. The section concludes with an account of experimental work on several ultrafast nonradiative relaxation processes of electronic excitations in solids.

5.2.1 Multiphoton Interactions

Modelocked lasers producing ultrashort light pulses offer several distinct advantages over other light sources for the observation of multiphoton transitions. In most materials very high light intensities are required for efficient multiphoton interaction. These intensities are readily achieved with picosecond light pulses of quite moderate total energy. Heating and material damage generally represent serious problems for longer pulses of comparable intensities, but these difficulties are often avoided with the use of very short pulses. Furthermore, multiphoton spectroscopy employing picosecond excitation makes it possible to readily distinguish between direct and stepwise multiphoton transitions by measuring the time response of the process. However, it has not yet been possible to fully utilize these principal advantages, because the frequency range and the tunability of the available picosecond light sources were limited. For instance, in two-photon spectroscopic work long duration laser pulses and flashlamps were mostly used, whereas multiphoton experiments involving picosecond pulses have so far been performed mostly at a few isolated frequencies. Detailed general reviews of two-photon spectroscopy are referenced under [5.29]. Here we will restrict ourselves to the discussion of multiphoton experiments involving ultrashort light pulses.

Multiphoton transitions can be detected directly by measuring the optical transmission of a sample at high intensities, or else indirectly by observing, for example, the photoluminescence or the photoconductivity resulting from multiphoton excitation.

Let us briefly discuss the direct method by considering the variation of the light intensity $I(z)$ in a material in which multiphoton transitions take place between two energy states with population N_1 (ground state) and N_2 (excited state). The change of the light intensity is described by

$$(1/I)dI/dz = -\beta_n I^{n-1} = -(N_1 - N_2)\sigma_n (I/\hbar\omega)^{n-1} \tag{5.22}$$

where β_n and σ_n are the n-photon absorption coefficient and the n-photon absorption cross section, respectively. The calculation of these coefficients by higher order perturbation theory has been discussed in detail in the literature [5.29].

The aim of multiphoton spectroscopy is, of course, to measure the magnitude and frequency dependence of these quantities experimentally. However, multiphoton transitions also provide an alternative method of material excitation. For example, large volumes of a medium can be excited readily by suitable adjusting the incident light intensity. The different selection rules of the higher order processes also permit us to obtain additional new information on electronic states not accessible by a single-photon interaction.

After the advent of picosecond light pulses much interest in two-photon absorption of fluorescent organic dyes was created by the work of *Giordmaine* et al. [5.30] who showed that the nonlinear excitation mechanism provided a convenient means for measuring the pulse duration by two-photon induced fluorescence (TPF – see Chap. 3 for details of this method). *Bradley* et al. [5.31] measured two-photon absorption cross section accurately in a variety of dyes under picosecond and nanosecond excitation. Typical values ranged from 10^{-51} to 10^{-49} cm^4s at a wavelength of 1.06 μm (rhodamine 6G and B, acridine red, disodium fluorescein, and DODCI). For these dyes the two-photon cross sections for picosecond and nanosecond excitation were the same within experimental error.

Numerous papers have been published on multiphoton interactions of picosecond pulses in solids, and very significant differences for picosecond and nanosecond light pulses were found. *Arsen'ev* et al. [5.32] have observed *photoluminescence* in several II-VI direct band gap semiconductors induced by two-photon and three-photon absorption of picosecond light pulses. For example, photoluminescence was generated in ZnS ($E_g = 3.6$ eV) by two-photon excitation at $\lambda = 0.53$ μm ($\hbar\omega = 2.34$ eV) and in CdS ($E_g = 2.42$ eV) by three-photon absorption at 1.06 μm ($\hbar\omega = 1.17$ eV). *Jayaraman* and *Lee* [5.33] observed *photoconductivity* in GaAs ($E_g = 1.4$ eV) produced by pulses at 1.06 μm. The change of conductivity varied directly with the square of the energy of the exciting pulses, suggesting free carrier generation via a two-photon process.

Three-photon absorption was measured in cadmium sulfide by *Jayaraman* and *Lee* [5.34] and by *Penzkofer* and *Falkenstein* [5.35]. In [5.34] picosecond pulses at 1.06 μm were used of produce photoconductivity in single crystals of CdS and in polycrystalline material. The conductivity due to incident pulses varied according to a power law with an exponent close to three, indicating excitation via three-photon transitions. *Penzkofer* and *Falkenstein* [5.35] measured the energy transmission for a single crystal of CdS for intensities between 10^9 and 4×10^{10} W/cm^2. In a 1 mm thick sample the energy transmission decreased significantly below the value given by the Fresnel reflection losses for intensities in excess of 10^{10} W/cm^2, falling to about 50% at 4×10^{10} W/cm^2. Careful measurements showed that the falloff of the transmission with incident intensity was faster than expected for a simple three-photon absorption process. However, by invoking free carrier absorption of the electron-hole plasma generated by the three-photon process the experimental data could be very well accounted for. The three-photon absorption coefficient was measured to be $\beta_3 = (1.1 \pm 0.3) \times 10^{-20}$ cm^3/W^2. This value is in good agreement with the one estimated in [5.34] from photoconductivity data.

Reintjes and *McGroddy* [5.36] investigated nonlinear absorption of 1.06 μm pulses in a 1-cm-thick single crystal of silicon at 20 K ($E_g \sim 1.2$ eV) and 100 K ($E_g \sim 1.19$ eV). They established that the ratio of transmitted optical energy to incident energy varied linearly with the incident intensity, as expected for a two-photon absorption process. From the slope of the line the two-photon absorption cross section β_2 was inferred to be 1.9×10^9 cm/W. In silicon, direct transitions to the conduction band minimum at the center of the Brillouin zone would require energies greater than 3.4 eV, which is much larger than twice the photon energy of 1.17 eV at 1.06 μm. *Reintjes* and *McGroddy* [5.36] attributed the observed optical nonlinearity to phonon-assisted, indirect two-photon transitions between the valence band and the conduction band minimum near the X-point of the Brillouin zone.

The indirect band gap of silicon at 100 K is very close to the single photon energy of the 1.06 μm pulses. There is a substantial probability for single photon transitions at the higher temperature. Under these conditions contributions to the optical nonlinearity are expected from stepwise processes whereby an electron is excited in the conduction band by absorption of a first photon, and a second photon is subsequently absorbed via the free-carrier absorption mechanism. These two different contributions were indeed observed experimentally, but they were separated by means of the excite-and-probe technique in which the absorption induced by an intense pump pulse is measured with the help of a weak probe pulse. Because the recombination time of the free carriers is long compared with the duration t_p of the light pulses, free carrier absorption persists even if the probe pulse is delayed with respect to the pump pulse. For zero delay time the combined effect of instantaneous two-photon absorption and free carrier absorption is seen, while for delay times $t_D > t_p$ only free carrier processes are observed. The probe pulse experiment showed that the contribution from two-step processes is smaller than that from the genuine two-photon process.

Bechtel and *Smith* conducted a very thorough study of two-photon absorption in variety of semiconducting crystals [5.37]. They used single picosecond pulses from a modelocked Nd-YAG laser for a direct determination of the two-photon absorption coefficient at $\lambda = 1.064$ μm. In this work the spatial and temporal structure of the laser pulses was analyzed carefully and taken into account for the evaluation of the two-photon absorption coefficient

Table 5.2

	GaAs	GaP	CdTe	CdSe	ZnTe
β_2 [cm/GW]	28 ± 5	0.2 ± 0.1	25 ± 5	30 ± 5	8 ± 4

from the measured (integrated) energy transmission of the samples. Also, these authors ascertained that possible complications from free carrier absorption (due to carrier generated via two-photon excitation) did not obscure their measurements. Table 5.2 shows the two-photon absorption coefficients for several III-V and II-VI semiconductors reported in [5.37]. These data are probably the most reliable to date. Most of the previous values obtained with longer laser pulses or with less well characterized light pulses were shown to be too high, in some cases by as much as an order of magnitude.

The first detailed investigation of the frequency dependence of two-photon absorption with picosecond light pulses was made by *Penzkofer* et al. [5.38]. They measured two-photon spectra in CdS single crystals (room temperature) using the excite-and-probe technique. A small volume of the sample was illuminated by a single very intense light pulse at 1.06 μm ($t_p = 6$ ps). A second weak picosecond probe pulse was used to measure the absorption induced by the pump pulse. The probe pulse was generated by a parametric process in a liquid cell filled with water [5.39]. The frequency spectrum of the parametrically generated light stretches continuously from the ir to the uv, yet the pulse had a duration similar to that of the intense exciting pulse, so that precise temporal overlap between pump and probe was readily achieved.

Two different types of experiments were performed. First, the incident and the transmitted probe pulses were detected at a fixed frequency with the help of spectrometers, and the variation of the crystal transmission at this frequency was measured as a function of the intensity of the exciting pulse (at 1.06 μm). Pump intensities as high as 10^{10} W/cm^2 were applied without causing material damage, and substantial absorption at the probe frequency was induced. For example, the probe transmission at 6000 Å was reduced by a factor of 20 in the presence of a 10^{10} W/cm^2 pump pulse. The strong two-photon interaction made it possible to determine accurate absolute values of the two-photon absorption coefficient from the measured variation of the induced absorption with pump intensity. It was ascertained that no significant nonlinear absorption occurred with the pump pulse alone.

In the second type of measurement the frequency dependence of the two-photon absorption coefficient was studied. For this purpose each of the two spectrometers measuring the incident and the transmitted probe spectrum, respectively, was equipped with a vidicon in conjunction with a multichannel analyzer, which made it possible to measure the complete induced absorption spectrum of the probe pulse from a single picosecond pulse.

The CdS two-photon absorption spectrum measured in this way is depicted in Fig. 5.14, plotting two-photon absorption cross section versus the combined photon energy of the pump pulse and the probe pulse for σ-polarized ($E \perp c$) and π-polarized ($E \| c$) probe light. Both cross sections increase rapidly at an energy of about 2.4 eV close to the band gap energy. The π cross section saturates at a value of about 4×10^{-49} cm^{4}s, whereas the σ-spectrum reaches a maximum of about 2×10^{-49} cm^{4}s near 3 eV. Note that even small differences between σ and π are resolved, e.g., a crossover of the spectra is seen at about 2.6 eV.

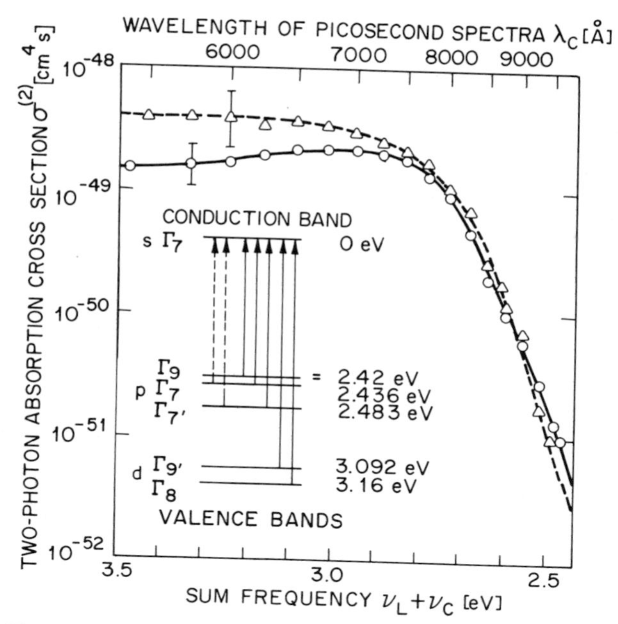

Fig. 5.14. Two-photon absorption cross section as a function of energy for CdS. Dashed curve (triangles): $E||c(\pi)$; solid curve (circles): $E\perp c(\sigma)$. The insert shows the band structure of CdS at the Γ point. Solid arrows and dashed arrows: Allowed two-photon transitions for σ and π polarization, respectively (after *Penzkofer* et al. [5.38])

The work of [5.38] demonstrates clearly that the picosecond techniques available to date offer a useful tool for measuring two-photon spectra accurately and quantitatively.

Multiphoton excitation can also be used to excite electrons from the bound states of a material to the free vacuum state, i.e., electrons can be knocked out of the material. The multiphoton photoelectric effect was demonstrated for the first time by *Sonnenberg* et al. [5.40]. Using picosecond pulses at 1.06 μm, *Burnham* [5.41] observed three-quantum photoelectric emission from a blue sensitive standard photocathode of an image orthicon. This effect was utilized to produce a direct oscilloscope display of the third-order autocorrelation function of the ultrashort pulses, i.e., the pulsewidth was measured with the help of this device (see Chap. 3 for the relevance of third-order autocorrelation). *Forkas* and *Horvath* [5.42] reported two-photon photoelectric emission from a Cs_3Sb photocathode when irradiated with picosecond pulses at 1.06 μm. From their work the emission coefficient of the cathode is estimated at roughly 4×10^{-14} Acm^2/W^2: a current density of 10^{-4} A/cm^2 was measured at an incident intensity of about 50 kW/cm^2. Measurements of the second order correlation function of modelocked pulses from an argon ion laser were reported by *Bennett* et al. [5.43]. They used a technique similar to that described in [5.41] but in their experiment the photoemission was due to a two-photon photoelectric effect.

Let us conclude the discussion of picosecond multiphoton interactions with a few remarks concerning multiphoton emission. This topic is relevant because of the intrinsic tendency of multiphoton stimulated emission to form extremely short pulses. It is apparent from (5.25) that light is amplified if a population inversion of levels ($N_2 > N_1$) exists. Note that the gain due to n-photon stimulated emission is proportional to I^{n-1}. Clearly, this dependence of gain on intensity leads to a very strong enhancement of any peaks in the light intensity, i.e., intensity maxima are amplified selectively. This is the opposite of the well-known pulse limiting action of a two-photon absorber. Laser action based on two-photon emission was originally proposed by *Sorokin* and *Braslau* [5.44]. Papers discussing the various aspects of multiphoton emission are listed under [5.45]. However, the experimental realization of two-photon lasers turned out to be very difficult. Probably the most serious complication is caused by the competition of anti-Stokes (electronic) Raman processes, which tend to quench the two-photon emission process very effectively. (For details see [5.45]).

5.2.2 Saturation of the Optical Absorption

In this section we discuss single photon transitions between the valence band and the conduction band in semiconductors. Nonlinearities of the optical absorption are caused by modifications of the thermal equilibrium distribution of electrons and holes. During the interaction with a strong light pulse the number of states in resonance with the light is reduced because the available initial and final states become crowded with electrons and holes, and this effect leads to a bleaching of the optical absorption. Two different situations should be distinguished. When the lifetime of the optically coupled states is short compared with the inverse pump rate (and the interband recombination time), then the optical nonlinearity can be described by a dynamic Burstein-Moss shift, i.e., by band filling [5.46]. With increasing optical excitation the quasi-Fermi levels of the electrons and holes are shifted into the conduction band and the valence band, respectively. For direct transitions saturation occurs when the energy separation of the quasi-Fermi levels approaches the photon energy [5.46].

On the other hand, when the excitation rates and the relaxation rates of the coupled states are comparable, the saturation results directly from the filling of the interacting states.

Very high concentrations of nonequilibrium carriers are readily generated by picosecond optical excitation. By studying saturated interband transitions with picosecond light pulses information may be obtained about the distribution of electrons and holes, and about the changes with time of the carrier distribution caused by relaxation processes or by diffusion and recombination.

As a first example let us discuss experiments in germanium involving indirect transitions from the valence band to the L-point conduction band minimum ($E_g \sim 0.66$ eV). *Kennedy* et al. [5.47] investigated nonlinear absorption of an 8 μm thick crystal wafer of intrinsic germanium at liquid nitrogen temperatures using single picosecond pulses at 1.06 μm. Two different types of experiments were performed. First, they measured the transmission of their sample as a function of the energy of the incident pulses. Second, they used the excite-and-probe technique to study the time dependence of the transmission by measuring the transmission of a weak probe pulse as a function of the arrival time after excitation by a strong pump pulse.

In the first experiment the measured transparency of the germanium sample increased from the small-signal value of about 8×10^{-5} to a maximum of 2×10^{-3} at an incident energy of $\simeq 3\times10^{-3}$ J/cm^2 (3000 MW/cm^2). The apparent decrease of the absorption coefficient by a factor of about 1.5 was interpreted as evidence for saturation of the interband transition.

Time-dependent measurements over a range of probe delay times from -8 ps to $+4$ ps revealed an increase in transmission over a very narrow time interval of only 4 ps, which appears to be even shorter than the duration of the pulses used in the experiment. The maximum probe pulse transmission at zero delay time was about three times the transmission ratio of the pump pulses. The authors of [5.47] considered the result of the time dependent experiment as evidence for intraband relaxation of the nonequilibrium carriers at a rate shorter than the pulse duration (believed to be about 5 ps).

More detailed investigations of the same transitions of Ge by *Shank* and *Auston* [5.48] and later by *Smirl* et al. [5.49] led to an alternative interpretation of the observed nonlinearity. By using a similar excite-and-probe technique, *Shank* and *Auston* carefully measured the probe pulse transmission over a much wider range of delay than described in [5.47]. It was shown that in addition to the fast rise and decay of the probe pulse seen by *Kennedy* et al. there was a distinct increase of the transmission over a much longer time scale (Fig. 5.15). The slow feature reached a maximum at a delay time of about 20 ps and decayed with a time constant of approximately 100 ps. The structure corresponding to the original feature seen by *Kennedy* et al. was shown to be an isolated, sharp spike centered around zero delay time having a width of only 2 ps. This is much less than the duration $t_p \sim 7$ ps of the pulses used in the experiment. In subsequent work, *Smirl* et al. [5.49] also confirmed the existence of a slowly varying bleached absorption in Ge.

These results were interpreted by *Shank* and *Auston* [5.48] as follows. The sharp spike is attributed not to carrier relaxation but to a parametric interaction between the strong pump pulse and the probe pulse. As a result of the parametric process, light is scattered from the pump beam to the probe beam, leading to an apparent increase in probe transmission. In the excite-and-probe technique the interacting beams typically propagate at a small angle (a few degrees) with respect to each other, and an optical grating may be formed due to the interference between pump and probe. The spatial modulation of the total light intensity is transformed into a corresponding spatial variation of the density of the generated electron-hole plasma. The incremental refractive index due to the plasma now forms a phase grating from which the pump light is scattered into the direction of the probe beam.

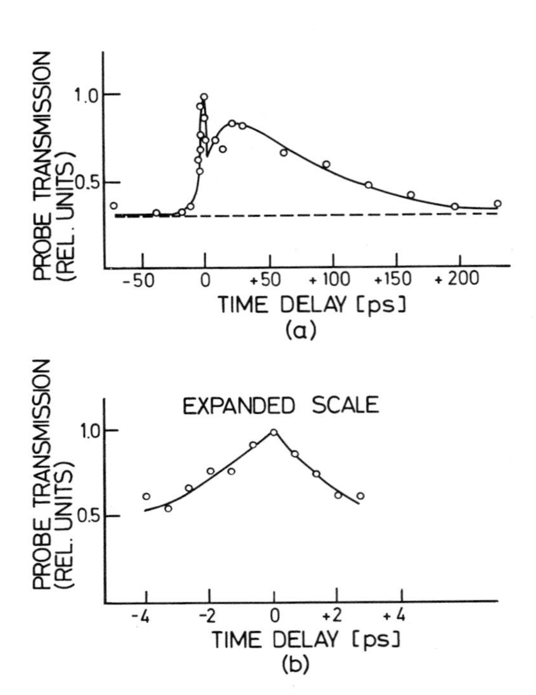

Fig. 5.15. Variation of the probe pulse transmission with delay time for germanium, showing a sharp spike near zero delay. The width of the spike is less than the pulse duration (after *Shank* and *Auston* [5.48])

The parametric scattering takes place only if the delay between probe and pump does not exceed the coherence length. Very often the output pulses from modelocked lasers are not transform-limited [5.50], i.e., the coherence time can be much shorter than the pulse duration, explaining why the sharp feature is narrower than the pulse width [5.48].

The slow feature in the variation of the probe transmission was tentatively attributed by *Shank* and *Auston* to bleaching of the optical absorption resulting from band filling, i.e., there is indication of a dynamic Burstein-Moss effect. For such a bleaching mechanism the rise of the transmission is expected to follow the integrated pulse intensity, i.e., the transmission risetime should equal the total pulse duration. The decay, on the other hand, is probably due to diffusion resulting in a decrease of the plasma density and a corresponding decrease of the quasi-Fermi level separation. Initially, the electron-hole plasma is generated in a very thin layer ($1/\alpha \simeq 10^{-4}$ cm) near the crystal surface, and subsequently the electrons and holes diffuse into the bulk of the material. The time scale for the carrier redistribution by diffusion processes is typically 100 ps (see Subsec. 5.2.3). Note, however, that the time variation in probe transmission is related to the true time variation of the absorption coefficient by

$$T(t) = \exp[-\alpha(t)L] = \exp[-\alpha_0 L]\exp[\Delta\alpha(t)L] \quad (5.23)$$

where α_0 is the unperturbed absorption coefficient, $\Delta\alpha$ is the induced variation, and L is the sample length. It is seen that $T(t)$ is related directly to the variation of, say, the level population only if $\Delta\alpha L$ is small compared with unity. On the other hand, if $\Delta\alpha L$ is comparable or larger than unity, the apparent aperture time obviously can be much shorter than the relaxation times of the processes causing the variation of $\Delta\alpha$.

These considerations also play a role in the interpretation of the recovery time of the optical absorption in the bleaching experiments of *Reintjes* et al. [5.51]. They investigated nonlinear optical absorption in epitaxial crystal films of $InAs_xP_{1-x}$ and Ga_xIn_{1-x} (direct band gap materials). The composition of these crystals was adjusted in such a way that the energy gap was about 10 meV less than the photon energy of the 1.06 μm picosecond pulse used for saturating the transitions. The dynamic Burstein-Moss shift is expected to lead to drastic bleaching effects at relatively low incident pulse energy, if the photon energy is close to the energy gap. In fact, increases in transparency by several orders of magnitude were observed in the experiment (room temperature). For example, for an essentially opaque ($T \sim 10^{-6}$) 25 μm thick crystal of $InAs_{0.19}P_{0.81}$ bleaching was observed to start at an incident energy of approximately 0.5 mJ/cm^2 ($\sim 25\ MW/cm^2$), and the transmission reached a peak of about 15% (including reflection losses) for a tenfold increase of the incident energy (5 mJ/cm^2).

The time dependency of the induced transmission was measured by means of the excite-and probe technique. It was found that the bleached state lasted for about 20 to 40 ps, depending on the initial amount of bleaching by the pump pulse (Fig. 5.16). The recovery times of the absorption could not be related directly to the material relaxation times, because in these experiments $\Delta\alpha L$ was large.

Experimentally the optical bleaching energy turned out to be about ten times higher than the values estimated theoretically on the basis of the band-filling model. *Reintjes* et al. [5.51] considered as a possible explanation of this discrepancy that the radiative carrier recombination is enhanced by stimulated emission. Band filling to such an extent that strong bleaching of the absorption is observed necessarily gives rise to very large optical gain. Stimulated emission could very effectively speed up the recombination of nonequilibrium carriers, particularly in direct band gap materials. This effect would lead to an increase of the energy required for optical bleaching.

Similar bleaching experiments at 1.06 μm in mixed compound semiconductors of mercury cadmium telluride, $Hg_{1-x}Cd_xTe$, were reported by

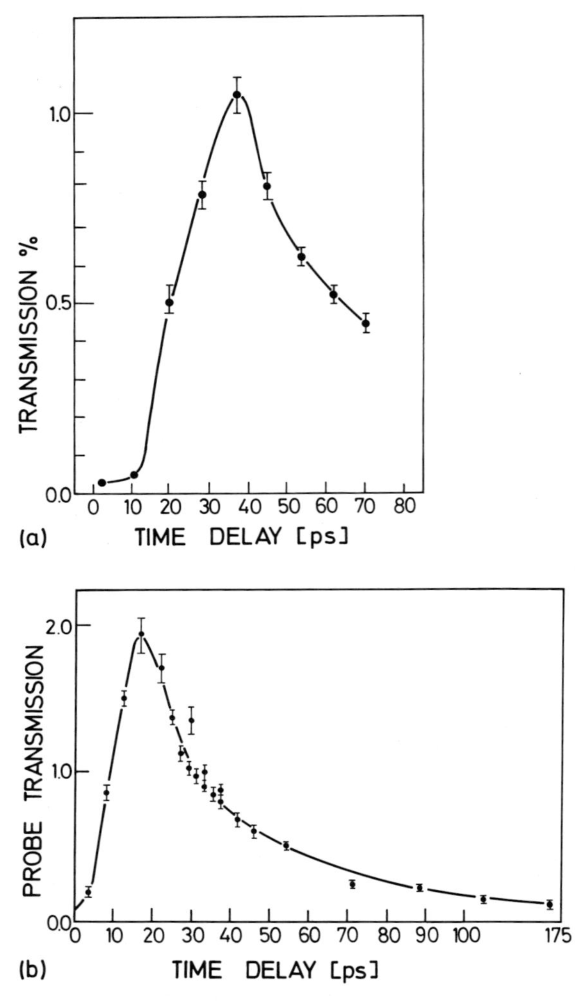

Fig. 5.16 a and b. Recovery of the optical absorption in a $Ga_{0.8}In_{0.2}As$ crystal after bleaching by a strong pump pulse at 1.06 μm. Variation of the probe pulse transmission with delay time for two different pump pulse intensities (after *Reintjes* et al. [5.51])

Matter et al. [5.52]. However, in this case the photon energy ($\hbar\omega = 1.17$ eV) was much higher than the band gap energy. For example, for a crystal composition corresponding to $x = 0.61$ the band gap is approximately 0.75 eV at room temperature. The transmission of a 5 μm thick sample of this composition increased from the low power value of 4×10^{-5} for incident energies less than 1 mJ/cm^2 to a maximum of approximately 10^{-3} at 100 mJ/cm^2. Comparing with the results for the III-V type crystal films of larger band gaps [5.51] it is seen that apparently saturation occurs only at much higher optical power.

For $x=0.4$ and $x=0.2$ corresponding, respectively, to an energy gap of 0.42 and 0.1 eV (at 297 K), no optical bleaching could be observed.

For the $x=0.61$ sample the recovery of the absorption was measured by using delayed, weak probe pulses (1.06 μm). Aperture times of the order of 10 to 20 ps were observed, but an interpretation in terms of carrier relaxation times was not given, nor did the authors of [5.52] discuss the mechanism of the observed optical saturation.

Saturable absorption of semiconductors was also observed at infrared wavelengths with subnanosecond pulses from CO_2 lasers, both at the fundamental (10.6 μm) [5.53] and at the second harmonic. A very interesting example was reported by *Nurmikko* [5.54]. He studied nonlinear absorption in indium antimonide using 400 ps pulses at 5.3 μm. At 110 K the band gap of InSb is 220 meV, which is only 14 meV less than the photon energy. Furthermore, the density of states near the conduction band edge is anomalously low because the effective mass of the conduction electrons is only 0.01 m_0. These two factors combine to give a very low energy threshold for optical bleaching via the Burstein-Moss effect. Experimentally substantial bleaching was observed for incident pulse intensities as low as a few kW/cm^2, e.g., in a 25 μm thick crystal wafer of InSb (110 K) a transparency increase by a factor of 10^3 was measured for 10 kW/cm^2 pulses.

5.2.3 High Density Electron-Hole Plasmas

In this section we will discuss picosecond experiments dealing with transport and recombination processes of electron-hole plasmas in semiconductors. Optical pulsed excitation offers the advantage that extremely dense electron-hole plasmas are readily generated in pure intrinsic semiconductor material permitting investigation of high density plasma phenomena under conditions of high degeneracy, yet the concentration of donors and acceptors can be very small (impurity effects are negligible).

When the photon energy exceeds the band gap energy, the incident radiation is absorbed in a very thin material layer near the surface; e.g., the depth of the layer is given by the reciprocal of the absorption coefficient, $1/\alpha \simeq 10^{-4}$ to 10^{-5} cm. The electrons and holes generated in that thin surface layer then rapidly diffuse into the bulk. *Auston* and *Shank* [5.55] investigated the dynamics of such a diffusion process when electrons and holes are excited by a powerful picosecond light pulse. In their experiment a 1.06 μm light pulse incident approximately normal to the $\langle 111 \rangle$ face generated an electron-hole plasma by interband absorption in an intrinsic single crystal of germanium. They used an ellipsometer arrangement (see Fig. 3.17) to measure the surface plasma density as function of time. A circularly polarized weak probe pulse, also at 1.06 μm, was reflected from the excited area of the crystal at approximately the principal angle ($\sim 70°$). The polarizing prism in front of the detector was adjusted for maximum extinction in the absence of the excitation pulse. The electron-hole plasma created by the excitation produced a change δn of the refractive index at the sample surface. As a result, the transmission of the ellipsometer increased by an amount proportional to the square of the fractional index change $\delta n/n$. The index change, on the other hand, was proportional to the plasma density; the reflected probe beam then measured the concentration of electrons and holes at the surface.

In the experiment the transmission of the ellipsometer is measured as a function of the arrival time of the probe pulse, and from these data the variation of the refractive index with time was computed (see Fig. 3.18). Excitation with 10 mJ/cm^2 produced a maximum fractional index change (decrease) of 5% which corresponds to a plasma density of about 10^{20}/cm^3. The results showed that following excitation the density at the surface decreased to half its peak value in about 30 ps. Carrier recombination can be neglected on this time scale, and *Auston* and *Shank* [5.55] attributed the observed decay to a dilution of the plasma by diffusion of electrons and holes out of the surface layer into the bulk. A value of $D=230$ cm^{2}s^{-1} for the ambipolar diffusion coefficient was obtained by fitting theoretical calculations of the diffusion process to the experimental data. The high density value of the diffusivity, $D=230$ cm^{2}s^{-1}, inferred in this way was 3.5 times larger than the low density ambipolar diffusion coefficient of germanium, $D=2D_eD_h/(D_e+D_h)=65$ cm^{2}s^{-1}. *Shank* and *Auston* were able to show that such an increase in diffusivity is expected for plasma densities of the order of 10^{20}/cm^3 as a consequence of the high degeneracy of the gas of free carriers.

In a different experiment *Auston* et al. [5.56] looked at recombination processes in a dense electron-hole plasma. Again the plasma was generated in a germanium crystal by absorption of a pump pulse at 1.06 μm (8 ps duration). The time variation of the plasma density was monitored with the help of delayed probe pulses at 1.55 μm which measured the excited state absorption due to the generated free carriers. The probe pulses were produced by stimulated Raman scattering in benzene. At 1.55 μm there is still some interband absorption in Ge, but the free carrier absorption coefficients of electrons and holes are quite large ($\sigma_e=4\times10^{-18}$ cm^2, $\sigma_h=3.3\times10^{-17}$ cm^2); at plasma densities of the order of 10^{20} cm^3 free carriers contribute significantly to the probe pulse absorption in the excited Ge sample. *Auston* et al. estimated a maximum plasma density of 3.4×10^{20} cm^{-3} for an exponential layer having a $1/e$ width of $1/\alpha=7.1\times10^{-5}$ cm. Under these conditions induced free carrier absorption decreases the probe transmission to 40% of its initial value.

The observed variation of the induced absorption as a function of time is shown in Fig. 5.17 for two different values of the initial plasma density. The following conclusions were drawn from the experimental results. The probe technique based on free carrier absorption measures the integrated carrier concentration along the direction of propagation of the probe beam, i.e., a mere redistribution of the excited electrons and holes by diffusion conserves the total number of carriers. Therefore, diffusion processes are expected not to affect the probe pulses (transverse diffusion can be neglected). The results then clearly represent evidence that the total concentration of electrons and

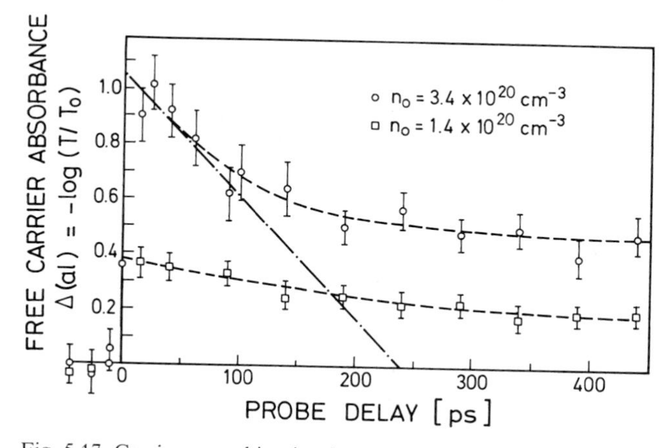

Fig. 5.17. Carrier recombination in a high density electron-hole plasma in germanium, measured by free carrier absorption of a picosecond probe pulse at 1.55 μm (after *Auston* et al. [5.56])

holes decreases with time, which led to the conclusion that a very fast recombination process exists. Comparing the decay curves for the two different initial plasma densities, it was noted that the recombination rate is faster at the higher density.

Auston et al. identified the recombination as an Auger process whereby an electron and a hole recombine transferring the recombination energy to a third electron or hole as kinetic energy. If equal concentration n of electrons and holes is assumed, the rate of change of the electron concentration (or hole concentration) is given by

$$\partial n/\partial t = -\gamma_3 n^3 \tag{5.24}$$

where γ_3 is the Auger recombination rate constant. From the observed initial decay rate of the free carrier concentration and from the known initial plasma concentration the Auger rate was calculated to be $\gamma_3 = 1.1 \times 10^{-31}$ cm^6 s^{-1}. It was estimated that at these high densities the Auger process indeed dominates the recombination; radiative processes contribute less than 1% to the total recombination rate.

Comparison of the two results in Fig. 5.17 showed that the Auger rate at the smaller density is about 1.8 times the high density Auger rate. *Auston* et al. suggested that the observed decrease of the Auger rate with increasing plasma concentration may be due to a limitation of the number of available final states for Auger processes for more strongly degenerate conditions at higher densities.

Electrical transport in a dense electron-hole plasma in silicon was the subject of a picosecond investigation reported by *Auston* and *Johnson* in [5.57]. The experimental technique made use of a microstrip transmission line (see Chap. 4) consisting of a high purity 0.5 mm thick silicon wafer with a uniform aluminum film ground plane on the bottom, and a 0.4 mm wide aluminum strip on the top, interrupted by a 2 mm gap. The structure formed a transmission line, having a characteristic impedance of 50 Ω. One section of the transmission line was charged to a known bias voltage while the section across the gap was connected via matched coaxial cables to a fast oscilloscope.

In the experiment the gap is made photoconductive by illuminating a well-defined area of the gap with a picosecond light pulse at 0.53 μm, generating a high concentration of electrons and holes in a very thin surface layer of the material. In this way the charged section is suddenly switched on by the light pulse, and an electrical signal travels down the transmission line. By measuring this electrical signal the effective conductance of the gap can be calculated. When the energy of the optical switching pulse and the illuminated volume is known, then the carrier mobility of the material forming the gap can be evaluated. With this technique one can measure mobilities at a very high carrier concentration of *intrinsic* material, in which impurity scattering can be neglected. The measurement is performed on such a short time scale that diffusion, trapping, and recombination can be ignored. The microstrip transmission line was used to investigate electron-hole scattering for carrier concentration ranging from $10^{15}/\mathrm{cm}^3$ to about $10^{20}/\mathrm{cm}^3$. Figure 5.18 shows the

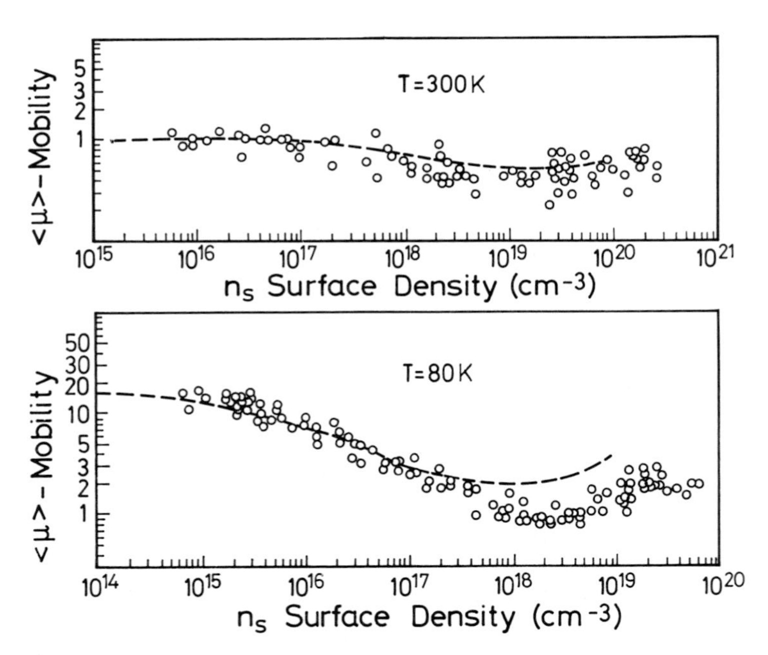

Fig. 5.18. Sum of electron and hole mobilities as a function of the carrier density. $\langle\mu\rangle$ is a spatial average over the density profile generated by the optical excitation. n_s is the initial carrier density at the surface. Dashed curve: theoretical calculations (after *Auston* and *Johnson* [5.57])

experimental result, plotting mobility (spatially averaged [5.57]) versus carrier density for two different temperatures, 80 K and room temperature. For both cases the experimental mobility is normalized to the room temperature (lattice) mobility. Figure 5.18 shows an interesting variation of the mobility with concentration. There is a steady decrease of the mobility as the carrier concentration is raised to densities of about $10^{18}/\text{cm}^3$. At 80 K the mobility reaches minimum at a value 15 times below the low density value. For still higher densities the mobility increased again.

According to *Auston* and *Johnson* [5.57] the mobility is reduced due to electron-hole scattering by the screened Coulomb interaction. With increasing concentration the Coulomb scattering rate is initially increased up to a point at which still higher carrier concentrations screen the Coulomb potential more effectively rather than to further enhance the scattering rate. The experimental results were in reasonably good agreement with mobility calculations based on the electron- hole scattering mechanism (dashed curves in Fig. 5.18).

5.2.4 Nonradiative Relaxation of Electronic Excitation

In this section we will discuss several examples of picosecond nonradiative relaxation of electronic excitations in solids. Let us begin with electronic relaxation of crystal impurities. The properties of impurity atoms play an important role in several branches of solid state physics. For example, many solid state lasers are based on electronic states that arise from the interaction of the impurities with the static crystal lattice. On the other hand, the dynamic interaction with the lattice leads to nonradiative relaxation processes, in which excited electronic states of impurities decay by emission of phonons. An extensive literature has been accumulated on nonradiative multiphonon transitions between electronic states of rare earth and transition metal impurity ions, yet very few experiments have been performed in the subnanosecond and picosecond time regime.

We will discuss now an example of picosecond multiphonon relaxation of copper impurities embedded in a lithium tantalate host crystal [5.58]. The electronic state of interest is the $^2T_2$ first excited state, which arises from the crystal field splitting of the ground state of the free Cu^{2+} ion. Transitions between the vibronic manifold corresponding to this excited state and the ^2E ground state give rise to a broad absorption band centered at approximately 1 μm. At shorter wavelengths the absorption spectrum of $LiTaO_3{:}Cu^{2+}$ is dominated by charge transfer of the Cu^{2+} ion, which in a sense can be regarded as photoionization of the impurity. The long wavelength cutoff of the charge transfer bands is at about 4200 Å. Energetically it is possible to reach the charge transfer bands from the excited $^2T_2$ state by absorption of light at 0.53 μm, the second harmonic of the neodymium laser. This situation affords a convenient means of measuring the lifetime of the $^2T_2$ band. The fundamental at 1.06 μm can be used to populate the $^2T_2$ states, whereas the changing number of excited ions can be readily monitored by measuring the charge transfer absorption with a delayed second harmonic probe pulse. A first measurement of the $^2T_2$ relaxation time at room temperature was performed by *Auston* et al. [5.58]. Subsequently, a detailed study of the relaxation of $LiTaO_3:Cu^{2+}$ including the temperature dependence of the relaxation time between 20 and 450 K was made by *von der Linde* et al. [5.59]. They found that for all temperatures the excited state (charge transfer) absorption cross section decreased exponentially with probe pulse delay time, indicating an exponential decay of the population of the $^2T_2$ state. The measured variation with temperature of the relaxation rate $1/\tau$ is shown in Fig. 5.19. The curve is characteristic of a thermally stimulated relaxation process. The relaxation rate is constant up to a temperature of about 100 K, but at higher temperatures the rate increases rapidly. Experimentally the lifetime τ of $^2T_2$ decreased from the low temperature limit of 450 ps to a value of only 10 ps at 450 K.

A simple model of vibronic coupling was used to interpret these results. In the configuration diagram shown in Fig. 5.20 the two electronic states $^2T_2$ and ^2E are represented by parabolas; the vertical distance between the minima corresponds to the electronic energy difference between $^2T_2$ and ^2E, whereas the horizontal shift of the vertices represents the change of the equilibrium position of the ions upon electronic excitation. The model assumes that after optical excitation along the upward vertical path very fast vibrational relaxation takes place so that the ions quickly reach quasi-thermal equilibrium near the $^2T_2$ vertex. The authors of [5.59] suggested that the lifetime of $^2T_2$ is determined by an internal conversion process in which the electronic energy of $^2T_2$ ($\hbar\Omega$) is converted into vibrational energy of the electronic ground state. Because $\hbar\Omega$ is much greater (typically thousands of cm^{-1}) than the phonon energies (hundreds of cm^{-1}), a large number of vibrational quanta is created by internal conversion, i.e., we are dealing with a multiphonon process. This excess vibrational energy is subsequently dissipated to the lattice very rapidly. In [5.59] a comparison was made of the measured picosecond relaxation rate with a theoretical model of internal multiphonon conversion published by *Englman* and *Jortner* [5.60]. In this paper the transition rate is written as a product of an electronic matrix element and Franck-Condon vibrational overlap functions. The Franck-Condon factors are expressed as a function of only two basic parameters, an effective vibrational frequency $\bar{\omega}$ and a coupling constant S_0. The latter, S_0, is the excess vibrational energy of the impurity after an electronic transition expressed as the effective number of phonons, i.e., $S_0 = E_v/\hbar\bar{\omega}$ (see Fig. 5.20).

The observed temperature dependence could be accounted for very well with values of $S_0 = 5$ and $S_0 = 490\ cm^{-1}$. These data suggest that the pure electronic energy difference between $^2T_2$ and ^2E is $\hbar\Omega \sim 8000\ cm^{-1}$. It follows that a luminescence band should be observed at a Stokes shift of $2\ E_v = 2\ S_0\hbar\bar{\omega} = 6000\ cm^{-1}$ (1.6 μm). By optically exciting the Cu^{2+} impurities with 1.06 μm light pulses a fluorescence band in the infrared at about 1.75 μm was indeed detected. The measured quantum efficiency of the fluorescence of a few times 10^{-6} (at ~ 20 K) was in fair agreement with the value estimated from the

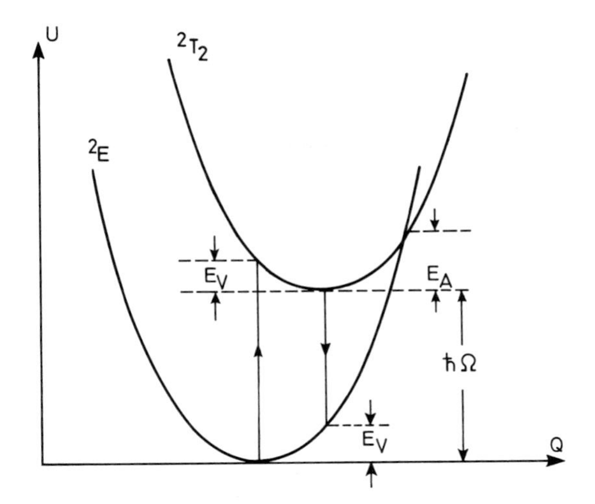

Fig. 5.19. Variation with temperature of the inverse lifetime of the $^2T_2$ excited electronic state of Cu^{2+} impurities in $LiTaO_3$. The curve represents a fit of the experimental data by the theory of [5.60] (after *von der Linde* et al. [5.59])

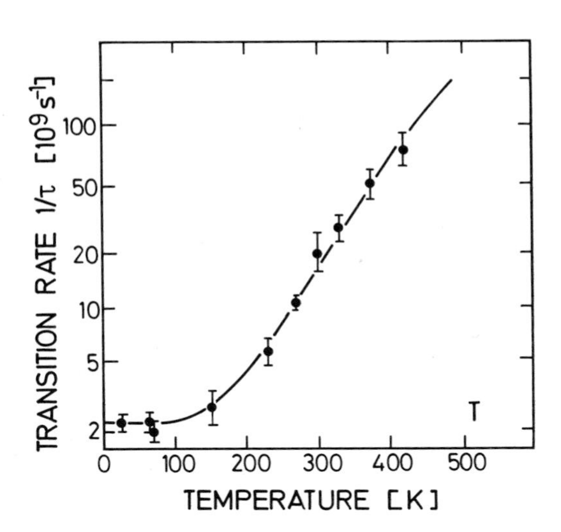

Fig. 5.20. Configuration diagram illustrating the model used in [5.59] for the interpretation of the relaxation of the $^2T_2$ state of Cu^{2+}. $\hbar\Omega$ is the electronic energy gap between the ground state (^2E) and the excited state ($^2T_2$). E_v is the excess vibrational energy, E_A is the activation energy (after *von der Linde* et al. [5.59])

oscillator strength of the $^2E - ^2T_2$ transition (integrated absorption coefficient) and the measured total relaxation rate at 20 K. Thus it appears that despite the simplifications involved, the model used in [5.59] gives a fairly consistent overall picture of the observed fast nonradiative relaxation of the excited copper impurities in $LiTaO_3$.

Next let us consider a nonradiative relaxation process in a quite different system, a tetracene crystal. The lowest excited electronic state in this organic molecular crystal is a triplet (Frenkel) exciton [5.61]. The singlet exciton state has an energy which is about 10% less than twice the energy of the triplet exciton. It is well known that two triplet excitons can fuse and form a singlet exciton, because there is approximate energy resonance for such a process. The reverse reaction is also possible and gives rise to a nonradiative exciton decay mechanism: a singlet exciton can split into a pair of triplet excitons. However, the latter process is associated with an activation energy because the singlet energy is *less* than the energy of two triplet excitons. Therefore it is expected that singlet exciton fission is negligible at sufficiently low temperature, but that fission may eventually contribute to the singlet exciton relaxation rate at high temperatures.

Alfano et al. [5.62] have reported time resolved measurements at room temperature of the singlet exciton fluorescence at 5800 Å of a tetracene crystal. They measured a significantly smaller lifetime than the value reported at low temperatures [5.63]. In their experiment the tetracene crystal was excited by frequency doubled pulse trains from a modelocked Nd-glass laser. A time resolution for the fluorescence measurement of about 8 ps was achieved by optically gating the fluorescence with the help of a CS_2 Kerr shutter controlled by the fundamental pulses (see Chap. 3). *Alfano* et al. showed that with picosecond excitation the singlet exciton fluorescence rises to a maximum within about 12 ps (determined by the excitation pulses) and then decays exponentially with a time constant of (145 ± 50) ps. By using the low temperature limit of the exciton lifetime in tetracene of 810 ps and assuming that decay by fission is the only temperature dependent relaxation process, *Alfano* et al. calculated the singlet exciton fission rate at room temperature to be $5.7 \times 10^9\ s^{-1}$, or a time constant for fission of 180 ps. On the basis of an activation energy of 1600 cm^{-1} [5.64] the high temperature limit of the fission rate should be $1.7 \times 10^{13}\ s^{-1}$, but the temperature dependence was not investigated in [5.62].

The final example of picosecond nonradiative relaxation is concerned with alkali halides. *Bradford* et al. [5.65] have studied the formation of lattice defects following interband electronic excitation in a KCl crystal. In alkali halides these processes represent an important nonradiative relaxation mechanism of electronic excitations. Here we consider a process in which the recombination energy of an optically excited electron-hole pair serves to knock a halide ion off its lattice site, leaving behind an anion vacancy occupied by an electron, which is an *F*-center, and a neutral interstital halide atom (a Frenkel defect). The latter can combine with another halide ion to form a negatively charged diatomic halide molecule, Cl_2^-, which is called an *H*-center [5.66].

F-centers possess characteristic broad absorption bands in the visible, and it is therefore possible to monitor the formation of these lattice defects by measuring the growth of the *F*-center absorption band during intense interband excitations.

In the experiment of *Bradford* et al. single picosecond pulses at $\lambda = 2660$ Å ($\hbar\omega = 4.66$ eV) were produced by frequency quadrupling the output of a mode-locked Nd:YAG laser. With these pulses the KCl crystal ($E_g \simeq 6.2$ eV) was excited via two-photon absorption at a temperature of 25 K. Because the *F*-band absorption of KCl peaks at about 2.34 eV, very close to the second harmonic of the YAG laser ($\lambda = 0.53$ µm, $\hbar\omega = 2.33$ eV), the build-up of *F*-center concentration can be monitored conveniently by measuring the absorption of second harmonic probe pulses. *Bradford* et al. used three probe pulses of different optical delay with respect to the uv pump pulse. The first probe pulse arrived at the sample well before the uv pump pulse, serving as a reference. A second probe pulse of variable delay interrogated the instantaneous transmission shortly after excitation, whereas the third probe pulse at a fixed delay of 10 ns measured the final value of the induced absorption long after the exciting uv pulses had passed. Experimentally, the ratio D_2/D_3 of the optical density seen by the second and third probe pulse is measured as a function of the delay time between the second probe pulse and the uv light. The experimental result is plotted in Fig. 5.21, showing a rise of the probe absorption to a maximum within about 40 ps. The experimental data were analyzed by assuming a gaussian shape for both the pump and the probe pulses. Further-

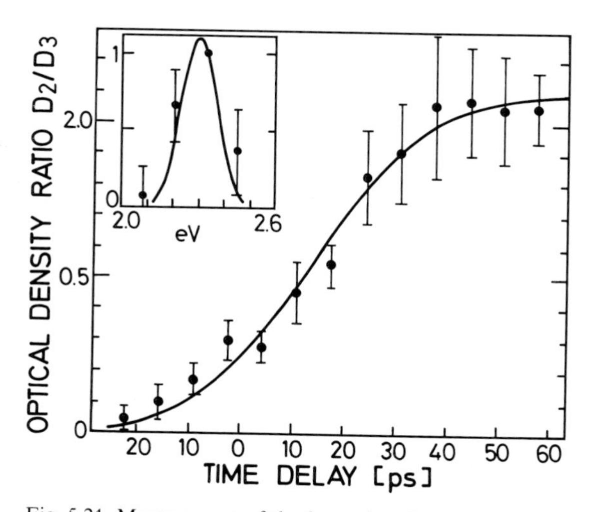

Fig. 5.21. Measurement of the formation time of *F* centers in KCl ($T \simeq 25$ K). Normalized optical density of the probe pulse at 0.532 µm versus delay time between the pump pulse at 0.266 µm and the probe pulse. From the observed rise of the probe pulse absorption a formation time of $\tau = 11$ ps was derived. The insert shows the shape of the induced absorption band measured 85 ps after excitation of the KCl crystal by the pump pulse (after *Bradford* et al. [5.65])

more, it was assumed that the buildup in time of the lattice defects follows an exponential law. The best fit of the calculated response with the experimental points was obtained with a value of 11 ps for the formation time of ground state *F*-centers. This result was considered to be accurate to within a factor of about 2, due to the assumptions made in the analysis.

Additional spectral measurements were made to confirm that the observed change of the crystal absorption was associated with *F*-centers. At a fixed delay of 85 ps the crystal was probed with interrogating pulses at four different closely spaced frequencies (generated by stimulated Raman scattering). The result shown in the insert of Fig. 5.21 is consistent with the well-known *F*-center absorption band, although it appears to be somewhat broadened.

An important conclusion from this experiment was that the observed growth of the *F*-centers is due to the formation of new Frenkel defects by the uv light, and not due to photoexcited electrons filling existing vacancies. The vacancy concentration of the samples was shown to be much smaller than the final *F*-center concentration obtained after uv exposure. Furthermore, it is known that *F*-centers in the first excited state are formed when electrons are trapped at halide vacancies. The long lifetime of the excited state ($\simeq 600$ ns) would cause a much slower rise time of the induced absorption, if the *F*-centers were generated via trapping of photoexcited electrons.

Comparing the *F*-center concentration produced in the experiment with the amount of absorbed uv light, *Bradford* et al. estimated a probability of about 15% for the formation of a lattice defect following the creation of an electron-hole pair by interband optical excitation.

5.3 Picosecond Spectroscopy of Excitons

In this section we will discuss picosecond experiments concerned with the interaction of excitons in semiconductors. Particularly at very high exciton concentrations, an amazing wealth of new phenomena was discovered in the past several years. Because many of these high density effects are of transient nature and occur on a subnanosecond time scale, picosecond optical techniques are a very useful tool for studying these effects.

Before we enter into the discussion of picosecond exciton experiments some elements of the physics of excitons [5.67] are very briefly outlined emphasizing aspects that are relevant to the investigations of II-VI type semiconductors discussed in this section.

5.3.1 Overview of the Properties of Excitons

Excitons in II-VI semiconductors belong to the category of weakly bound Wannier-Mott excitons [5.68]. It is well known that many properties of these excitons can be understood by regarding the exciton as a hydrogenic system in the effective mass approximation [5.69]. Due to the attractive Coulomb force between electrons and holes an electron-hole pair can form a bound hydrogen-like system, the exciton. Some of the phenomena to be discussed can be understood from this point of view. For example, it has been suggested that two excitons can form an excitonic molecule [5.70] as a result of the covalent attraction between the excitons, very much like the formation of a covalently bound H_2 molecule out of two hydrogen atoms. The excitonic molecule will play a crucial role in many experiments discussed in this section.

There are, of course, a number of significant differences between an electron-proton system and an exciton. First, the ratio of the masses of the constituents differs by many orders of magnitude from that of hydrogen. Second, the concept of effective masses for electrons and holes in a semiconductor crystal implies the possibility of an anisotropy of the masses [5.71]. It follows from band structure theory that the effective mass exhibits tensor properties [5.71]. A very good example is the mass of a hole in cadmium sulfide, which is 5 and 0.7 m_e [5.72] for propagation parallel and perpendicular to the c-axis, respectively. These tensor properties are quite important for treating an excitonic molecule in CdS and similar II-VI type semiconductors.

Note also that the mass of an excitonic molecule formed from two excitons would be three to four orders of magnitude smaller than the mass of a H_2 molecule. An interesting consequence of the very small molecular mass is the possibility that the repulsive part of the molecular interaction due to the Pauli exclusion principle dominates over both the exchange attraction and the van der Waals forces [5.73]. If the net interaction between molecules is repulsive, a gas of excitonic molecules does not form a liquid or a crystal as does H_2 gas. However, the excitonic molecular system can be regarded as a boson gas with a weakly repulsive interaction. In such a system Bose-Einstein condensation [5.74] can occur when the system is cooled below a characteristic transition temperature. In the condensed phase a large fraction of the molecules accumulates in a state of zero momentum, $k=0$. The possibility of Bose-Einstein condensation of excitonic molecules is a major topic discussed in this section.

The picture of excitons as hydrogen-like atomic systems is quite helpful in visualizing the nature of these objects and in understanding some of their interactions such as the formation of exciton complexes and the binding of excitons to impurities. However, the interaction of excitons with light, which is of crucial importance in our discussion, can only be described correctly by the polariton concept [5.75].

In the usual scheme of interaction of light with atomic energy levels weak coupling is assumed, and the energy resides either in the electromagnetic field (as a photon) or in the atom (as material excitation). A transition between these two situations is called a radiative transition. From this point of view a clear distinction is made whether the energy of the coupled system is electromagnetic or materialistic at any one time, and coupling manifests itself only via energy transfer between field and matter (radiative transitions).

This approach does not provide a satisfactory description when the coupling is stronger. The polariton concept takes into account more accurately the coupling between the electromagnetic field and the material. A direct consequence of the polariton theory is the fact that we can no longer use the concept of radiative transitions as defined above. Instead, the energy propagates in the form of a coupled wave of partly electromagnetic and partly materialistic character, the polariton wave.

The new normal mode arising from the coupling exhibits a characteristic dispersion law $\omega(k)$, the well-known polariton dispersion curve [5.75]. As an example, the exciton-polariton dispersion for the A_1 exciton in CdS is depicted in Fig. 5.22. The dotted lines show the dispersion relations for the two uncoupled subsystems, i.e., the light wave and the exciton. The mixed character of the coupled mode is quite pronounced in the region of the crossover region of the dotted lines. On the other hand, in the asymptotic regions of very large and very small wavevector the coupled wave shows dominantly light and exciton character, respectively.

Let us discuss the meaning of radiative transitions in the framework of polariton theory [5.75]. Assume that exciton-polaritons are excited in a crystal. When the polariton waves arrive at the crystal-vacuum boundary, light is emitted because part of the polariton energy leaks from the crystal and propagates in free space as light waves. Similarly, when light is incident on a crystal, part of the optical energy will be transformed and propagate into the crystal as a polariton wave. Absorption of light from this point of view occurs because there are various damping mechanisms whereby polariton energy is dissipated in the material [5.75], e.g., by emission of phonons.

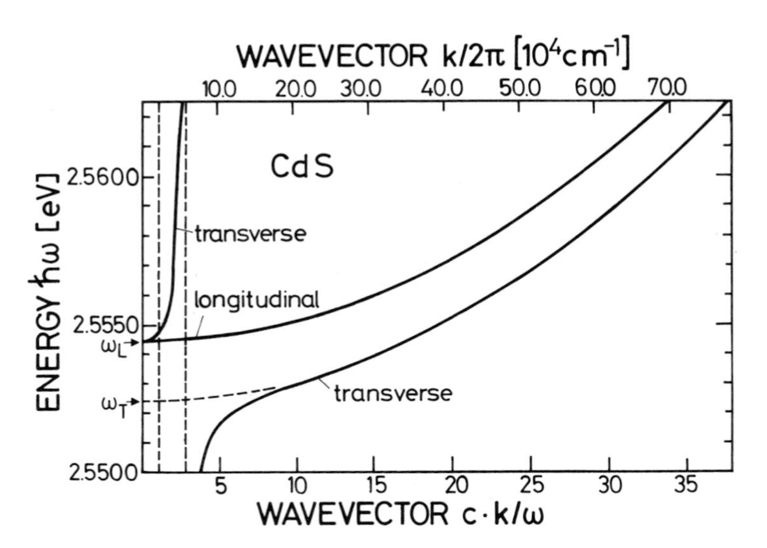

Fig. 5.22. Part of the A_1 exciton-polariton dispersion curve of cadmium sulfide. ω_L and ω_T are the longitudinal and the transverse exciton frequencies, respectively. The dashed curves show the dispersion relations for light and excitons neglecting the coupling. (After Wiesner [5.76])

The principles of the interaction between excitons and light are of paramount importance for the interpretation of the picosecond optical experiments to be discussed later.

Before we conclude this overview of exciton properties we present a typical emission spectrum of CdS in the exciton regime. Figure 5.23 shows the luminescence of a CdS single crystal of high purity taken at very low excitation intensity at a temperature of 1.6 K [5.76]. Note that the two dominant sharp features at 2.5357 eV and 2.5458 eV are due to impurity effects and not due to intrinsic crystal properties. In fact, the intrinsic luminescence from the $n=1$ state of the A exciton at 2.552 eV [5.76] is very much weaker. Exciton spectra are frequently dominated by impurity effects even if the impurity concentration is very small. The explanation is the giant oscillator strength effect which is typical of exciton systems [5.77]. In the spectrum of Fig. 5.23 the bands

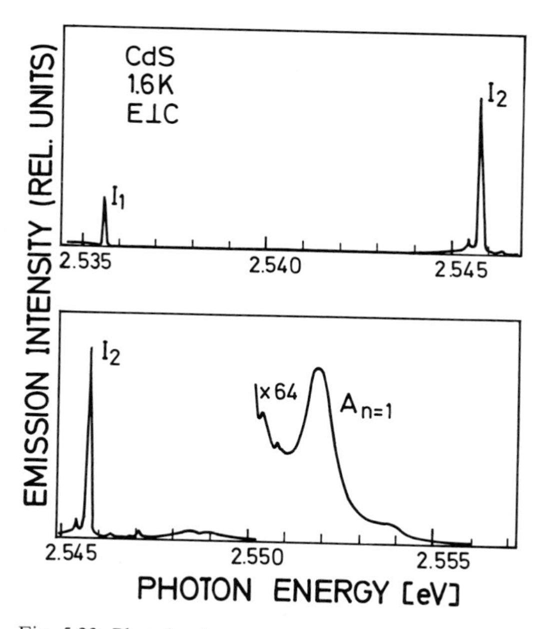

Fig. 5.23. Photoluminescence spectra of CdS, showing the I_1 and I_2 bands which are due to excitons bound to neutral donors and neutral acceptors, respectively. Note the expanded scale for the intrinsic A_1 luminescence (after *Wiesner* [5.76])

denoted I_1 and I_2 are due to excitons bound to neutral acceptors and neutral donors, respectively [5.78]. The dominance of impurity effects has often led to uncertainties in the interpretation of spectra, and this problem will also arise in our discussions of picosecond exciton experiments.

This concludes the brief introduction of excitons. We shall now discuss the kinetics of intrinsic A_1 excitons at low concentrations.

5.3.2 A_1-Exciton Bottleneck

Heim and *Wiesner* have investigated the relaxation kinetics of free A_1-excitons in CdS on the lower branch of the dispersion curve [5.79]. Fig. 5.24 depicts a portion of the CdS emission spectrum showing the intrinsic A_1 feature which is of interest here. The relevant part of the polariton dispersion curve and the reflectivity are also included in the figure. The weak luminescence peak at 2.554 eV has been attributed to the fact that the optical reflectivity (at the crystal vacuum boundary) exhibits a minimum at just this energy. The dominant luminescence peak at 2.552 eV, on the other hand, originates from polaritons belonging to the "knee" area of the dispersion curve. Its strength was thought to be related to a bottleneck situation for the relaxation of polaritons in this part of the dispersion curve, resulting in a pile-up of population at energies corresponding to the knee. This effect has been theoretically predicted by *Toyozawa* [5.81] and experimentally confirmed by *Heim* and *Wiesner* by means of picosecond time-resolved spectroscopy [5.79]. In their experiment free electrons and holes were generated via interband transitions using pulses of 200 ps duration from a modelocked argon laser operating at $\lambda = 4529$ Å. After relaxation of the electrons and holes from the continuum (by emission of LO phonons) the lower branch of the dispersion curve is populated with exciton-polaritons. By using delayed coincidence photon counting [5.82] they measured the buildup and the decay of the luminescence from the various parts of the polariton branch, and with the help of deconvolution techniques [5.79] they achieved a time resolution of approximately 200 ps. Their result is reproduced in Fig. 5.25. Curve *b* shows the measured luminescence decay time

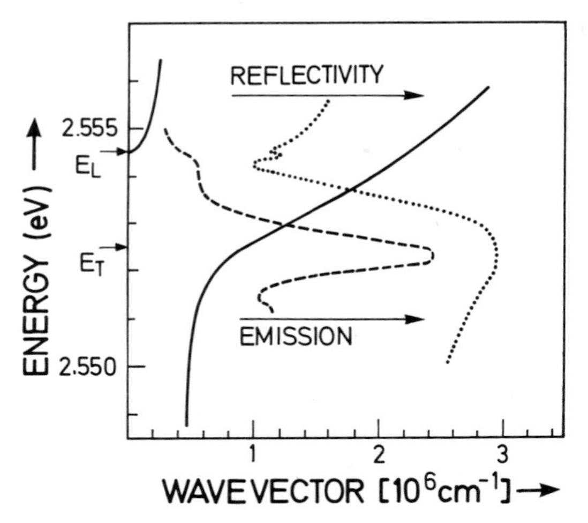

Fig. 5.24. Part of the CdS exciton-polariton dispersion curve (full line). The dashed curve shows the intrinsic A_1 exciton luminescence, and the dotted curve represents the optical reflectivity (after *Gross* et al. [5.70])

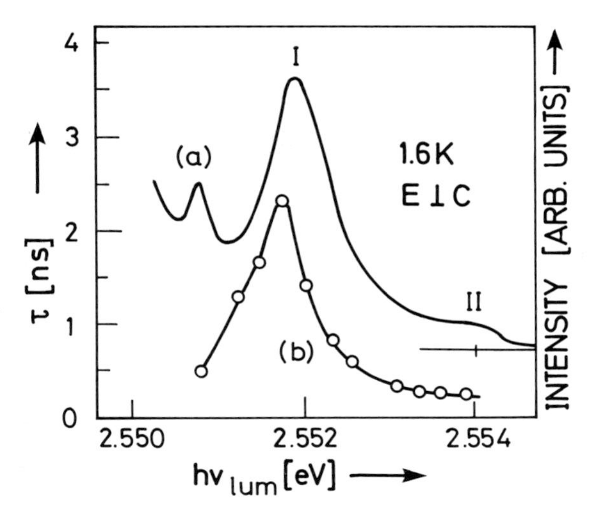

Fig. 5.25. Variation of the luminescence intensity (a) and of the luminescence decay time (b) with energy for cadmium sulfide (after *Heim* and *Wiesner* [5.79])

as a function of photon energy. A maximum decay time of 2.3 ns was found for polaritons at an energy of 2.552 eV corresponding to peak I in curve *a* of Fig. 5.25. Going to lower energies the polariton lifetime decreases rapidly, whereas at higher energies the lifetime approaches the resolution limit of 200 ps at the position of peak II in curve *a*.

For an interpretation of their results *Heim* and *Wiesner* considered the various population and depopulation processes in three different regimes of the dispersion curve: range *A* far above the knee at energies greater than 2.553 eV; range *B* slightly above the knee at approximately 2.552 eV; range *C* at energies less than 2.551 eV (small wavevectors).

Relaxation along the dispersion curve takes place by emission of acoustic phonons: low energy portions of the polariton curve are populated by acoustic relaxation from more energetic portions. An important characteristic of this process is a decreasing relaxation rate with smaller wavevectors [5.83]. Radiative and nonradiative decay processes also contribute to the depopulation. The radiative decay rate varies with energy in accordance with the reflectivity exhibiting a minimum in regime *B*, whereas the nonradiative rate was assumed to be smaller and energy independent [5.79].

The bottleneck situation arises because regime *B* is rapidly populated from *A* by effective phonon emission, but the relaxation rate from *B* to *C* is slow because of the decreasing efficiency of acoustic phonon relaxation at smaller wave vectors. In addition, the radiative losses in *B* are minimal.

All these factors cooperate to produce a long decay time and a pileup of population in range *B*, as observed in the experiment. On the other hand, it follows that the decay time in *A* is fast since both the acoustic and the radiative decay mechanisms are efficient in A. Similarly, shorter decay times are expected as one moves down the polariton curve into C because of increasing radiative losses.

The investigations of *Heim* and *Wiesner* have revealed an interesting energy dependence of the exciton lifetime. The experimental observations are in qualitative agreement with the predicted existence of a bottleneck near the knee of the polariton dispersion curve. However, we note that in their interpretation *Heim* and *Wiesner* did not take into account the presence of the triplet exciton states just above the knee. It is conceivable that triplet-singlet relaxation could also contribute to the observed apparent lifetime increase [5.84].

5.3.3 Excitons Luminescence Spectra at High Density

In the preceding paragraph we were concerned with excitons under the conditions of low excitation intensity of about 1 W/cm^2 producing exciton concentrations of the order of 10^{13}/cm^3 [5.79]. Exciton densities in excess of 10^{17}/cm^3 are readily generated with more powerful laser pulses. At such high exciton concentrations many new interaction channels open up, notably collisions between excitons, and formation of exciton complexes. These phenomena show up, for example, as new features in the luminescence spectra. Before proceeding to the discussion of picosecond experiments involving high density excitons we will review the changes in the luminescence spectra that occur when the exciton concentration is increased.

Figure 5.26 presents photoluminescence spectra [5.85] from a CdS crystal at 1.8 K taken at various excitation intensities from a nitrogen laser ($\lambda = 3371$ Å). For comparison the bottom curve displays a low intensity spectrum (Hg lamp excitation). For the laser excited spectra the rate of photon absorption per unit volume ranges from about 10^{18} to 10^{19} cm^{-3} s^{-1}, and the exciton concentrations are many orders of magnitude larger than those dealt with in the preceding paragraph. It is seen that the spectra change drastically as the pump intensity is increased. We are interested in the two major new features denoted M and P, at 2.545 and 2.527 eV, respectively. The P band [5.86] has been attributed to inelastic exciton-exciton scattering. An example of such a collisional process is indicated schematically in Fig. 5.27. Two exciton-polaritons collide producing a continuum excitation, i.e., a free electron-hole pair and a polariton belonging to the lower end of the dispersion curve. This polariton propagates freely because it is photon-like, and it can leak from the crystal at the boundaries.

Neglecting the kinetic energy of the two original excitons the energy of the photon produced in the collision is approximately

$$\hbar\omega_{\mathrm{p}} = 2E_{\mathrm{x}} - E_{\mathrm{G}} = E_{\mathrm{x}} - B_{\mathrm{x}} \tag{5.25}$$

where $E_{\mathrm{G}} = 2.582$ eV and $B_{\mathrm{x}} \simeq 30$ meV are the energy gap and the A exciton

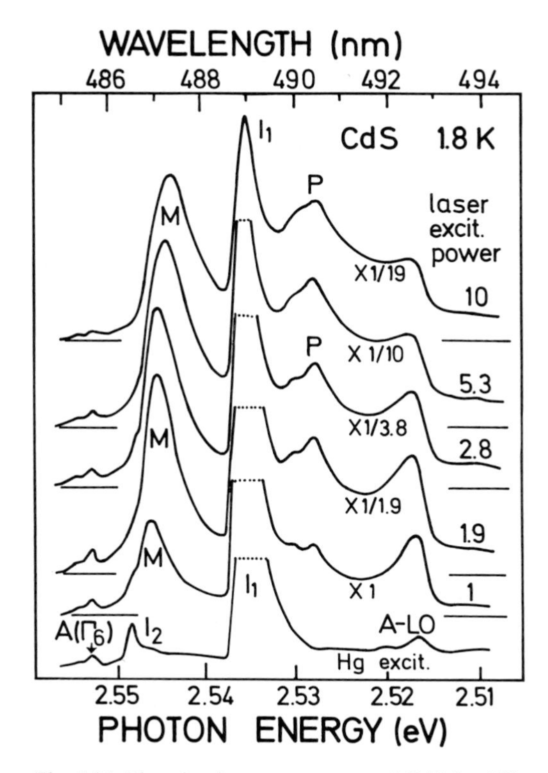

Fig. 5.26. Photoluminescence spectra of CdS for different excitation intensities (nitrogen laser at 337.1 nm). The top curve corresponds to 40 kW/cm^2. The bottom trace shows a low intensity spectrum taken with mercury lamp excitation (after *Shionoya* et al. [5.85])

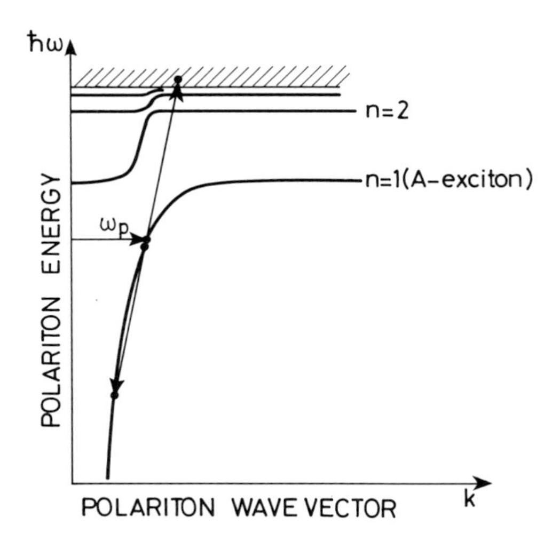

Fig. 5.27. Schematic representation of an exciton-exciton inelastic collision process of the type responsible for the *P* band emission. One of the collision partners is scattered out of the exciton-like regime into the photon-like regime of the dispersion curve

binding energy for CdS, respectively, and $E_x = E_G - B_x$ is the A_1 exciton energy neglecting the kinetic energy. Thus the emission due to inelastic collisions of excitons is expected at an energy of about $\hbar\omega_p \simeq 2.522$ eV.

Next we discuss the M band feature. This band has been interpreted by *Shionoya* et al. [5.85] as evidence for the formation of excitonic molecules [5.87], or biexcitons. It was argued [5.85] that the most probable radiative annihilation process for an excitonic molecule is a decay leaving a free exciton and a photon (a photon-like polariton). If $E_M = 2E_X - B_M$ is the energy of the excitonic molecule, where B_M is the molecular binding energy, then the energy $\hbar\omega_M$ of the photon originating from the radiative annihilation of a molecule is approximately

$$\hbar\omega_M = E_M - E_X = E_X - B_M \tag{5.26}$$

where we have again neglected the kinetic energy of the exciton and of the excitonic molecule. As discussed by *Akimoto* and *Hanamura* [5.88] the molecular binding energy is expected to be about 5 meV for CdS. Thus the emission due to excitonic molecules will occur at $\hbar\omega_M \simeq 2.547$ eV.

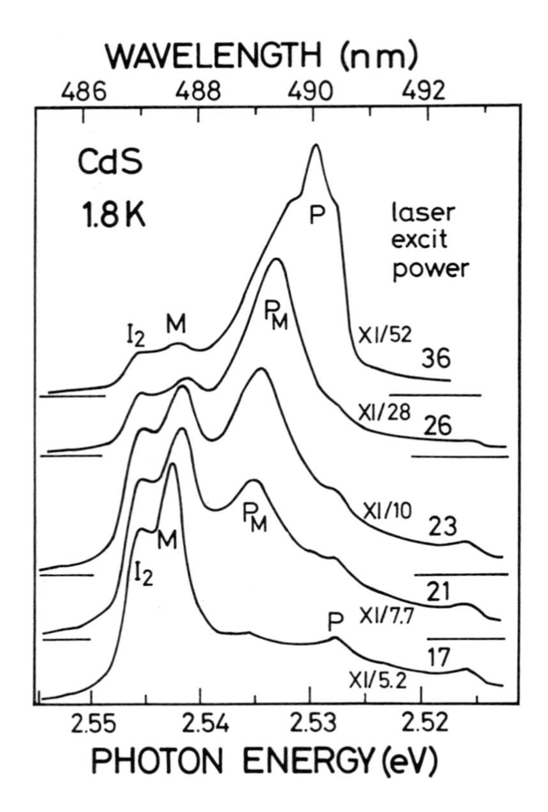

Fig. 5.28. Photoluminescence spectra at very high nitrogen laser excitation. The top curve corresponds to 140 kW/cm^2 (after *Saito* et al. [5.89])

At still higher excitation power another new luminescence band at about 2.535 eV was observed by *Saito* et al. [5.89]. This new structure is denoted by P_M in the spectra shown in Fig. 5.28. *Saito* et al. attributed the P_M band to inelastic collisions between excitonic molecules, and, according to their interpretation, two molecules collide producing three free excitons and a photon-like polariton. Again, the photon-like polariton propagates through the crystal and is detected. The energy $\hbar\omega_{P_M}$ of the photon emitted during this collision is given by

$$\hbar\omega_{P_M} = E_x - 2B_M \qquad (5.27)$$

when the kinetic energy is neglected.

It is interesting to note that the I_1 band in Fig. 5.28 is absent whereas it is clearly seen at all excitation levels in Fig. 5.26. Apparently a different CdS crystal, possibly with lower acceptor concentration or with a different degree of compensation was used in the experiments of [5.89]. The strong I_1 band would have obstructed the detection of the P_M band in the sample used in [5.85]. This situation stresses the importance of a proper sample characterization and the difficulties that arise in the comparison of spectra from different samples (see Subsec. 5.3.6).

5.3.4 Dynamics of Excitons at High Density

Having surveyed high density exciton spectra, we return to the discussion of picosecond experiments. *Kuroda* et al. [5.90] reported the observation of Bose condensation in cadmium selenide when a high concentration of excitons was generated by two-photon absorption of picosecond light pulses at 1.06 μm. During an effort to support their conclusion detailed studies of the dynamic behaviour were performed of high density excitons in CdSe. Excitons were excited by band-to-band transitions using ultrashort light pulses, and the time evolution of the various luminescence and absorption bands was measured on a picosecond time scale.

One obvious reason for choosing CdSe for these studies was the availability of a convenient source of picosecond light pulses at a frequency suitable for the generation of excitons in that material, i.e., the modelocked Nd:glass laser. In CdSe the intrinsic A_1-exciton polariton band is located at $\hbar\omega_{A_1} = 1.825$ eV ($\lambda = 6793$ Å); for intense excitation new luminescence bands similar to those observed in CdS are also found in CdSe: the M band at $\hbar\omega_M = 1.819$ eV ($\lambda = 6815$ Å), the P_M band at $\hbar\omega_{p_M} = 1.816$ eV ($\lambda = 6827$ Å), and the P band at $\hbar\omega_p = 1.812$ eV ($\lambda = 6842$ Å). In a first set of measurements [5.91] the time dependence of the luminescence from M, P_M and P bands was examined. A CdSe crystal immersed in liquid helium ($T = 1.8$ to 4.2 K) was excited by the second harmonic of pulse trains from a passively modelocked neodymium glass laser. The luminescence was focused into a monochromator after passing through a CS_2 Kerr shutter gated by the fundamental pulse trains at $\lambda = 1.06$ μm.

Time and frequency resolved luminescence spectra were thus measured point by point by varying the delay time between the pulses at the second harmonic ($\lambda = 0.53$ μm) and the gating pulses, and by changing the frequency setting of the monochromator. Each data point thus obtained represents an average over the individual pulses in the train and over several different pulse trains (laser shots). The time resolution was approximately 10 ps, and the frequency resolution corresponded to 10 Å [5.91].

Experimental results of *Kuroda* et al. [5.91] are reproduced in Fig. 5.29. From part *a* of this figure we see that the *M* band intensity rises rapidly, reaching a maximum at a delay time of about 100 ps after excitation and decaying with a time constant of approximately 100 ps. The light intensity observed at a delay of about 400 ps was thought to be due to the short wavelength wing of the *P* band and not to *M* band luminescence.

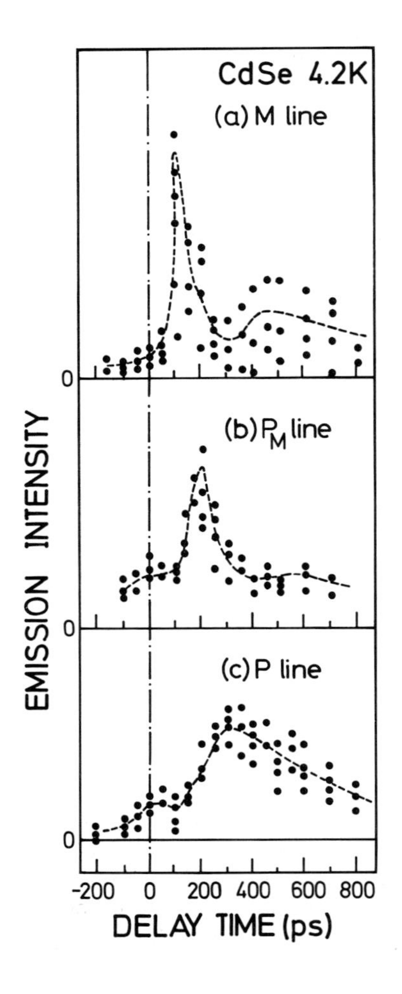

Fig. 5.29. Experimentally measured time dependence of the luminescence of the various bands in cadmium selenide (after *Kuroda* et al. [5.91])

Figure 5.29b depicts the time behavior of the P_M band with a maximum at about 200 ps and a decay time similar to that of the M band. The P luminescence is the latest to reach a maximum at 400 ps, and it differs from the other two bands by a relatively slow decay with a time constant of about 20 ps.

In a second set of measurements information about the formation and the decay of the free A_1-exciton was obtained. These observations are based on the concentration-dependent shift of the A_1 free exciton band [5.92]. In these experiments a high concentration of excitons was generated by two-photon absorption of picosecond pulse trains at 1.06 μm. The A_1 absorption band was probed by using "white" continuum picosecond pulses [5.93] generated by self-phase modulation of the second harmonic pulses in a slab of BK7 optical glass. The time dependence of the shift of the A_1 absorption peak was measured by varying the delay between the pump pulses at 1.06 μm and the continuum probe pulses. It was found that the A_1 peak shifted to higher energies when the band was probed under conditions of high exciton concentration [5.91]. The energy shift reached a maximum of about $\Delta E = 3$ to 4 meV at a delay of 20 ps after excitation. Thereafter the A_1 peak moved back to its normal position at 1.825 eV within approximately 50 ps.

Figure 5.30 qualitatively shows the observed time behavior of the A_1 absorption band, as well as that of the various luminescence bands discussed

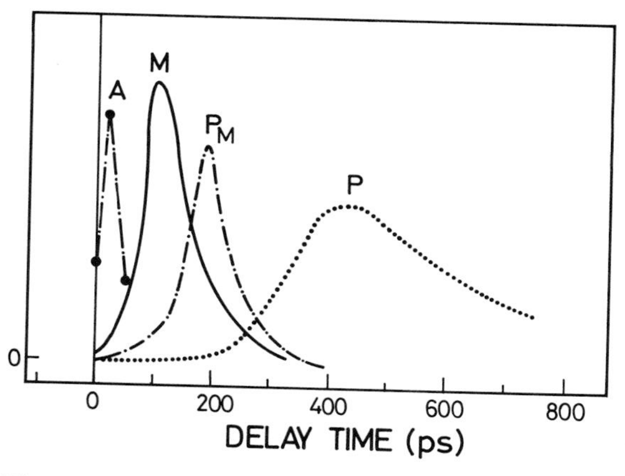

Fig. 5.30. Qualitative time behavior of the luminescence bands and of the A_1 exciton absorption peak (after *Kuroda* et al. [5.91])

above. The bulk of experimental results was interpreted by *Kuroda* et al. [5.91] as follows. We begin with the A_1 absorption band. *Hanamura* [5.92] has predicted a shift of the A_1 band to higher energies as the total concentration of excitons increases. However, for a consistent interpretation of the time dependence of the A_1 shift the authors of [5.91] had to assume that the shift depends also on the concentration of excitons in the state $n=1$, and not just on the total concentration of all species of excitons. Theory predicts [5.94] that A type excitons in $n=2$ (A_2 excitons) are generated very rapidly from the continuum electron-hole excitations. The A_1 excitons are formed from A_2 excitons with a time constant of about 10^{-11} s [5.91]. Thus, the shift in time of the A_1 peak was interpreted as an indication that the A_1 excitons are produced within 20 ps after band-to-band excitation. The rise of the M luminescence at about 100 ps was regarded as evidence for preferential formation of excitonic molecules from free A_1 excitons. In the opinion of the authors of [5.91] the rise of the P_M band closely following the M band represents further proof of a rapid increase of the concentration of excitonic molecules at a delay time of 100 ps. Because each molecule-molecule collision contributing to the P_M band leaves three free excitons, the exciton concentration increases rapidly at about 300 ps, and as a result the rate of inelastic exciton-exciton collisions increases, explaining the maximum of the P band at approximately 400 ps.

Note that a similar time behavior of the P band had already been observed in the earlier work of *Figueira* and *Mahr* [5.95], although these authors interpreted their results as possible evidence of a Mott transition. The essential point of the time-resolved experiments according to the authors of [5.91] is the fact that picosecond excitation apparently produces a high concentration of excitonic molecules very shortly after excitation. Only much later do those exciton decay processes occur that give rise to "hot" free excitons of large kinetic energy, such as the inelastic exciton-exciton collisions responsible for the P band.

It is concluded [5.91] that with picosecond pulse excitation a situation highly favorable for the observation of Bose condensation is produced, and in the next section we will discuss an attempt to observe Bose condensation of excitonic molecules in CdSe.

5.3.5 Bose Condensation of Excitonic Molecules

Since excitonic molecules can be approximately regarded as bosons, it has been suggested that a gas of excitonic molecules can undergo Bose condensation when cooled below a characteristic transition temperature. According to *Hanamura* the transition temperature T_c is given by [5.87]

$$T_c = 2\pi \frac{\mu}{2M} [Na_0^3/\zeta(3/2)]^{2/3} B_x \tag{5.28}$$

where μ and M are the reduced and the total mass of the molecule, respectively, ζ is the Zeta-function, a_0 is the exciton Bohr radius, and N is the concentration of excitonic molecules.

It is seen from (5.28) that a lower limit of the molecular concentration for Bose condensation will exist, if the crystal temperature is held fixed. On the other hand, an upper concentration limit for the existence of a Bose condensed phase is imposed by the critical density for the Mott transition [5.96]. If the number of molecules exceeds the threshold value for the Mott transition, the excitonic molecules break up, and a metal-like electron-hole liquid is formed. For excitonic molecules the Mott density is given by

$$N_{\mathrm{Mott}}=\tfrac{1}{2}(2a_0)^{-3}. \tag{5.29}$$

It turns out that at liquid helium temperature the concentration range for the existence of a Bose condensate is rather narrow in materials like CdS and CdSe.

It is characteristic of the Bose condensate that below the transition temperature a large fraction of the molecules is concentrated in the state of zero momentum, $k=0$. According to *Hanamura* [5.87], the Bose condensate can be readily detected by measuring the M band luminescence, because the same radiative decay channels as above T_c are also operative in the condensed phase, i.e., the molecules decay leaving free excitons and photons. However, the energy $\hbar\omega_c$ of the photons emitted during annihilation of a molecule in the condensed phase has a well-defined value

$$\hbar\omega_c=E_x-B_M \tag{5.30}$$

because the kinetic energy is zero. Thus, if Bose condensation takes place in a gas of excitonic molecules, a drastic change of the shape of the M band is expected, i.e., a large fraction of the emission should occur in the form of a sharp line at precisely $\hbar\omega_c$ [5.87].

In search of evidence for Bose condensation *Kuroda* et al. [5.90] carefully investigated the M band luminescence of a CdSe crystal. The CdSe sample was excited via two-photon absorption with picosecond light pulses ($\lambda=1.06$ μm) at temperatures between 1.8 and 4.2 K. The authors of [5.90] claimed that they have indeed succeeded in detecting a very sharp feature in the M-band luminescence at an energy of $\hbar\omega_c=1.8195$ eV. Their experimental result is reproduced in Fig. 5.31, showing a detail of the M luminescence with a spike at 1.8195 eV superimposed on a strongly fluctuating background. It was pointed out in [5.90] that the spike was seen only for a very narrow range of excitation intensity, and it was argued that this behavior is expected because of the limited concentration range for Bose condensation [see (5.28) and (5.29)]. The broad, fluctuating background in Fig. 5.31 was attributed to the variation of the excitation intensity along the pulse trains. Only for a few pulses in the train will the concentration of excitonic molecules lie in the narrow margin for Bose condensation, while for the remaining pulses the concentration will be either too low or too high. Consistent with the theoretical expectation it was observed that the intensity range over which the sharp spike could be detected decreased significantly as the temperature of the CdSe sample was raised from 1.8 to 4.2 K.

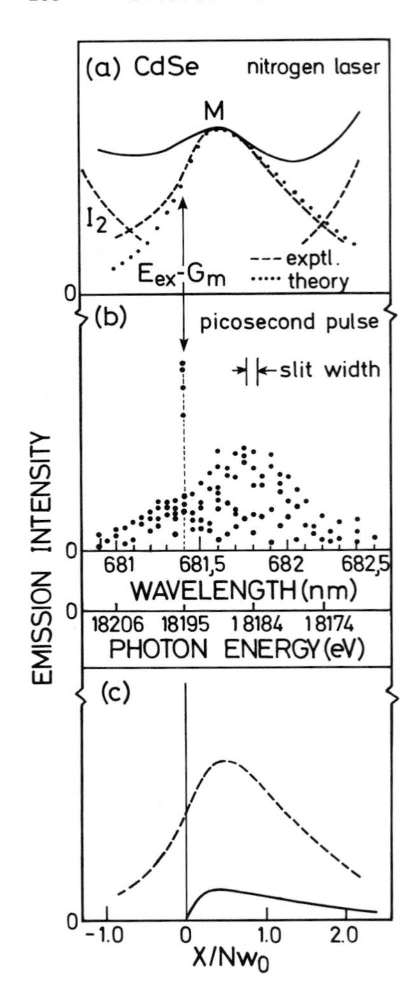

Fig. 5.31. Detail of the photoluminescence spectrum of cadmium selenide showing the M band (a) for nitrogen laser excitation, (b) for excitation via two-photon absorption by a train of picosecond pulses 1.06 μm, (c) theoretical calculations of the luminescence spectrum for the Bose condensed state of the excitonic molecules (full curve) and for the non-condensed state at higher temperatures (dashed curve). $X = E_x - B_M - \hbar\omega$; $Nw_0 = 1.8$ meV (after *Kuroda* et al. [5.90])

5.3.6 Discussion of the High Density Exciton Experiments

The evidence for Bose condensation of a gas of excitonic molecules was essentially based on the following principal experimental facts [5.90, 91].

1) From time-resolved luminescence measurements it was concluded that only with picosecond excitation could one hope to create a sufficiently high density of excitonic molecules before the temperature of the exciton system would be raised above the Bose transition temperature by energy dissipation through the various decay mechanisms.

2) With picosecond excitation a sharp luminescence line was observed at an energy position consistent with radiative decay of an excitonic molecule of

zero momentum. No such feature was seen with nanosecond laser pulse excitation [5.90].

3) The sharp luminescence line existed only within a very narrow range of excitation intensity, in agreement with the theoretically predicted limited concentration range over which Bose condensation will occur.

Although all these experimental observations are qualitatively consistent with the picture of Bose condensed excitonic molecules, other interpretations of the results have not been definitely ruled out, and a detailed quantitative analysis is still lacking.

Johnston and *Shaklee* [5.97] estimated the temperature changes following photoexcitation of excitons. They came to the conclusion that under the conditions of the experiment of *Kuroda* et al. [5.90] excitonic molecules cannot be generated at densities less than the Mott density and temperatures less than the Bose transition temperature, i.e., these two conditions cannot be satisfied simultaneously. They pointed out that a minimum amount of energy of

$$\Delta E = 4\hbar\omega - 2E_x - B_M \tag{5.31}$$

will be deposited in the crystal per excitonic molecule created, when two-photon excitation by interband transitions at a frequency ω is used. For CdSe and $\hbar\omega = 1.17$ eV ($\lambda = 1.06$ μm) (5.31) gives $\Delta E = 1$ eV. It is easy to show that when such a large energy is to be dissipated the final temperature will be lower than the Bose transition temperature only for molecular concentrations well above the Mott density. According to *Johnston* and *Shaklee* the condition $T < T_c$ and $N < N_{\text{Mott}}$ cannot be simultaneously satisfied even if one were to generate excitonic molecules by direct excitation of the A_1 band, with a dissipation energy as small as $\Delta E = B_M = 5$ meV for CdSe. Therefore it is concluded that excitonic molecules would break up and form a metallic electron-hole liquid before Bose condensation could take place.

Note that the calculation of *Johnston* and *Shaklee* is very conservative. They chose 100% efficiency for the formation of excitonic molecules, and they computed the rise of the *lattice* temperature. Furthermore, their estimate was based on excitation by a single picosecond pulse, whereas the experiment of *Kuroda* et al. [5.90] was performed with pulse trains of about 30 individual pulses of 7 ns separation. Under these conditions accumulative heating is expected, because heat transport can be neglected on the time scale of the pulse train. Thus, the actual rise of the exciton temperature will be larger than the temperature rise estimated by *Johnston* and *Shaklee*.

Obviously the question of experimental evidence for Bose condensation in [5.90] depends crucially on the interpretation of the various luminescence bands as given in [5.85] and [5.89] (see Subsec. 5.3.3). For example, it has been questioned by *Dite* et al. [5.98] whether the M band represents sufficient proof of the existence of excitonic molecules in CdS, and, similarly, whether the P_M band is indeed due to inelastic collisions between excitonic molecules.

In [5.85] and [5.89] the assignment of the M and P_M bands was mainly based on the following arguments.

1) The position and the shape of both the M and the P_M band could be explained by using the concept of excitonic molecules and collisions among molecules.

2) The M band was found to be independent of the intensity of the I_1 and I_2 luminescence suggesting that the M band is of intrinsic origin and not related to impurities.

3) The intensity of M and P_M grew faster than linear when the excitation power was increased. A nonlinear dependence on excitation is indeed expected if these bands involve interactions between several excitons. It was claimed that the nonlinear behavior was not the result of stimulated emissions [5.89].

The observations reported by *Dite* et al. [5.98] differ from the results of [5.85] and [5.89] in several points. *Dite* et al. did observe different behavior of the P_M band luminescence in CdS depending on the sample purity and the degree of compensation. For example, a distinct P_M band was seen in high purity samples (less than $5 \times 10^{15}/\text{cm}^3$ donors and acceptors) even though there was no M band at excitation power densities of a few kW/cm^2. On the other hand, no P_M band could be found in highly compensated CdS crystals at excitation levels approaching the damage threshold.

Although stimulated emission was ruled out in [5.89], *Dite* et al. presented strong experimental evidence that stimulated emission plays a role for both the M and the P_M bands. At a fixed pump intensity these authors varied the length of the strip of pumped material, and they were able to show that the intensity and shape of the luminescence bands strongly depended on the strip length in a manner consistent with the assumption of gain due to stimulated emission of a few hundred cm^{-1}.

Dite et al. offered an alternative explanation of the M and P_M bands as acoustic phonon sidebands of the I_2 and the I_1 bound exciton zero phonon line, respectively. It was pointed out that despite the low concentration of donors and acceptors large gains are expected as a consequence of the giant oscillator strength effects [5.77]. The observation that not the I_1 and I_2 bands but only their sidebands exhibited stimulated emission was explained by pointing out the low threshold for stimulated emission of a vibronic transition due to the analogy of vibronic lasing with a four level lasing mechanism.

We conclude this section by mentioning a result reported recently by *Švorec* and *Chase* [5.99]. They have used two-photon spectroscopy in an attempt to establish the existence of excitonic molecules in CdS. These experiments were based on a prediction of *Hanamura* [5.100] who showed that excitonic molecules can be generated directly by simultaneous absorption of two light quanta of energy $\hbar\omega = E_x - B_M/2$. The two-photon absorption coefficient due to this process is expected to be very strong because of the combined effect of resonance enhancement and giant oscillator strength [5.77]. Therefore two-photon absorption should be a useful tool for detection of excitonic molecules.

In their experiment *Svorec* and *Chase* did indeed observe a strong two-photon peak at 2.5457 eV, and one might be tempted to interpret this observation as proof of excitonic molecules. However, these authors came to the conclusion that the two-photon feature at 2.5457 eV is not due to excitonic molecules but more likely due to interference of contributions to the interband two-photon absorption from different intermediate states.

5.4 Summary

Since 1965 when ultrashort laser pulses became available for the first time the techniques for the production of picosecond light pulses and for the detection of very rapid phenomena created by them have progressed to a point where it is now possible to obtain reliable, quantitative data from picosecond optical experiments. In this chapter we have discussed the application of picosecond techniques in a wide variety of investigations.

Section 5.1 described how the various vibrational relaxation processes in liquids and crystals can be directly measured with a time resolution as short as 10^{-12} s. The picosecond Raman probe techniques of Section 5.1 yielded accurate values of optical phonon lifetimes, molecular dephasing times, and energy relaxation times of individual normal vibrational modes of molecules. These techniques also permit measurement of vibrational decay routes and study of rapid vibrational energy transfer in molecular liquids.

Clearly we are just at the outset of studying material vibrations by means of picosecond light pulses. Various expansions of the present efforts may be anticipated for the future. For example, in most of the previous work the vibrational modes of interest were excited by single frequency stimulated Raman scattering. With this mode of excitation just the very strongest Raman active vibration out of the total vibrational spectrum of a system can be studied. This limitation may be overcome by using a double pulse technique in which the difference frequency of two synchronized picosecond pulses is tuned to any desired Raman active mode of a system. Suitable pulses may be obtained from a pair of tunable dye lasers or parametric devices synchronously excited by a common modelocked pump laser. On the other hand, tunable infrared picosecond pulses will probably complement Raman excitation in the near future.

The variety of experiments discussed in Section 5.2 is an indication of the many opportunities for picosecond optical experimentation in solid state physics, particularly in semiconductor physics. In the discussion of multiphoton absorption it was shown how the use of ultrashort light pulses overcame the discrepancies of some earlier work by eliminating competing processes that obstructed multiphoton measurements with longer pulses. Picosecond techniques now yield quantitative values of multiphoton absorption cross sections and accurate multiphoton spectra.

The investigations of optical bleaching of interband transitions in semiconductors (Subsect. 5.2.2) are interesting because eventually new information on ultrashort relaxation times of photoexcited hot carriers may be obtained from such experiments. However, so far it has not yet been possible to unambiguously identify the mechanisms responsible for the optical bleaching effects in the various semiconductor materials. The interpretation of the measured recovery times of the optical absorption in terms of fundamental relaxation processes is difficult because optical saturation in semiconductors may be accompanied by other competing processes, for example, parametric scattering, free carrier absorption, and stimulated emission. Further work is required to gain a better understanding of these phenomena.

Subsection 5.2.3 described interesting new experiments on very dense electron-hole plasmas generated by picosecond optical excitation. Novel information on diffusion, conduction, and recombination of a very dense gas of free carriers was obtained. A unique advantage of these experiments lies in the fact that the measurements are performed in pure semiconductor materials permitting investigation of intrinsic material properties which are often obscured by impurity effects in conventional experiments with doped material.

On the other hand, studying the interaction of free carriers with impurities is of course a very important research topic of its own. For instance, the various nonradiative recombination mechanisms of electrons and holes are still not very well understood. Multiphonon nonradiative relaxation involving impurity levels as intermediate steps has been considered as a possible effective recombination process. This mechanism is similar to the electronic relaxation process via multiphonon emission discussed in Subsection 5.2.4 for ionic impurities in an insulator.

Section 5.3 finally dealt with diverse picosecond phenomena involving the formation, relaxation and interaction of excitons following interband optical excitation by picosecond light pulses. Time-resolved observations of exciton luminescence spectra revealed interesting dynamic processes on a subnanosecond time scale. The measurements were interpreted in terms of scattering and decay processes of exciton complexes. The picosecond exciton work culminated in the claim of evidence of Bose condensation of excitonic molecules. However, it appears that further work is needed to support or rule out this interpretation.

In conclusion, this survey emphasized the interdisciplinary character of picosecond spectroscopy and indicated many interesting applications of ultrashort light pulses. It became apparent that some applications have already matured in such a way that useful quantitative information not previously available can now be obtained. In other areas just a few initial experiments have been reported to date. It is to be expected that in the years to come these more recent efforts will expand, and that other interesting scientific applications of picosecond techniques will be found.

References

5.1 G. Placzek: *Marx Handbuch der Radiologie*, vol. VI/2, p. 205 (1934)
5.2 N. Bloembergen: Am. J. Phys. **35**, 989 (1967);
M. Maier, W. Kaiser: In *Laser Handbook*, vol. 2, ed. by F. T. Arecchi, E. O. Schulz-Dubois (North Holland, Amsterdam 1972)
5.3 N. Bloembergen: *Nonlinear Optics* (W. A. Benjamin, New York 1965)
5.4 J. A. Giordmaine, W. Kaiser: Phys. Rev. **144**, 676 (1966)
5.5 D. von der Linde, A. Laubereau, W. Kaiser: Phys. Rev. Lett. **26**, 954 (1971)
5.6 F. De Martini, J. Ducuing: Phys. Rev. Lett. **17**, 117 (1966)
5.7 A. Laubereau, D. von der Linde, W. Kaiser: Phys. Rev. Lett. **28**, 1162 (1972)
5.8 J. Brandmüller, K. Buchardi, H. Hacker, H. W. Schrötter: Z. Angew. Phys. **22**, 117 (1967)
5.9 A. Laubereau, G. Wochner, W. Kaiser: Phys. Rev. **A 13**, 2212 (1976)
5.10 A. Laubereau: Chem. Phys. Lett. **27**, 600 (1974)
5.11 S. F. Fischer, A. Laubereau: Chem. Phys. Lett. **35**, 6 (1975)
5.12 A. Laubereau, G. Wochner, W. Kaiser: Opt. Commun. **17**, 91 (1976)
5.13 M. Matsuoka, H. Nakatsuka, J. Okada: Phys. Rev. **A 12**, 1062 (1975)
5.14 R. R. Alfano, S. L. Shapiro, Phys. Rev. Lett. **26**, 1247 (1971)
5.15 K. Park: Phys. Lett. **22**, 39 (1966)
5.16 A. Laubereau, D. von der Linde, W. Kaiser: Phys. Rev. Lett. **27**, 802 (1971)
5.17 A. K. McQuillan, W. R. L. Clements, B. P. Stoicheff: Phys. Rev. **A 1**, 628 (1970);
F. Stenman: J. Appl. Phys. **40**, 4164 (1969);
S. A. Solin, A. K. Ramdas: Phys. Rev. **B 1**, 1687 (1970)
5.18 A. Laubereau, G. Wochner, W. Kaiser: Opt. Commun. **14**, 75 (1975)
5.19 M. Born, K. Huang: Dynamical Theory of Crystal Lattices (Oxford University Press, London 1962)
5.20 A. Laubereau, D. von der Linde, W. Kaiser: Opt. Commun. **7**, 173 (1973)
5.21 A. Mooradian, G. B. Wright: Solid State Commun. **4**, 431 (1966);
A. S. Barker: Phys. Rev. **165**, 917 (1968)
5.22 R. R. Alfano: J. Opt. Soc. Amer. **60**, 66 (1970);
R. R. Alfano T. G. Giallorenzi: Opt. Commun. **4**, 271 (1971);
P. G. Harper: Opt. Commun. **1 D**, 68 (1974)
5.23 S. Ushioda, J. D. McMullen: Solid State Commun. **11**, 299 (1972)
5.24 A. Laubereau, A. Seilmeier, W. Kaiser: Chem. Phys. Lett. **36**, 232 (1975)
5.25 R. R. Alfano, S. L. Shapiro: Phys. Rev. Lett. **29**, 1655 (1972)
5.26 A. Laubereau, G. Kehl, W. Kaiser: Opt. Commun. **11**, 74 (1974)
5.27 A. Laubereau, L. Kirschner, W. Kaiser: Opt. Commun. **9**, 182 (1973)
5.28 P. R. Monson, S. Patumtevapibal, K. J. Kaufmann, G. W. Robinson: Chem. Phys. Lett. **28**, 312 (1974)
5.29 J. M. Worlock: In *Laser Handbook*, vol. 2, ed. by F. T. Arecchi, E. O. Schulz-Dubois (North Holland, Amsterdam 1972);
V. I. Bredikhin, M. D. Galanin, V. N. Genkin: Sov. Phys. Uspekhi **16**, 299 (1973)
5.30 J. A. Giordmaine, P. M. Rentzepis, S. L. Shapiro, K. W. Wecht: Appl. Phys. Lett. **11**, 216 (1967)
5.31 D. J. Bradley, M. H. R. Hutchinson, H. Koetser, T. Morrow, G. H. C. New, M. S. Petty: Proc. R. Soc. London A **328**, 97 (1972)
5.32 V. V. Arsen'ev, V. S. Dneprovskii, D. N. Klyshko, L. A. Sysoev: Sov. Phys. JETP **33**, 64 (1971)
5.33 S. Jayaraman, C. H. Lee: Appl. Phys. Lett. **20**, 292 (1972)
5.34 S. Jayaraman, C. H. Lee: J. Appl. Phys. **44**, 5480 (1973)
5.35 A. Penzkofer, W. Falkenstein: Opt. Commun. **17**, 1 (1976)
5.36 J. F. Reintjes, J. McGroddy: Phys. Rev. Lett. **30**, 901 (1973)
5.37 J. H. Bechtel, W. L. Smith: Phys. Rev. B **13**, 3515 (1976)
5.38 A. Penzkofer, W. Falkenstein, W. Kaiser: Appl. Phys. Lett. **28**, 319 (1976)

5.39 A. Penzkofer, A. Laubereau, W. Kaiser: Phys. Rev. Lett. **31**, 863 (1973)
5.40 H. Sonnenberg, H. Heffner, W. Spicer: Appl. Phys. Lett. **5**, 95 (1964)
5.41 D. C. Burnham: Appl. Phys. Lett. **17**, 45 (1970)
5.42 G. Y. Forkas, Z. Gy. Horvath: Phys. Stat. Sol. (a) **3**, K 29 (1970)
5.43 W. R. Bennett, D. B. Carlin, G. J. Collins: IEEE J. **QE-10**, 97 (1974)
5.44 P. Sorokin, N. Braslau: IBM J. Res. Dev. **8**, 177 (1964)
5.45 S. Yatsiv, M. Rokni, S. Barak: Phys. Rev. Lett. **20**, 1282 (1968);
R. L. Carman: Phys. Rev. A **12**, 1048 (1975);
V. S. Letokhov: JETP Lett. **7**, 221 (1968)
5.46 P. D. Dapkus, N. Holonyak, Jr., R. D. Burnham, D. L. Keune: Appl. Phys. Lett. **16**, 93 (1970)
5.47 C. J. Kennedy, J. C. Matter, A. L. Smirl, H. Weiche, F. A. Hopf, M. O. Scully: Opt. Commun. **16**, 118 (1976)
5.48 C. V. Shank, D. H. Auston: Phys. Rev. Lett. **34**, 479 (1975)
5.49 A. L. Smirl, J. C. Matter, A. Elci, M. O. Scully: Opt. Commun. **16**, 118 (1976)
5.50 D. von der Linde: IEEE J. **QE-8**, 328 (1972)
5.51 J. F. Reintjes, J. C. McGroddy, A. E. Blakedee: J. Appl. Phys. **46**, 879 (1975)
5.52 J. C. Matter, A. L. Smirl, M. O. Scully: Appl. Phys. Lett. **28**, 507 (1976)
5.53 A. F. Gibson, C. A. Rosito, C. A. Raffo, M. F. Kimmit: Appl. Phys. Lett. **21**, 356 (1972)
5.54 A. V. Nurmikho: Opt. Commun. **16**, 365 (1976)
5.55 D. H. Auston, C. V. Shank, Phys. Rev. Lett. **32**, 1120 (1974)
5.56 D. H. Auston, C. V. Shank, P. LeFur: Phys. Rev. Lett. **35**, 1022 (1975)
5.57 D. H. Auston, A. M. Johnson: to be published
5.58 D. H. Auston, A. M. Glass, A. A. Ballmann: Phys. Rev. Lett. **28**, 897 (1972)
5.59 D. von der Linde, D. H. Auston, A. M. Glass, K. F. Rodgers: Solid State Commun. **14**, 137 (1974)
5.60 R. Englman, J. Jortner: Mol. Phys. **18**, 145 (1970)
5.61 C. E. Swenberg, N. E. Geacintov: In *Organic Molecular Photophysics*, vol. 1, ed. by J. B. Birks (John Wiley & Sons, London 1973)
5.62 R. R. Alfano, S. L. Shapiro, M. Pope: Opt. Commun. **9**, 388 (1973)
5.63 A. W. Smith, C. Weiss: Chem. Phys. Lett. **14**, 507 (1972)
5.64 C. E. Swenberg, W. T. Stacy: Chem. Phys. Lett. **2**, 237 (1968)
5.65 J. N. Bradford, R. T. Williams, W. L. Faust: Phys. Rev. Lett. **35**, 300 (1975)
5.66 W. B. Fowler, Ed.: *Physics of Color Centers* (Academic Press, New York 1968)
5.67 R. S. Knox: In *Theory of Excitons, Solid State Physics*, Suppl. 5, ed. by F. Seitz and D. Turnbull (Academic Press, New York 1957)
5.68 G. H. Wannier: Phys. Rev. **52**, 191 (1937);
N. F. Mott: Trans. Faraday Soc. **34**, 500 (1938)
5.69 G. Dresselhaus: J. Phys. Chem. Solids **1**, 14 (1956);
W. Kohn: In *Solid State Physics*, vol. 5, ed. by F. Seitz, D. Turnbull (Academic Press, New York 1957)
5.70 M. Lampert: Phys. Rev. Lett. **1**, 450 (1958)
5.71 See, for example, J. M. Ziman: *Electrons and Phonons* (University Press, Oxford 1963)
5.72 J. J. Hopfield, D. G. Thomas: Phys. Rev. **122**, 35 (1961)
5.73 E. Hanamura: In *Springer Tracts in Modern Physics*, vol. 73: Excitons at High Density, ed. by H. Haken, S. Nikitine (Springer, Berlin, Heidelberg, New York 1975) p. 43
5.74 J. M. Blatt, K. W. Boer, W. Brandt: Phys. Rev. **126**, 1691 (1962)
5.75 J. J. Hopfield: Phys. Rev. **112**, 1555 (1958)
5.76 P. Wiesner: Thesis, Technical University Stuttgart (1974)
5.77 E. I. Rashba: In *Springer Tracts in Modern Physics*, vol. 73: Excitons at High Density, ed. by H. Haken, S. Nikitine (Springer, Berlin, Heidelberg, New York 1975) p. 150
5.78 D. G. Thomas, J. J. Hopfield: Phys. Rev. **128**, 2155 (1962)
5.79 U. Heim, P. Wiesner: Phys. Rev. Lett. **30**, 1205 (1973);
P. Wiesner, U. Heim: Phys. Rev. B **11**, 3071 (1975)
5.80 E. Gross, S. Permogarov, V. Travnikov, A. Selkin: Solid State Commun. **10**, 1071 (1972)
5.81 Y. Toyozawa: Prog. Theoret. Phys. Suppl. **12**, 112 (1959)

5.82 L. M. Bollinger, G. E. Thomas: Rev. Sci. Instr. **32**, 1044 (1961)
5.83 A. A. Demidenko: Sov. Phys. Sol. State **5**, 2074 (1964)
5.84 S. Suga, K. Cho, P. Hiesinger, T. Koda: J. Luminescence **12/13**, 209 (1976)
5.85 S. Shionoya, H. Saito, E. Hanamura, O. Akimoto: Solid State Commun. **12**, 223 (1973)
5.86 D. Magde, H. Mahr: Phys. Rev. Lett. **24**, 890 (1970)
5.87 For a review of excitonic molecules see, for example, E. Hanamura: Optical Properties of Solids; New Developments, ed. by B. O. Seraphin (North Holland, Amsterdam 1976)
5.88 O. Akimoto, E. Hanamura: J. Phys. Soc. Japan **33**, 1537 (1972); O. Akimoto: J. Phys. Soc. Japan **35**, 973 (1973)
5.89 H. Saito, S. Shionoya, E. Hanamura: Solid State Commun. **12**, 227 (1973)
5.90 H. Kuroda, S. Shionoya, H. Saito, E. Hanamura: J. Phys. Soc. Japan **35**, 534 (1973)
5.91 H. Kuroda, S. Shionoya: J. Phys. Soc. Japan **36**, 476 (1974)
5.92 E. Hanamura: J. Phys. Soc. Japan **29**, 50 (1970)
5.93 See, for example, D. H. Auston: Chapter 4 of this book
5.94 M. Lax: J. Phys. Chem. Solids **8**, 66 (1959)
5.95 J. F. Figueira, H. Mahr: Solid State Commun. **9**, 679 (1971)
5.96 N. F. Mott: Phil. Mag. **6**, 287 (1961)
5.97 W. D. Johnston, Jr., K. L. Shaklee: Solid State Commun. **15**, 73 (1974)
5.98 A. F. Dite, V. I. Revenko, V. B. Timofeev, P. D. Altukhov: JETP Lett. **18**, 341 (1974)
5.99 R. W. Svorec, L. L. Chase: Solid State Commun. **17**, 803 (1975)
5.100 E. Hanamura: Solid State Commun. **12**, 951 (1973)

6. Picosecond Relaxation Processes in Chemistry

K. B. Eisenthal

With 16 Figures

One of the basic questions in chemistry today is concerned with the degradation of energy in a molecular system. The time-dependent redistribution of energy among the various degrees of freedom within a molecule on excitation to some excited state and the interactions and energy exchange of the excited molecule with surrounding molecules and external fields are of fundamental importance to a description of molecular phenomena. It is the competition between the various dissipative pathways which determines whether light is emitted or nonradiative physical and chemical processes dominate in the degradation of energy and materials in the system of interest.

With the application of picosecond lasers to studies of physical and chemical processes, a revolutionary breakthrough has occurred in the investigation of ultrafast molecular processes. It is in the picosecond time domain that processes such as electron and proton transfer, energy degradation, and chemical changes often occur. The ultrafast time resolution afforded by picosecond lasers makes it possible to determine the evolution of a molecular system, perturbed by a picosecond laser pulse, through its various energy states and transient chemical species. Picosecond laser studies of the orientational and translational motions of molecules, as well as the energy-exchanging processes and chemical reactions occurring in liquids, are now providing new insights, not only into the nature of these fundamental processes, but into the nature and properties of the liquid state itself.

In this review the topics to be covered include: intermolecular energy transfer, molecular orientational relaxation in liquids, photodissociation and the cage effect in liquids, electron transfer phenomena, and internal conversion and intersystem crossing in molecules. Ground state vibrational decay processes and relaxation phenomena in biological systems are covered in Chapters 5 and 7, respectively.

6.1 Intermolecular Energy Transfer

The nonradiative transfer of electronic energy between molecules can take place over a considerable range of spatial separations and orientational configurations [6.1–6]. The efficiency of the excitation transfer process is dependent on the nature of the intermolecular interactions, the molecular separations and relative orientations, the extent of energy overlap and the lifetime of the excited donor molecule. For example a dipole-dipole coupling mechanism can lead to transfer over distances of the order of 100 Å as opposed to an exchange coupling which is short range and can extend to distances of the order of 10 Å. The donor and acceptor states involved in the transfer process can be excited- or ground-electronic states of various multiplicities containing variable amounts of vibrational excitation. This "diffusion" of energy from some initial distribution of excited and unexcited molecules through a variety of alternative distributions plays a vital role in a variety of physical, chemical and biological processes [6.7–11], such as the quenching of the donor fluorescence and the appearing of new emission bands, and the initiating of chemical reactions by transfer to a reactive molecule or site or the reverse process. "Diffusion" has been used also to assist in the mapping of distances between chromophores in biological molecules.

Prior to the development of picosecond lasers, studies of the dynamics of energy transfer were restricted to times of the order of 10^{-8} s and longer and were generally limited to light-emitting molecules. Thus, the nature of the laws governing the transfer could not be examined in the subnanosecond time scale.

6.1.1 Singlet-Singlet Transfer

One approach to the study of singlet-singlet energy transfer uses the linear polarization of the picosecond exciting light to induce an orientational anisotropy in the distribution of ground and excited molecules [6.9, 10]. Molecules whose transition moments have a large component along the polarization direction of the exciting picosecond light are preferentially excited. Thus, the formerly isotropic system becomes anisotropic, i.e., more excited molecules are oriented parallel to the field direction, and hence more unexcited molecules are oriented with their transition moments perpendicular to the field direction. In a rigid environment the induced anisotropy can only relax by unimolecular decay of the excited molecules and by energy transfer. In a system where the donor and acceptor molecules are of the same species, the randomization results from transfer between molecules of different orientations. On the other hand, in a mixed system the anisotropy in donor orientations can also relax by transfer between donor and acceptor molecules regardless of their mutual orientation. The decay of the anisotropy can be monitored by measuring the polarization dependent change in the donor absorption with time. The absorption is greater for probe light polarized perpendicular rather than parallel to the polarization of the exciting light because of the relative depletion of ground-state molecules in what we call the "parallel configuration". By probing at successively later times after excitation the decay of the dichroism due to the continued effects of energy transfer and the unimolecular excited state decay can be determined. Measurement of the latter quantity in the absence of transfer is then combined with the measured decay of the dichroism to obtain the energy-transfer dynamics.

To determine the nature of the transfer it is advantageous to study a two-component system, i.e., distinct donor and acceptor molecules versus a one-component system for which the acceptor molecules are the ground-state donor molecules. For the two-component system the acceptor molecules are randomly distributed both in orientation and distance with respect to the anisotropically distributed excited donor, and hence the donor decay function $\bar{w}(t)$ can be readily calculated. In the one-component system, although the distribution in distances is random, the distribution in orientations is perturbed by the excitation pulse [6.11–15]. Furthermore, in the two-component system only one transfer step from the donor to acceptor need be considered, assuming that vibrational relaxation in the excited acceptor molecule is rapid compared with excitation transfer. In the one-component system several steps may be necessary before randomization has occurred and would therefore have to be included in any theoretical treatment. The decay of the anisotropy induced by the picosecond excitation pulse is obtained by measuring the ratio of the parallel and perpendicular transmissions of an attenuated picosecond probe beam as a function of time. The ratio of the components of the transmitted beam at a time t after the excitation pulse is found to be given by [6.10],

$$\frac{I_{\parallel}(t)}{I_{\perp}(t)}=\frac{I_{\parallel}(0)}{I_{\perp}(0)}\exp\left[\alpha\bar{w}(t)l\right]$$

where α is a constant, $\bar{w}(t)$ is the donor decay function, and l is sample path length. If the coupling of the excited donor and ground state acceptor molecules is by a dipole-dipole interaction, then for a random distribution of acceptors $\bar{w}(t)$ is given by

$$\bar{w}(t)=\exp\left[-\frac{t}{\tau}-0.845\pi^{\frac{1}{2}}\sqrt{2/3}\,N_{\mathrm{A}}\left(\frac{R_0}{R_{\mathrm{g}}}\right)^3\left(\frac{t}{\tau}\right)^{\frac{1}{2}}\right]$$

where τ is the donor lifetime in the absence of transfer, R_0 is the critical transfer distance and is a measure of the strength of the dipole-dipole coupling, and $N_{\mathrm{A}}/(4/3)\pi R_{\mathrm{g}}^3$ is the concentration of acceptors. With rhodamine 6G as the donor and malachite green as the acceptor it is seen from Fig. 6.1 that the dipole-dipole interaction is a good description of energy transfer up to the earliest time measured, i.e., 20 ps. The critical transfer distance, R_0, is found to be about 53 Å which is in good agreement with the value calculated from the spectra (48 Å). It should be noted that in experiments of this type one must consider the possibility of stimulated emission caused by the probe pulse and amplified spontaneous emission processes. For the rhodamine 6G molecule the excited state vibrational relaxation time is known to be complete at the time of the earliest measurement (20 ps) [6.16–18]. The power densities used in the excitation pulse must be adjusted to avoid the complications of amplified spontaneous emission. In addition if the full pulse train is used rather than a single pulse, the possibility of buildup effects must be considered.

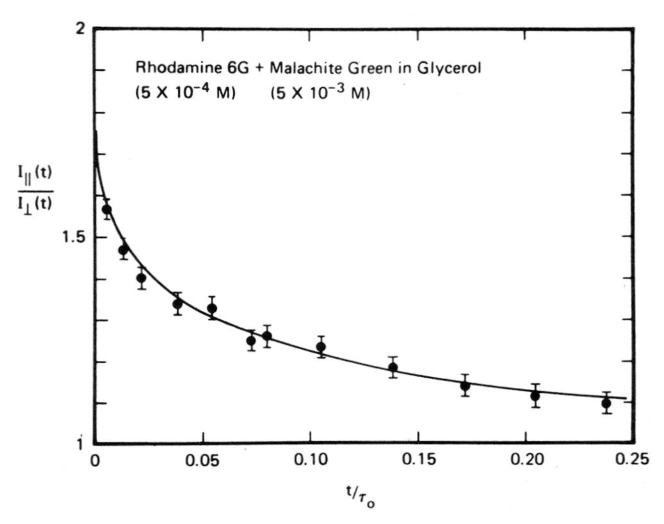

Fig. 6.1. Probe transmission ratio $I_{\|}/I_{\perp}$ as a function of time after the excitation pulse. τ_0 is the excited state donor lifetime in the absence of acceptors. Solid line is theoretical fit indicating that the Förster treatment for singlet-singlet transfer is adequate up to the earliest times measured, i.e., about 20 ps (after [6.10])

6.1.2 Triplet-Triplet Transfer

Since triplet-triplet energy transfer occurs via an exchange interaction, it is a short-range process and generally occurs between neighboring molecules, perhaps as far as 10–15 Å. Most studies in solution are therefore limited by the diffusion of the donor and acceptor molecules to some neighboring or near-neighboring molecular configurations. If the concentration of acceptor molecules is sufficiently high, then the rate of transfer is too rapid to study by conventional flash photolysis methods. Furthermore, earlier studies covered a time domain which was long compared to vibrational relaxation times. Thus the initial and final states of the donor-acceptor pair which were monitored were thermally equilibrated. Information on the vibronic energy in the excited donor prior to transfer and the distribution of vibronic energy in the ground state donor and excited acceptor after transfer is thus not obtainable. Since the energy transfer process is dependent on the energy distribution of the interacting states, and also on the rate of vibrational relaxation, both intramolecularly as well as intermolecularly, via coupling to the surrounding molecules, this short time information is of key importance.

With the application of picosecond laser methods, we are now gaining some insight into the role of vibrational energy distribution in triplet excitation transfer processes [6.19]. Some of the donor-acceptor pairs which have been studied include benzophenone as the donor and cis-piperylene, trans-piperylene, and 1-methylnaphthalene as the acceptors. In these investigations the solvent was composed of the acceptor molecules and hence translational molecular diffusion was not the rate-determining step. For some systems rotational motion may be of importance in satisfying orientational requirements of the donor-acceptor pair for energy transfer [6.20–21]. The method used for determining the triplet-triplet dynamics was to excite the donor to an excited singlet state with a picosecond pulse and to monitor the donor triplet population with a picosecond probe pulse at a wavelength corresponding to a donor triplet-triplet absorption. In this way both the buildup of the donor triplet due to intersystem crossing from the donor singlet and the decay of the donor triplet due to energy transfer to the acceptor molecule are obtained. See Table 6.1 and Fig. 6.2.

A small but definite difference is found in the rate of energy transfer from benzophenone to cis- versus trans-piperylene with the cis form being the faster one. The electronic contribution to the transfer rate is expected to be the same for the cis and trans forms since their electronic structures are essentially the same. The difference in the observed transfer is attributed to the different vibrational overlap functions (Franck-Condon factors) for the two forms. This difference in the Franck-Condon factors could arise from the roughly 500 cm^{-1} difference in the trans- versus cis-triplet energies. In this interpretation the final vibrational energy distribution would not be the same for the two pairs. In addition the rapid rate of energy transfer observed (10 ps) is in the time domain of vibrational relaxation, and thus the transfer might occur from a thermally nonequilibrated triplet benzophenone. This latter point is used in part to explain the slower rate of energy transfer from benzophenone to 1-methylnaphthalene (20 ps). It is estimated that the Franck-Condon factors are poorer in this pair than for the piperylenes, and thus donor vibrational relaxation can compete with triplet energy transfer. In this way a potentially favorably channel for energy transfer (i.e., from vibrationally excited triplet benzophenone) is reduced in importance, and the energy transfer rate is slower. However, it is also clear that the differences in the electronic contributions to the transfer can be different for the piperylenes and 1-methylnaphthalene, and thus could contribute to the observed difference in transfer rates.

Table 6.1. Buildup and decay times of triplet state donor species following singlet excitation in acceptor solvents (after [6.19])

Donor concentration (mole/liter^{-1})	Donor (solute)	Acceptor (solvent)	k^{-1} ($1-e^{-1}$ buildup time, ps)	k_Q^{-1} (e^{-1} decay time, ps)
0.5	Benzophenone	*cis*-Piperylene	7 ± 3	9 ± 3
0.5	Benzophenone	*trans*-Piperylene	7 ± 3	11 ± 3
0.5	Benzophenone	1-Methylnaphthalene	$<$ ca. 10	20 ± 5
0.01	1-Nitronaphthalene	*cis*-Piperylene	8 ± 3	678 ± 136

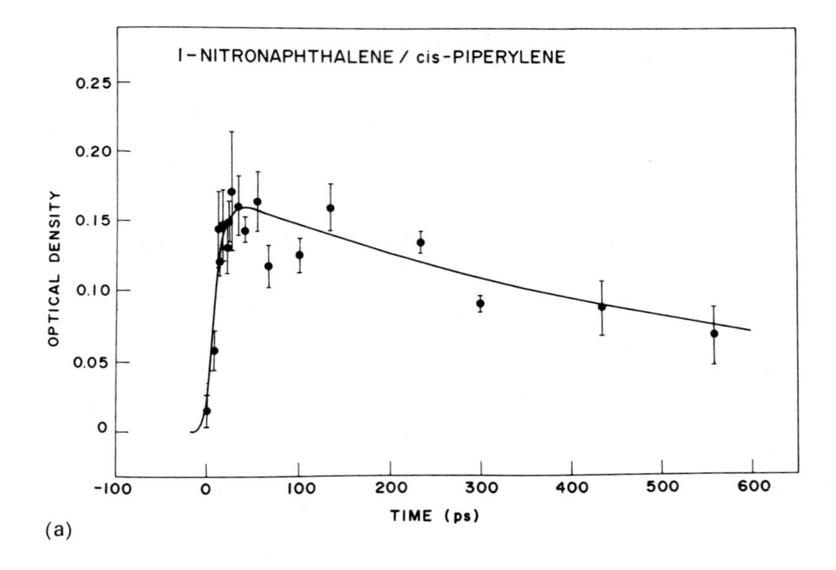

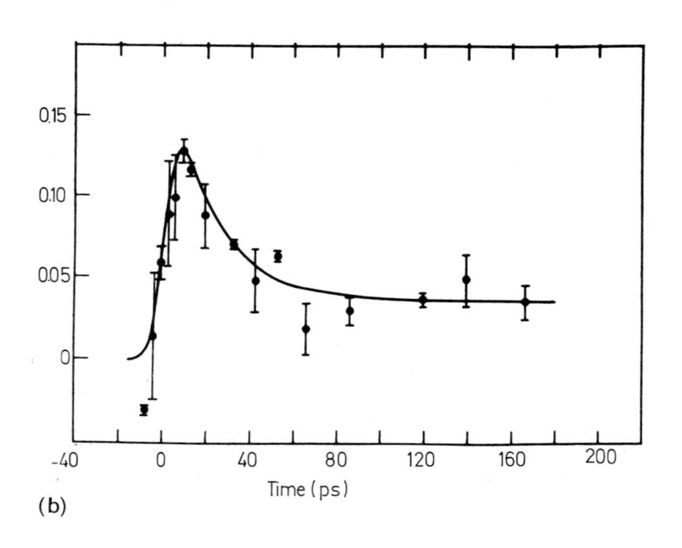

Fig. 6.2. (a) Optical density at 530 nm due to triplet 1-nitronaphthalene formation and decay as a function of delay time following singlet state excitation at 345.5 nm as measured in 0.01 mole/liter of 1-nitronaphthalene in *cis*-piperylene. Risetime indicates rapid triplet buildup while slow decay shows the inefficiency of triplet transfer to the solvent (after [6.19]). (b) Optical density at 530 nm due to triplet benzophenone formation and decay as a function of delay time following singlet state excitation at 354.5 nm as measured in 0.50 mole/liter of benzophenone in 1-methylnaphthalene. Rapid decay demonstrates efficient triplet-triplet transfer process in this solvent (after [6.19])

In the 1-nitronaphthalene-cis-piperylene system the triplet transfer rate (680 ps) is "certainly" slower than the vibrational relaxation rate. The transfer is thus from a thermally equilibrated donor molecule. In fact the transfer must be thermally assisted since the cis-piperylene acceptor triplet energy is roughly 1000 cm^{-1} above the triplet energy of the donor, 1-nitronaphthalene. This thermally assisted model is consistent with other studies of the intensities of donor phosphorescence as a function of temperature [6.22].

In another type of experiment the dynamics of triplet energy transfer in crystals have been investigated. The fluorescence risetime and decay in a tetracene crystal excited by a picosecond pulse (at 0.53 μm) has been used to estimate the incoherent hopping rate of triplet excitons. [6.23] See Fig. 6.3. Tetracene is a rather interesting system in that one channel for the decay of the singlet exciton is via fission into two triplet excitons. The rate of this process is dependent, in part, on the velocity at which the newly formed triplet excitons separate and thus avoid a geminate recombination process leading to the initial singlet exciton. This triplet energy hopping or transfer from neighbor to neighbor is estimated to be at a rate greater than 10^{13} s^{-1}. Using an average lifetime for the triplet exciton of 100 μs and an average jump distance of 7 Å, a diffusion length of the order of microns is obtained.

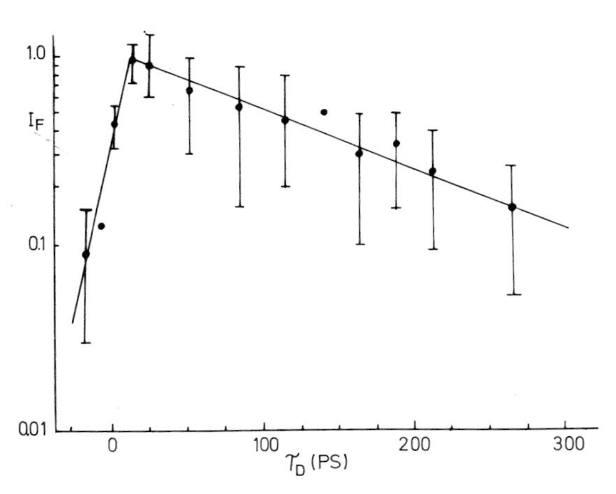

Fig. 6.3. Fluorescence from a tetracene crystal as a function of delay time measured at 580 nm for multiple pulse excitation. Decay time is 145 ps (after [6.23])

6.2 Orientational Relaxation of Molecules in Liquids

The torques experienced by a molecule in a liquid due to thermal molecular motions lead to a randomization of the molecule's orientation with time. The dynamics of this molecular rotation, dependent on anisotropic intermolecular forces, contain information on the microscopic structure of the liquid. Although a number of methods have been used to study orientational relaxation processes, the exciting feature of the recent applications of picosecond laser techniques is that rotational motions are measured directly in the time domain. A number of different and complementary picosecond laser approaches have been developed to study the rotational behavior of molecules. In one method, which uses the optical Kerr effect, the birefringence induced by an intense picosecond pulse is monitored with an attenuated picosecond pulse as a function of time. This method is most appropriate for studies of pure liquids and highly concentrated mixtures. A second approach monitors the decay of the dichroism induced by picosecond laser excitation of solute molecules present at low concentrations in the solution being investigated. The dichroism decays as the solute molecules rotate and thus transform the orientational distribution from an anisotropic to an isotropic one. The third method involves the creation of a transient grating by the intersection of two coherent light pulses in the liquid of interest. Time resolved measurements can be obtained by monitoring the decay of the induced diffraction pattern with a probe-light pulse.

6.2.1 Optical Kerr Effect

The optical Kerr effect results from an intensity-dependent change in the refractive index induced by an intense light pulse propagating through a material. The optical Kerr effect was first observed in a number of liquids using a Q-switched nanosecond ruby laser [6.24, 25]; this followed a theoretical prediction and treatment of the phenomenon [6.26–30]. The use of picosecond lasers to induce a significant nonlinear refractive index in liquids has made it possible to measure the rotational motion of molecules in liquids [6.31], to investigate the short term nature of optical self-trapping in liquids [6.32], and to study a variety of ultrafast processes with a laser-generated ultrafast light gate [6.31].

Due to the polarization of the optical field, the change in the refractive index parallel to the beam polarization (assuming a linearly polarized beam) can be different from the change in the perpendicular directions. Thus an isotropic medium, such as a liquid, can be made anisotropic, and thus birefringent. In a liquid composed of anisotropic molecules, such as carbon disulfide (CS_2), we can view a major contribution to the refractive index difference as resulting from the partial alignment of the CS_2 molecules along the optical field direction. The light pulse induces a dipole in the CS_2 molecules. The induced dipole interacts with the light field and leads to a torque on the dipole which tends to orient the long axis of the CS_2 molecule along the polarization direction of the light field. This induced anisotropy in the orientation of the molecules produces a phase difference as the probe light propagates through the liquid, between the components polarized parallel versus perpendicular to the initial intense picosecond light pulse, Thus the probe pulse polarized at 45° to the initial pulse experiences a phase difference ϕ

between its parallel and perpendicular components in a path length l at the wavelength λ (in vacuum) and is given by

$$\phi = \frac{2\pi}{\lambda}(n_{\parallel} - n_{\perp})l.$$

When the relaxation of the induced refractive index change (birefringence) is comparable to the pulsewidth, the expression [6.33] for the birefringence is

$$\Delta n = n_{\parallel} - n_{\perp} = \frac{n_2}{2\tau} \int_{-\infty}^{t} |E(t')|^2 e^{-(t-t')/\tau} dt'$$

where absolute value $|E|^2$ is the intensity of the initial picosecond pulse and n_2 is the orientational contribution to the nonlinear refractive index. Up to this point in our discussion we have considered only the orientational contribution to the induced birefringence. There can also be an electronic contribution to the nonlinear refractive index which will have a relaxation time far shorter than the picosecond light pulse [6.31, 34–37]. The birefringence due to the electronic part of the nonlinear refractive index will thus only last as long as the initial pulse duration, i.e., it will "instantaneously" follow the excitation pulse in time.

Returning to the experimental arrangement, which consists of the liquid placed between crossed polarizers, we find that the probe light polarized at 45° to the strong excitation pulse, can only pass through the system when the liquid is birefringent. Thus by measuring the transmitted probe intensity as a function of time relative to the initial pulse we obtain the decay of the induced birefringence. The decay of the birefringence at times subsequent to the initial pulse is due to orientational relaxation processes whereas the change in birefringence during the time period when the probe and initial pulses overlap in time can have contributions which are both electronic and orientational in origin. It has been suggested that for CS_2 the major contribution to the nonlinear refractive index change is due to orientational effects [6.31] whereas for nitrobenzene, electronic contributions should also be considered [6.37].

The first picosecond observations of optically induced birefringence were made in the liquids CS_2, nitrobenzene, methylene diiodide, and dichloroethane at room temperature [6.31, 32]. See Fig. 6.4. For nitrobenzene a decay time of the birefringence of 32 ps was obtained [6.31]. With the other liquids the decay of the induced birefringence was rapid compared with the duration of the light pulses used (5 ps) and thus yielded the response time of the system. The method used was to induce the birefringence with the intense 1.06 μm output from a modelocked Nd:glass laser and to probe with frequency-doubled light at 0.53 μm. Although the result obtained for nitrobenzene is somewhat lower than the orientational relaxation times obtained from depolarized Rayleigh scattering – 36 ps [6.38], 39 ps [6.39], and 50 ps [6.40] – it does indicate

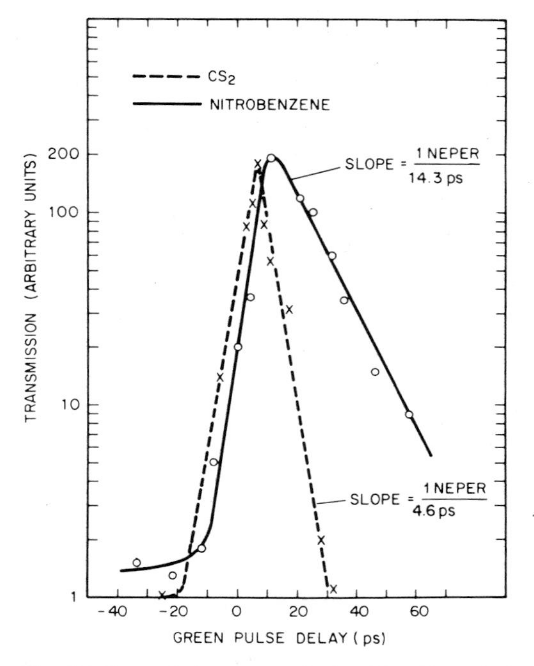

Fig. 6.4. Transmission curves obtained for CS_2 and nitrobenzene using an optical-gate technique. The curves have been normalized to the same height for easier comparison. Each point is average of about 6 measurements. Decay of nitrobenzene birefringence is 32 ± 6 ps (after [6.31])

that the primary contribution to the decay of the optically induced birefringence is orientational in nature. Picosecond measurements on nitrobenzene carried out in various other laboratories are found to be in good agreement, i.e., they range from 27 ps [6.41] to 32 ps [6.42]. It has been suggested that the lower values found for the time constant of the birefringent decay as compared with the orientational relaxation values obtained from light scattering in a variety of liquids (nitrobenzene, *m*-nitro-toluene and various liquid mixtures) arise from the coupling of the orientational motions with shear modes [6.42]. This explanation, though interesting, has not yet been clearly established and requires further study. It has, however, been demonstrated that the rotational motion of nitrobenzene cannot be described by very small angular jumps characteristic of Debye-type rotational diffusion nor by very large jumps. This result has been established by comparing the optically induced birefringence or light scattering results with dielectric relaxation measurements [6.41].

For the case of the liquid CS_2, picosecond measurements using a high repetition rate rhodamine 6G dye laser yielded a value for the optical Kerr effect relaxation of 2.1 ps [6.43]. This is a good agreement with light scattering results of 1.96 ps [6.44] and thus indicates that for CS_2 the decay of the induced birefringence is due to orientational relaxation. Studies of a variety of other liquids including bromobenzene, toluene, iodomethane and mixture of CS_2 and CCl_4 indicated fair agreement between the orientational relaxation and the macroscopic viscosity as given by the Debye relation, i.e., the relaxation time scales linearly with the viscosity [6.41]. For the case of mixtures of CS_2 and CCl_4 the scaling appears to be linear but not over the full concentration range studies. Furthermore great care should be exercised in mixtures since the contributions of both components to the induced birefringence must be included. Correspondingly, if one seeks to obtain relaxation times from light scattering, then the background scattering due to the solvent must be considered also.

Picosecond laser methods have recently been used to induce relaxation processes in liquid crystals [6.45]. This work follows earlier work on liquid crystals which used *Q*-switched lasers [6.46]. Direct measurements of the orientational relaxation have yielded information about the phase transition and temperature dependence of the viscosity coefficient for the liquid crystal p-methobenzilidene p-m-butylaniline [6.45].

6.2.2 Induced Dichroism Method

Unlike the induced birefringence method previously described, the induced dichroism method [6.47] is suitable for studies of solute rotational motions at low concentrations, and can be carried out in any solvent into which the solute can be introduced. The optical Kerr method requires high concentrations, at least several percent, of the species of interest whereas the induced dichroism method is limited to low concentrations of the solute molecules. In the latter method the rotational of the individual solute molecules are obtained, whereas in the induced birefringence method the concentrations of solute molecules is so high that solute-solute interactions as well as the solute-solvent interactions must be considered. As a further point, the dichroism method is applicable to molecules which have absorptions at frequencies corresponding to the frequency of the picosecond light pulse. The birefringence method can be applied to any liquid which has an optical Kerr constant sufficiently large to yield an induced birefringence. These two methods can be viewed as complementary in the systems amenable to study.

The principal idea of the dichroism method is to induce an anisotropy in the orientational distribution in the excited and ground state populations with an intense picosecond pulse [6.47]. The return of the system to an isotropic state is monitored with an attenuated picosecond pulse as a function of time. The induced orientational anisotropy results from the preferential excitation of those molecules whose absorption axes have a large component along the polarization direction of the exciting light field. Thus, there are more excited molecules having their absorption axes parallel to the field and correspondingly more unexcited molecules with their transition moments ($S_0 \rightarrow S_1$) oriented perpendicular to the field direction. Due to this nonuniform distribution the absorption of the probe pulse will depend on its direction of polarization, i.e., the system is dichroic. The difference in absorption of probe light polarized parallel and perpendicular to the polarization of the excitation pulse will decay in time as a function of excited state lifetime, solute concentration, and solution viscosity. For example, in a fluid environment the anisotropic orientational distribution can transform to an isotropic one via the rotational motion of the molecules. At low concentrations in a highly viscous medium the rotations are frozen out, and the anisotropy decays as the excited molecules return to the ground state. At high concentrations the anisotropic distribution can also decay in time by intermolecular energy transfer between molecules of differing orientations. Clearly, for the study of rotational motions low concentrations are necessary to avoid the complicating effects of energy transfer.

The method of induced dichroism with picosecond laser pulses has been used to study the orientational relaxation of rhodamine 6G in a variety of solvents [6.47, 48]. In particular, the effects of solute-solvent hydrogen bonding interactions on the rotational motion of the solute rhodamine 6G were investigated in a series of solvents [6.48].

As shown in Fig. 6.5, a linear relation between the measured relaxation times and the measured solution viscosities for the series chloroform, formamide and the alcohols from methanol through octanol was obtained. The observed linear scaling is in agreement with the Debye-Stokes-Einstein hydrodynamic model for the case in which the hydrodynamic volume V of the rotating particle is a constant, i.e.,

$$\tau_{\mathrm{OR}} = \frac{\eta V}{KT}$$

where η is the shear viscosity. This agreement with the hydrodynamic model is rather surprising since the volumes of the hydrogen bonded complexes should vary considerably through the series from methanol, to octanol. Furthermore one might expect since the strengths of hydrogen-bonding interactions of rhodamine 6G with chloroform and methanol are different, that the relaxation times would not be equal even though the viscosities are the same. Similar arguments can be applied to the results obtained in the liquids formamide and 1-pentanol. To explain the apparent insensitivity of the orientational relaxation times to the volumes of the hydrogen bonded complexes and the strengths of hydrogen bonding interactions, it was proposed that the rotational motion of the complex cannot be described as that of a rigid particle. Allowing some flexibility in the hydrogen bond, it was then assumed that rotational motion of the solute, of the order of several degrees with respect to the hydrogen bond, does not introduce any significant strains in the solute-solvent hydrogen bond. However, on larger rotations the resulting strains in the hydrogen bond should be reflected in the solute rotations. This effect of strain on the solute rotation is most likely minimized by the rapid forming and breaking of the hydrogen bond with the same or other surrounding

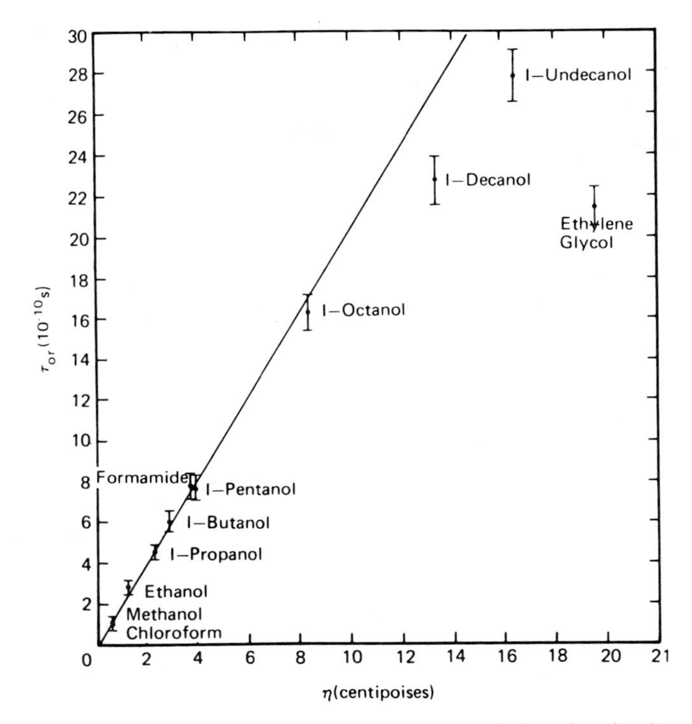

Fig. 6.5. Orientational relaxation time versus solution viscosity for rhodamine 6G in various solvents. Linear dependence is expected from Debye-Stokes-Einstein model (after [6.48])

solvent molecules. Thus by combining flexibility in the hydrogen bond with the dynamic process of bond forming and breaking, it is seen that the rotational motion of the solute could be roughly the same whether it is hydrogen bonded or not. Thus the hydrodynamic volume is unchanging through the series of liquids studied, and linear scaling of τ_{OR} with η becomes plausible. Furthermore, if the time variation in the torques responsible for the solute rotation, due to interactions with the surrounding solvent molecules, is primarily dependent on solvent-solvent interactions (which in turn determines the solution viscosity), then the linear scaling of τ_{OR} vs η found in the rhodamine 6G-solvent systems is not surprising. The observed deviation from linearity of τ_{OR} vs η in the liquids 1-decanol and 1-undecanol is probably due to the breakdown of the continuum hydrodynamic model since the solvent molecules are larger than the solute molecule. The sharp deviation observed in the ethylene glycol solvent is though to be due to the extensive solvent-solvent aggregation via hydrogen bonding. If the solute does not experience the full frictional effects of this aggregation, then τ_{OR} will be faster than the measured viscosity would lead one to expect. The measured orientational relaxation time would thus deviate below the τ_{OR} vs η line which was found to be the case in this system.

6.2.3 Transient Grating Method

Using the rather novel approach of a transient grating technique, the rotational relaxation times and fluorescence lifetimes of rhodamine 6G in several alcohols have been obtained [6.49]. Frequency-doubled modelocked Nd:YAG laser pulses of about 60 ps duration and wavelength 5300 Å were used to excite the rhodamine 6G molecules. Attenuated pulses at the same wavelength were used to probe the decay of the induced grating by the time-dependent changes in the probe-light diffraction pattern. The grating which results from strong absorption at the antinodes of the colliding excitation beams decays as the intensity of absorption decreases due to rotational motions, i.e., the orientational anisotropy decays, and also as the excited molecules return to the ground state. See Fig. 6.6. The rotational motion times were found to be in agreement within experimental error, with the values obtained from the induced dichroism method [6.48].

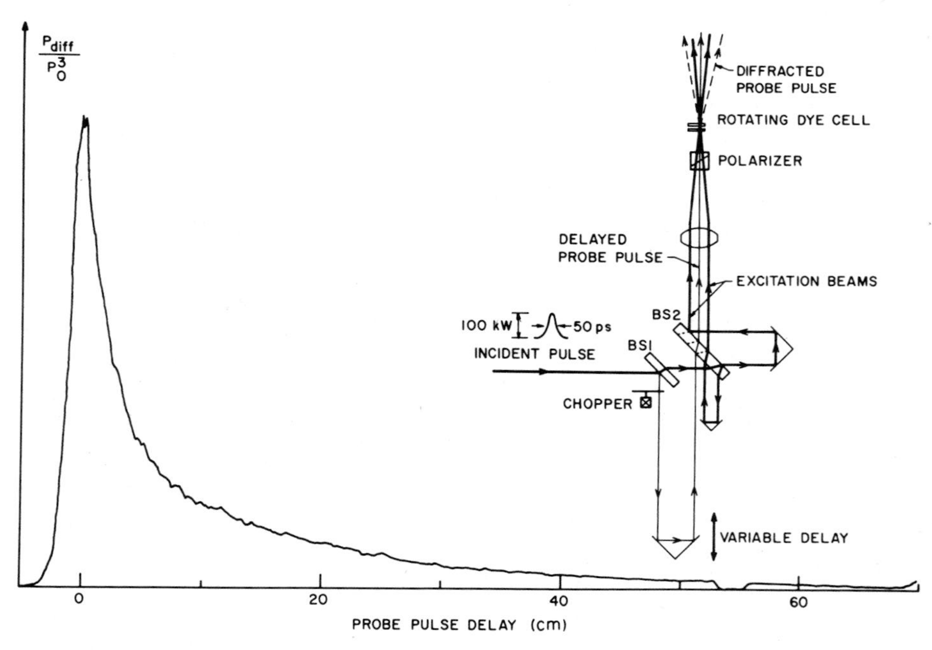

Fig. 6.6. Schematic of transient grating apparatus and typical result for 10^{-4} M solution of rhodamine 6G in methanol (after [6.49])

6.3 Photodissociation and the Cage Effect

In a gas phase dissociation, such as $AB \rightarrow A + B$ the probability of the original fragments A and B re-encountering one another before reacting with A and B fragments from other molecular dissociations in the system is close to zero. However, in a liquid the original fragments A and B are surrounded by solvent molecules which interfere with their escape. Consequently, there is a significant probability for A and B to re-encounter and recombine to produce the original parent molecule AB. The chemistry following a dissociation event is therefore dependent on the relative probability of 1) recombination of the original fragments which results in no net chemical change and 2) their escape and subsequent reaction with other species present in the system. The enhanced probability of recombination of the original fragments, referred to as the cage effect [6.50], is dependent on the kinetic energy of the fragments and on the nature of the fragment-solvent interactions, e.g., the efficiency of momentum transfer. Although there has been much discussion of primary and secondary cage effects in the literature, it is perhaps more useful to consider all recombinations of original partners as due to the cage effect. The arbitrary separations of the cage effect into primary and secondary processes is questionable, since the cage is not a static structure as originally postulated, i.e., fragments bouncing around in a "rigid" solvent environment. The description of the cage effect must therefore be of a dynamic nature, dependent on the translational motions of both the fragment and solvent molecules.

Due to the generation of the A and B fragments by a dissociative process, the distribution of A and B is not, at least initially, spatially random. Since A and B molecules in the early time domain are more likely to be near to each other, there are local concentration gradients in the solution, and a conventional kinetic treatment cannot be used to describe the dynamics of the geminate (i.e., pairwise recombination of original fragments) and nongeminate recombination processes. In either a concentration diffusion or random flights description at least two problems come to mind. One is that the diffusion coefficient in the usual description is assumed to be independent of the separation of A and B. This may be incorrect for the processes considered here, since A and B are within a few molecular diameters, at most, in the early stages of the reaction. Second, the motions of A and B may be correlated and not describable by a random walk, since the motion of one fragment influences the motion of the solvent molecules, which in turn can effect a drag on the other fragment. Furthermore, in these processes it may not be accurate to describe the solvent as a continuous and isotropic medium, and therefore motions in certain directions with certain displacement sizes may be favored.

Since the processes involved in "cage"-effect reactions are in a time domain beyond the scope of conventional dynamic methods, the extensive studies of these phenomena have heretofore been investigated by indirect and time independent methods. For example, information on the quantum yields for dissociation obtained from scavenger experiments have provided valuable insights into the nature of the "cage" processes [6.51]. However, with these methods no measurements of the dynamics of the geminate recombinations were possible, and thus the time scale for the geminate processes, i.e., whether these processes were 10^{-13} s or 10^{-11} s or 10^{-9} s, remained unknown. To determine the nature of "cage" effect reactions, it is necessary to obtain information on the early time motions of the fragments since this is the key to the partitioning between geminate and nongeminate recombinations.

The system selected for study using a picosecond laser was I_2, since it is a simple molecule of great interest, and a great deal is known about its spectroscopic properties and chemistry. Two different solvents, namely, hexadecane and carbon tetrachloride, were used in the experiments [6.52]. The system was pulsed with an intense 5300 Å picosecond light pulse (half width ~5 ps). At this wavelength I_2 is excited to the $^3\Pi_{0_u^+}$ $(v' \approx 33)$ state. (A small fraction of I_2 molecules are excited to the $'\Pi_u$ state and directly dissociate).

The I_2 molecules in the $^3\Pi_{0_u^+}$ state undergo a collisionally induced predissociation leading to a pair of ground-state $^2P_{3/2}$ iodine atoms. The iodine atoms can geminately recombine or can escape and subsequently react with iodine atoms produced elsewhere in the liquid. The population of I_2 molecules was monitored with a weak 5300 Å picosecond light pulse from times prior to the strong excitation pulse up through 800 ps after the excitation pulse. The strong excitation pulse depopulates a good fraction of the ground state I_2 molecules and thus results in an increase in the transmission of the probe pulse. As the iodine atoms recombine, the population of absorbers (iodine molecules) increases, and therefore the transmission of the probe pulse decreases. In this way the recombination dynamics of the iodine atoms were followed by monitoring the time-dependent population change of I_2 molecules.

In Figs. 6.7, 8 it is seen that the transmission of the probe light increases to a peak value at about 20–25 ps after the strong excitation pulse in both the CCl_4 and hexadecane solvents. The transmission reaches a stable value in both solvents at about 800 ps. The residual difference in absorption between the long time values (800 ps) and the initial absorption $(t<0)$ is due to those iodine atoms which have escaped their original partners. The iodine atoms which have escaped will recombine at much later times $(>10^{-8}$ s) with iodine atoms from other dissociation events, i.e., the nongeminate recombinations. The dynamics of the geminate recombination (the cage effect) and the escape of fragments leading to the nongeminate recombination processes are directly seen in Figs. 6.7, 8. The geminate recombination times are about 70 ps in hexadecane and 140 ps in carbon tetrachloride. From the time scale of these geminate recombinations, it seems unlikely that a description of the cage effect in terms of a static solvent cage would be physically reasonable. To describe the dynamics of the recombination, Noyes treatment based on a random flight model was used [6.53, 54]. The theoretical curves, the solid and dashed lines, are shown in Figs. 6.7, 8. In comparing the theory to experimental results it was found that if the theoretical curve was adjusted to fit the long time behavior, where one would expect the random-walk description to be most valid, then the agreement with the early time behavior was poor. The lack of agreement between theory and the experimental results can be due to the crudeness of the theory, e.g., assuming one distance between the iodine atoms on thermalization rather than a distribution of distances, or to the more fundamental issues mentioned earlier, namely, the correctness of a simple random-walk description for atoms within several Angstroms of each other.

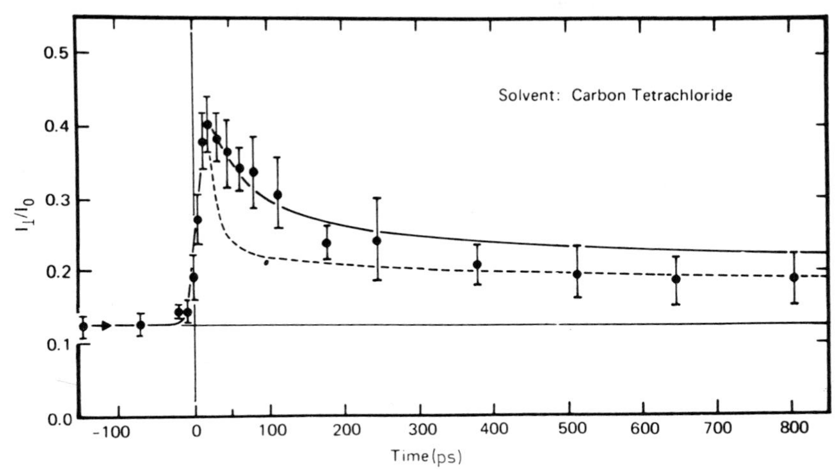

Fig. 6.7. Probe transmission $I_\perp/I_0$ as a function of time after excitation for iodine in CCl_4. Dashed and solid curves are the decay functions calculated from Noyes random flight model (after [6.52])

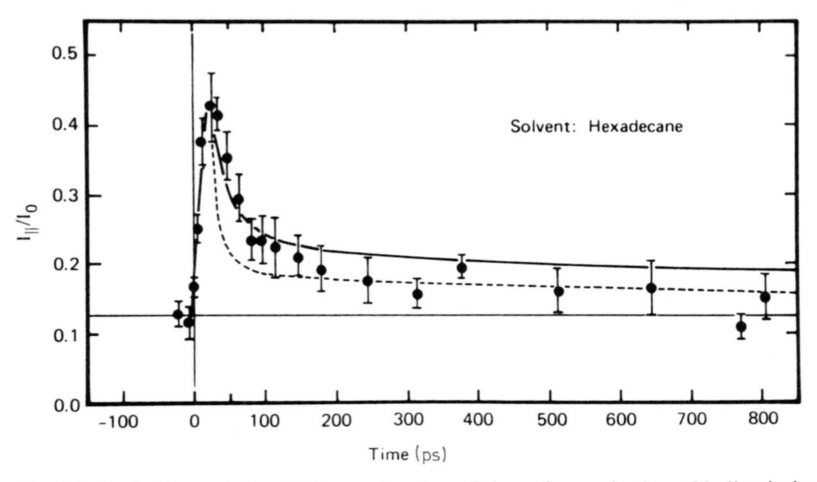

Fig. 6.8. Probe transmission $I_\parallel/I_0$ as a function of time after excitation of iodine in hexadecane. Dashed and solid curves are the decay functions calculated from Noyes random flight model (after [6.52])

Information on the dynamics of a collision-induced predissociation process which generates the iodine atoms from the excited bound $^3\Pi_{0_u^+}$ state was also obtained from these experiments. The observed peak in the probe transmission occurs at a time (20–25 ps) significantly after the decay of the excitation pulse (full width less than 8–10 ps) in both the CCl_4 and hexadecane solvents. Therefore, the continual rise in the transmission of the probe light after the excitation pulse cannot be due to the further depopulation of the ground state iodine molecules by the excitation pulse. The rise time of the transmission can be explained by assuming that the probe light can be absorbed not only by ground state but also the $^3\Pi_{0_u^+}$ excited molecules. Photodissociative recoil studies [6.55] on I_2 show that I_2 in the $^3\Pi_{0_u^+}$ state absorbs light at 0,53 μm. Thus, the probe is monitoring the change in both the ground and excited molecular iodine populations. It was therefore concluded that after the excitation pulse reduces the ground-state population, the subsequently observed increase in the probe transmission was due to the decay of the excited iodine molecules into iodine atoms. This experiment thus provided the first direct observation of the dynamics of a collision-induced predissociation in the liquid state. A rate constant of about $10^{11}\ s^{-1}$ was obtained. This is about 10^5 larger than the spontaneous predissociation process observed in I_2 at low pressures.

6.4 Electron Transfer Processes

6.4.1 Electron Photoejection and Solvation

The dynamics of electron localization in a solvent yields fundamental information on the structure and energies of the solvated electron, ionic aggregates in liquids, and the structure and relaxation properties of the liquid itself. Pulse radiolysis methods, some with a time resolution of 20 ps, have been used to study the kinetics of solvation, and conventional flash photolysis methods have been used in investigations of the solvation and electron photoejection processes [6.56–74]. With the use of picosecond lasers in flash photolysis studies, new information on these phenomena has been obtained [6.75–80].

Generation of an excess electron in a liquid can be achieved in a variety of ways including pulse radiolysis [6.56–59, 81], ejection from a field emission tip [6.82], or from a photocathode [6.83], as well as solute photoionization processes. The relaxation steps involved in the localization of the quasi-free ejected electron [6.84, 85] (provided the localized state is energetically stable) are radiationless transitions to localized states but in a nonequilibrium solvent environment. This is followed by solvent relaxation which leads to an equilibrium localized ground state. If the localization kinetics can be viewed in terms of these "separable" consecutive events, then it should be recognized that the later process involving solvent equilibration is directly related to the solvent dielectric relaxation time whereas the earlier radiationless processes are not.

In studies of the dynamics of electron solvation in water, picosecond laser techniques were employed to obtain time-resolved spectra [6.75, 76]. A frequency-quadrupled Nd:glass laser pulse at 2650 Å was used to photoeject an electron from ferrocyanide ion in water and the subsequent solvation steps were monitored in the 3100–9000 Å range with a picosecond continuum. It was found that there is an initial absorption at 1.06 μm which develops in about 2 ps, and which is followed by the "normal" hydrated electron absorption after 4 ps. It is not known whether the excitation of the ferrocyanide ion at 2650 Å leads to "direct" ejection of the electron or whether a metastable state of longer lifetime is produced which then leads to the ionization step. Thus the 2-ps segment might contain both the ionization step and the initial relaxation of the quasi-free electron to a solvent nonequilibrated localized state. In any event the results are consistent with the model proposed for the dynamics of the localization of the ejected electron.

Bleaching expierments on Na-NH_3 solutions indicated that the excited state relaxation time of the solvated electron is ~ 0.2 ps [6.79]. This value was obtained from the change in optical density following excitation by assuming a saturation expression relating optical density to relaxation time. This approach to estimating the relaxation time was used since the decay time was beyond the time-resolution capabilities of the 8 ps duration light pulses used in these experiments. The relaxation of 0.2 ps found in NH_3 is rapid compared with the initial relaxation step of the quasi-free electron in water and alcohols as obtained from laser and pulse radiolysis experiments. This suggests that the relaxation process being studied by the bleaching experiments in ammonia involves transitions between bound states of the solvated electron and does not involve the generation of a quasi-free electron followed by its subsequent localization.

From conventional flash photolysis studies of ionic aggregates such as the sodium salt of tetraphenylethylene dianions (T^{2-}, $2Na^+$), information on the structure of the ionic aggregates, and their electron ejection and transfer processes have been obtained [6.86]. In particular it was concluded that photoejection in the solvent tetrahydrofuran, (THF) involved the (T^{2-}, $2Na^+$) aggregate and not the partially dissociated (T^{2-}, Na^+) species. In addition significant differences were observed in the solvent THF and dioxane. It was suggested that electron ejection was not occurring in dioxane but rather an intramolecular electron transfer from an excited state of the aggregate to a sodium cation was taking place. Direct evidence for this scheme was, however, not obtained, possibly due to the rapidity of the back-electron transfer process relative to the time resolution capabilities of the flash-photolysis system. Unfortunately, follow-up studies using picosecond lasers did not resolve this particular point, i.e., whether in dioxane autoionization is unimportant or the back-electron transfer is too fast for resolution by picosecond methods [6.78]. However, evidence was found that demonstrated marked differences in the relaxation processes in the solvents tetrahydrofuran versus dioxane following picosecond excitation. The bleaching of the (T^{2-}, $2Na^+$) absorption at

4880 Å following excitation with a 5300 Å pulse was found to last 10 ps in dioxane, as opposed to tetrahydrofuran, where the bleached state lasted for several nanoseconds. The results in tetrahydrofuran, in which electron ejection does occur, are rather interesting in that the absorption of the $T\cdot^-$ radical ion was not observed. The appearance of this radical, which absorbs at 6600 Å, is anticipated since, on photoejection of an electron from the (T^{2-}, $2Na^+$) aggregate, the ($T\cdot^-$, $2Na^+$) species should be produced. It has been suggested that the $T\cdot^-$ radical ion initially produced is distorted and in the time scale covered by the experiment (60 ps) the absorption at 6600 Å, characteristic of the $T\cdot^-$ ground state, does not have sufficient time to appear [6.78]. Further experiments are necessary to clarify this interesting explanation.

6.4.2 Excited-State Charge Transfer Complexes

In an earlier section we discussed excited-state interactions which quenched the donor fluorescence by a dipole-dipole energy transfer mechanism. Another important class of excited-state interactions which quenches the fluorescence of an excited molecule involves the transfer of charge rather than the transfer of energy. This transfer of an electron from a ground-state donor molecule D to an excited-state acceptor molecule A* produces a new species in low dielectric solvents, namely an excited-state charge transfer complex or exciplex $(A^-—D^+)^*$ [6.87]. Depending on the molecules involved, the excited molecule can be the electron acceptor or donor.

The charge transfer interaction not only quenches the emission from A* but gives rise to a new emission in low dielectric solvents characteristic of the exciplex $(A^-—D^+)$, produces ion radicals in high-dielectric solvents, provides new pathways for energy degradation, and changes the chemistry of the system. The physical and chemical nature of these diverse processes has been extensively studied since the discovery of excited-state charge transfer complexes [6.87–91]. However, with the application of picosecond laser methods, new insights into this fundamental problem have been achieved.

A key aspect of the formation of an exciplex is the diffusion of a ground-state donor and an excited-state acceptor to a separation at which electron transfer can be effected. Since the transfer occurs via an exchange type coupling, it is expected to be short range in nature. A determination of the key parameters in the electron-transfer process in dependent on adequate theories of diffusion controlled reactions for describing the kinetics of the electron-transfer processes [6.92, 93].

Conventional kinetic treatments assume that the reactivity of a molecule does not change in any time interval subsequent to the start of the reaction. Thus, a time-independent rate constant would be adequate to describe the dynamics of the reaction. In a highly reactive system this description is inadequate since the spatial distribution of molecules is not an equilibrium distribution and is thus changing in time. This leads to a rate "constant" which is also changing in time. To adequately test diffusion theory and the expected transient behavior, it is necessary to determine the full time characteristics of the chemical reaction.

Picosecond laser studies of anthracene serving as the acceptor and N, N-diethylaniline as the donor have provided new insights into the electron transfer process and diffusion controlled reactions [6.94–96]. A frequency-doubled picosecond pulse at 3472 Å was used to excite anthracene to S_1, and the fundamental picosecond pulse at 6943 Å was used to monitor the appearance of a new absorption resulting from the electron transfer step [6.94, 96]. It was found that the experimental results were in excellent agreement with a theoretical treatment which included all transient terms (see Fig. 6.9). There had been some speculation that all transient terms were not necessary for a description of the dynamics for times longer than 10 ps. This was not found to be the case; the full transient description was necessary to conform with the experimental results. The two parameters which appear in the theory could be extracted with some confidence, namely, the distance R at which the transfer occurs, and a rate constant k for the reaction if D and A* were in an equilibrium distribution having an average distance R. The values obtained are $R=8$ Å and $k=10^{11}$ liter/mole-s. The value for R is in good agreement with that obtained from fluorescence quenching studies [6.88]. The same values for R

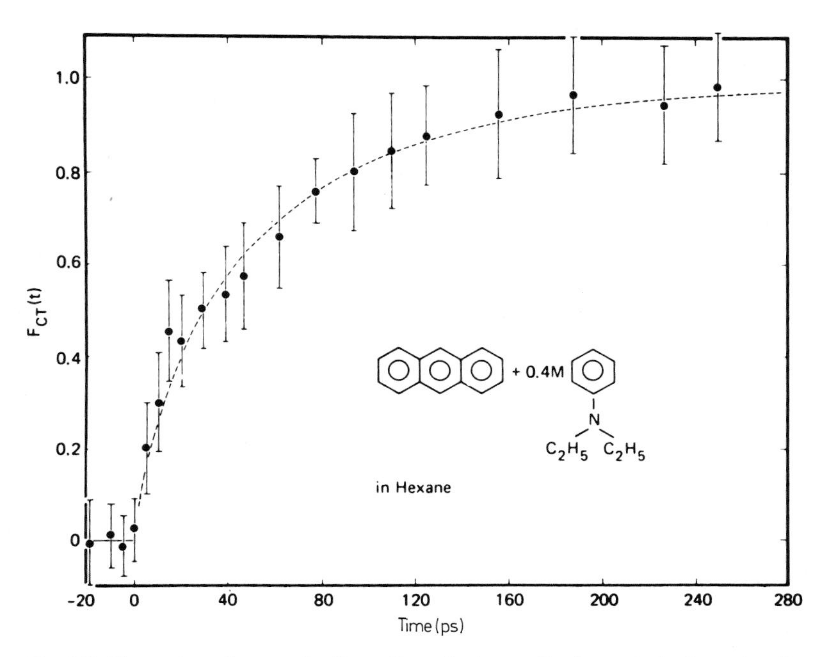

Fig. 6.9. Charge-transfer complex formation, F_{CT} (normalized population), vs. time for anthracene plus diethylaniline in hexane. The dashed line is the theoretical curve. Good agreement is obtained when including all transient terms, or equivalently, the time-changing character of the rate "constant" is thus established (after [6.96])

and k were obtained independently for all donor concentrations ranging from 0.1 M to 1.0 M in the solvent used, hexane.

However in 3 M N, N-diethylaniline or neat N, N-diethylaniline the formation of $(A^- - D^+)^*$ showed no transient behavior. The kinetics followed an exponential time-dependence characteristic of a bimolecular process with a time-independent rate constant of 10^{11} s^{-1}. At these high donor concentrations, excited A* molecules have donor molecules as immediate neighbors. Thus it appears that translational motions as contained in the diffusion treatment are not of key importance. In acetronitrile which has a large dielectric constant, $\varepsilon \sim 37$ at 25 °C, the radical ions $A\cdot^-$ and $D\cdot^+$ are produced. The dynamics of electron transfer at high concentrations of the donor are found to be essentially the same in hexane and in acetonitrile.

In addition to the distance requirements for excited state electron transfer, there can also be orientational restrictions on the transfer process. One way to gain insight into these latter requirements is to study intramolecular electron transfer between a donor and an excited acceptor hooked together by methylene groups, i.e., $A-(CH_2)_n-D$ [6.89, 90]. To this end, picosecond laser studies of the anthracene$-(CH_2)_3-$N, N-dimethylaniline system were carried out [6.95]. As in the free system the A moiety was excited with a 3472 Å pulse and the electron transfer step monitored with a 6943 Å pulse. The behavior in polar solvents such as acetonitrile and methanol was found to be considerably different from that observed in nonpolar solvents such as hexane [6.97]. In nonpolar solvents the initial rapid charge transfer step is rapid and then either levels off or increases very slowly after about 40 ps [6.97]. However, electron transfer does not occur for all of the $A^*-(CH_2)_3-$DMA molecules in the system. This is thought to be due to a distribution of ground state geometries. In other words, molecules which are in the "appropriate" configuration can undergo exciplex formation whereas molecules which are in the "wrong" configurations (e.g., extended form) cannot achieve the appropriate geometry within the lifetime of $A^*-(CH_2)_3-$DMA to effect electron transfer.

On the other hand, the observation that electron transfer occurs for almost all $A^*-(CH_2)_3-$DMA molecules in polar media can be due either to favorable molecular configurations for electron transfer in polar media, or due to the relative shifting of the $A^*-(CH_2)_3-$DMA and $\cdot A^-(CH_2)_3-DMA^{\dot{+}}$ energy surfaces leading to an enhanced electron transfer probability for an "extended" configuration in polar media. In polar media it is also found that there is a fairly rapid decay (probably back-electron transfer) though slower than the initial transfer step. This back transfer is not observed in the nonpolar media, at least for times up to 1 ns. For the polar media the fairly rapid decay process may involve formation of the acceptor triplet in the back transfer step. It should be noted that the center-to-center separation in an extended form is about 4–5 Å. From the studies of the free donor and acceptor systems at donor concentrations of 3 M or higher it is known that the electron transfer is completed in about 20–25 ps. At the distances separating the hooked-together donor and acceptor, one would thus expect the transfer to be completed in this same time period (20–25 ps) rather than the longer times observed. It is concluded that the differences are due to the less-than-favorable geometries achievable in the hooked-together molecule.

6.5 Picosecond Measurements of Internal Conversion and Intersystem Crossing

In this section we shall review measurements on internal conversion and intersystem crossing, with special emphasis on recent measurements and techniques. Internal conversion refers to the process in which an excited singlet state converts to a lower energy singlet state nonradiatively with the energy disappearing in the form of heat, or to the process in which an excited triplet state converts to a lower triplet state by a nonradiative deactivation. Intersystem crossing involves the nonradiative transition between states of differing multiplicity. Picosecond pulse techniques afford special opportunities for studying internal conversion and intersystem crossing because many organic molecules possess lifetimes of the order of a nanosecond and less because of these processes. For these complex organic molecules the molecular interactions responsible for internal conversion and intersystem crossing are not well understood. While it is true that spectral data can sometimes provide information, many molecules do not exhibit any noticeable emission, and a systematic study of short-lived "nonemissive" species affords the possibility of determining the underlying mechanisms through direct testing of the models for internal conversion and intersystem crossing. Moreover, because large populations of excited states may be generated rapidly with picosecond pulses, one may, besides establishing the nature of the electronic transitions, identify higher excited states via their absorption or emission spectra. In addition, one can investigate the effects of the environment on rapid relaxation processes and the transfer of electronic energy between molecules. From an applied point of view, information derived from studies of these complex molecules can be used as a guide for synthesizing dyes for a variety of reasons including their use for modelocking elements or for scintillator materials with rapid response.

Two picosecond techniques have been mainly used to measure rapid energy transfer in large molecules: an absorption technique and a fluorescence emission method. In the former, the sample is prepared with an exciting pulse and the absorption or transmission of the sample is probed with a weak interrogation pulse at varying delay times. If a molecule emits sufficient fluorescence, rapid transfer can also be investigated by observing the emission with either a picosecond-resolution optical gate or streak camera, or, when extremely high resolution is not required, with conventional electronics. Let us begin by discussing some absorption experiments because, historically, such measurements were accomplished first.

6.5.1 Absorption Measurements of Internal Conversion and Intersystem Crossing

The first general technique for measuring ultrashort nonradiative transitions with picosecond pulses [6.98, 99] used 1.06-μm intense pulses from a mode-locked Nd:glass laser to excite dye molecules to a higher electronic state and much weaker pulses, derived from the intense pulses, for probing the transmission of the sample before, during, and after excitation by the intense pulses. Because the transmission depends upon the number of dye molecules in the ground state, the transmission reaches a maximum when the probe pulse coincides in time with the intense exitation pulse, after which the transmission returns to a minimum at later delay times as the excited state molecules return to the ground state. A measure of the recovery time then gives the lifetime of the excited state. For Eastman-9740 dye the recovery time was in the neighborhood of 8 to 25 ps [6.98] and about 6 ps for Eastman-9860 dye [6.99]. Because in both experiments the transmission recovered rapidly to its value prior to excitation, these results are of significance to the chemist because they represent the first attempts to measure rapid internal conversion directly by using picosecond pulses. The rapid recovery of Eastman-9860 dye has been subsequently observed by others [6.100, 101].

Similar probe techniques were applied to the measurement of the internal conversion rate from the second excited singlet (S_2) state to the first excited singlet state (S_1) in a cresyl violet sample [6.102]. The internal conversion rate between the second and first excited singlet states in molecules such as cresyl violet cannot be obtained from fluorescence or phosphorescence yields because emission from the second excited singlet state is difficult to observe, so that picosecond techniques are essential. Second-harmonic pulses at 347.2 nm derived from the modelocked ruby laser were used to excite cresyl violet in solution to the second excited singlet state. The buildup of the population in the first excited singlet state was then measured by probing the gain with a weak beam at 694.3 nm, which corresponds to the long-wavelength side of the fluorescence spectrum emitted by the first excited singlet state. It was found that the gain of the probe increased after excitation and an S_2-to-S_1 internal-conversion time of about 30 ps for cresyl violet in methanol and of about 56 ps for cresyl violet in ethylene glycol was obtained [6.102]. Because the viscosities of the solutions are 0.58 and 19.9 cp, respectively, and because the internal conversion rates are only moderately different, it was concluded that the environment has affected the decay rate, although not as much as expected if the conversion were due to collisional deactivation. The more rapid rate measured in the methanol solution may be indicative of the less-rigid structure surrounding the molecule in the lower-viscosity solution.

Another interesting measurement of rapid internal conversion was carried out on the dye molecule [6.103] crystal violet, $[(CH_3)_2NC_6H_4]_3C^+$. The structure of this molecule is known to be D_3 propeller shaped three-dimensionally, with the phenyl rings rotated 32° from the central plane. The molecule, although exhibiting intense visible absorption bands, is almost completely nonfluorescent with a quantum yield, Q, below 10^{-4}. The crystal violet molecules were excited with an intense 530 nm pulse promoting the molecules from S_0 to S_1, and a very weak interrogating pulse was used to probe the return of the molecules to the ground state. Because the recovery was very rapid and complete within 100 ps, it was concluded that the rapid internal conversion process was being measured. By choosing a series of solvents that covered a viscosity range from 0.01 to 120 p, it was shown that the ground state recovery time varies as $\eta^{1/3}$, where η is the viscosity of the solvent.

Solvent effects other than viscosity are judged to be relatively unimportant. A model developed [6.104] for a series of triphenylmethane dyes predicts a viscosity dependence for the quantum yield of $Q = C\eta^{2/3}$. In this model absorption of light produces a Franck-Condon vertical excited state with the phenyl rings still at a ground state equilibrium angle θ_0. The rings then rotate toward a new equilibrium angle θ, and the nonradiative deactivation of the excited state depends upon $(\theta - \theta_0)^2$. It was further assumed that the radiative rate is independent of θ. A new model must probably be developed to explain the new data because the present model predicts the same $\eta^{2/3}$ dependence for the lifetime as well as for the quantum yield, whereas the measured dependence for the lifetime is $\eta^{1/3}$.

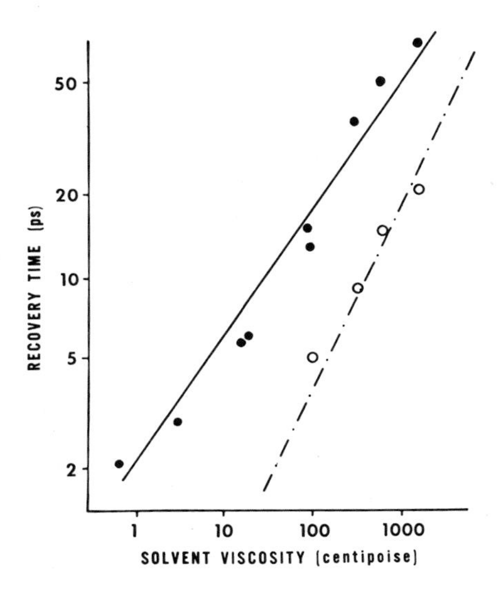

Fig. 6.10. Recovery time of malachite green in glycerol after excitation to first singlet state with 0.5 ps pulse. Experimental results consists of a long-term recovery of absorption that could be described by a simple exponential, and by a more rapid, partial recovery. Thus two exponential time constants can be plotted as a function of solvent viscosity. Solid points represent longer time constants, and open circles correspond to faster, initial time constants. Longer lifetimes follow approximately $\eta^{1/2}$ dependence (after [6.105])

Further measurements on another triphenylmethane dye, malachite green, have been made using the superior resolution available from a modelocked cw dye laser [6.105]. With 0.5 ps excitation pulses it was found that the recovery time is only 2.1 ps for malachite green in methanol and that the decay is exponential. By studying the recovery time in a number of solvents it was determined that there was always a long-term recovery, whereas in the higher viscosity solvents, there was also an initial, more rapid, partial recovery. The data could be fit with recovery functions of the form $R(t)=\exp(-t/\tau_1)+a\cdot\exp(-t/\tau_2)$ (see Fig. 6.10). Since the fast initial recovery is in agreement with the S_1 lifetime calculated from the quantum efficiency, it was suggested that on a short time scale molecules in S_1 rapidly convert to a highly energetic level of S_0, giving rise to a partial recovery of the absorption. Subsequently, this hot distribution in the ground state relaxes, giving rise to the slower rate observed for complete recovery of the absorption. The longer lifetimes were found to depend on the viscosity of the solvent approximately as $\eta^{1/2}$, but the $1/e$ point of the total recovery curves varies closely as $\eta^{1/3}$ in agreement with the previously described work [6.103].

Rapid intersystem crossing has been measured for a number of molecules in a variety of solvents [6.106, 107]. The mechanisms and dynamics of triplet formation in both benzophenone and nitronaphthalene were studied [6.106, 107] by exciting the molecules with a pulse at 354.5 nm to produce excitation in higher singlet states, and then probing the formation of the lowest triplet state by means of an absorption at 530 nm known to originate because of triplet-triplet absorption. Results demonstrating rapid formation of benzophenone triplets in benzene and ethanol solutions are shown in Fig. 6.11. It was found that the buildup times for triplet formation for benzophenone in ethanol and for benzophenone in benzene are 16.5 ± 3 and 30 ± 5 ps, respectively [6.106], whereas the triplet-formation time for benzophenone in *n*-heptane is 8 ± 2 ps [6.107] (see Fig. 6.12). These results, as well as several other results for 1- and 2-nitronaphthalenes in solution [6.106, 107], are summarized in Table 6.2.

There are two basic pathways by which the initially excited singlet state S_i may be converted to the T_1 state: 1) The S_i state may either vibrationally relax to other singlet vibronic states, which then transfer their energy by intersystem

Table 6.2. Mean lifetimes for triplet buildup following 3545 Å excitation

Molecule	Solvent	$\tau=k^{-1}$ (ps)
1-Nitronaphthalene	Ethanol	8 ± 2
1-Nitronaphthalene	Benzene	12 ± 2
2-Nitronaphthalene	Ethanol	22 ± 2
2-Nitronaphthalene	Benzene	10 ± 2
Benzophenone	Ethanol	16.5 ± 3
Benzophenone	Benzene	30 ± 5
Benzophenone	*n*-Heptane	8 ± 2

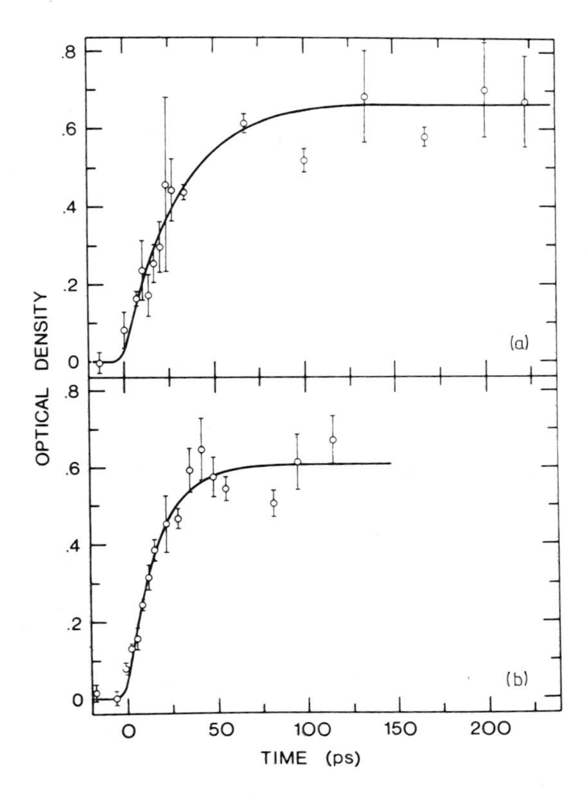

Fig. 6.11 a and b. Optical densities at 530 nm due to triplet benzophenone formation as a function of delay following singlet state excitation as measured in (a) 0.40 mole/liter benzophenone in benzene, and (b) 0.47 mole/liter of benzophenone in ethanol. Build-up time is 30 ps in benzene and 16.5 ps in ethanol (after [6.106])

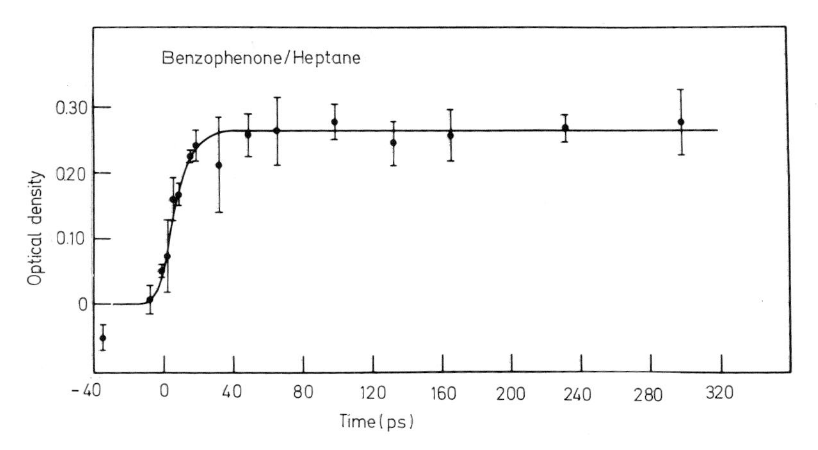

Fig. 6.12. Optical densities at 530 nm due to triplet benzophenone formation as a function of delay following singlet state excitation as measured in *n*-heptane solvent. Energy remains in triplet state for a long time after excitation. Result should be compared with Fig. 6.2 where triplet-triplet transfer occurs depleting triplet state (after [6.107])

crossing to a vibronically excited triplet state that, in turn, then relaxes to the T_1 state; or 2) intersystem crossing to a vibronically excited triplet state may occur immediately and is then followed by nonradiative relaxation to the T_1 state. Because the nonradiative steps may be comparable in time to the intersystem crossing rate, the aforementioned [6.106, 107] results place a lower limit on the intersystem crossing rate. In other studies it has been reported that intersystem crossing depends on the wavelength of excitation into the $1n\pi^*$ state of benzophenone [6.108–111].

These phenomena may be quite complex, if vibrational relaxation is occurring on a time scale comparable to that of intersystem crossing. The solvent-dependent risetimes indicate that conventional singlet-triplet mixing appears to be an insufficient explanation because this mechanism is quite sensitive to the spacing between the singlet and triplet states. It has been suggested [107] that the variations in buildup time may be due to the effectiveness of different solvents in relaxing the excited benzophenone molecules vibrationally to or from singlet levels that are strongly coupled to the triplet manifold.

The very fast triplet buildup times in the nitronapthalenes are difficult to reconcile with the low triplet yields that have been reported. All these new measurements are providing a deeper insight into rapid nonradiative processes in the condensed phase, but many more experiments are required for a complete understanding of these complicated phenomena.

In an extension of the previously quoted work [6.107] the internal conversion between excited electronic states of the molecule 4-(1-naphthylmethyl) benzophenone in a benzene solution have been undertaken [6.112]. In these experiments the benzophenone molecule and the methylnapthalene molecule are connected by a sigma type chemical bond so that energy may be transferred from one part of the connected molecule to the other. As in the aforementioned experiments, the 0.2 M sample of 4-(1-naphthylmethyl)benzophenone is excited to the S_1 state by a pulse at 353 nm, and the formation of a triplet state is probed by the absorption of the sample at 530 nm. An initial absorption with a lifetime of 10 ps is accompanied by the development of a much weaker absorption with a decay time greater than the longest delay time used in the experiment. The short-time and long-time components were interpreted as originating from triplet-triplet absorptions from the benzophenone and 1-methylnaphthalene parts of the molecule, respectively. The initial rapid decay of the absorption of the double molecule can be contrasted against a very slow decay of about 1 ns observed for an equimolar solution of benzophenone and naphthalene in benzene with each solute at 0.2 M to simulate the double molecule experiments at 0.2 M. Thus the contribution to quenching by nearest neighbors is small. The results appear to support a model in which the wavefunctions of the low-energy states of the double molecule are approximately products of naphthalene- and benzophenone-like, single-excitation functions with the small interaction between the chromophores providing the coupling.

A more general extension of the probe technique has been used by several groups to measure internal conversion and intersystem crossing. A pulse at a given "wavelength" excites the sample, but the probe pulse consists of a picosecond continuum that spans the entire visible region and sometimes extends beyond. These very broad continua, whose generation by nonlinear optical techniques is described in Chapter 4, were first used [6.36, 113–115] to monitor inverse Raman spectra and their use suggested for picosecond flash-photolysis experiments. These continua make it possible to systematically study the picosecond transient behavior of molecules, which are weak emitters.

Experiments using these flash-photolysis techniques have helped to resolve discrepancies in the literature on the lifetime of DODCI dye [6.116]. In further studies [6.117] the effect of the solvent on the decay kinetics of bis-(4-dimethyl-aminodithiobenzil)-Ni(II), or BDN for short. BDN is a nickel complex which absorbs in the infrared and is difficult to study by emission spectroscopy because it has a low quantum yield and because the emission probably extends well into the infrared. BDN was excited with a single intense pulse at 1.06 μm, and excited-state absorption was then monitored with a picosecond continuum; a broad absorption band was discovered at 510 nm, which appeared promptly upon excitation. By monitoring this excited-state absorption as a function of delay time, it was found that the absorption band decays exponentially and that the transient bleaching at 1.06 μm recovered in about the same time as the excited state absorption in a number of solvents. The excited state absorption lifetime was measured to be 220 ps for BDN in iodoethane; 3.6 ns for BDN in 1,2-dichloroethane; 2.6 ns for BDN in 1,2-dibromoethane; and 9 ns for BDN in benzene. An external heavy-atom effect apparently leads to the more rapid recovery in the halogenated solvents. Analysis indicates that the excited-state absorption represents the lifetime of a state, or states, in which the $3b_{2g}$ orbital is occupied.

Porphyrin molecules have also been studied by excitation with a 530 nm pulse and the evolution of the absorption spectra probed with picosecond resolution by means of a continuum [6.118]. For octaethylporphinatotin (IV) dichloride [(OEP)$SnCl_2$] the absorption spectrum of the excited-singlet state and the decay of the S_1 state and the growth of the T_1 triplet state have been observed. The time-resolved absorption spectra of (OEP)$SnCl_2$ are shown in Fig. 6.13. The absorption spectra due to the S_1 singlet state and the T_1 triplet state as deduced from such measurements is shown in Fig. 6.14. The spectrum labelled S_1 decays in about 500 ps, and the spectrum represented by T_1 appears in about the same time. By analysis it was found that the quantum yield for triplet formation is 0.8 ± 0.008 for (OEP)$SnCl_2$, and from the known fluorescence quantum yield of about 0.01 the quantum yield for internal conversion from the S_1 to the S_0 state was deduced to be about 0.19. The difference between the absorption spectrum of the first excited-singlet state and that of the first excited-triplet state in a porphyrin molecule has also been obtained using these methods. For many of these experiments picosecond flash photolysis is used to locate a number of new absorption bands and then standard techniques,

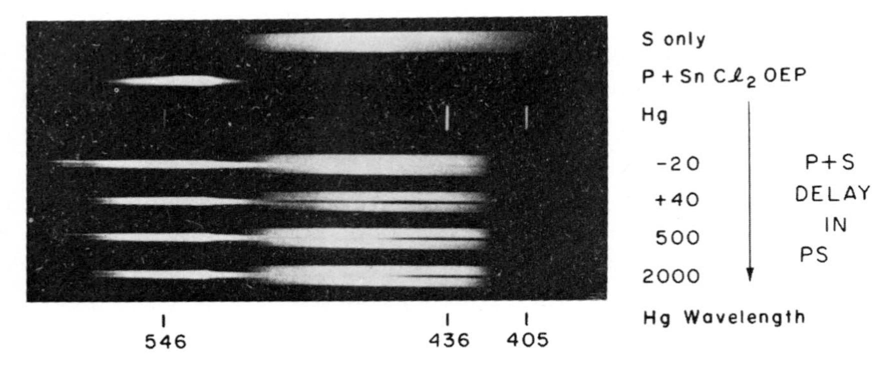

Fig. 6.13. Flash photolysis with picosecond pulses can locate the position of new transient bands. Calibration and transient absorption spectra of (OEP)$SnCl_2$ in 1,2-dichloroethane at room temperature shown here: (1) probe pulse S only; (2) pump pulse P and sample; (3) Hg calibration; (4) through (7) have probe pulse S, pump pulse P, and sample. The time delay between the arrival of P and S at the sample is indicated on the right in ps. The photolysis excites only the central band. Note how absorbence from excited singlet state appears just after excitation as dark absorption at about 470 nm, and then shifts toward the blue ($\simeq 430$ nm) due to triplet formation (after [6.118])

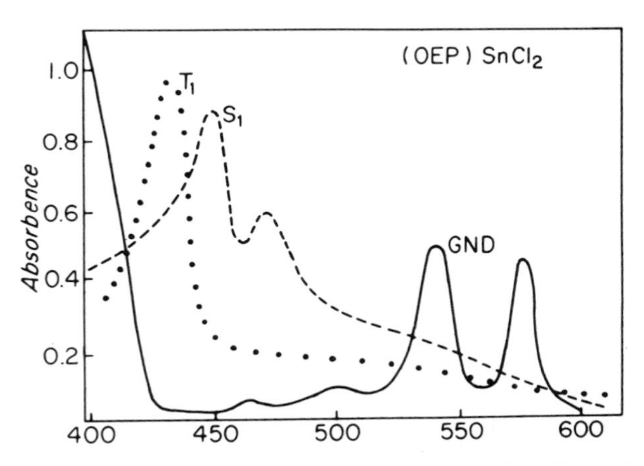

Fig. 6.14. Ground state absorbence of (OEP)$SnCl_2$ (solid line). Schematic absorbence of first excited singlet S_1 (dashed line) and of first excited triplet T_1 (circles) as derived from flash photolysis and probe experiments (after [6.118])

such as optical multichannel analyzer measurements, are used to observe changes of optical density at a particularly interesting wavelength.

In other experiments using the picosecond continua, two cyanine dyes, cryptocyanine (1,1′-diethyl-4,4′-carbocyanine iodide) and DTTC (3,3′-diethyl-2,2′-thiatricarbocyanine iodide), were excited with 694.3 nm pulses and the evolution of transient absorption bands examined [6.119]. A new absorption band, which decayed in 90 ± 30 ps was detected in DTTC at 525 nm. Since the ground state was also observed to recover in about this time, the newly observed band was attributed to a transition between excited-singlet states. Measurements of the recovery time for the ground state of cryptocyanine were found to be consistent with the results of earlier workers [6.120].

The lifetime of DODCI has also been measured, though [6.121–123] under high-power conditions. Picosecond flash-photolysis methods have also been used to measure the $S_n \leftarrow S_1$ and $T_n \leftarrow T_1$ absorption spectra of anthracene in solution [6.124]. The anthracene is excited with pulses at 347.2 nm, and a continuum covering the entire region from 390 to 920 nm is used to probe the transient spectra. At short delay times of 250 to 300 ps, a strong absorption band at 600 nm corresponding to the $S_n \leftarrow S_1$ transition is observed, while at much longer delay times, 4 to 5 ns, a strong transition is found at 420 nm which is assigned to $T_n \leftarrow T_1$.

Lifetimes of the excited states of a number of transition metal compounds have been established [6.125] using probe techniques and picosecond flash photolysis. Lifetimes in the subnanosecond range were established for some nonluminescent compounds of iron and ruthenium. For a number of transition metal complexes the interstate nonradiative processes such as intersystem crossing were found to be extremely fast.

6.5.2 Emission Measurements of Internal Conversion and Intersystem Crossing

One of the first applications of the modelocked laser was its use in measuring the nanosecond fluorescence decay times of dye molecules in solvents. In an early technique the samples were excited with either the fundamental or the second harmonic of pulses generated by a modelocked ruby laser, and the fluorescence was detected with a planar diode detector used in conjunction with a travelling-wave oscilloscope [6.126, 127]. In these experiments both the ultrashort pumping pulses for minimizing deconvolution problems as well as the high-intensity picosecond pulse source were used. Prior to these measurements more conventional schemes employed either nanosecond flashlamp pumping or more cumbersome fluorometry techniques. These experiments [6.126, 127] were still limited, however, to a resolution of about 0.5 ns.

The first studies in which fluorescence phenomena were detected with "true" picosecond resolution used the optical gate to study the emission from two dyes, DDI and cryptocyanine, both dissolved in methanol [6.120]. The samples were excited with pulses at 530 nm, and the emission was sampled at variable delay times by means of an ultrafast shutter operated by intense pulses at 1.06 μm. An excellent example of the results obtained with the optical gate technique is shown in Fig. 6.15. As can be seen, the fluorescence from DDI in methanol decayed in 14 ± 3 ps, while that of cryptocyanine decayed in 22 ± 4 ps. This ultrafast recovery time in cryptocyanine has since been verified [6.119, 128, 129]. The ultrashort fluorescence decay time demonstrates the rapid internal conversion of the first excited singlet state to the ground state.

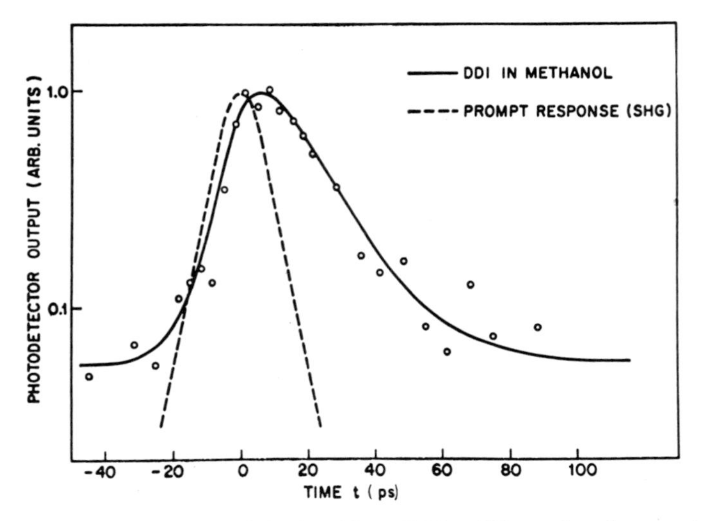

Fig. 6.15. Fluorescence light signal from DDI at 750 nm plotted versus time *t*, *t* being measured relative to the arrival time of 530 nm pulses which excite the fluorescence. The dotted line is a fit to the data obtained when system "views" the green light pulses. The width of this prompt curve, 14 ps at half-height, reflects the width of the gating 1060 nm pulses (8 ps) and the width of the 530 nm pulses (6 ps) (after [6.120])

Using an optical gate technique, measurements of the fluorescence decay time for erythrosin in solution have been obtained [6.130–132]. For erythrosin in water the measured decay times were found to be 90 ps [6.130], 110 ± 20 ps [6.132], and 57 ± 6 ps [6.131]. Since the quantum yield is 0.02 for erythrosin in water, the higher values should probably be preferred. The fluorescence rise-time is "instantaneous" as shown in two of the studies [6.131, 132], though other work [6.130] mistakenly identified their prompt risetime as a slow one because of the long delay between the peak of the fluorescence curve and the calibrated zero time. These apparent delays originate because of the continuous accumulation of excited state singlets produced by the wings of the pulse. The fluorescence lifetimes of a number of fluorescein derivatives have also been measured [6.132]. The lifetimes of fluorescein (Fl), eosin ($FlBr_4$), and erythrosin (FlI_4), were measured to be 3.6, 0.9, and 0.11 ns, respectively. The data are in excellent agreement with quantum-yield predictions, and the decrease in lifetime observed upon addition of heavy halogen atoms has been taken to be consistent with heavy atom enhanced intersystem crossing and with published triplet quantum yields.

A precautionary note should be added for measurements of fluorescence lifetimes. At high excitation intensities, nonlinear processes such as stimulated emission may occur. Stimulated emission effects were pointed out in two early studies [6.133, 134]. These effects lead to a nonexponential decay of the fluorescence and to a shortening of the lifetime. Other common effects, which must be considered, are the formation of transient species when an entire train of pulses is used to excite fluorescence, and concentration quenching. A fairly common effect is that leftover triplet states modify the lifetime for singlets produced by later pulses in the train. Such problems have led to a wide range of estimates for the lifetime of DODCI, for example, where the correct fluorescence lifetime of about 1.2 ns has now finally been firmly established by a number of investigators [6.43, 116, 135–138]. Some of the problems encountered in making lifetime measurements are discussed in Chapter 3.

In a number of experiments a streak camera has been used in conjunction with picosecond excitation to examine fluorescence emission with picosecond resolution. Emission from a number of dyes under modelocking conditions (see Chap. 2) [6.139, 140], fluorescence from DODCI [6.138] (see Chap. 3), and emission from dye vapors, from photosynthetic samples (see Chap. 7), and scintillator materials [6.141–143], have been obtained in this way. In the dye-vapor experiments [6.141] dimethyl POPOP and perylene molecules were excited in the gaseous phase with picosecond pulses at 353 nm. Then, using a streak-camera detection technique, the risetime of the fluorescence was found to be ≤ 20 ps for dimethyl POPOP and ≤ 30 ps for perylene. Since the vapors were at 300 °C, the collision rate between dye molecules could have been no more than 10^8/s. A dye molecule in the liquid phase would collide with the neighboring solvent molecules at a rate of about 10^{12}/s. Since fluorescence is commonly observed from the ground vibrational state of the first excited-singlet state, the rapid risetime in the vapor was interpreted as direct evidence supporting rapid internal relaxation of the molecules, even in the absence of collisions. The ability of large dye molecules to relax internally independent of their surroundings is a manifestation of the fact that such molecules have a large number of vibrational and rotational modes so that the probability of mutual interactions among the modes is very high. These interactions reduce the lifetime of any particular level, because any excess energy provided by the excitation wavelength can be rapidly redistributed over the large ensemble of densely packed levels. In addition to the experimental progress due to picosecond lasers, there have been important theoretical advances, in recent years, in the description of radiationless transitions and optical coupling in complex molecules [6.144–151].

With regard to the rapid relaxation of large dye molecules, two innovative techniques have been introduced for measuring vibrational relaxation times in large dye molecules [6.16, 17]. In the first technique molecules are excited from the ground state S_0 to the first excited-singlet state S_1 by photons of frequency ω_1. Then the molecules decay to the ground vibrational state of S_1. This decay can be followed by measuring the gain produced by a probe pulse of frequency ω_2, which is usually chosen to correspond with the red end of the fluorescence band. In the second technique ground-state vibrational relaxation can be monitored by exciting the molecules to S_1, employing a suitable delay to allow the molecules to relax to the ground state of S_1, stimulating these molecules to relax to an upper vibrational state of S_0 by an intense beam at frequency ω_2, and then probing the vibrational relaxation by the return of absorption at ω_1 caused by return of the population to the ground state. Because the pulsewidths were of the order of 5 to 6 ps, and the vibrational relaxations were very fast, deconvolution of the pulse shape functions was a difficult matter, but it was concluded that the vibrational relaxation times were about several picoseconds in the rhodamines with an error bar also of several picoseconds. With pulses of 6 ps duration it is not an easy matter to determine the zero time to within a couple of picoseconds.

Just how fast large dye molecules can relax in the excited state is still an open question. Recent results have provided a partial answer to this question [6.18]. Using subpicosecond pulse excitation, it has been possible to apply a previously used technique [6.16] with greater precision. Rhodamine 6G and rhodamine *B* molecules were excited in a number of solvents up to a higher electronic state with a subpicosecond pulse (0.9 ps) at 307.7 nm, and the gain of the emission was then probed with a subpicosecond pulse (0.9 ps) at 615 nm. Gain measurements show that the fluorescence emission begins promptly, as shown in Fig. 6.16. After deconvolution, the risetime is less than 0.2 ps, the resolution of the apparatus. The result is at first somewhat surprising because in order for the excitation to reach the first excited-singlet state, the molecules must relax through an ensemble of levels with a total energy gap of ~14000

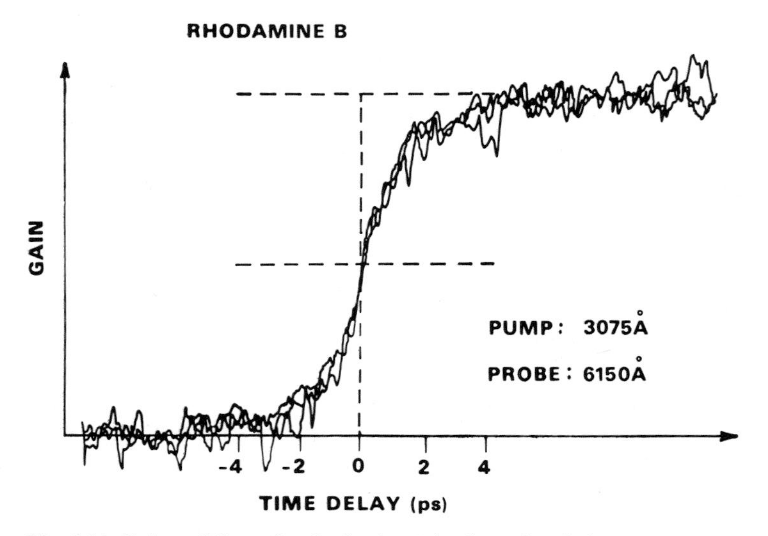

Fig. 6.16. Gain at 615 nm for rhodamine *B* in glycerol and also a low viscosity solvent after excitation with subpicosecond pulse at 307.5 nm. Third curve is a prompt curve. Similar results were obtained for rhodamine 6G. Results clearly demonstrate the rapidity of the decay of excess energy to the ground state of first excited singlet state. Risetime is, after deconvolution, less than 0.2 ps (after [6.18])

cm^{-1}. Though for these complex molecular systems other possible explanations present themselves, such as emission from higher vibronic levels of the first excited-singlet state, though such emission has rarely been observed, the following interpretation [6.18] seems to be indicated: deactivation of the excited state of rhodamine 6G is exceedingly fast.

The rapid de-excitation of an excited electronic state has also been obtained by determining the risetime of the spontaneous fluorescence intensity for a number of dyes [6.152]. Dyes such as rhodamine B, rhodamine 6G, and erythrosin B in such solvents as water, ethanol, and methanol were excited with 530 nm, 10 ps pulses and the emission detected with an optical shutter. The observed risetime corresponded to a delay of less than 1 ps, consistent with later work [6.18].

The risetime of Stokes-shifted spontaneous fluorescence has also been measured for several dyes using a train of 355 nm pulses generated by a mode-locked Nd:glass laser to excite the solutions and an optical Kerr cell to detect the emission near 460 nm [6.153]. Delay times of less than 1 ps, 3.2 ps and 4.3 ps were observed for dimethyl POPOP in ethanol, esculin in ethanol and 1,1,4,4-tetraphenylbutadiene in cyclohexane, respectively.

Previously, it had been claimed that the time- and frequency-resolved vibrational relaxation in an excited state of rhodamine 6G had been determined [6.154]. It was observed upon pumping rhodamine 6G with intense 530 nm pulses of several picosecond duration that intense stimulated emission occurred many picoseconds after the zero time of the apparatus. It was therefore concluded that the vibrational relaxation time was about 6 ps. It is now clear that the onset of stimulated emission in this experiment is governed more by parameters such as the integral of the excitation pulse-shape function, which is closely related to the number of molecules in the excited state, and by the stimulated emission cross section, and is indirectly related to any vibrational relaxation time.

Paradoxical results also had been obtained in other experiments [6.155]. It was found that the spectral emission from a sample of rhodamine 6G in glycerol takes the shape of a Lorentzian with a 12 nm halfwidth at 559 nm a few picoseconds after excitation, and then evolves into a skewed gaussian shape with 37-nm halfwidth centered about 569 nm at 66 ps after excitation [6.155]. The red-shift and spectral broadening are interpreted as follows: upon excitation the dipole moment of the excited state differs from that of the ground state, so that the solvent molecules surrounding an excited state rhodamine molecule are not in equilibrium. Initially the spectrum represents the emission from the vibrationally relaxed rhodamine molecules, while at later times the spectrum shifts to the red on account of reorientations of the solvent molecules driven by their interactions with the new dipole moment, and because of reorientations of substituent groups on the rhodamine 6G molecules as well. The reported [6.155] shift of wavelength toward the red would appear to imply that the cross section for gain at 615 nm as previously measured [6.18, 152] would tend to increase over a $\sim$66 ps or longer period. But for rhodamine 6G in glycerol, no slow increase was observed [6.18], but rather a risetime of ≤ 0.2 ps. These seemingly contradictory results are difficult to reconcile and must be clarified by further experiments. It is possible, though unlikely, that measurements taken in the same fashion as [6.155] will show that the spectrum at 615 nm maintains a constant spectral emission cross section as a function of delay time after excitation. It is likely that effects due to pulse train excitation or some other such complication might explain some of the more puzzling results.

Solvent reorientation about molecules with a large dipole moment have also been reported [6.156–158]. The fluorescence risetime from the S_2 and S_1 states of *p*-dimethylaminobenzonitrile (DMAB) in ethanol and methylcyclohexane have been measured [6.156] by means of an optical-gate technique. In the polar solvent the ^1L_a emission rises to a constant level at 40 to 50 ps after climbing from zero intensity at $t=0$. The onset of the ^1L_b emission for the sample in methylcyclohexane is about 7 to 20 ps. According to this study [6.156] the ^1L_a state of DMAB contains a large admixture of an intramolecular charge transfer state characterized by a large dipole moment ($\sim$23 D), and so the long risetime represents orientational reorganization of the ethanol molecules surrounding the giant dipole. In the nonpolar solvent where such an interaction does not occur, the risetime, according to these workers, is governed by the vibrational activation. A possible difficulty with these interpretations is that the excitation pulses are at 265 nm and their durations are unknown because of experimental difficulties at this wavelength. Such pulses are generated by passage of 530 nm pulses through either KDP or ADP crystals and the pulses may be stretched to 50 ps duration in even normal-size crystals. These effects originate because of the narrow phase-matching window of most nonlinear crystals in the ultraviolet [6.159, 160]. Careful deconvolution of the pulse shape is necessary for interpreting experimental results. The dependence of group dispersion effects on sample path lengths has also recently been analyzed [6.161]. Pulse stretching effects can be especially large in the ultraviolet region and can be very important in long sample cells. According to this analysis, the results with 265 nm pulses could be due to dispersion effects.

An early work on vibrational relaxation in an excited electronic state of azulene was found to be of great interest [6.162]. Recent attempts in various laboratories to reproduce these results have been unsuccessful [6.163–165]. It is therefore necessary that the exact conditions of the different experiments be carefully delineated and that further work be carried out to resolve this discrepancy. Some measurements on radiative processes in benzophenone in solution and in the vapor phase have been reported [6.166–169] as well as results for 3,4 benzpyrene and naphthalene [6.170–172]. Further work on relaxation processes in azulene has also been carried out [6.173–177]. Especially noteworthy are the theoretical analyses of the decay of highly excited electronic states [6.172].

It has been shown that the decay of benzophenone in the vapor phase is nonexponential and in the microsecond range [6.166, 167]. The nonexponential decays have been verified but a somewhat different interpretation given, in terms of triplet states [6.178]. These later workers indicate that oscillations observed in the tail of the decay previously observed [6.167] are probably experimental artifacts and not quantum beats.

In another interesting application of picosecond fluorescence techniques emission originating from upper singlet states by using a two-photon absorption technique has been observed [6.179]. Ordinarily these emissions cannot be observed because, except in rare cases, fluorescence is emitted from higher singlet states with a very low quantum efficiency. Two-photon excitation allows the excitation of levels that are spectroscopically forbidden by one-photon excitation. Two-photon excitation also permits a more uniform spatial distribution of excited molecules in a sample which is strongly one-photon absorbing, e.g., a solid, concentrated solution, or a very intense transition. Furthermore, states that are one-photon allowed but hidden by stronger overlapping transitions can be amenable to detection by two-photon absorption. In addition direct scattering processes which would tend to obscure any weak emission are avoided with two-photon absorption techniques. By irradiating samples with a combination of the wavelengths 1060 nm, 530 nm, and 354 nm generated from the fundamental or harmonics of the Nd:glass laser, spectra for the excited states of such dyes as rhodamine 6G perchlorate in 2-propanol, rhodamine B in 2-propanol, and acridine red in 2-propanol were obtained. From the intensity of typical spectra, an upper limit of 10^{-4} can be placed on the fluorescence quantum efficiency from the upper states, and the lifetime of these states is estimated to be less than 5 ps.

Acknowledgement. The author acknowledges the support of the U.S. Army Office of Research and NIH (National Heart and Lung Institute) and the extensive assistance and advice of Dr. *S. L. Shapiro*.

References

6.1 G. Carios, J. Franck: Z. Physik **17**, 202 (1923)
6.2 Th. Forster: Ann. Physik **2**, 55 (148)
6.3 Th. Forster: Z. Naturforsch. **4 A**, 321 (1949)
6.4 D. L. Dexter: J. Chem. Phys. **21**, 836 (1953)
6.5 M. D. Galanin: J. Exptl. Theor. Phys. USSR **28**, 485 (1955)
6.6 A. N. Terenin, V. L. Ermolaev: Dokl. Akad. Nauk SSSR **85**, 547 (1952)
6.7 F. Wilkinson: In *Advances in Photochemistry*, Vol. 3, ed. by W. A. Noyes, G. S. Hammond, J. N. Pitts, Jr. (Wiley Interscience, New York 1964) pp. 241–268
6.8 R. G. Bennett, R. E. Kellogg: In *Progress in Reaction Kinetics*, Vol. IV, ed. by G. Porter (Pergamon, London 1966) pp. 215–238
6.9 K. B. Eisenthal: Chem. Phys. Lett. **6**, 155 (1970)
6.10 D. Rehm, K. B. Eisenthal: Chem. Phys. Lett. **9**, 387 (1971)
6.11 G. Weber: Trans. Faraday Soc. **50**, 552 (1954)
6.12 A. Ore: J. Chem. Phys. **31**, 442 (1959)
6.13 K. B. Eisenthal, S. Siegel: J. Chem. Phys. **41**, 652 (1964)
6.14 K. B. Eisenthal, S. Siegel: J. Chem. Phys. **42**, 2494 (1965)
6.15 I. Z. Steinberg: J. Chem. Phys. **48**, 2411 (1968)

6.16 D. Ricard, H. Lowdermilk, J. Ducuing: Chem. Phys. Lett. **16**, 617 (1972)
6.17 D. Ricard, J. Ducuing: J. Chem. Phys. **62**, 3616 (1975)
6.18 E. P. Ippen, C. V. Shank: unpublished results
6.19 R. W. Anderson, Jr., R. M. Hochstrasser, H. Lutz, G. W. Scott: J. Chem. Phys. **61**, 2500 (1974)
6.20 J. K. Roy, M. A. El-Sayed: J. Chem. Phys. **30**, 3442 (1964)
6.21 K. B. Eisenthal: J. Chem. Phys. **50**, 3120 (1969)
6.22 W. R. Anderson, Jr., R. M. Hochstrasser: unpublished results
6.23 R. R. Alfano, S. L. Shapiro, M. Pope: Opt. Commun. **9**, 388 (1973)
6.24 G. Mayer, F. Gires: Compt. end. Acad. Sci. (Paris) **258**, 2039 (1964)
6.25 P. D. Maker, R. W. Terhune, C. M. Savage: Phys. Rev. Lett. **12**, 507 (1964)
6.26 A. D. Buckingham, J. A. Pople: Proc. Phys. Soc. **A68**, 905 (1955)
6.27 A. D. Buckingham: Proc. Phys. Soc. **A68**, 910 (1955)
6.28 A. D. Buckingham: Proc. Phys. Soc. **B69**, 344 (1956)
6.29 S. Kielich: Acta Phys. Pol. **30**, 683 (1966)
6.30 S. Kielich: *Dielectrics and Related Molecular Processes,* Vol. I (Chem. Soc. of London, London 1972) p. 192
6.31 M. A. Duguay, J. W. Hansen: Appl. Phys. Lett. **15**, 192 (1969)
6.32 F. Shimizu, B. P. Stoicheff: IEEE J. **QE-5**, 544 (1969)
6.33 R. A. Fisher, P. L. Kelley, T. K. Gustafson: Appl. Phys. Lett. **14**, 140 (1969)
6.34 R. G. Brewer, C. H. Lee: Phys. Rev. Lett. **21**, 267 (1968)
6.35 A. P. Veduta, B. P. Kirsanov: Sov. Phys. JETP **27**, 736 (1968)
6.36 R. R. Alfano, S. L. Shapiro: Phys. Rev. Lett. **24**, 592 (1970)
6.37 K. Sala, M. C. Richardson: Phys. Rev. **A12**, 1036 (1975)
6.38 G. R. Alins, D. R. Bauer, J. I. Brauman, R. Pecora: J. Chem. Phys. **59**, 5310 (1973)
6.39 G. I. A. Stegeman, B. P. Stoicheff: Phys. Rev. **A7**, 1160 (1973)
6.40 V. S. Starunov, E. V. Tiganov, I. L. Fabelinskii: Sov. Phys. JETP Lett. **4**, 176 (1966)
6.41 G. Mourou, M. M. Malley: Opt. Commun. **13**, 412 (1975)
6.42 P. P. Ho, W. Yu, R. R. Alfano: Chem. Phys. Lett. **37**, 91 (1976)
6.43 E. P. Ippen, C. V. Shank: Appl. Phys. Lett. **26**, 62 (1975)
6.44 S. L. Shapiro, H. P. Broida: Phys. Rev. **154**, 129 (1967)
6.45 J. R. Lalanne: Phys. Lett. **51 A**, **74** (1975)
6.46 G. K. L. Wong, Y. R. Shen: Phys. Rev. Lett. **30**, 895 (1973)
6.47 K. B. Eisenthal, K. H. Drexhage: J. Chem. Phys. **51**, 5720 (1969)
6.48 T. J. Chuang, K. B. Eisenthal: Chem. Phys. Lett. **11**, 368 (1971)
6.49 D. W. Phillion, D. J. Kuizenga, A. E. Siegman: Appl. Phys. Lett. **27**, 85 (1975)
6.50 J. Franck, E. Rabinowitch: Trans. Faraday Soc. **30**, 120 (1934)
6.51 F. W. Lampe, R. M. Noyes: JACS **76**, 2140 (1954)
6.52 T. J. Chuang, G. W. Hoffman, K. B. Eisenthal: Chem. Phys. Lett. **25**, 201 (1974)
6.53 R. M. Noyes: J. Chem. Phys. **22**, 1349 (1954)
6.54 R. M. Noyes: JACS **78**, 5486 (1956)
6.55 G. E. Busch, R. T. Mahoney, R. I. Morse, K. R. Wilson: J. Chem. Phys. **51**, 837 (1969)
6.56 M. J. Bronskill, R. K. Wolff, J. W. Hunt: J. Chem. Phys. **53**, 4201 (1970)
6.57 M. S. Matheson, W. A. Mulac, J. Rabani: J. Phys. Chem. **67**, 2613 (1963)
6.58 L. I. Grossweiner, H. I. Joschek: Advan. Chem. **50**, 279 (1965)
6.59 G. Czapski, H. A. Schwarz: J. Phys. Chem. **66**, 471 (1962)
6.60 W. L. Waltz, A. W. Adamson, P. D. Fleischauer: JACS **89**, 3923 (1967)
6.61 P. Bennema, G. J. Hoytink, J. M. Luprinski, L. J. Oosterhoff, P. Sellier, J. D. Van Voorst: Molec. Phys. **2**, 431 (1959)
6.62 J. Joussot-Dubien, R. Lesclave: J. Chem. Phys. **61**, 1631 (1964)
6.63 W. A. Gibbons, G. Porter, M. I. Savadatti: Nature **206**, 1355 (1965)
6.64 K. D. Cadogan, A. L. Albrecht: J. Chem. Phys. **43**, 2550 (1965)
6.65 C. R. Goldschmidt, G. Stein: Chem. Phys. Lett. **6**, 299 (1970)
6.66 G. Kenney-Wallace, D. C. Walker: Ber. Bunsenges. Phys. Chem. **75**, 634 (1971)
6.67 M. Ottolenghi: Chem. Phys. Lett. **12**, 339 (1971)

6.68 M. Shirom, G. Stein: J. Chem. Phys. **55**, 3379 (1971)
6.69 J. Jortner, M. Ottolenghi, G. Stein: JACS **85**, 2712 (1963)
6.70 G. J. Hoytink, P. J. Zandstra: Molec. Phys. **3**, 371 (1960)
6.71 J. D. Van Voorst, G. J. Hoytink: J. Chem. Phys. **42**, 3995 (1965)
6.72 J. Eloranta, H. Linschitz: J. Chem. Phys. **38**, 2214 (1963)
6.73 M. Fisher, G. Rämme, S. Claesson, M. Swarc: Chem. Phys. Lett. **9**, 306 (1971)
6.74 G. Makkes Van der Deyl, J. Dousina, S. Speiser, J. Kommandeur: Chem. Phys. Lett. **20**, 17 (1973)
6.75 P. M. Rentzepis, R. P. Jones, J. Jortner: Chem. Phys. Lett. **15**, 480 (1972)
6.76 P. M. Rentzepis, R. P. Jones, J. Jortner: J. Chem. Phys. **59**, 766 (1973)
6.77 T. L. Netzel, P. M. Rentzepis: Chem. Phys. Lett. **29**, 337 (1974)
6.78 W. S. Struve, T. L. Netzel, P. M. Rentzepis, G. Levin, M. Swarc: JACS **97**, 3310 (1975)
6.79 D. Huppert, W. S. Struve, P. M. Rentzepis, J. Jortner: J. Chem. Phys. **63**, 1205 (1975)
6.80 D. Huppert, P. M. Rentzepis: J. Chem. Phys. **64**, 191 (1976)
6.81 L. M. Dorfman: In *Investigations of Rates and Mechanisms of Reactions,* Vol. VI, Part II, ed. by G. G. Hammes (Wiley-Interscience, New York 1974) pp. 463–519
6.82 B. Halpern, R. Gommer: J. Chem. Phys. **43**, 1069 (1968)
6.83 M. A. Woolf, G. W. Rayfield: Phys. Rev. Lett. **15**, 235 (1935)
6.84 J. Jortner: Ber. Bunsenges. Phys. Chem. **75**, 696 (1971)
6.85 D. Copeland, N. R. Kestner, J. Jortner: J. Chem. Phys. **53**, 1189 (1970)
6.86 G. Levin, S. Claesson, M. Swarc: JACS **94**, 8673 (1972)
6.87 H. Leonhardt, A. Weller: Ber. Bunsenges. Phys. Chem. **67**, 791 (1963)
6.88 A. Weller: In *5th Nobel Symposium,* ed. by S. Claesson (Interscience, New York 1963) p. 413
6.89 E. A. Chandross, H. T. Thomas: Chem. Phys. Lett. **9**, 393 (1971)
6.90 T. Okada, T. Fujita, M. Kubota, S. Masaski, N. Mataga, R. Ide, Y. Sakata, S. Misumi: Chem. Phys. Lett. **14**, 563 (1972)
6.91 M. Ottolenghi: Accts. Chem. Res. **6**, 153 (1973)
6.92 R. M. Noyes: Progr. Reaction Kinetics **1**, 129 (1961)
6.93 K. Gnadig, K. B. Eisenthal: unpublished results
6.94 T. J. Chuang, K. B. Eisenthal: J. Chem. Phys. **59**, 2140 (1973)
6.95 T. J. Chuang, R. J. Cox, K. B. Eisenthal: J. Chem. Phys. **62**, 2213 (1975)
6.96 T. J. Chuang, K. B. Eisenthal: J. Chem. Phys. **62**, 2213 (1975)
6.97 W. R. Ware, J. S. Novros: J. Phys. Chem. **70**, 3246 (1966)
6.98 J. W. Shelton, J. A. Armstrong: IEEE J. QE-**7**, 696 (1967)
6.99 R. I. Scarlet, J. F. Figueira, H. Mahr: Appl. Phys. Lett. **13**, 71 (1968)
6.100 M. M. Malley, P. M. Rentzepis: Chem. Phys. Lett. **3**, 534 (1969)
6.101 M. R. Topp, P. M. Rentzepis, R. P. Jones: J. Appl. Phys. **42**, 3415 (1971)
6.102 C. Lin, A. Dienes: Opt. Commun. **9**, 21 (1973)
6.103 D. Magde, M. W. Windsor: Chem. Phys. Lett. **24**, 144 (1974)
6.104 T. Förster, G. Hoffman: Z. Physik Chem. NF **75**, 63 (1971)
6.105 E. P. Ippen, C. V. Shank, A. Bergman: Chem. Phys. Lett. **38**, 611 (1976)
6.106 R. M. Hochstrasser, H. Lutz, G. W. Scott: Chem. Phys. Lett. **24**, 162 (1974)
6.107 R. W. Anderson, R. M. Hochstrasser, H. Lutz, G. W. Scott: Chem. Phys. Lett. **28**, 153 (1974)
6.108 A. Nitzan, J. Jortner, P. M. Rentzepis: Chem. Phys. Lett. **8**, 445 (1971)
6.109 P. M. Rentzepis: Science **169**, 239 (1969)
6.110 P. M. Rentzepis, C. J. Mitschele: Anal. Chem. **42**, 20 (1970)
6.111 P. M. Rentzepis, G. E. Busch: Molec. Photochem. **4**, 353 (1972)
6.112 R. W. Anderson, R. M. Hochstrasser, H. Lutz, G. W. Scott: Chem. Phys. Lett. **32**, 204 (1975)
6.113 R. R. Alfano, S. L. Shapiro: Phys. Rev. Lett. **24**, 584 (1970)
6.114 R. R. Alfano, S. L. Shapiro: Phys. Rev. Lett. **24**, 1217 (1970)
6.115 R. R. Alfano, S. L. Shapiro: Chem. Phys. Lett. **8**, 631 (1971)
6.116 D. Magde, M. W. Windsor: Chem. Phys. Lett. **27**, 31 (1974)
6.117 D. Magde, B. A. Bushaw, M. W. Windsor: Chem. Phys. Lett. **28**, 263 (1974)
6.118 D. Magde, M. W. Windsor, D. Holten, M. Gouterman: Chem. Phys. Lett. **29**, 183 (1974)
6.119 H. Tashiro, T. Yajima: Chem. Phys. Lett. **25**, 582 (1974)

6.120 M. A. Duguay, J. W. Hansen: Opt. Commun. **1**, 254 (1969)
6.121 G. E. Busch, R. P. Jones, P. M. Rentzepis: Chem. Phys. Lett. **18**, 178 (1973)
6.122 G. E. Busch, K. S. Greve, G. L. Olson, R. P. Jones, P. M. Rentzepis: Chem. Phys. Lett. **33**, 412 (1975)
6.123 G. E. Busch, K. S. Greve, G. L. Olson, R. P. Jones, P. M. Rentzepis: Chem. Phys. Lett. **33**, 417 (1975)
6.124 N. Nakashima, N. Mataga: Chem. Phys. Lett. **35**, 487 (1975)
6.125 A. D. Kirk, P. E. Hoggard, G. B. Porter, M. G. Rockley, M. W. Windsor: Chem. Phys. Lett. **37**, 199 (1976)
6.126 A. S. Pine: J. Appl. Phys. **39**, 106 (1968)
6.127 M. E. Mack: J. Appl. Phys. **39**, 2483 (1968)
6.128 G. Mourou, G. Busca, M. M. Denariez-Roberge: Opt. Commun. **4**, 40 (1971)
6.129 J. Fouassier, D. Lougnot, J. Faure: Chem. Phys. Lett. **30**, 448 (1975)
6.130 R. R. Alfano, S. L. Shapiro: Opt. Commun. **6**, 98 (1972)
6.131 G. Mourou, M. M. Malley: Opt. Commun. **11**, 282 (1974)
6.132 G. Porter, E. S. Reid, C. J. Tredwell: Chem. Phys. Lett. **29**, 469 (1974)
6.133 M. E. Mack: Appl. Phys. Lett. **15**, 166 (1969)
6.134 H. E. Lessing, E. Lippert, W. Rapp: Chem. Phys. Lett. **7**, 247 (1970)
6.135 H.-J. Cirkel, L. Ringwelski, F. P. Schäfer: Z. Phys. Chem. NF **81**, 158 (1972)
6.136 C. V. Shank, E. P. Ippen: In *Topics in Applied Physics,* Vol. 1, ed. by F. P. Schäfer (Springer Berlin, Heidelberg, New York 1973) p. 141
6.137 C. V. Shank, E. P. Ippen: Appl. Phys. Lett. **26**, 62 (1975)
6.138 J. C. Mialocq, A. W. Boyd, J. Jaraudias, J. Sutton: Chem. Phys. Lett. **37**, 236 (1976)
6.139 E. G. Arthurs, D. J. Bradley, A. G. Roddie: Chem. Phys. Lett. **22**, 230 (1973)
6.140 E. G. Arthurs, D. J. Bradley, P. N. Puntambekar, I. S. Ruddock, T. J. Glynn: Opt. Commun. **12**, 360 (1974)
6.141 S. L. Shapiro, R. C. Hyer, A. J. Campillo: Phys. Rev. Lett. **33**, 513 (1974)
6.142 S. L. Shapiro, V. H. Kollman, A. J. Campillo: FEBS. Lett. **54**, 358 (1975)
6.143 A. J. Campillo, R. C. Hyer, S. L. Shapiro: Nucl. Instr. Meth. **120**, 533 (1974)
6.144 G. W. Robinson, R. P. Frosch: J. Chem. Phys. **38**, 1187 (1964)
6.145 W. Siebrand: J. Chem. Phys. **44**, 4055 (1966)
6.146 B. R. Henry, M. Kasha: Ann. Rev. Phys. Chem. **19**, 161 (1968)
6.147 M. Bixon, J. Jortner: J. Chem. Phys. **48**, 715 (1968)
6.148 J. Jortner, R. S. Berry: J. Chem. Phys. **48**, 2757 (1968)
6.149 K. Freed, J. Jortner: J. Chem. Phys. **50**, 2916 (1969)
6.150 S. Fischer, E. W. Schlag: Chem. Phys. Lett. **4**, 393 (1969)
6.151 K. G. Spears, S. A. Rice: J. Chem. Phys. **55**, 5561 (1971)
6.152 G. Mourou, M. M. Malley: Chem. Phys. Lett. **32**, 476 (1975)
6.153 J. Covey, M. M. Malley, G. Mourou: to be published
6.154 P. M. Rentzepis, M. R. Topp, R. P. Jones, J. Jortner: Phys. Rev. Lett. **25**, 1742 (1970)
6.155 M. M. Malley, G. Mourou: Opt. Commun. **10**, 323 (1974)
6.156 W. S. Struve, P. M. Rentzepis, J. Jortner: J. Chem. Phys. **59**, 5014 (1973)
6.157 W. S. Struve, P. M. Rentzepis: J. Chem. Phys. **60**, 1533 (1974)
6.158 W. S. Struve, P. M. Rentzepis: Chem. Phys. Lett. **29**, 23 (1974)
6.159 J. Comly, E. Garmire: Appl. Phys. Lett. **12**, 7 (1968)
6.160 R. C. Miller: Phys. Lett. **26 A**, 177 (1968)
6.161 M. R. Topp, G. C. Orner: Opt. Commun. **13**, 276 (1975)
6.162 P. M. Rentzepis: Chem. Phys. Lett. **2**, 117 (1968)
6.163 J. P. Heritage: to be published
6.164 M. M. Malley, G. Mourou: unpublished
6.165 P. R. Monson, G. W. Robinson: unpublished
6.166 G. E. Busch, P. M. Rentzepis, J. Jortner: Chem. Phys. Lett. **11**, 437 (1971)
6.167 G. E. Busch, P. M. Rentzepis, J. Jortner: J. Chem. Phys. **56**, 361 (1972)
6.168 A. Nitzan, J. Jortner, P. M. Rentzepis: Chem. Phys. Lett. **8**, 445 (1971)
6.169 P. M. Rentzepis, G. E. Busch: Molec. Photochem. **4**, 353 (1972)

6.170 P. Wannier, P. M. Rentzepis, J. Jortner: Chem. Phys. Lett. **10**, 102 (1971)
6.171 P. Wannier, P. M. Rentzepis, J. Jortner: Chem. Phys. Lett. **10**, 193 (1971)
6.172 A. Nitzan, J. Jortner, P. M. Rentzepis: Proc. Roy. Soc. Lond. A **327**, 367 (1972)
6.173 P. M. Rentzepis: Chem. Phys. Lett. **3**, 717 (1969)
6.174 P. M. Rentzepis: Science **169**, 239 (1970)
6.175 P. M. Rentzepis, J. Jortner, R. P. Jones: Chem. Phys. Lett. **4**, 599 (1970)
6.176 D. Huppert, J. Jortner, P. M. Rentzepis: Chem. Phys. Lett. **13**, 225 (1972)
6.177 E. Drent, G. Makkes van der Deijl, P. J. Zandstra: Chem. Phys. Lett. **2**, 526 (1968)
6.178 R. M. Hochstrasser, J. E. Wessel: Chem. Phys. Lett. **19**, 156 (1973)
6.179 G. C. Orner, M. R. Topp: Chem. Phys. Lett. **36**, 295 (1975)

7. Picosecond Relaxation Measurements in Biology

A. J. Campillo and S. L. Shapiro

With 24 Figures

The application of picosecond techniques to biological measurements is in its infancy. In the past three years numerous groups throughout the world have begun to apply picosecond measurements to photosynthetic problems and to study important biological molecules such as hemoglobin, rhodopsin, and DNA. Because initial research in this area appears to be quite productive, we can look forward to a rapid expansion of effort.

There are a number of reasons why picosecond technology offers special advantages in the biological area. Complex macromolecules are amenable to picosecond studies because energy-transfer processes are often rapid; because the number of possible excitation modes is large, which results in many paths of decay; because energy transfer and migration times in such molecules offer the opportunity, in principle, of determining structural spacings; because the close proximity of molecules in complex systems, such as photosynthetic systems, leads to rapid exciton transfer and so offers the opportunity for studying energy transport mechanisms; and because atoms, molecules, or molecular side chains attached to macromolecules often rearrange themselves as a fundamental part of a biophysical process in a time interval which may occur within picoseconds.

This chapter is divided into four parts. The first section deals with rapid photosynthetic processes, the area of biology which has received the most attention from picosecond experimentalists; the second section describes a measurement in hemoglobin made with the powerful, new, cw modelocked dye-laser techniques; the third section discusses measurements on molecules which play an important role in the visual process; and the fourth describes measurements in DNA.

7.1 Photosynthesis

7.1.1 Primary Events in Photosynthesis

All life on earth depends directly or indirectly upon photosynthesis. The exact manner by which plants convert solar energy into useful chemical energy is still an unsolved riddle. Clearly, an understanding of this very important process would have many far-reaching practical implications in our present energy- and food-conscious world. For example, the development of novel solar-energy conversion systems as alternative energy sources and the increase of agricultural output through the use of additives for higher photosynthetic efficiency would have a profound impact on our standard of living. Because the photosynthetic process relies intimately on the absorption of light, it is no surprise that the laser would be an invaluable tool in unravelling its mysteries.

What, then, occurs during photosynthesis? In green plants, for instance, there are special subcellular organelles called chloroplasts, usually ellipsoidal and about 5–10 μm in length, in which photosynthesis occurs. Photosynthesis can be summarized by the deceptively simple equation

$$H_2O+CO_2 \xrightarrow{\text{light}} (CH_2O)+O_2\,, \qquad (7.1)$$

that is, water, carbon dioxide, and light are utilized by the chloroplasts to manufacture oxygen and carbohydrates. The light quanta provide energy in excess of the necessary 112 kcal/mole CO_2 to drive the process while the chloroplast, consisting of pigments, proteins, and enzymes, acts as a catalyst. There has been considerable evidence accumulated over the past forty years that the photosynthetic apparatus consists of units within the chloroplasts. *Emerson* and *Arnold* [7.1] using repetitive pulses of exciting light of variable intensity and frequency established that such a "unit" composed of 2500 chlorophyll (Chl) molecules acting in cooperation, evolved one O_2 molecule. Since four hydrogen atoms are transferred from two H_2O to CO_2 to evolve one O_2, it became convenient to consider a smaller subunit, called the photosynthetic unit (PSU), accomplishing each hydrogen transfer using about 600 Chl molecules. In green plants there are two photosystems, PS I and PS II, which are in close proximity and which act in cooperation to transfer one hydrogen atom and each absorbing a quanta of light. Thus, in general, a minimum of eight quanta is necessary to generate O_2 and to reduce one CO_2 molecule. In green plants these reactions need the cooperation of 40 to 400 chlorophyll a molecules and of additional accessory antenna pigments to harvest the light photons and to convey the electronic excitation to a trapping site, called a reaction center. The reaction center and its associated antenna pigments form a photosynthetic unit, the smallest entity capable of photosynthesis. Photosystem I, the smaller of the two units, produces upon light absorption both a strong reductant and a weak oxidant, whereas System II, upon absorption of a photon produces a strong oxidant and a weak reductant. Electron flow from the weak reductant to the weak oxidant is coupled to the conversion of adenosine diphosphate and inorganic phosphate to adenosine triphosphate (ATP). ATP assists the strong reductant to reduce CO_2 to a carbohydrate (CH_2O), and the strong oxidant oxidizes H_2O to O_2. Photosynthetic bacteria have only one photosystem, which closely resembles that of PS I.

The following equations summarize the process of photosynthesis in higher plants:

$$R+h\nu \rightarrow R^* \qquad (7.2)$$

$$R^* + \text{Chl a} \rightarrow \text{Chl a}^* + R \tag{7.3}$$

$$\text{Chl a}^* + \text{Chl a} \rightarrow \text{Chl a} + \text{Chl a}^* \tag{7.4}$$

$$\text{Chl a}^* + P \rightarrow \text{Chl a} + P^* \tag{7.5}$$

$$4(P^*X) \rightarrow 4(P^+X^-) \tag{7.6}$$

$$4P^+ + 2H_2O \rightarrow 4P + O_2 + 4H^+ \tag{7.7}$$

$$4X^- + 4Y \rightarrow 4X + 4Y^- \tag{7.8}$$

$$4H^+ + 4Y^- \rightarrow 4YH \tag{7.9}$$

$$4YH + CO_2 \rightarrow CH_2O + H_2O + 4Y \tag{7.10}$$

Equation (7.2) describes the absorption of light by a pigment molecule R in the chloroplast and (7.3) the nonradiative transfer from the pigment to a chlorophyll a molecule. Actually a series of transfers occurs between the same and different pigments first. Equations (7.4) and (7.5) describe the process of energy migration whereby the absorbed excitation is transferred a distance to a reaction center in much the same manner a bucket brigade passes water to a fire. Equation (7.6) describes the reduction of P, the primary electron donor, and the oxidation of X, the primary electron acceptor. These first few steps ending in (7.6) are generally referred to as the realm of "physical" photosynthesis while the remaining chemical equations involving CO_2, O_2, H_2O, CH_2O and a secondary electron acceptor Y are considered the realm of "chemical" photosynthesis.

The equations merely represent a summary of what is occurring. In reality numerous additional intermediate steps occur. Indeed, there are also a number of exceptions to the above summary. For example, besides green plants there exist species of bacteria which are incapable of oxidizing H_2O but substitute hydrogen donors instead, H_2A'. As a result, no O_2 is evolved in this form of photosynthesis. (For the interested reader who would like a more general description of photosynthesis, the following review works are recommended: [7.2–7.6].)

In this chapter, we will confine our discussion to the physical or *primary* aspects of photosynthesis. These "primary events" occur on a picosecond time scale. Here the picosecond light source can be an invaluable tool. Absorption of a laser photon by one of the antenna molecules sets off a complex sequence of events leading to the reduction of an acceptor in just a few hundred picoseconds. How this is brought about can be directly deduced using techniques of picosecond chronoscopy. Fluorescence decay times from specific molecules give direct information as to how long the molecule remains in a specific state before transferring its energy internally or to a neighbor. Similarly, photoinduced absorbance changes provide a tipoff to the photochemical fate of various oxidized species. We discuss the picosecond measurements in the sequence that the actual events occur in photosynthetic systems. In the photosynthetic unit light is first absorbed by the pigment molecules, the energy is funneled by exciton migration to the reaction center, and finally the energy is stored by means of new states formed in the reaction center and used for driving the photosynthetic process. Subsections 7.1.2, 7.1.3, and 7.1.4 discuss measurements on the component pigment molecules, on exciton migration, and on the newly formed states in the reaction center, respectively.

7.1.2 The Pigments

The entire photobiological range is geared to accept the sun's radiation. As *Wald* [7.7] points out, the sun emits 75% of its radiant energy between 300 and 1100 nm. The pigment molecules which absorb radiation in this wavelength region and then transport the energy to locations where it can be converted by photosynthesis are the subject of this section. The study of energy transfer between chlorophyll molecules in solution is of special importance because in photosynthetic systems energy is transferred from chlorophyll molecule to chlorophyll molecule until it arrives at what is most probably a special chlorophyll complex in the reaction center. Besides chlorophyll, which is so important for energy transfer, we will briefly discuss the carotenoids which play a role in photosynthetic transfers. Because pigment molecules interact in a complex fashion, and because the transfer rates may be as short as a few picoseconds, picosecond spectroscopy can yield specific information as to their structural positions with respect to one another and as to the exciton interactions responsible for energy transport.

Regardless of the wavelength at which a plant is excited in the visible region, most of its fluorescence is emitted between 670 and 740 nm. This band is that of chlorophyll a, and, because it is so dominant, it is thought that the other pigment molecules transfer their energy to chlorophyll a. Thus chlorophyll a (which is found in all higher plants and all simple algae, and which also appears in a modified form as bacteriochlorophyll a in bacteria) plays a central role in energy transfer. The molecular structure of chlorophyll is shown in Fig. 7.1. The chlorophyll "head", composed of four pyrrol rings connected by CH links, forms a porphin ring about the central magnesium atom and together with the side chains constitute a porphyrin structure. The porphin ring is responsible for the light-absorption and energy-transfer characteristics. Note that the porphyrin structure with an iron atom replacing the magnesium is present in heme, $C_{34}H_{32}N_4O_4Fe$, of the blood's hemoglobin (see Fig. 7.17). The porphyrins have a conjugated structure, that is, they possess a regular alternation of single and double bonds. Conjugated systems are found in many dye molecules and, in porphyrin, give rise to mobile electrons associated with the porphin ring structure as a whole. In chlorophyll these π electrons cause strong absorption bands in the visible and ultraviolet regions of the spectrum.

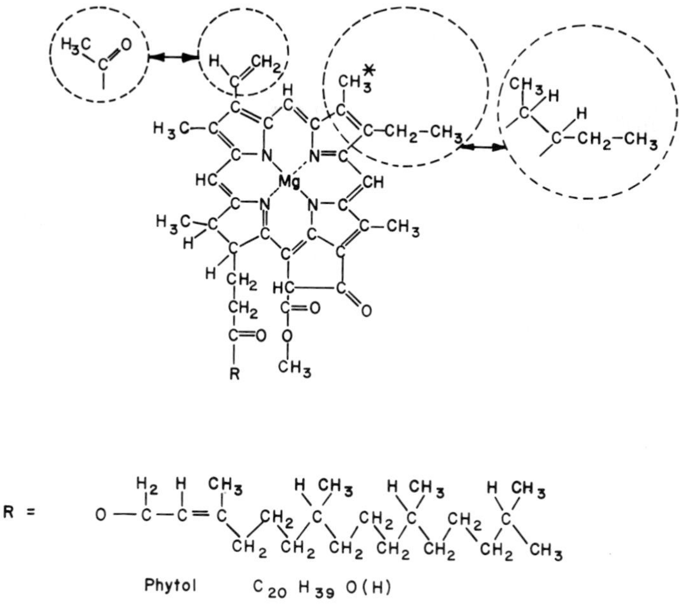

Fig. 7.1. Schematic diagram of chlorophyll a molecule. Chlorophyll b differs only in the replacement of the CH_3 group by a CHO group at the position marked by an asterisk. Bacteriochlorophyll a is formed by replacing the two dotted sections by the sections shown. The porphyrin ring is responsible for many of chlorophyll's spectral properties. The phytol tail represented by R connects to porphyrin head at position shown

The porphyrin head is squarish and compact, about 15.5 by 15.6 by 3.7 Å across. This shape allows the molecules to bunch together for efficient exciton transfer. The central magnesium atom in the porphyrin structure is bonded to four nitrogen atoms. Suspected mechanisms for coordinating chlorophyll molecules assume their bonding with this central magnesium atom [7.8]. The chlorophyll "tail" consists of a long phytol chain ($C_{20}H_{39}OH$) which, except for a single double bond, would be saturated. While this phytol tail is nonpolar and hydrophobic, the porphyrin head is polar because it contains the magnesium atom which tends to be positively charged and hence has an affinity for water and is hydrophilic. These properties determine the structural interactions of chlorophyll with its environment.

The structure of chlorophyll b is almost identical to that of chlorophyll a, with a single CHO bond replacing a CH_3 bond in the position marked by an asterisk in Fig. 7.1. The molecular weight of water-free chlorophyll a ($C_{55}H_{72}O_5N_4Mg$) is 893, whereas that of water-free chlorophyll b is 907; both are good-sized dye molecules. The absorption characteristics for chlorophylls a and b are shown in Fig. 7.2a. There are two strong absorption bands in chlorophyll a, the first singlet located at about 665 nm and the other band located at 435 nm, often referred to as the Soret band. Although there is little apparent light absorption between 460 and 600 nm, chlorophyll and other pigment molecules are present in plants in sufficient concentration so that a substantial fraction of light in this spectral region is absorbed. Among these are the carotenoids, found in many plant leaves, which absorb between 420 and 480 nm, and the phycobilins, found in blue-green, brown, and red algae, which absorb between 490 and 654 nm. Chlorophyll is transparent between 700 and 3000 nm, and this property perhaps provides the mechanism for preventing overheating. The exact positions of the bands are sensitive to

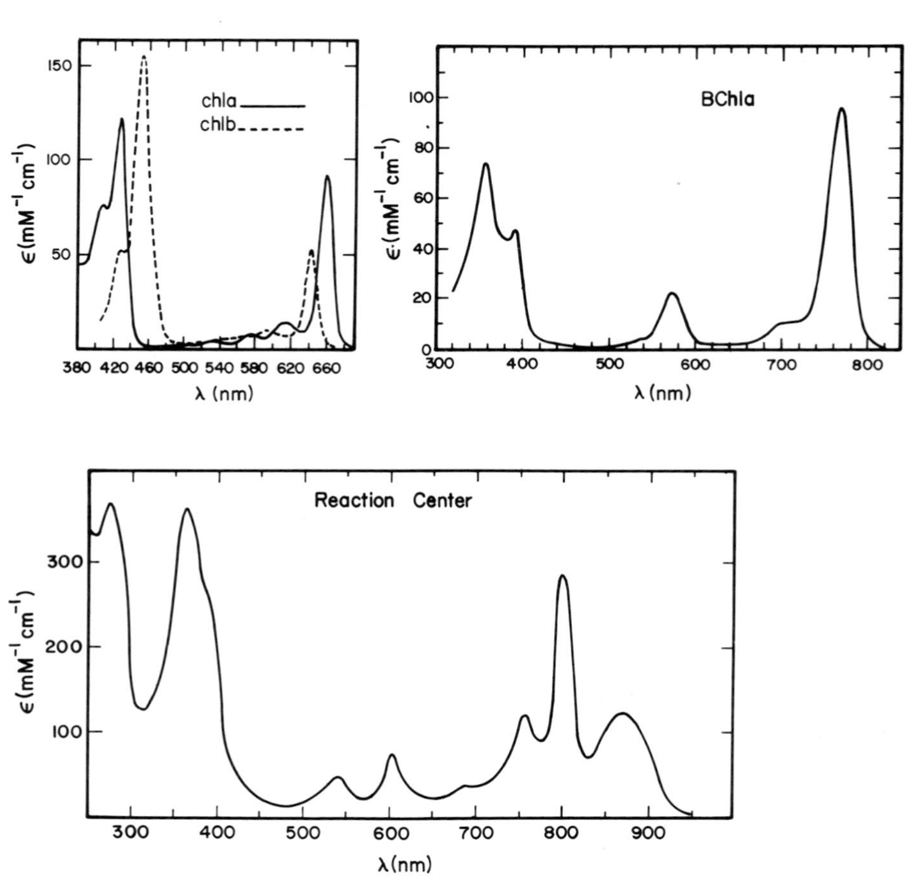

Fig. 7.2. *Top* left: Absorption spectra of chlorophylls a and b in ethyl ether. The chlorophyll a Soret band appears at 430 nm and the first excited singlet at 660 nm. This molecule fluoresces at 680 to 740 nm (after *Zscheile* and *Comar* [7.9]). *Center* right: Absorption spectrum of bacteriochlorophyll a in ether solution (after *Philipson* and *Sauer* [7.10]). *Bottom:* In bacterial cells molecules fluoresce at about 920 nm. Absorption spectrum of reaction centers from *R. spheroides*. Very weak fluorescence is detected at about 920 nm (after *Reed* [7.11])

whether the solution is wet or dry—trace amounts of water have a strong influence.

Bacteriochlorophyll a, present in purple and green bacteria, differs from chlorophyll a by the presence of four additional hydrogen atoms and the absence of just one more double bond. With these simple changes, the spectrum shifts to the red. The main absorption band of bacteriochlorophyll a is at 770 nm, as shown in Fig. 7.2b. However, in bacterial cells, the bands are shifted to between 800 and 890 nm and they fluoresce at about 920 nm.

Because of the large size of the chlorophyll molecule and because of structural considerations, it is doubtful that the molecule can undergo rotation in a cell. In a solution, the molecule can rotate, but, because of its large size, quite slowly; from the Debye equation the rotation time in typical solvents can be estimated to be >100 ps. Because the most interesting regions for investigating chlorophyll interactions are at high concentration where excitation-transfer times are considerably shorter than 100 ps, and because of structural considerations, it would probably be only of indirect interest to apply picosecond spectroscopy techniques to a measurement of the rotational lifetime in solution.

Other important biological pigments include the carotenoids which are conjugated polyenes and are structurally quite similar to the retinals discussed in Section 7.3. All-trans-β carotene is thought to absorb between 250 and 400 nm and to emit in the range of about 350 to 650 nm [7.12]. The carotenoids, such as β carotene, act as accessory pigments in plants absorbing the light and transferring the energy to the chlorophylls. Plants seem to be able to adjust the amount of accessory pigments to suit the light level involved. Carotenoids may act as adjustable neutral filters at high light levels in order to protect the plant.

Picosecond measurements can be applied to study pigment molecules in at least three ways: Measurement of radiative and nonradiative loss of energy of independent molecules in solution; measurement of excitation transfer between different molecules in solution by interactions at a distance; and measurement of aggregation properties of pigments in certain solvents.

Campillo et al. [7.13] first used streak-camera techniques for biological studies. They studied α and β carotenes *in vitro* at high concentrations and high excitation intensities and found surprising decay times ($\simeq 50$ ps). Since the quantum efficiency for fluorescence is expected to be less than 2×10^{-3} [7.12], and since they used 530 nm excitation, results are probably due to either multiple photon absorption to a higher singlet, such as S_2, to excited state-excited state interactions, or to impurity transfer processes. The excited state properties of the polyenes, such as the carotenoids and retinals, are still controversial [7.12]. In fact, one cannot completely rule out the possibility that the observed spectral emission, thought to come from the carotenes, is due to an impurity [7.12, pp. 47 and 48].

Chlorophyll a and b fluorescence lifetimes have been obtained with picosecond pulse excitation by *Kollman* et al. [7.14], by *Shapiro* et al. [7.15], and by *Paschenko* et al. [7.16–17]. Samples are excited with a 530 nm pulse train derived from the second-harmonic generation in KDP produced upon passage of the pulse train at 1060 nm emitted by a modelocked Nd:glass laser. Fluorescence emission is collected onto the slit of a streak camera. When a train of pulses excites a sample, the fluorescence produced by one of the pulses can be examined by triggering the camera to sweep during the interval when the

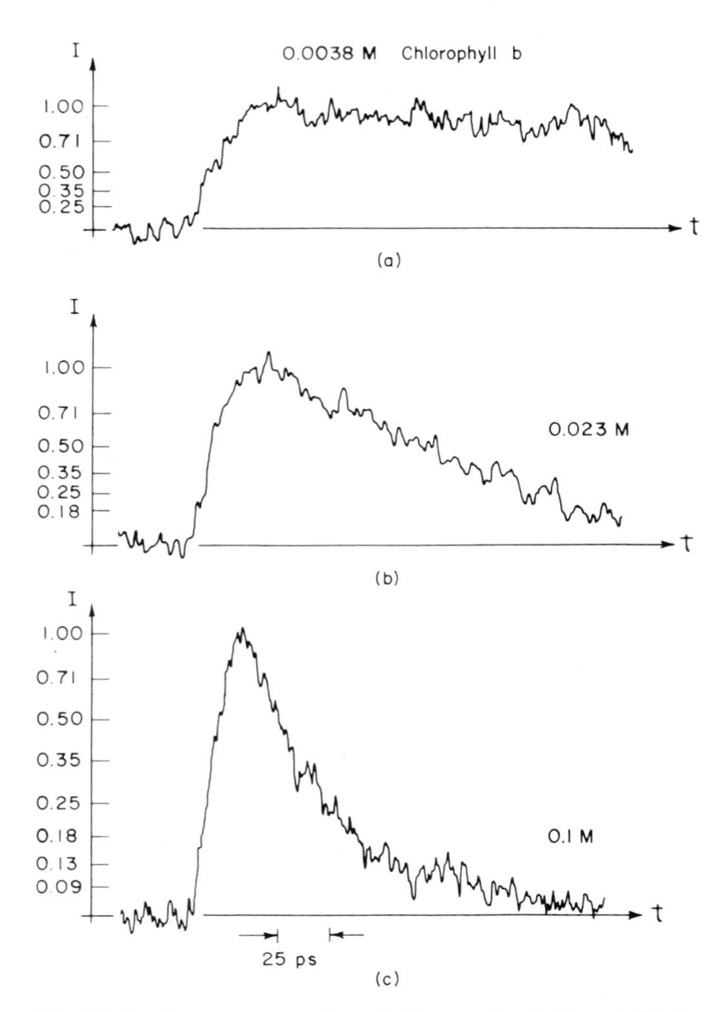

Fig. 7.3. Densitometer traces of streak photographs of chlorophyll b fluorescence in chloroform showing concentration quenching phenomena. Parts (a), (b) and (c) show lifetimes of 572, 120 and 15 ps, respectively. The lifetime follows a R^6 dependence between donor and acceptor, consistent with a Förster energy transfer mechanism. At low concentrations lifetime can be as long as 5 ns in certain dry solutions (from *Shapiro* et al. [7.15])

fluorescence produced from such a pulse is arriving at the streak tube. Experimental results showing the dependence of the lifetime of chlorophyll b on the concentration in a chloroform solution are shown in Fig. 7.3. Each curve represents a densitometer trace of a streak photograph. At low concentrations the lifetime is of the order of a nanosecond, and decreases with concentrations approaching those of live cells, 0.1 M [7.18], to about 15 ps for chlorophyll b. In certain very dry solutions the chlorophyll lifetime approaches 5 ns at very low concentrations [7.19]. The fluorescence risetimes are less than 10 ps, and careful calibration indicates that the fluorescence decay is not exponential.

Similar results were obtained for chlorophyll a in solution. Results such as these are explainable by concentration quenching, a term applied to the phenomenon whereby an increase in concentration leads to a decrease in quantum efficiency. Such phenomena were first studied in chlorophyll solutions by *Watson* and *Livingston* [7.20]; we will briefly summarize the theory describing such a phenomenon before explaining these lifetime measurements.

The modern quantum-mechanical theory describing the transfer of electronic excitation energy between similar molecules was formulated by *Förster* in 1948 [7.21]. Excellent reviews of the theory are found in *Förster* [7.22] and with regard to chlorophyll in *Knox* [7.23–25]. The theory describes the migration of electronic energy away from an excited molecule and was originally formulated to explain the strong depolarization properties of certain dye molecules in solution, including the chlorophyll molecules.

Förster [7.21] showed that the rate of excitation transfer and the critical molecular-separation distance for such transfer during the excitation lifetime were, in certain cases, calculable from the absorption and emission spectra, and that this form of transfer explained the depolarization measurements. Dye molecules in solution are usually treated by assuming the "very-weak-coupling" case [7.22]. This approximation is appropriate for systems that at high concentration have little or no alterations in the absorption spectra but have quite different luminescence properties. For such systems the essence of Förster's calculation is as follows: Let a donor molecule D and an acceptor molecule A be separated by a distance R. If molecule D is then excited while molecule A remains in the ground state, the probability that molecule D will transfer its electronic excitation energy to A will be proportional to interaction terms of the form $U_{\mathrm{AD}} = \langle \phi_{\mathrm{A}}^{*} \phi_{\mathrm{D}} | V_{\mathrm{DA}} | \phi_{\mathrm{D}}^{*} \phi_{\mathrm{A}} \rangle$ squared, where V_{DA} represents the interaction between the two states. Terms which enter into the resonance energy U are the transition-charge densities, which can be expanded into a point multipole series. The first term is the dipole-dipole term so that, ignoring higher order terms,

$$U = \frac{1}{n^2 R^3} \boldsymbol{m}_{\mathrm{D}} \cdot \boldsymbol{m}_{\mathrm{A}} - \frac{3}{R^5} (\boldsymbol{m}_{\mathrm{D}} \cdot \boldsymbol{R}) (\boldsymbol{m}_{\mathrm{a}} \cdot \boldsymbol{R}) \tag{7.11}$$

and

$$U^2 = \frac{\kappa^2 |\boldsymbol{m}_{\mathrm{D}}|^2 |\boldsymbol{m}_{\mathrm{A}}|^2}{n^4 R^6}, \tag{7.12}$$

where κ is a numerical factor depending on the direction of the dipole moments of the two molecules ($\kappa^2 = 2/3$ for random orientations). Integrating over all the initial states Förster arrives at an expression for the transfer rate k, which is related to the absorption and emission spectra

$$k = \frac{9\kappa^2 (\ln 10) c^4}{128\pi^5 n^4 N' \tau} \frac{1}{R^6} \int f_D(\nu) \varepsilon_A(\nu) \frac{d\nu}{\nu^4}, \tag{7.13}$$

where $\varepsilon_A(\nu)$ represents the molar decadic coefficient, $f_D(\nu)$ represents the fluorescence quantum spectrum normalized to unity on a frequency scale, τ represents the fluorescence lifetime, N' is the number of molecules per millimole (6.02×10^{20}), and n is the refractive index. The equation is often written in the following simplified form

$$k = \frac{1}{\tau} \left(\frac{R_0}{R} \right)^6, \tag{7.14}$$

where R_0, as calculated from the numerical factors above, represents the molecular separation at which the radiative losses are equal to the nonradiative losses due to the dipole-dipole interaction, and is determined by the overlap of the absorption and fluorescence spectra. *Stryer* and *Haugland* [7.26] have shown that the dependence of k with R goes as $1/R^6$ in their experiments with variable length "stick" molecules.

The expression for k must be integrated over all neighboring molecules [7.27] to obtain the time dependence of the excited-state population as measured in an experiment with picosecond pulses. The excited-state population decays nonexponentially as

$$N_D^* = N_D(t=0) \exp\left[-t/\tau - N_v (\pi/\tau)^{\frac{1}{2}} R_0^3 t^{\frac{1}{2}}\right], \tag{7.15}$$

where $N_D(t=0)$ is the excited-state population immediately after excitation and N_v is the number of acceptor molecules per unit volume. The physical basis for the nonexponential decay is that those molecules which are close to each other transfer their excitation energy rapidly because of the R^{-6} dependence of k, while those far away from each other transfer their energy much more slowly. Of course, the molecules simply lose their energy exponentially when they do not interact as at low concentrations. The expressions given by (7.13–15) are derived assuming several conditions are met: First, the concentration must be low enough that the spacing between molecules does not become comparable to their dimensions; when the electronic clouds overlap, for example, higher order multipole terms must be considered (see *Chang* [7.28]). Second, the incident light intensities must be low enough so that interactions between excited-state donors, excited-state acceptors, or between excited-state donors and excited-state acceptors are not possible (see *Fröhlich* and *Mahr*

[7.29]). The orientations of the molecules are assumed to be random and the molecules are allowed to rotate rapidly during the lifetime of the excited state —the case of frozen molecular orientations such as those might occur in a viscous liquid has been calculated by *Steinberg* [7.30]). Fourth, thermal equilibrium of the vibrational motion is assumed; thus at the point at which theory predicts a transfer time comparable to the molecular collision time, about 10^{-13} s, the theory breaks down. Fifth, the donors and acceptors are randomly distributed within a given volume; also, the theory is invalid if the interaction is so strong that energy is transferrred in times shorter than the vibrational period of the molecules. *Rehm* and *Eisenthal* [7.31] experimentally verified (7.15) (see Chap. 6).

Förster's formulae can also be applied to the case where the donor and acceptor molecules are of the same type. This case is particularly important in photosynthesis because of the high concentration of chlorophyll a in the cell. The overlap of the emission and absorption bands for this case, though not nearly as great as for two different dye molecules with nearly perfect spectral overlap, can nonetheless lead to the rapid transfer of electronic excitation with rates approaching 10^{12}/s at high concentrations. As an example consider the respective roles played by chlorophyll b and chlorophyll a in a plant. Because the overlap between the emission spectrum of chlorophyll b and the absorption spectrum of chlorophyll a is nearly perfect, excitation transfer from chlorophyll b to chlorophyll a occurs with a quantum efficiency near unity and consequently fluorescence from chlorophyll b is very weak *in vivo;* excitation transfer between chlorophyll a molecules also occurs rapidly until eventually the excitation is quenched at the reaction center. Unfortunately, neither lifetime nor quantum-efficiency measurements can reveal whether excitation transfer takes place between like molecules in a liquid; this inability is due to the fact that lifetime and quantum-efficiency measurements can in no way discriminate the new state from the old if an excited-state molecule transfers its energy to a like neighbor. Equation (7.15) will not account for the fluorescence decay curve when all of the molecules are the same. For such a case R_0 can be determined from depolarization data and compared against the value calculated from the absorption and fluorescence spectra. *Knox* [7.23, 24] has reviewed all the known measurements for chlorophyll a and has concluded that the average experimental value for R_0 is in the neighborhood of 70 Å, but that the difference between the smallest and largest measurements leads to an uncertainty of a factor of 50 in the transfer rate k.

To explain the picosecond fluorescence lifetimes obtained at high concentrations (Fig. 7.3), a different type of acceptor molecule must have been present in the solution because, otherwise, the lifetime could not change with concentration so dramatically. There are at least four types of acceptors that might quench the luminescence: 1) formation of dimers or higher oligomers in chlorophyll solutions at high concentrations; 2) an impurity molecule in the solution to which the excitation is transferred; 3) transfer to another chlorophyll molecule which is in an excited state – the excess energy is liberated as heat; and 4) the possibility that previous pulses in the train leave some chlorophyll molecules in the triplet state so that singlet-triplet fusion is the quenching mechanism. The second mechanism is unlikely because the main impurities in such a solution would most likely be the pheophytins which, however, are present in much too low concentrations and are much too similar to the chlorophylls to produce any lifetime shortening. The third mechanism, excited state-excited state interactions, has been observed in the context of excited *F*-center interactions in KI by *Fröhlich* and *Mahr* [7.29]. However, this mechanism is strongly dependent on intensity, and no such strong dependence is observed experimentally. The singlet-triplet interaction mechanism has been postulated by *Rahman* and *Knox* [7.32] to be quite effective in chlorophyll solutions. There does not appear to be sufficient intensity to obtain a triplet population that could produce such short lifetimes, but to rule out singlet-triplet quenching with certainty, the experiment must be repeated with single-pulse excitation.

The most likely mechanism for explaining the picosecond fluorescence lifetime in solutions is the formation of dimers and higher oligomers at high concentrations. Concentration quenching – which does not require a population of impurities, excited states, or triplets – is known to occur in most liquids and in the chlorophylls in particular [7.20]. It has now been fairly well established that chlorophyll aggregation occurs in certain chlorophyll solutions [7.8]. A dimer concentration of about 0.03 *M* for a nominal concentration of 0.1 *M* could lead to excitation-transfer times in the 10 ps range.

The functional aspect of this aggregation mechanism remains unclear, although the structure of chlorophyll has been known since the early part of the century. Chlorophyll in certain solutions tends to aggregate and to form different types of chlorophyll complexes. In polar-electron-donor solvents such as methanol, acetone, and pyridine, a strong absorption band occurs at about 665 nm, the exact location depending slightly on the solvent, whereas in dry nonpolar solvents a new, strong absorption band occurs at about 680 nm. This new band is strongly dependent on the concentration and has been interpreted as due to dimer and oligomer formation. The absorption band in nonpolar solvents resembles one in plants, and has led to the hypothesis that antenna chlorophyll molecules form one gigantic oligomer cluster for collecting light quanta [7.8].

In addition to the band at 680 nm, there is a weaker *in vivo* absorbance band in plants near 700 nm, which was discovered by *Kok* [7.33]. This band contains a contribution from about 1% of the chlorophyll in the plant, and *Katz* and coworkers [7.8] have postulated that the band is due to a second kind of dimer, one that contains two chlorophyll molecules bound together by a water molecule; hence, the name "special pair" is often given to this form of dimer. The addition of a small amount of water has a profound effect upon the spectrum of chlorophyll in solution. The magnesium atom in the center of the porphyrin ring seems to play an important role in the structure of both the ordinary and the special pair dimer. From the infrared spectra *Katz* et al. [7.8] have interpreted the ordinary dimer as being formed due to the interactions of the keto $C=O-Mg$ on one chlorophyll with another chlorophyll molecule. From proton magnetic resonance and carbon-13 magnetic resonance they have concluded that the chlorophyll-water adduct is connected by a single water molecule with the hydrogen atoms of the water molecule bound to the oxygen molecules in the "V-th ring" of the chlorophyll molecule, and the oxygen atoms of the water molecule bound to the magnesium atom of a second chlorophyll molecule. Furthermore, vapor-phase osmometry measurements by *Ballschmiter* et al. [7.34] indicate that in a 0.1 *M* solution of chlorophyll a in the solvent *n*-hexane, clusters of as large as twenty or more chlorophyll molecules can be found in the dry solution. At high concentrations the monomeric form becomes vanishingly small in hexane, and the spectrum becomes dominated by the band at 680 nm. The resemblance of *in vivo* absorption spectra with that of the chlorophyll oligomers has been taken as good experimental support for the view that the antenna pigment molecules are really oligomers, $(Chl_2)_n$. This is consistent with the "pebble mosaic" model of the PSU discussed in Subsection 7.1.3.

Another recent model by *Fong* [7.35, 36] based on quantum mechanical calculations leads to the suggestion that the reaction center may contain a C_2 symmetrical structure of chlorophyll molecules bound together by water molecules. The exciton interactions within such a $(Chl-H_2O)_2$ complex may then lead to a triplet state which is long-lived and therefore may act as a trap.

Beddard and *Porter* [7.37] have proposed that another form of dimer is responsible for the quenching. In polar solvents, where the formation of stable dimers is less likely, they propose that a pair of chlorophyll molecules whose separation is less than a critical distance form a trap. When one pair of molecules within this critical distance is excited, the pair interacts to form an excimer or other excited complex. They have estimated that at a concentration of 0.1 *M* a suitable high fraction of chlorophyll molecules would fall within a 10 Å distance to lead to strong quenching.

The formation of dimers and of higher oligomers in solution complicates the interpretation of lifetime measurements. The number of donor and acceptor species is a strong function of concentration, as are the types of acceptors. Maintaining water-free solutions is therefore a necessity. Nor are the fluorescence properties of oligomers well characterized. On the other hand, picosecond spectroscopy forms a viable alternative method for investigating both the formation of oligomers in solution and the energy transfer between monomers and oligomers. Besides lifetime measurements, the quantum efficiency can also be measured in a number of solvents as a function of concentration, or excitation intensity. This latter method is discussed in Subsection 7.1.3 and holds promise for studying oligomers because if more than one chlorophyll molecule is excited per oligomer, the quantum efficiency should drop due to exciton-exciton annihilation processes, thereby allowing an estimate of the size distribution of the oligomers.

More theoretical work must also be done to understand the interactions of chlorophyll molecules in solution. A most interesting case, that of multiple transfer between identical molecules before trapping, has not received much attention. A recent paper by *Craver* and *Knox* [7.38] treats the case where the molecules shares its excitation with both the nearest and next nearest neighbors and they find excellent agreement with experiment. In Förster's theory [7.21] only the nearest neighbor transfer is considered, while an extension by *Ore* [7.39] allows transfer to the next nearest neighbor but does not allow for the return of the excitation to the original donors from the next nearest neighbor. A comparison of the different excitation transfer theories may be found in *Knox* [7.40].

7.1.3 Exciton Migration

After an excited pigment molecule transfers its energy to Chl a, the excitation is, on the average, still quite some distance from a reaction center. If photosynthesis relied on a one-step Förster transfer from the excited Chl a to the reaction center, the transfer process would be inefficient because of the slow transfer times involved. However, if the excitation hops randomly between neighboring Chl a molecules, theory shows that the transfer becomes rapid and efficient. This random transfer process is believed to be the one to occur and is usually referred to as "exciton migration".

The term "exciton" used here is somewhat more restrictive than that commonly used in solid-state physics. A more general definition of a biological exciton, as used, e.g., by *Knox* [7.23, 24] and *Pearlstein* [7.41], refers to the specific case where the electron and a hole occupy the same site. However, under the coherent exciton concept a "site" might include any array of chemically similar molecules, with the excitation not necessarily confined to only one of these molecules [7.42]. This definition includes the conventional Frenkel or localized exciton as a limiting case. A justification for considering only the case of electron and hole occupying the same site within biological systems is offered by the fact that widely separated charges would usually interact too weakly to be treated as a single excitation.

We will not attempt to summarize the voluminous theoretical work on the biological exciton but refer the interested reader to an excellent review by *Knox* [7.24]. In this section we outline key theoretical approaches and certain necessary physical concepts that relate to pertinent picosecond experimental work.

In Subsection 7.1.1 we saw that photosynthesis requires the cooperation of two photosystems, each containing 40 to 400 chlorophyll molecules. To formulate an appropriate theory of energy migration we must know in some detail the topology of the chlorophyll matrix. When the concept of the PSU was first introduced, the concentration and optical properties of the chlorophyll within it were known, but little about its supposed structure. At that time *Franck* and *Teller* [7.43] asked if it was possible for an exciton to migrate in an efficient enough manner to ensure small radiative losses. A random-walk calculation along a one-dimensional chlorophyll chain showed that it simply took too long for the "walk". It is now known that the dimensionality of the calculation is very important, and for a given PSU size, two- and three-dimensional random walks are much faster and indeed consistent with the efficient energy migration process of photosynthesis. For example the number of steps $\langle n \rangle$ required for trapping in a PSU composed of N chlorophyll molecules goes as N^2 for a linear chain, $N \log N$ for a square lattice, and N for a cubic lattice [7.44–46].

Since a multidimensional model for the PSU is most appropriate, we will review our present knowledge of the chloroplast structure to gain some insight as to the possible structure. The chloroplast is a subcellular, usually saucer-shaped, organelle, 5 to 10 μm in diameter and 1 to 2 μm thick. By using phase-contrast microscopy one can discern dark bodies, called grana, with a diameter ranging from 3000 to 5000 Å. Nearly all the chlorophyll is found in the grana (concentration, 0.2 *M*). Electron microscopy reveals that the internal membranes of chloroplasts are flattened, hollow sacs, called thylakoids, arranged in stacks to form the grana. The thylakoids are embedded in the cytoplasm of the chloroplast, called stroma. With increased magnification, electron microscopy shows repetitive structures, called quantasomes, on the surfaces of the thylakoids. The quantasomes, measuring about 200 by 150 by 100 Å, are arranged like cobblestones and are, according to the "pebble-mosaic model" [7.47], to be identified with the photosynthetic unit. This model acquires its name from the similarity in appearance between mosaic floors found in the Far East and the arrangement of the quantasomes. The thylakoid membrane consists of these globular quantasomes on both sides of a thin central lipid layer. It is thought that the quantasomes on the inner and on the outer thylakoid surfaces are associated with PS II and PS I, respectively, with the lipid layer acting as both a support and a bridge between the two photosystems. A popular hypothesis due to *Thornber* [7.48, 49] envisions the quantasome (PSU) as a sack-like structure tightly filled with a variety of protein building blocks ("pebbles"). The most abundant of the protein pebbles consist of a chlorophyll or chlorophyll/pigment antenna protein, each containing about 10 chlorophyll or pigment molecules. The function of these specialized antenna proteins is to collect sunlight efficiently and to transfer the quanta to another protein structure called the reaction center. In addition to these pebbles there are the cytochromes, ferredoxin, and a variety of other specialized protein pebbles, all believed to be held together by a "lipid" glue.

Because the above picture of the PSU is fairly complicated, theoreticians have necessarily employed nonrigorous approaches or have assumed simple models. All usually consider "weak", or Förster, coupling between neighboring chlorophylls [7.22]. Although no doubt the excitons are initially coherent, interactions with vibrational modes after the first few picoseconds result in a loss of coherency. Thus, there is some justification in considering excitons to be localized [7.50], hopping or diffusing about. Most theoreticians have treated them accordingly, but it should be mentioned that picosecond techniques may allow us to observe the coherent properties of such excitons. The earliest approach involved solution of a diffusion equation of the form

$$\frac{\partial \rho(\boldsymbol{r},t)}{\partial t} = D\nabla^2 \rho(\boldsymbol{r},t) - \frac{1}{\tau}\rho(\boldsymbol{r},t) + S(\boldsymbol{r},t) - \sum_i k_i(\boldsymbol{r}-\boldsymbol{r}_i)\,\rho(\boldsymbol{r},t) \tag{7.16}$$

where D is a diffusion constant, ρ is an exciton probability density, τ is a characteristic lifetime in the absence of traps, $S(\boldsymbol{r},t)$ is a source term, and $\sum_i k_i(\boldsymbol{r}-\boldsymbol{r}_i)\,\rho(\boldsymbol{r},t)$ represents a trapping term. The diffusion coefficient can be related directly to the optical properties of the pigment molecules [7.21], and the equation can be solved in a straightforward manner subject to specific boundary conditions [7.51].

Another straightforward approach involves the use of the Stern-Volmer equation. A typical equation for excitons might have the form

$$\frac{d[B^*]}{dt} = CI - \frac{[B^*]}{\tau} - k[B^*]\,[X] \tag{7.17}$$

where $[B^*]$ is the exciton density, $[B]$ the chlorophyll concentration is $\gg [B^*]$, I is the excitation intensity, CI represents a volume rate of exciton production, $[X]$ is the density of traps, and k is a rate constant representing trapping of the excitons. With this approach *Vredenberg* and *Duysens* [7.52] derived an expression for the fluorescence yield

$$\phi_\mathrm{f} = \frac{\tau/\tau_0}{1+\tau k[X]}\,. \tag{7.18}$$

This approach verifies qualitatively many experimentally observed features. In the presence of long (>10 ns) "saturating" light flashes, the reaction centers bleach, so $[X]$ decreases, and from (7.18) we see that ϕ_f increases. Although the Stern-Volmer and the diffusion-equation approach are extremely useful for predicting features such as these, the many assumptions involved lead to rather imprecise lifetime and quantum-efficiency estimates.

A more precise approach involves following the time dependence of the excitation of the molecules over simple systems by generating a set of equations that accounts for all molecules, for all possible decay paths, and for all intermolecular transfers. This approach pioneered by *Bay* and *Pearlstein* [7.53] involves the use of a random-walk picture and extensive digital-computer use. In their calculations they assumed a model consisting of 300 chlorophyll a molecules, spaced 17 Å apart, arranged in a cubic lattice bounded by an oblate spheroid surface with dimensions similar to those observed in electron-microscope pictures of quantasomes. The localized exciton is imagined to hop from site to site in a random walk until it is irreversibly trapped at a reaction center located at the center of the lattice. Only the six nearest neighbor hops are considered. Because the Förster transfer time to any one neighbor is about 4 ps, a hopping time of 0.7 ps is inferred. By solving a set of master equations subject to these conditions they calculate an average trapping time of 86 ps ($\phi_f \sim 0.57\%$). Although their model is simple, *Bay* and *Pearlstein* [7.53] established a number of techniques for treating a Förster random-walk picture of exciton migration.

The random-walk approach has been extended by *Knox* [7.46] and involves an application to the Förster master equations of the inverse-matrix method of *Montroll* and *Shuler* [7.54]. This technique is discussed at length by *Knox* [7.24], *Pearlstein* [7.55], and *Hemenger* et al. [7.56]. A matrix G represents all decay and transfer processes and appears in the master equations

$$\frac{\partial \rho_i}{\partial t} = -\sum_{j=1}^{N} G_{ij}\rho_j + E_i \tag{7.19}$$

where ρ_i is the probability that molecule i is excited, N is the number of molecules in the PSU, and E_i is an external excitation rate of the molecule i. Off-diagonal matrix elements represent the rate of transfer from one molecule to another, whereas diagonal elements include monomolecular decay rates, $1/\tau_i$, and outgoing transfer rates. It can be shown that the fluorescence yield is given by

$$\phi_f = \sum_{j=1}^{N} \sum_{k=1}^{N} \frac{1}{\tau_{0j}} (G^{-1})_{jk} f_k \tag{7.20}$$

where G^{-1} is an inverse matrix, f_k is the fraction of light absorbed by molecule k, and τ_{0j} are natural radiative lifetimes.

By using this approach, *Knox* [7.23] calculated the trapping time for an infinite two-dimensional square array with one trap for every 25 sites. An interesting aspect of this calculation involves the effect of the specific orientation of the dipole moments of the chlorophyll a molecules at the lattice sites. Knox has shown that the trapping times can be altered by more than an order of magnitude by assuming different orientations. For example, parallel dipoles resulted in a 35 ps estimate, whereas a perpendicular interlocking arrangement (see Fig. 7.4) results in a trapping time of 830 ps. If the molecules are densely packed, monopole effects become important and also greatly affect the trapping time.

Another important consideration is the role of pigment heterogeneity. Recently, *Swenberg* et al. [7.57] have pointed out that the presence of other pigments can very substantially affect both the chlorophyll a fluorescence yield and the trapping time, because the pigments, having higher energy levels, effectively act as obstacles or "antitraps". Because migration takes place only between chlorophyll a molecules, the presence of the antitraps limits the possible migration routes and directs the energy transfer. This effect is related

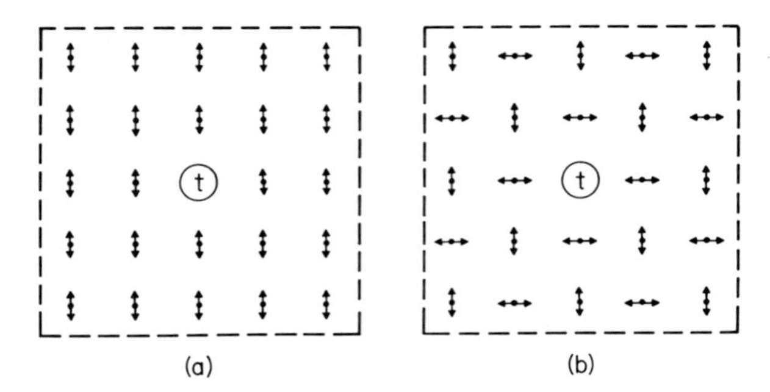

Fig. 7.4 a and b. Importance of molecular orientation in migration process is demonstrated in calculation by *Knox* [7.23] for the two-dimensional lattices depicted. Arrows show dipole orientations of chlorophyll molecules, and t represents a trap. Choosing an appropriate Förster coupling strength, the exciton migration time is 35 ps in (a), but 830 ps in (b). Orientations within photosynthetic unit are not known

to the well-known solid-state physics phenomenon of "percolation" (*Shante* and *Kirkpatrick* [7.58]). At a critical concentration of antitraps the lattice begins to "percolate", that is, to form directed rather than random pathways to the reaction center. Using (7.20) and assuming a 5×5 square array, *Swenberg* et al. [7.57] demonstrated that fluorescence yields may vary by over an order of magnitude with variations in the chlorophyll a to chlorophyll b ratio. They believe this explains the experimentally observed low fluorescence efficiency of PS I and the higher efficiency of PS II because most accessory pigments are present in the latter.

It is obvious from the previous discussion that a certain amount of experimental evidence must be fed back into the theoretical formulation. Experimental parameters, such as R_0, and topological properties of the PSU can, in principle, be determined from the time evolution of the intensity and polarization of fluorescence from *in vivo* pigment molecules. Prior to the development of picosecond techniques, measurements involved the use of nanosecond-duration excitation flashes (e.g., hydrogen lamps) and nanosecond resolution detectors. After deconvolution of the detector response and excitation pulse-shape functions, the resolution was necessarily limited to a few tenths of a nanosecond.

Another earlier approach involves the use of "phase fluorometry". This powerful technique is based upon the delay between absorption and emission associated with the finite lifetime of the fluorescence. If the excitation source is modulated at an angular frequency ω, then the phase delay δ of fluorescence with respect to excitation is given by $\tan\delta = \omega\tau$, where τ is the fluorescence decay time. This technique is described in detail in *Müller* et al. [7.59] and in *Borisov* and *Godik* [7.60]. Although this method does not necessarily involve the use of picosecond sources, it is nevertheless capable of inferring subnanosecond phenomena. *Borisov* and *Godik* [7.61], using a phase fluorometer modulated at a rate of 12.3×10^6 Hz, have demonstrated resolutions as short as 50 ps. The major disadvantage of this technique manifests itself when the observed lifetime has a nonexponential character, as in the case of the two-photosystem fluorescence of higher plants. For such a case data reduction can be tedious and in some cases ambiguous. However, before the advent of recently developed picosecond methods, phase fluorometry was the mainstay for short-lifetime measurements in photosynthesis and was applied to a number of *in vivo* samples. By using these methods *Borisov* and *Godik* [7.61] and *Borisov* and *Il'ina* [7.62] obtained most interesting results. Exciton lifetimes of 30 to 70 ps have been observed in a number of photosynthetic bacteria and in PS I of higher plants. These experiments clearly demonstrated for the first time that the exciton-migration processes could be effected on a picosecond time scale. Previous measurements, usually involving higher plants, measured the much longer lifetimes associated with PS II. For example, an intensity dependent lifetime measurement for PS II is described by *Müller* et al. [7.63].

To our knowledge, the earliest experiment identifiable with the use of modelocked laser techniques in biophysics was performed by *Merkelo* et al. [7.64]. The excitation source consisted of a modelocked He:Ne laser which emitted a continuous train of subnanosecond pulses. The repetitive nature of the excitation allowed the use of appropriate signal-averaging techniques and led to a resolution of 80 ps. Fluorescence from pigments *in vitro* and *in vivo* was examined with two techniques. The first consisted of a photomultiplier and sampling oscilloscope and provided a direct measurement, whereas the second involved a phase-delay measurement of the fluorescence with respect to the incident laser beam. *In vitro* studies of chlorophyll b and phycocyanin at low concentrations yielded fluorescence decay times of 3.87 ± 0.05 and 1.14 ± 0.01 ns, respectively, whereas *in vivo* chlorophyll a fluorescence from *Chlorella pyrenoidosa* decayed in 1.4 ± 0.05 ns.

A modified version of this approach, also by *Merkelo* et al. [7.65], is shown in Fig. 7.5. By passing modelocked pulses through a spinning half-wave plate the anisotropic emission pattern is effectively rotated, and a detector observes fluorescence with a polarization alternately parallel and perpendicular to the electric vector of the exciting light. In this manner, the depolarization time of luminescence could be determined by using a phase-delay technique. Phase differences as small as 10 μrad could be detected. A drawback of this method is the heavy reliance on analytical processing of the data. The time evolution of each component is nonexponential [7.65], and extensive analysis is required. By using this method *Merkelo* et al. [7.65] present fluorescence data for the depolarization time of chlorophyll b in ethanol as a function of temperature.

Much higher resolutions are possible with modelocked lasers in conjunction with such components as the optical gate of *Duguay* and *Hansen* [7.66]. This approach is employed by *Seibert* et al. [7.67] who use 1060 nm pulses from a Nd:glass laser for producing the birefringence in the gate, and 4 ps, 530 nm pulses to excite an *in vivo* escarole sample. By optically delaying

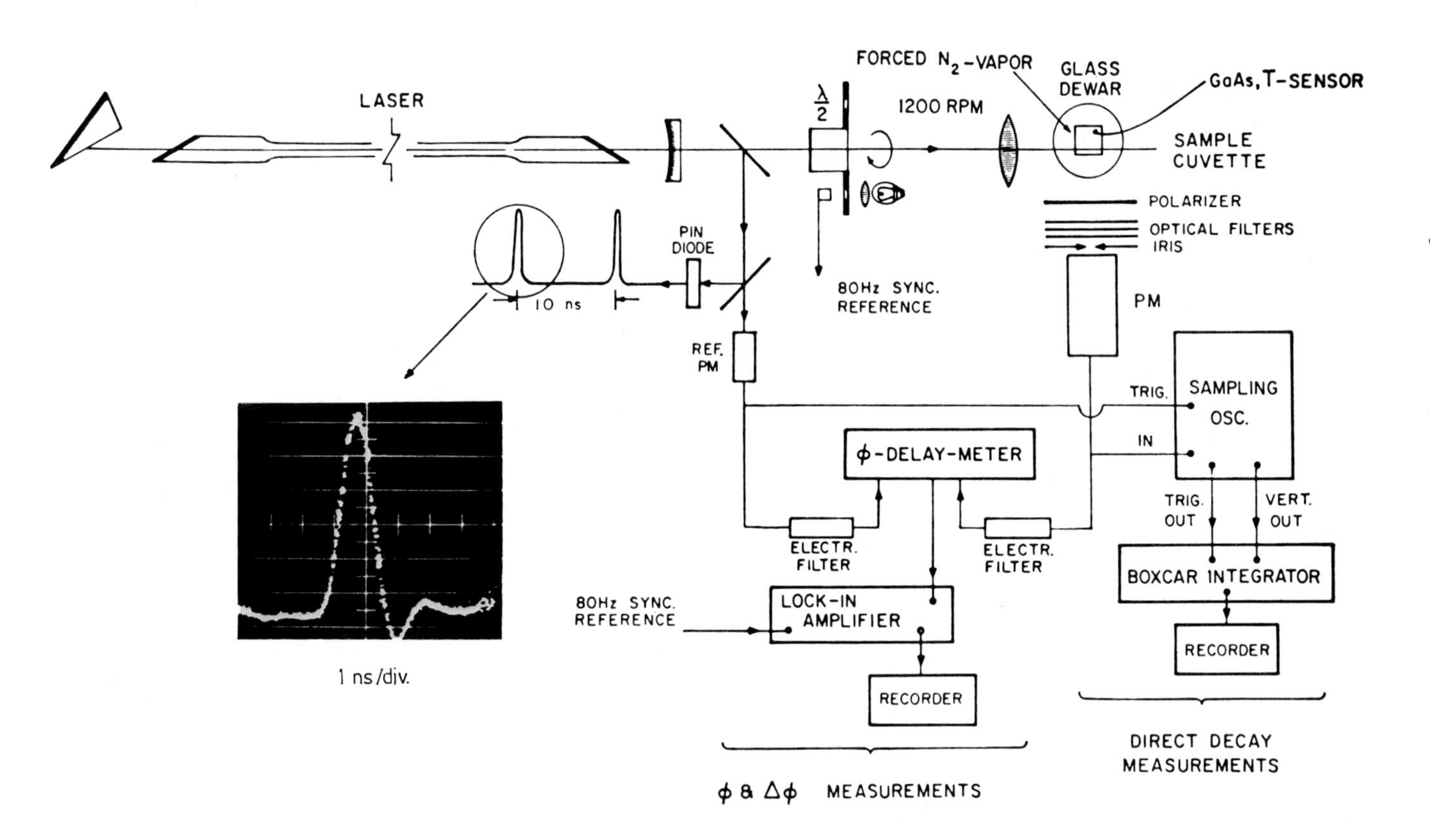

Fig. 7.5. Schematic of experimental apparatus of *Merkelo* et al. [7.65] for measuring depolarization time of luminescence. The repetitive nature of the modelocked He:Ne laser excitation source and signal-averaging techniques made possible time resolutions as short as 80 ps

the time of arrival of the gating pulses, Seibert et al. found that the chlorophyll a fluorescence decays in 320 ± 50 ps.

This study was repeated by *Seibert* and *Alfano* [7.68] who used an identical technique. Their data for isolated spinach chloroplast fluorescence at 685 nm show three prominent features: two maxima at 15 and 90 ps and an intervening dip at 50 ps. They believe these features originate from PS I (lifetime 10 ps) and PS II (lifetime 210 ps) and interactions between the two photosystems. *Yu* et al. [7.69] have examined particles rich in PS I and PS II and report lifetimes of 60 and 200 ps, respectively, somewhat consistent with results from whole chloroplasts.

An important consideration here and in a number of experiments, which will soon be described, are the very high excitation intensities required. The danger exists of multiply exciting an individual photosynthetic unit, resulting in exciton decay times that reflect these artificial conditions. Besides, when using a Duguay gate, a full modelocked pulse train is often used to enhance the signal-to-noise ratio. Earlier pulses in the train may create long-lived intermediaries, e.g., triplet states, or alter the initial condition of the traps and thereby affect the observed decay time. Quite aware of these possibilities, *Seibert* et al. [7.67] measured the quantum efficiency of fluorescence as a function of excitation intensity. They observed no variation with intensity and concluded they were operating at low enough intensities for reliable measurements. However, a recent result by *Campillo* et al. [7.70] suggests that the constant quantum efficiency and lifetime are caused by the use of a full train of high-power modelocked pulses. With single-pulse operation the quantum efficiency decreases with increasing intensity, consistent with an excited-state annihilation process. Many early results, such as those by *Seibert* and *Alfano* [7.68] and *Yu* et al. [7.69], as well as some later ones, will probably require reinterpretation due to the use of high-intensity pulse trains. Before discussing this point further, we shall describe other picosecond exciton-lifetime measurements.

The full power of picosecond technology has only recently begun to be brought to bear on the problem of *in vivo* energy migration with the use of ultrafast (1 to 10 ps resolution) streak cameras. This technique, described in detail in earlier chapters, provides a simpler experimental setup with increased sensitivity. Using this technique *Shapiro* et al. reported *in vivo* results [7.14, 15]. In these studies second-harmonic pulses derived from a modelocked Nd:glass laser excited samples of the algae *Chlorella p.* and *Anacystis n.* A microdensitometer trace of a photograph of the resultant streak representing the 690 nm chlorophyll a fluorescence from *Anacystis* is shown in Fig. 7.6. Somewhat surprisingly, the lifetimes were very short, that of *Anacystis n.* Being 80 ps. These lifetimes were much shorter than expected from species containing PS II and differed with earlier quantum-efficiency measurements. Furthermore, unpublished work by *Campillo* and *Shapiro* [7.71] indicated that numerous species from algae to higher plants all had lifetimes in the 40 to 100 ps range, a most unexpected result.

Independently, another group, *Paschenko* et al. [7.16, 17], began employing streak cameras in conjunction with modelocked ruby and Nd:glass laser pulse excitation. A densitometer trace showing one of their results is displayed in Fig. 7.7. More recently, a third group, *Beddard* et al. [7.72], repeated many of the measurements of *Shapiro* et al. [7.15] and verified the very short lifetimes. Unfortunately, in all these initial streak-camera studies the researchers used a modelocked train to excite the sample, usually observing fluorescence from a single pulse near the middle of the train.

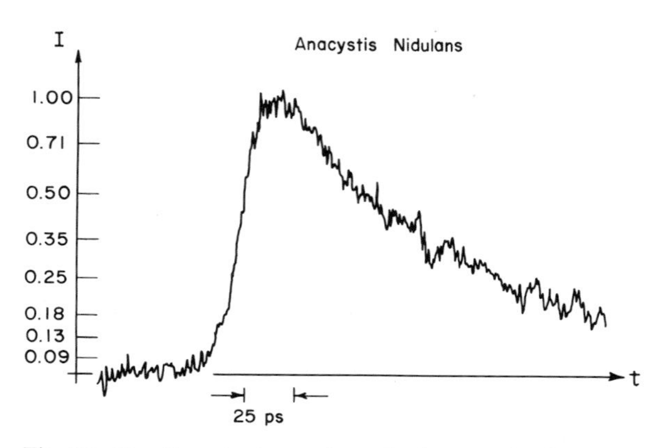

Fig. 7.6. Densitometer trace of streak photograph of chlorophyll *a* fluorescence from *Anacystis nidulans*. The very short observed lifetime (80 ps) is consistent with singlet-triplet annihilative processes; see text (from *Shapiro* et al. [7.15])

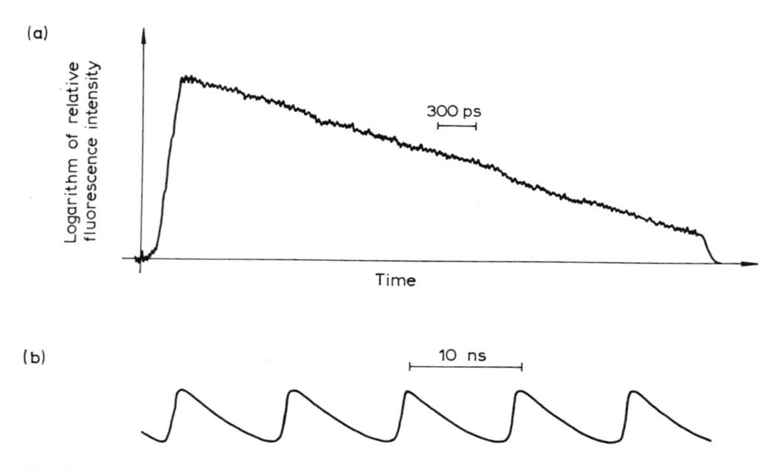

Fig. 7.7. (a) Densitometer trace of streak photograph of fluorescence from chlorophyll *a* in ethanol solution. (b) Oscilloscope display of fast diode output showing same result as in (a) from *Paschenko* et al. [7.17])

The observation of very short and comparable lifetimes of widely varying species led researchers to consider the likelihood of nonlinear optical effects. Such a possibility would be expected in *in vivo* experiments because the PSU antenna system acts both as an efficient collector of light and as a funnel to bring the excitation to the reaction center. This same funneling action might easily bring two or more excitons within close proximity of each other where they might interact. The strongest evidence against these effects, however, was the linear proportionality of fluorescence yield with intensity observed by *Seibert* et al. [7.67]. Two other groups, *Campillo* and *Shapiro* [7.71] and *Beddard* et al. [7.72], further verified that the fluorescence caused by one of the pulses of a modelocked train had a decay time that was roughly constant over a decade variation in excitation intensity.

Nevertheless, a reinvestigation was prompted by the recent results of *Mauzerall* [7.73]. He showed that the quantum efficiency of fluorescence emission from *Chlorella decreases* at higher pumping intensities for 7 ns excitation pulses, and he interpreted the decrease in terms of excited state interactions within multitrap domains. The excitation pulses in his experiment were three orders of magnitude longer than those used typically in picosecond lifetime studies, and it was not clear that this effect was operable on the picosecond time scale if, for instance, it were due to triplet-exciton collisions.

It is also known that under certain long-flash-saturating light conditions the quantum efficiency increases with intensity, a point discussed earlier. Consequently, *Campillo* et al. [7.70] repeated this measurement with a single pulse selected from a modelocked Nd:YAG laser by means of a longitudinal-mode *KD*P* Pockels cell developed by *Hyer* et al. [7.74] at intensities comparable to previous picosecond measurements. Results (see Fig. 7.8), show a decrease of quantum efficiency with intensity, in agreement with the results of

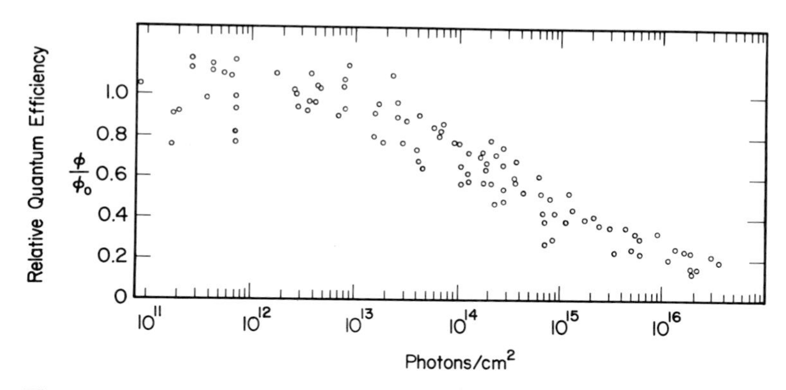

Fig. 7.8. Single pulse excitation of *Chlorella* shows decrease in quantum efficiency with energy density due to "multiple hits" of the photosynthetic unit and the resultant exciton annihilative processes (from *Campillo* et al. [7.70])

Mauzerall [7.73]. As can be seen in Fig. 7.8 the quantum efficiency is relatively constant at low energy densities, but begins to decrease at an energy density of about 10^{13} photons/cm^2.

The experimental quantum-efficiency results are consistent with a non-linear optical effect caused by interactions of the excited-state population. This effect would occur when more than one photon excited a photosynthetic unit leading to the simultaneous presence of more than one excited-state chlorophyll molecule per photosynthetic unit. Then excited-state singlets could also interact with one another leading to singlet-singlet annihilation [7.75, 76], the excess energy possibly disappearing in the form of heat or in the formation of triplets. Because the fluorescence intensity is proportional to the population of excited singlet chlorophyll molecules, the diminution of singlets by such processes would lead to a shortening of the lifetime and a decrease in quantum efficiency. Such interactions have been noted previously by *Fröhlich* and *Mahr* [7.29], in connection with exciton migration in crystals, and by *Mauzerall* [7.73]. In *Chlorella* the total cross section of absorption is thought to be of the order of 100 Å^2 at 530 nm, so that at an energy density of about 10^{13} photons/cm^2 the quantum efficiency should begin to decrease due to "multiple hits" of the photosynthetic units.

It is remarkable that the results are so similar even though the excitation pulse of *Campillo* et al. [7.70] was 350 times shorter than that of *Mauzerall* [7.73]. This similarity suggests that the curve for the quantum efficiency as a function of energy density is time independent for times shorter than 7 ns. Apparently, triplets do not play a major role in single-pulse experiments because several nanoseconds are usually required to build a significant population, and this would be inconsistent with the picosecond results. Interaction of singlets with charge states in the reaction center can be ruled out by recent experiments which show similar effects in reaction centerless chromatophores [7.77]. Thus, the singlet-singlet annihilation processes appear to offer a plausible explanation for the decrease in quantum efficiency.

Such an explanation is consistent with the observed functional form of the quantum efficiency with intensity according to the theoretical considerations of *Mauzerall* [7.73], *Goad* and *Gutschick* [7.78], and *Swenberg* et al. [7.79]. These researchers have used a Stern-Volmer equation approach with a term $-\gamma[B^*]^2$ added to reflect the singlet exciton annihilative processes and have averaged over independent PSU's with an assumed Poisson distribution of multiple excitations to show that a slow decrease in quantum efficiency is to be expected. It is important to note here that the manner in which the excitons interact reflects strongly the network of chlorophyll interactions within the PSU. Thus, measurements such as shown in Figs. 7.8 and 7.10 will eventually shed new insight into the topology of the PSU. Initially, *Mauzerall* [7.73] has attempted to correlate the shape of the curve shown in Fig. 7.8 with the existence of multiple trap domains. *Goad* and *Gutschick* [7.78] have attempted to theoretically determine an appropriate distribution of PSU sizes while *Swenberg* et al. [7.79] obtain an adequate fit with a "lake" model.

Recently, *Campillo* et al. [7.80] have demonstrated directly in the time domain the existence of these processes. Figure 7.9 shows the observed variation in lifetime with intensity for single pulse excitation in *Chlorella*. The lifetime was observed to vary by more than an order of magnitude with a decade change in intensity. This is consistent with a singlet-singlet annihilative rate constant, γ, of about 10^{-8} cm^3/s. This established, without doubt, the existence of these processes at intensities exceeding 10^{14} photons/cm^2.

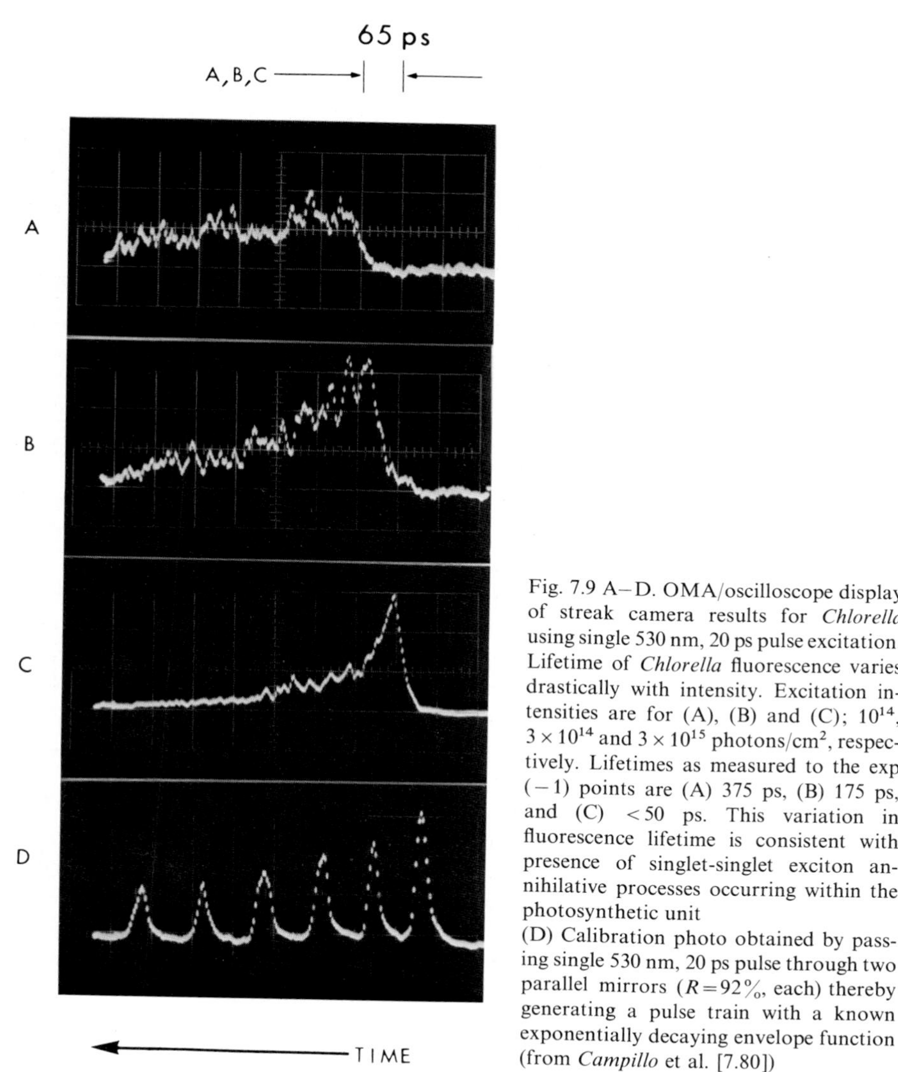

Fig. 7.9 A–D. OMA/oscilloscope display of streak camera results for *Chlorella* using single 530 nm, 20 ps pulse excitation. Lifetime of *Chlorella* fluorescence varies drastically with intensity. Excitation intensities are for (A), (B) and (C); 10^{14}, 3×10^{14} and 3×10^{15} photons/cm^2, respectively. Lifetimes as measured to the exp (-1) points are (A) 375 ps, (B) 175 ps, and (C) <50 ps. This variation in fluorescence lifetime is consistent with presence of singlet-singlet exciton annihilative processes occurring within the photosynthetic unit
(D) Calibration photo obtained by passing single 530 nm, 20 ps pulse through two parallel mirrors ($R=92\%$, each) thereby generating a pulse train with a known exponentially decaying envelope function (from *Campillo* et al. [7.80])

A Stern-Volmer equation describing singlet-singlet annihilation is simply written, $dn_s/dt = -kn_s - \gamma_{ss}n_s^2$, valid for $t > 20$ ps. Here $n_s = [B^*]$ is the singlet excited state population and is proportional to the observed fluorescence, k is a characteristic decay rate constant related to trapping, intersystem crossing, and radiative losses ($k = 1/\tau$), and γ_{ss} is a bimolecular rate constant for the singlet-singlet annihilation process. The solution for n_s is nonexponential [7.79]; however, the initial decay rate can be estimated from the following Hartree approximation [7.81]: $dn_s/dt \approx -kn_s - (\gamma_{ss}\bar{n}_s)n_s$. Thus, the initial decay rate observed using a streak camera technique is $k_{obs} \approx k + \gamma_{ss}\bar{n}_s$ where $\bar{n}_s$ is the initial density of excited states. To estimate γ_{ss} one must have specific information on the PSU geometry, e.g., dimensionality, etc. It is convenient to define $\Gamma \equiv \gamma_{ss}\bar{n}_s/I$, where I is the excitation intensity and Γ is independent of PSU geometry. With this simple formulation *Campillo* et al. [7.80] obtain a reasonable fit to their data. They find that $\Gamma \simeq 1.5 \times 10^{-5}$ cm^2/photons-s and $k = 1.25 \times 10^9$ s^{-1}, and hence $\tau \approx 800$ ps for *C. pyrenoidosa*. An alternative approach in estimating τ involves using both streak camera data and the observed quantum efficiencies of [7.70]. Assuming the relationship $\tau \approx \tau(\mathrm{I})/[\phi(\mathrm{I})/\phi_0]$ is valid at intensities $< 10^{14}$ photons/cm^2, then τ is estimated to be about 535 ps. The Hartree approximation probably yields an estimate that is high due to its neglect of the presence of a distribution of multiple excitations within the individual independent PSUs. The quantum efficiency approach is suspected to give a low estimate because of the possibility of background radiation. Nevertheless, an average of these two approaches yields $\tau \approx 650 \pm 150$ ps. This value agrees well with earlier low intensity measurements and those estimated using quantum efficiencies, but is much longer than previous picosecond studies.

The very long lifetime observed in *C. pyrenoidosa* may be indicative of pigment heterogeneity which limits the excitation transfer paths to the reaction center [7.57]. Heterogeneity may also cause the observed slow decrease of the quantum efficiency with intensity by effectively dividing the photosynthetic unit into regions of varying size with the consequence that the onset for exciton-exciton annihilation processes differs for these regions. Recent work by *Breton* and *Geacintov* [7.82] attempted to verify this concept. They have used multiple picosecond pulse and microsecond excitation to show that the quantum efficiency, as a function of pulse intensity, begins to decrease at lower intensities for PS I than for PS II. Note that since PS I is smaller than PS II, one would reach the opposite conclusion without the heterogeneity consideration. PS I being more homogeneous has a larger bimolecular rate constant γ_{ss}. However, with single picosecond pulse excitation they observe that PS I and PS II have identical quenching curves, indicating more complexity than just heterogeneity.

All previous fluorescence lifetime determinations made by picosecond pulse excitation on *in vivo* samples described earlier used energy densities that were typically in the range 0.1 to 10 mJ/cm^2 where singlet-singlet annihilation must be occurring. For example, *Beddard* et al. [7.72] use $\mathrm{I} > 10^{16}$ photons/cm^2;

Yu et al. [7.69] and *Seibert* and *Alfano* [7.68] use $I > 6 \times 10^{14}$ photons/cm^2. Thus, their data must also reflect these processes. Furthermore, in these studies the sample has been excited by pulse trains of at least 100 pulses. The fact that the observed lifetimes are constant with intensity under these conditions suggests another complicating process.

Campillo et al. [7.80] propose that the observed lifetime under multiple pulse excitation is determined by singlet-triplet fusion processes. Because triplets, which evolve from singlets by intersystem crossing, live longer [7.83] ($\tau_T = 30$ ns *in vivo*) than the time interval between pulses in the train (e.g., $\Delta t = 6.7$ ns [7.72]), the triplet population would increase from pulse to pulse until a steady-state value N_{Tss} is established. The triplet-triplet fusion rate [7.84] is thought to be less than 10^{-11} cm^3/s so that the steady-state triplet population is limited by singlet-singlet and singlet-triplet interactions at high intensities. Triplets especially are very effective quenchers of subsequently generated singlet excitons. For example, with as few as one triplet state per reaction center, N_{Tss} will be about 10^{18}/cm^3. *Rahman* and *Knox* [7.31] have estimated the singlet-triplet fusion rate ($S_1 + T_1 \rightarrow T_2 + S_0$, $T_2 \rightarrow T_1 +$ heat) to be 6.2×10^{-9} cm^3/s for chlorophyll a in *vivo*. On this basis, it is plausible to expect singlets to be annihilated by triplet excitons in roughly 60 ps. Initially, the decay rate would be somewhat faster due to singlet-singlet annihilative processes, but the singlet-triplet processes should eventually dominate.

At high intensities N_{Tss} is limited through a combination of singlet-singlet and singlet-triplet annihilation processes, eventually becoming independent of I. In this "saturated" regime, an appropriate Stern-Volmer equation for n_s is simply expressed as $dn_s/dt = -\gamma_{ST} N_{Tss} n_s$ and has a nearly *exponential* decay rate $\gamma_{ST} N_{Tss}$ *independent of I*. This picture is remarkably similar to Fig. 7.6 and the observations of [7.15, 72], where the singlet state decay constant $\gamma_{ST} N_{Tss}$ is observed to be in the range 10^{10} to 3×10^{10} s^{-1} with pulse train excitation. Certainly these annihilative processes occur at these intensities, and singlet-triplet processes offer an alternative interpretation for many previous experimental observations.

Although this nonlinear effect has apparently led several research groups astray, its occurrence is, in any event, fortuitous. As a result, a powerful new technique to probe exciton interactions within the PSU is provided. Among the many parameters measurable with this new tool are the PSU total absorbance cross section and the number of chlorophyll a molecules participating in the exciton migration process. Further information on pigment heterogeneity and topological features of the PSU can also be deduced.

Figure 7.10 shows quantum efficiency data as a function of intensity for 920 nm bacteriochlorophyll fluorescence for three mutants of the photosynthetic bacterium, *Rhodopseudomonas spheroides,* obtained by *Campillo* et al. [7.77]. The curves of the species with reaction centers are shifted from the PM-8 curve towards the higher intensities because the presence of reaction center traps tends to reduce the effective exciton density within the PSU. PM-8 is similar to the Ga mutant, but *lacks* reaction centers.

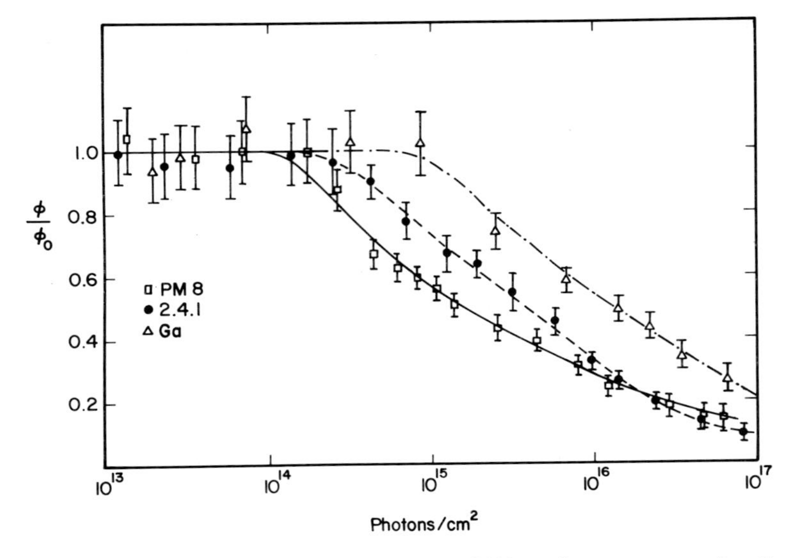

Fig. 7.10. Measurement of quantum efficiency of 920 nm fluorescence as a function of excitation intensity for single 20 ps, 530 nm pulse excitation of chromatophores of mutants of *R. spheroides*. Differences in curves reflect topology of photosynthetic unit; see text (from *Campillo* et al. [7.77])

It is clear from the above considerations that single-pulse excitation and lower intensities are necessary to obtain correct lifetime estimates. Sensitivity can be much improved with the use of a silicon-vidicon/optical-multichannel-analyzer arrangement coupled to a streak camera. For example, an Electro-photonics S-20 streak camera with modified $f/1$ input optics allows recording of fluorescence from *Chlorella* at a streak rate of 500 ps/cm (10 ps resolution) with a single $1-\mu$J excitation pulse at 10^{14} photons/cm^2. Although this intensity is still not low enough in the *Chlorella* system (see Fig. 7.8) it is, however, appropriate for any system whose $\phi/\phi_0\, vs\, I$ curve has been obtained and found to break downward at intensities greater than 10^{14} photons/cm^2, as is the case of the Ga, 2.4.1 and PM-8 mutants of *Rhodopseudomonas spheroides* shown in Fig. 7.10. The lifetimes of these mutants have been found by *Campillo* et al. [7.77] to be 100 ± 20, $\simeq 100$, and 1100 ± 300 ps, respectively. The R-26 carotenoidless mutant has a lifetime of 300 ps. Note the long lifetime in PM-8 in comparison to the mutants which possess reaction centers.

In many photosynthetic systems high resolutions are not required. In these cases the streak rate can be decreased and a much lower excitation intensity employed. It is indeed likely that improvements in streak cameras and alternative techniques will increase the sensitivity. The use of a continuous train of subpicosecond dye-laser pulses, as described by *Ippen* and *Shank* in an earlier chapter is most interesting and has a number of advantages. Clearly, the very low intensities and short durations of the pulses are among them.

The creation of long-lived intermediate states from previous pulses is avoided through the use of fast-flowing sample cells or the use of a low repetition rate cavity dumping scheme as used by *Shank* et al. (see Subsect. 7.2.4).

7.1.4 Reaction Center Oxidation

We have seen in previous sections how quanta may be transported from antenna chlorophyll molecules to a specialized chlorophyll complex called the reaction center. There the primary photochemical reaction takes place and consists of the transfer of an electron from a chlorophyll complex, P, to an acceptor, X. In the following we shall discuss experiments performed on reaction centers of a photosynthetic bacterium, *Rhodopseudomonas spheroides*. Bacteria are often used because of their relatively simple structure and because of the ease with which reaction-center complexes can be removed with detergents from chromatophores. Also, they are easily grown and quite reproducible in many of their properties; this is often not true for green plants. In particular, a carotenoidless mutant strain, R26, is often employed. The absence of carotenes and of their associated absorption bands greatly simplifies interpretation of the data. It is believed that in this species the reaction-center protein consists of four bacteriochlorophyll molecules, of two bacteriopheophytins, of a ubiquinone, and of a nonheme iron moiety [7.85–87].

Figure 7.11 shows what is believed to be the sequence of transitions required to accomplish the primary photochemical reaction. This figure is based upon a series of experiments performed by two groups that used picosecond tech-

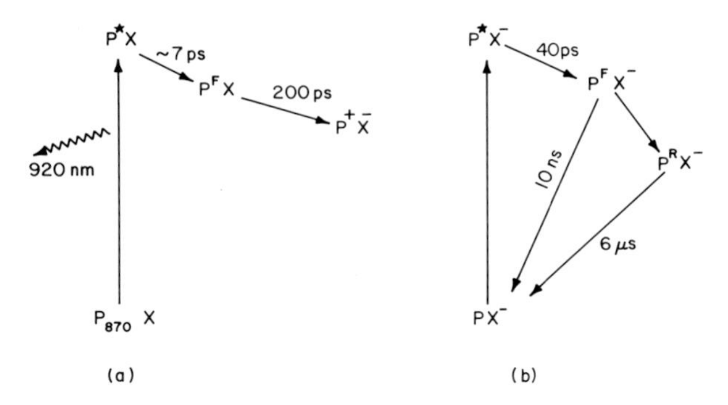

Fig. 7.11 a and b. Sequence of molecular transitions in reaction centers of *R. spheroides* as determined by picosecond experiments. (a) $P_{870}X$ is the ground state of the reaction center complex. This complex receives energy via a Förster energy migration process and P*X most likely represents the first excited singlet state. Fluorescence from P*X at 920 nm due to transitions back to the ground state is weak due to the very fast transfer (<7 ps) to the intermediate state P^FX. After another 200 ps, an electron is transferred to the primary electron acceptor, X. The electrochemical potential energy produced by this charge transfer is the driving potential by which all further chemical and physical changes in photosynthesis proceed. (b) Shows sequence of transitions when X is chemically reduced, thereby blocking the electron transfer from P^F to X. P^F is observed to relax to the ground state in 10 ns and to a triplet state P^R

niques, to be described in the following paragraphs. The ground state of the bacteriochlorophyll reaction center complex, shown in Fig. 7.11, is called P_{870} because of the prominent absorption band at 870 nm. This band overlaps strongly with the fluorescence band of the bacteriochlorophyll antenna molecules and so an efficient nonradiative transfer takes place. P* is thought to be the first excited state which quickly relaxes (<10 ps) to form the intermediary P^F. After another 200 ps, tell-tale signs clearly indicate that an electron transfer has occurred resulting in the formation of P^+X^-. The electrochemical potential energy produced by this charge transfer is the driving potential by which all further chemical and physical changes in photosynthesis proceed. Unfortunately, the exact nature of these states and of X is still an unsettled matter. There are a number of possibilities for the molecular species whose identity is P^+X^-: At present, the most likely assignment appears to be $(\mathrm{BChl\ BChl})^+\ (\mathrm{QFe})^-$, where an electron is transferred from a bacteriochlorophyll complex to a ubiquinone-iron complex.

Note that whatever the process an intermediate state P^F is apparently present. Its existence was suspected because of an anomaly between the photochemical and fluorescence quantum yields. Recently nanosecond excitation experiments by *Parson* et al. [7.88] performed with the primary processes blocked by chemical means have added evidence for the existence of an intermediate. By using 20 ns ruby laser flashes, the presence of two transient states was detected from absorbance changes: P^F, with a lifetime of about 10 ns, and P^R, with a lifetime of 6 μs (see Fig. 7.11b). The strongest evidence for the existence of the intermediate state P^F comes from the direct picosecond measurements of absorbance and spectral changes by *Kaufmann* et al. [7.89] and by *Rockley* et al. [7.90].

The first picosecond experiments on R26 reaction centers were performed by *Netzel* et al. [7.91]. By using an optical probe technique and a setup similar to that shown in Fig. 7.12, they studied absorption changes in the 864 nm band and found that the band, peculiar to this type of reaction center, bleaches within 10 ps of excitation with a 530 nm pulse. This interesting result clearly shows that a primary step in photosynthesis had been observed. However, it is not clear whether this observation represents anything but direct evidence of the very strong interactions between the bacteriopheophytins, originally excited by the 530 nm pulses, and the bacteriochlorophyll (P_{870}). It is suspected that the 864 nm band would bleach with the formation of P*.

In principle, by extending the studies of *Netzel* et al. [7.91] over a wider spectral range, it should be possible to deduce the sequence of events in the reaction centers. What, then, are some of the spectral characteristics of R26 reaction centers? The strongest absorption bands occur at 530, 600, 760, 800, and 865 nm; those at 600, 800, and 865 nm are thought to be due to bacteriochlorophyll absorption, whereas those at 530 and 760 nm may belong to bacteriopheophytin. Chemical oxidation of the reaction-center-bacteriochlorophyll complex is known to bleach the bands at 600 and 865 nm and to cause a small blue shift of the 800 nm band. It is to be expected that picosecond

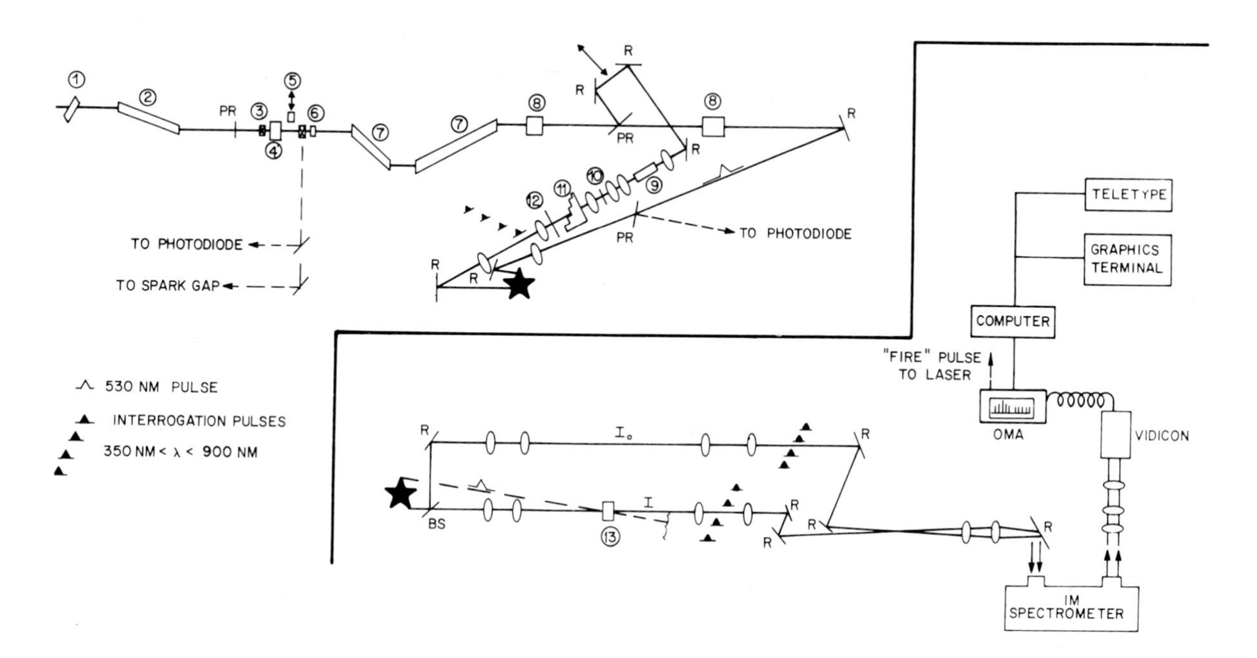

Fig. 7.12. Schematic of experimental apparatus of *Kaufmann* et al. [7.89]. Pulse from laser (2) is selected and amplified. A 530 nm pulse generated by second-harmonic generation (8) traverses one path and excites the sample at (13). A second pulse reflected at PR position generates a continuum at liquid cell at (9). The continuum passes through an echelon (11) forming a sequence of pulses which probe the absorption of the sample at set delay times. Spectrum is detected by spectrometer and compared against set of reference pulses I_0 which bypass the sample

optical excitation should also cause these changes. By using picosecond techniques, *Kaufmann* et al. [7.89] and *Rockley* et al. [7.90] found independently that these changes occur very rapidly. There are apparently no obvious optical changes that would correspond to the reduction of the primary electron acceptor, X.

The experimental arrangements of these two groups are quite similar, utilizing an optical probe technique (see Chap. 3). Fig. 7.12 shows a schematic of the setup of *Kaufmann* et al. [7.89]. A single 8 ps pulse is selected from the pulse train of a modelocked Nd:glass oscillator, is further amplified, and is split into two beams. The first beam, after frequency doubling in KDP, is used to excite the sample via the bacteriopheophytin absorption band at 530 nm, whereas the second beam generates superbroadening [7.92–95] in a liquid cell by self-phase modulation processes [7.96]. The resulting picosecond burst of white light [7.92–95] passes through an echelon to form a series of interrogation pulses separated from each other by known delays. These interrogation pulses are themselves split into two beams, I and I_0. The first pulse of the I-echelon train is coincident with the excitation pulse upon the sample so that the I-pulses monitor the absorbance changes in the sample, whereas the I_0-beam bypasses the sample and provides a reference. Both sets of pulses passed through a spectrometer and were imaged separately onto a silicon surface in such a way that a one-to-one correlation existed between the spatial position of each pulse and the time delay after excitation. The dual-beam arrangement permitted accurate adjustment for shot-to-shot variations of both the laser and the continuum-probing pulse. An SSRI optical multichannel analyzer was used to store the resulting signals from the vidicon and to display the resultant intensity patterns on an oscilloscope. Optical density changes as small as 0.03 could be measured.

The experimental setup of *Rockley* et al. [7.90] differed from that of *Kaufmann* et al. [7.89] in not using an echelon, the data being accumulated as a function of time delay from a single interrogation pulse on a shot-by-shot basis. Variations in the data between shots could be minimized by using the following technique. The interrogation beam was allowed to overfill the sample, whereas the excitation pulse illuminated only a band near the center of the sample. The interrogation pulse thus passed through both the center of the sample and the unexcited regions above and below, allowing a measurement of the absorbance differences caused by the excitation pulse. After traversing the sample, the interrogation pulse passed through a spectrograph and onto a silicon vidicon coupled to an OMA.

As the results of both groups are quite similar, we will outline here only those of *Rockley* et al. [7.90]. Upon excitation with 530 nm pulses the absorption spectrum of the P^F state appears immediately, within 10 ps. After several hundred picoseconds, the state decays, and the spectrum of the oxidized species P^+ becomes evident. These phenomena are shown in Figs. 7.13 through 7.15. Fig. 7.13 shows the absorption changes from 450 to 900 nm at 20 and 240 ps after picosecond pulse excitation, while Fig. 7.14A shows the absorption spectrum accompanying the formation of P^F. This spectrum, which can be compared with that in Fig. 7.13, was obtained by chemically blocking the final electron-transfer reaction as in Fig. 11 B and by excitation with 20 ns pulses from a ruby laser. The spectrum is very similar to that of Fig. 7.13 which is obtained 20 ps after excitation over the regions of the two figures which overlap, so the initial state must be P^F. The spectrum measured 240 ps after excitation has all the characteristics expected of P^+X^-. The decay rate of P^F, 246 ps (see Fig. 7.15), is the same as the rate of buildup of the oxidized state P^+. This fact would support the contention that P^+ forms directly from P^F. Thus P^F is an intermediate in the transformation of PX to P^+X^-.

Both groups verified that the bleaching of the 864 nm band occurs rapidly, as originally observed by *Netzel* et al. [7.91] *Rockley* et al. [7.90] also found that a bleaching of the 800 nm band occurs rapidly and decays prior to the conversion of P^FX to P^+X^-. They suggested that this represents another early step in the primary photochemical reaction. This possibility merits further investigation. One would initially expect that P^F might be simply the lowest

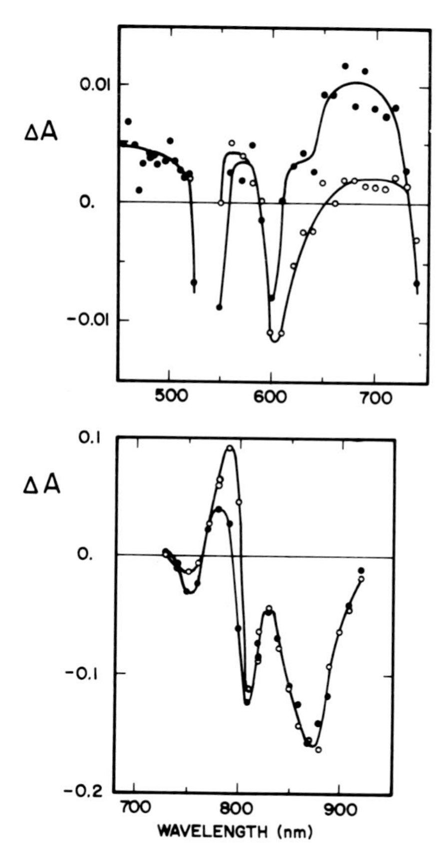

Fig. 7.13. Absorbance changes in *R. spheroides* reaction centers obtained 20 ps (●) and 240 ps (○) after picosecond pulse excitation. Figure shows spectrum of P^F converting to P^+, the radical cation. P^+ formation indicated by bleaching of absorption bands at 600 and 870 nm and blue shift of band at 800 nm (from *Rockley* et al. [7.90])

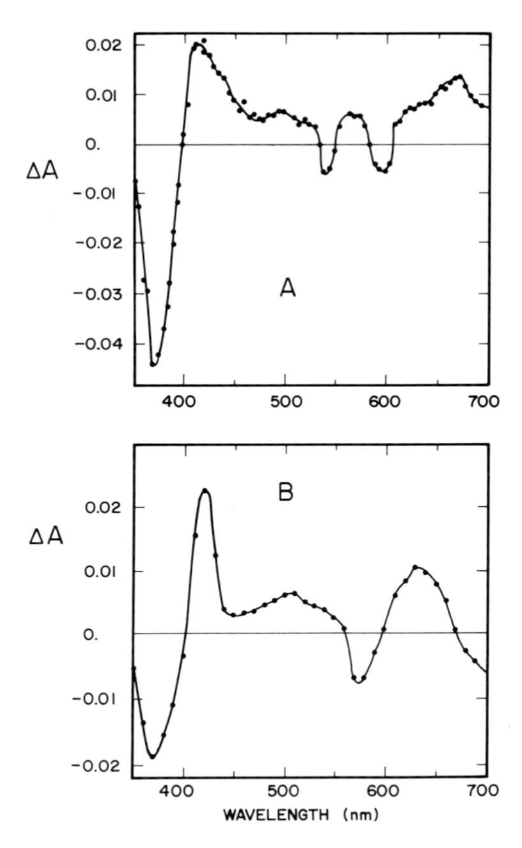

Fig. 7.14. (A) Absorbance changes accompanying the formation of the state P^F obtained by exciting chemically reduced reaction centers (i.e., $P_{870}X^-$) with 20 ns flashes. Spectrum closely matches that obtained 20 ps after excitation shown in Fig. 7.13 over region of overlap. Initial absorption changes observed in the picosecond experiments are therefore interpreted as due to the transient formation of the P^F state. (B) Calculation showing expected absorbance changes accompanying the one electron oxidation of bacteriochlorophyll to the cationic free radical and the reduction to the anionic free radical. Calculation predicts many features found in the picosecond experiments. Bleaching at 600 nm and new band at 420 nm are due mainly to formation of cation (Figure from *Rockley* et al. [7.90])

excited singlet, P*. However, measurements of the fluorescence yield of the 920 nm radiation emanating from P* result in a lifetime estimate of 20 to 40 ps with X in the reduced state, and of only 7 to 10 ps when X is not reduced. The lifetimes of P^F are 10 ns and 200 ps under these conditions. The blue shift of the 800 nm band accompanying the formation of P^F suggests electron transfer and may involve the migration of an electron from one of the four bacteriochlorophyll molecules to another. To test this possibility *Rockley* et al. [7.90] performed the calculation shown in Fig. 7.14 B. This plot is obtained by summing absorbance changes accompanying the one-electron oxidation of bacteriochlorophyll to the cationic free radical in CH_2Cl_2 and those accompanying the one-electron reduction to the anionic free radical in dimethyl formamide. This figure demonstrates than an anion-cation biradical could account for the spectral properties of P^F. As can be seen, the calculation predicts many of the observed spectral changes. However, it does not account for the bleaching of the absorption bands at 530 and 760 nm.

Recently, *Dutton* et al. [7.97] have extended the investigation of R26 reaction centers to the 1250 nm absorption band. This band is not apparent

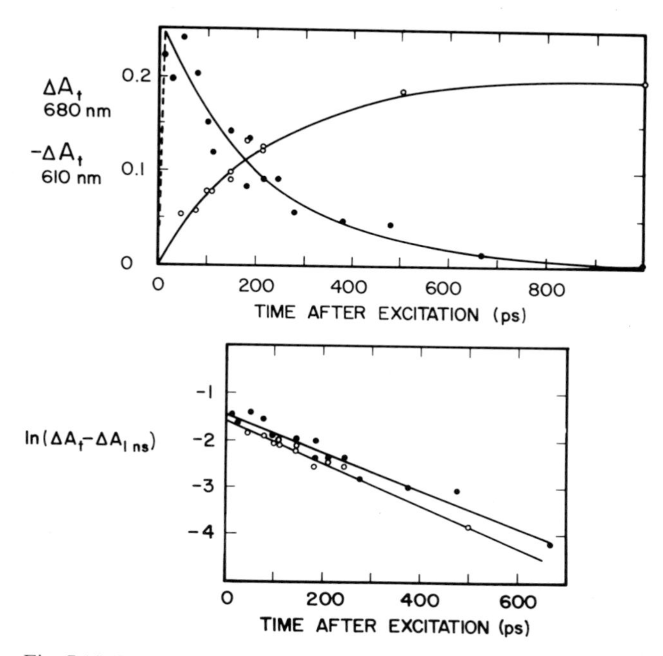

Fig. 7.15. Rate of formation of P^+ (○) measured from absorbance decreases at 610 nm, and rate of disappearance of P^F (●) measured from absorbance decrease at 680 nm after picosecond pulse excitation. Lower figure shows decrease of absorbance with time on semilog plot. P^F disappears in 246 ± 16 ps and P^+ forms in about 220 ± 14 ps. Kinetics indicate that P^+ therefore forms directly from P^F (from *Rockley* et al. [7.90])

in the spectrum of monomeric BChl in organic solution. When the reaction centers are treated with ferricyanide, resulting in formation of an oxidized dimer, an absorption at 1250 nm appears. Consequently, they assigned this absorbance band exclusively to the oxidized reaction-center dimer. By using the picosecond probe technique, they observed that the 1250 nm band forms within 10 ps after exciting the neutral reaction center with a 530 nm flash. Thus, apparently, the oxidized BChl dimer is present long before the formation of the final P^+X^- state. They suggest that the oxidized cation radical of the BChl dimer, represented as [BChl$\overset{+\cdot}{—}$BChl], is part of the intermediate state P^F.

Some of their data are summarized in Fig. 7.16. These data were obtained with a setup similar to an arrangement to be described in Section 7.3, except that a germanium diode replaced the vidicon. The data shown in this figure were accumulated with the reaction centers initially prepared in three ways: The circles represent data with reaction centers in their neutral form, [BChl——BChl]X; the darkened squares refer to data with the primary electron acceptor chemically reduced, [BChl——BChl]X^-; and the triangles refer to reaction centers in the ferricyanide-oxidized form, [BChl$\overset{+\cdot}{—}$BChl]X.

When prepared initially in the neutral form, the RC 1250 nm band appears within 10 ps and spectrally resembles that due to [BChl$\overset{+\cdot}{—}$BChl]; but when prepared with ferricyanide and subjected to 530 nm flashes, the reaction centers show no absorbance changes. This fact further strengthens the argument that [BChl$\overset{+\cdot}{—}$BChl] formation proceeds promptly from the neutral reaction center upon excitation. Appropriate to the conclusion that the formation of P^F does not involve photoreduction of X is the observation that chemical reduction of X does not prevent the appearance of this band after flash excitation.

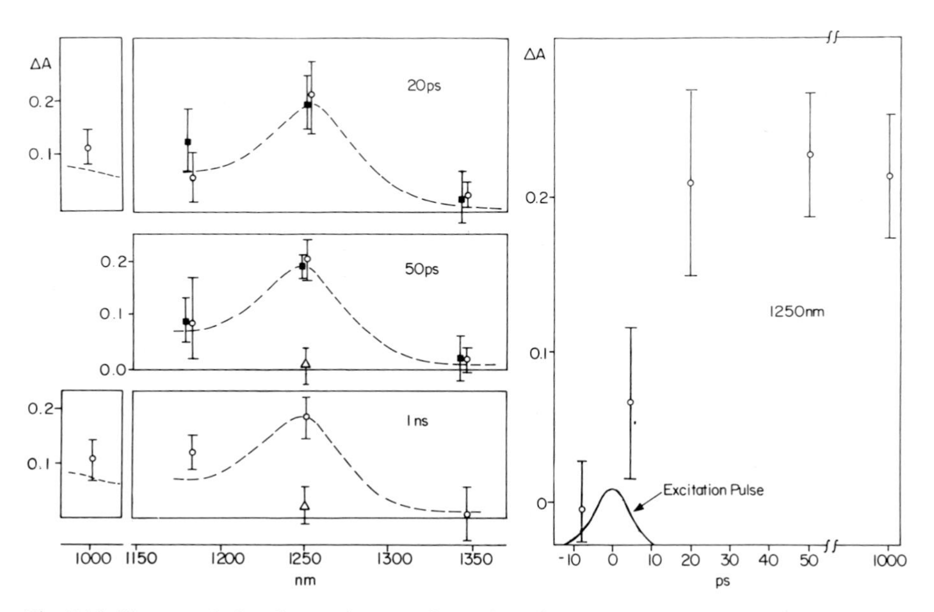

Fig. 7.16. Picosecond absorbance changes of *R. spheroides* reaction centers. Circles, squares and boxes refer to initial state of the reaction center: In the neutral form, (BChl——BChl) X=(○); in the ferricyanide oxidized form (BChl$\overset{+\cdot}{—}$BChl) X=(△) and with X reduced, (BChl——BChl) X^-=(■). On the left are shown data taken at various times after the flash. The right shows the rapid formation of (BChl$\overset{+\cdot}{—}$BChl) X (from *Dutton* et al. [7.97])

Because an electron has left the BChl dimer within 10 ps but does not appear on X until some time later, *Dutton* et al. [7.97] have suggested that the P^F state could be a composite of the oxidized BChl dimer and of another reduced component which they symbolize as I for "intermediate". They have attempted to identify I and have presented spectroscopic evidence against BChl. However, the possibility that I might be identified with BPh is strong. *Fajer* et al. [7.98] have calculated the expected ΔA spectrum of [BChl——BChl] BPh⟶[BChl$\overset{+\cdot}{—}$BChl] $BPh^{-\cdot}$ and have found that it closely resembles that shown in Figs. 7.13 and 7.14. Thus, in summary, *Dutton* et al. identify the states appearing in Fig. 7.11 as

$$P_{870}X = [\text{BChl} \text{——} \text{BChl}\,I]X \tag{7.21}$$

$$P^{*}X \ = [\text{BChl} \overset{*}{\text{——}} \text{BChl}\,I]X, \tag{7.22}$$

$$P^{F}X \ = [\text{BChl} \overset{+\cdot}{\text{——}} \text{BChl}\,I^{-}\cdot]X, \tag{7.23}$$

$$P^{+}X^{-} = [\text{BChl} \overset{+\cdot}{\text{——}} \text{BChl}\,I]X^{-}. \tag{7.24}$$

As is evident from this short discussion of reaction-center oxidation, much is still in an unsettled state. It would be desirable to extend these studies to reaction centers of other species and those containing carotenoids. In this context we should mention the experiment of *Leigh* et al. [7.99]. They have studied absorbance changes due to the endogenous bulk carotenoids in chromatophores of *Rhodopseudomonas spheroides* Ga mutant using a setup nearly identical to that shown in Fig. 7.12. Rapid formation (within 10 ps) of a 515 nm in the Ga mutant is observed and attributed to carotene response. As a similar band is absent in the R26 mutant, they interpret these observations as the response of the carotenoids to charge separation in the reaction center within 10 ps of excitation. Because these experiments were performed in chromatophores at very high flash intensities, multiple hits most certainly occurred within the antenna system. Thus, the preceding discussion of exciton annihilation in Subsection 7.1.3 is pertinent here. It is possible that the carotenoids respond to a higher level metastable state of the antenna bacterio-chlorophyll complex caused by multiphoton processes at such high excitation intensities.

As mentioned previously, the possibility of X being identified with a ubiquinone and I being identified with BPh is strong. As further proof that the primary electron acceptor is ubiquinone (associated with iron), *Kaufmann* et al. [7.100] performed an interesting experiment demonstrating the necessity of ubiquinone participation for a successful completion of photochemical events in the reaction centers of *R. spheroides*. The effects of ubiquinone can be tested by extracting the ubiquinone from the reaction centers and then reconstituting the reaction centers by adding ubiquinone. Upon 530 nm excitation, formation of the $P^{F}X$ state in the untreated sample is immediate and is subsequently followed by formation in 100 to 200 ps of the $P^{+}X^{-}$ state. However, for reaction centers lacking ubiquinone, the $P^{+}X^{-}$ state is not formed; the decay time is longer than 600 ps, consistent with expected decay time of 10 ns. Finally, upon reconstituting the reaction centers with the addition of ubiquinone, results are again similar to those of untreated reaction centers. It is clear that reaction centers depleted of ubiquinone behave as if there were no primary electron acceptor, confirming the important role of ubiquinone and providing further evidence for the identification of X with ubiquinone.

Recently, *Müller* et al. [7.101] have synchronized a nanosecond xenon flash with a picosecond excitation source. Following excitation of the sample, the temporal evolution of the spectral absorption can be photographed on a single shot with a streak camera. The provision of a reproducible, time-tested standard source for picosecond flash-photolysis experiments should greatly simplify the absorption measurements on biological samples.

7.2 Measurements with Hemoglobin

7.2.1 Properties of Hemoglobin

Hemoglobin (Hb) is the molecule within red blood cells responsible for transporting oxygen from the lungs to the rest of the body, and also for assisting with the transport of carbon dioxide back to the lungs. The molecular structure of mammalian hemoglobin was pieced together by *Perutz* et al. [7.102–104] after a careful analysis of x-ray photographs of the crystalline form. Hemoglobin is a protein consisting of four polypeptide chains and four heme groups with a combined molecular weight of 64500. Each chain closely resembles the structure of myoglobin and contains one heme group. Hemoglobin contains a pair of alpha chains with 141 amino-acid residues, and a pair of beta chains with 146 amino-acid residues. Together, the four chains form a tetrahedral array with each of the four heme groups widely separated. They are spaced from 25 to 40 Å apart and located near the surface of the molecule. Hemoglobin reacts with oxygen to form oxyhemoglobin (HbO_2) and with carbon monoxide to form carbon monoxyhemoglobin (HbCO). A very important structural transformation occurs when hemoglobin reacts with oxygen: the two beta chains move about 7 Å closer together while the two alpha chains hardly move with respect to one another. The heme groups in the alpha chain are separated by about 25 Å from those in the beta chain in both Hb and HbO_2; in the two alpha chains they are spaced 35 to 36 Å apart in both molecules, but in the two beta chains they are spaced by about 33.5 Å in HbO_2 and by about 39.9 Å in Hb. This transformation resembles breathing, except that the molecule contracts when it accepts oxygen and expands to release it. The ability of the hemoglobin to combine reversibly with oxygen is a property of the ferrous iron atom in the heme groups, but only when the group forms a complex with globin.

Once an oxygen molecule attaches itself to one of the heme groups in hemoglobin, it becomes easier for the three other hemes to take up oxygen. When three of the four heme groups have captured an oxygen molecule, the affinity of the fourth group becomes very strong. Conversely, when the molecule releases three of the four oxygen molecules, it becomes very easy for the fourth heme group to release its oxygen. These heme-heme interactions, however, are not well understood, and the elucidation of their cooperative mechanisms is therefore a primary target for investigators. Though x-ray techniques are very valuable for determining structural details, picosecond techniques can complement the x-ray studies by providing information on the dynamics of the structural transformations.

7.2.2 Optical Properties of Hemoglobin

The heme groups are responsible for the optical properties of hemoglobin. Their structure is very much like that of the chlorophylls. This was first recognized by *Hoppe-Seyler* [7.105, 106] who degraded chlorophyll with an alkaline treatment and found that its absorption spectrum resembled that of heme with the iron atom removed. An exhaustive treatment of the structural similarities of heme and chlorophyll may be found in *Marks* [7.107]. The framework of both molecules is a porphyrin structure; that of heme, shown in Fig. 7.17, is quite similar to that of chlorophyll, which is shown in Fig. 7.1. Where the magnesium atom in chlorophyll appears to coordinate the chlorophyll molecules, the iron atom at the center of heme performs the biological function of collecting oxygen. Similarities in the absorption spectra of the two can be seen by comparing Fig. 7.18 (heme) with Fig. 7.2a (chlorophyll a). As remarked earlier, the mobile π electrons in the conjugated porphin ring are responsible for the visible transitions in both compounds. Like chlorophyll, heme possesses two intense absorption bands, the Soret band at 430 nm and a second band at 555 nm [7.108, 109]. The absorption spectra of HbO_2 and HbCO are quite similar, with HbO_2 possessing a doublet at 541 and 577 nm, and HbCO having a similar doublet at about 540 and 569 nm. The differences in absorption spectra between Hb and HbO_2 are sufficient to produce a notable change in color: oxygen-free hemoglobin makes venous blood appear purple, whereas oxyhemoglobin makes arterial blood appear scarlet. These absorption differences are due to interactions of the electronic distribution within the porphyrin ring, with the surroundings of the heme group. In particular, the interaction of oxygen with the hemoglobin complex may cause the iron atom, which is supposedly displaced 0.75 Å out of the plane of the porphyrin ring

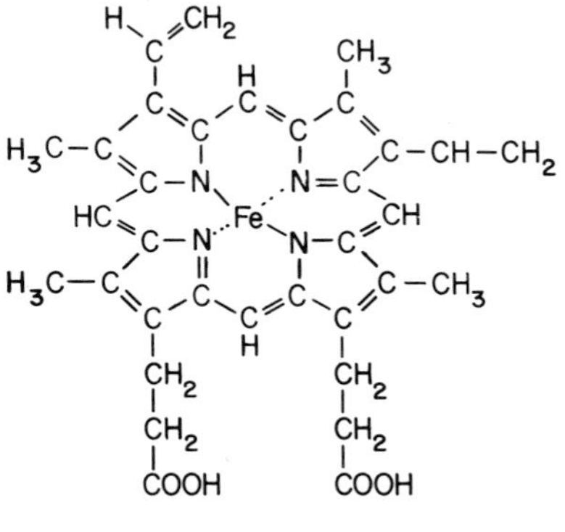

Fig. 7.17. Structure of heme. Oxygen and carbon monoxide bond to the iron atom out of the plane of of the porphyrin ring

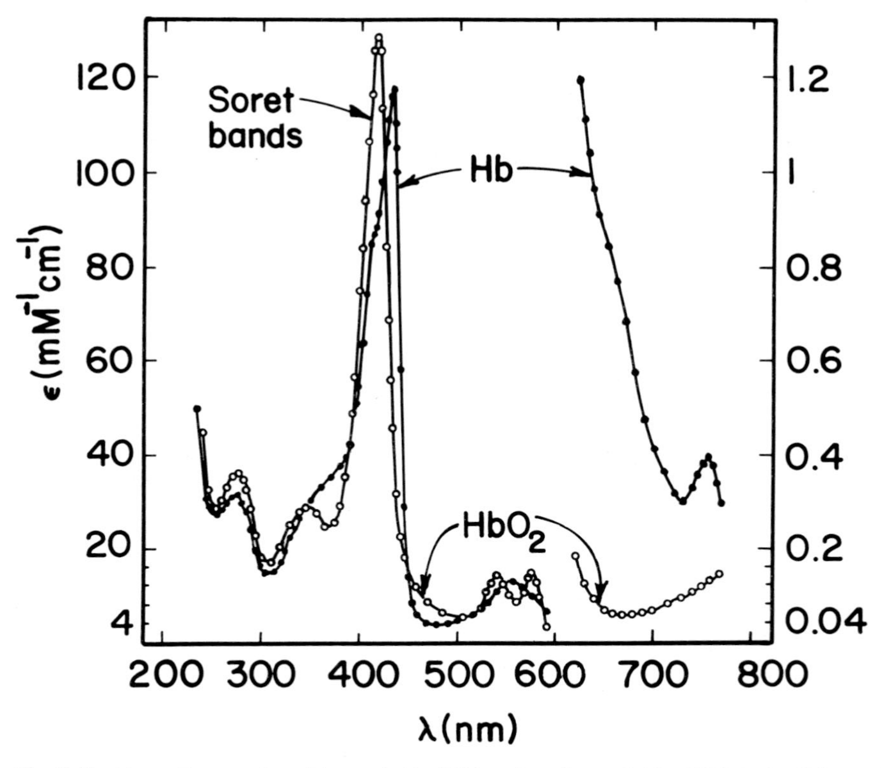

Fig. 7.18. Absorption spectra of hemoglobin (Hb) and oxyhemoglobin (HbO_2). At 615 nm the absorption of Hb is about six times greater than for HbO_2 or for HbCO. Scale at right is expanded by one hundred because of the weak absorption of the hemoglobin compounds at longer wavelengths (after *Sidwell* et al. [7.108])

toward the proximal histidine, to be moved back into the plane. As a result, the proximal histidine is thought to move 0.8 to 0.9 Å away from the porphyrin plane. Such motions are thought to be the triggering steps for the cooperative binding of oxygen by hemoglobin.

7.2.3 Photodissociation Properties of HbCO and HbO_2

The binding energies per molecule of oxygen and carbon monoxide to hemoglobin are $\simeq 0.8$ and $\simeq 0.9$ eV, respectively [7.110]. Consequently, hemoglobin has an affinity about 300 times higher for carbon monoxide than for oxygen, and this is the reason why carbon-monoxide poisoning occurs so easily. The binding mechanisms of the two molecules to the iron atom are thought to be grossly different [7.110]. The photodissociative properties of HbCO and HbO_2 are very different too: it has been known since 1895 [7.111] that while

$HbCO$ is easily dissociated with light in the range 2 to 4.8 eV, HbO_2 is quite difficult to dissociate. The quantum yield for the dissociation of myoglobin and carbon monoxide is near unity over the excitation range 280 to 546 nm. Because the aromatic side chains absorb most of the light at 280 nm, excitation transfer to the heme must have occurred from the rest of the protein molecule. *Stryer* [7.112] has shown that this transfer is consistent with Förster's theory from considerations of the distances between the amino acids and the heme.

The quantum yield for the dissociation of $HbCO$ appears to be more complicated than for myoglobin and carbon monoxide. The quantum yield for photodissociation for different heme proteins varies considerably [7.110]. The quantum efficiency depends on protein concentration, solvent, and structure of the protein even in nonassociating systems. The quantum yield for horse $HbCO$ was measured to be 0.25 by *Bücher* and *Negelein* [7.113] suggesting an inverse dependence on the number of heme groups present per molecule. More recent measurements [7.110] indicate quantum yields in the range of 0.4 to 0.7. The quantum yield for dissociation of HbO_2 of ~ 0.008 [7.111] is negligible by comparison.

7.2.4 Picosecond and Subpicosecond Measurements in Hemoglobin Compounds

An elegant study of the photolysis of hemoglobin complexes has been performed by *Shank* et al. [7.114]. Their experiment attacks the important process of cooperation in hemoglobin by means of sophisticated new techniques. In fact, their experiment may be assisting in the identification and probing of the dynamics of the very first triggering step in the cooperation process. The recently developed subpicosecond cw dye laser discussed in Chapter 2 and the associated techniques discussed in Chapter 3 are used for these hemoglobin experiments. *Shank* et al. [7.114] rely in part on the stronger absorption cross section of hemoglobin as compared to HbO_2 and $HbCO$ at 615 nm (see Fig. 7.18). A single subpicosecond pulse at 615 nm (2.0 eV), which is selected from a cw modelocked dye laser, is used to excite the samples, and a much weaker delayed image of the pulse is used to probe the induced absorption change as a function of delay time. The repetition rate (100 μs) is low enough to allow for complete recovery of the sample, but high enough to allow for extensive signal averaging. Results for the induced absorption as a function of delay time for both $HbCO$ and HbO_2 are shown in Fig. 7.19. In the $HbCO$ sample an increase in absorption is observed immediately after excitation. The rising absorption is shown by the dashed line, and the risetime is less than 0.5 ps (the dashed curve corrects an artifact introduced by the coherence between the pump and probe beams). For HbO_2 the induced absorption rises abruptly and then decays in about 2.5 ps.

Shank et al. [7.114] interpret the phenomena as follows: For $HbCO$, which is known to easily dissociate into Hb and CO, the increase of absorption must be due to the formation of Hb which has a much stronger absorption coefficient at 615 nm than $HbCO$. The dissociation is further indicated by the

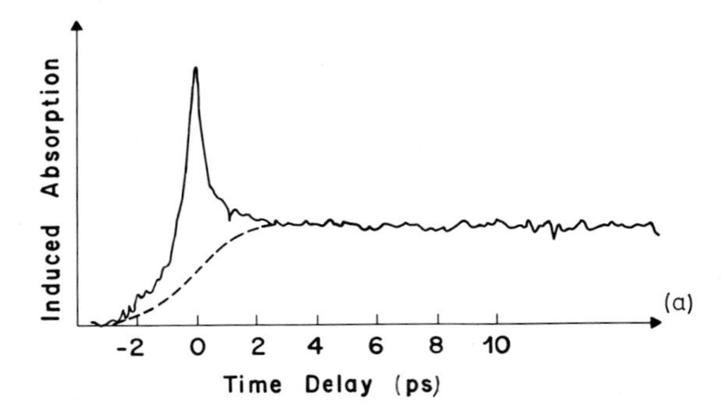

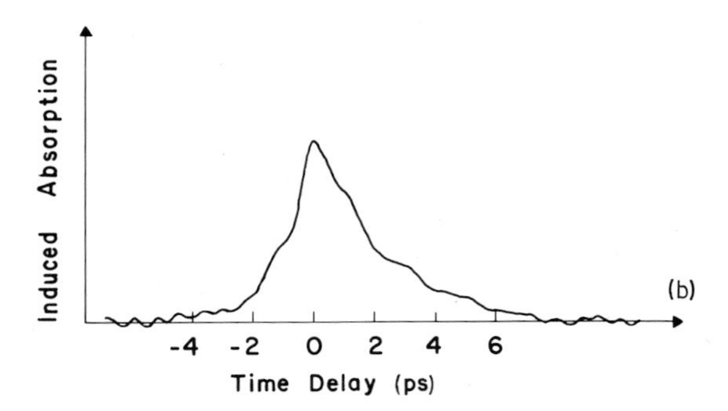

Fig. 7.19. (a) Increase in the induced absorption of HbCO sample follows excitation by subpicosecond pulse selected from a cw modelocked dye laser. The induced absorbance and slow recovery indicate HbCO dissociation. (b) Upon excitation with a subpicosecond pulse at 615 nm, the induced absorption of HbO_2 rises rapidly and decays in 2.5 ps (after *Shank* et al. [7.114])

fact that the induced absorption decays very little in the first microsecond after excitation. The high quantum yield for photodissociation must be due to the ultrafast dissociation process rather than to the inability of HbCO to release its excitation energy in nondissociative processes or else the risetime would be much longer. Surprising is the fact that the dissociation time is so brief, because the iron atom and the nearby histidine complex are supposedly required to be displaced considerably. *Shank* et al. [7.114] therefore suggest that the displacements involved in the formation of complexes may not be as large as is generally assumed; they further point out that their suggestion is consistent with recent x-ray diffraction studies on a particular type of mutant HbCO molecule known as HbCO Kansas.

The results for HbO_2 are interpreted as due to a short-lived excited state whose absorption is greater than that of the ground state. This interpretation is supported by measurements with an Hb sample where the results were similar to those obtained with the HbO_2 sample. Thus, the low quantum yield for photodissociation of HbO_2 must be due to a dissociation rate that is much slower than the 2.5 ps nondissociative recovery and cannot be due to the rapid combination of initially photolyzed molecules. The promise offered by cw modelocked dye-laser techniques in the biological area is evident.

7.3 The Visual Molecules

Studies in a third key area in biology using picosecond techniques are leading to a better understanding of the mechanism of vision. The pressing problem in this area is the identification of events that occur on the molecular level in the visual process. The very fast first steps, including the absorption of light and its retention for driving the visual process, resemble in many ways photosynthetic processes. Indeed, some of the molecules partaking in both processes are similar. A class of molecules, the carotenoids, are present in most normal plants, and as *Wald* [7.115, 116] first showed, vitamin A, which is structurally a double carotene, is present in the eye. However, he also points out that such an analogy is superficial [7.7], for although light is absorbed in the primary steps of photosynthetic and visual processes, the biological purposes are quite different. In photosynthesis the light is absorbed and the excitation energy is transferred efficiently and utilized for driving the entire photosynthetic process. In vision, on the other hand, an incoming photon triggers the response of nerves that have been set to discharge by internal chemical processes. Efficient energy use in vision does not appear to be of prime importance, although the eye is sensitive enough to detect a few photons only. As we have seen earlier, the efficient transfer of energy between chlorophyll molecules is an integral part of the photosynthetic process, whereas such energy-transfer processes have never been found [7.117], for example, between rhodopsin molecules that are so essential to human vision.

Though many rapid molecular steps in the visual process have been identified, many more probably remain to be discovered and elucidated. Picosecond lifetime measurements can help to establish the nature of the known transitions and present evidence for new transitions which often reflect structural rearrangements within molecules.

It has been demonstrated that four key molecules, at least, are essential to human vision [7.118–121]: rhodopsin, responsible for night vision, found in the rods of the eye; and three additional pigments, responsible for color vision, found in the cones of the eye. Because of the difficulties in isolating the three color-vision pigments, presumably due to their scarcity and the fact that they are all intermixed in the cones of the fovea, picosecond studies have concentrated on rhodopsin, which can be extracted from the rods much more easily. Indeed, there is no safe recipe for determining whether one has rhodopsin, although the two main criteria for identification are the extinction coefficient and the spectral position of the maximum absorption band.

Rhodopsin is a carotenoid protein with a molecular weight of about 40000 and if its shape were spherical it would have a corresponding diameter of about 40 Å. During the visual process, a photon is absorbed by a molecule such as rhodopsin which, after a series of rapid and slow steps, eventually decomposes into free retinal and opsin. At a later time, these two substances reform into rhodopsin to repeat the visual process. Somewhere during these multiple steps, ions, which are released through chemical reactions, cross the membranes containing the rhodopsin sacks and trigger nerve responses to the brain. In the visual process, at least the first three steps are expected to have transition times in the nanosecond and picosecond range. These first three steps are thought to be fast because spectral analyses based on the work of *Wald* and coworkers [7.118], among others, have shown that the rhodopsin molecule upon absorption of light undergoes a small structural change. This change takes place mainly on the retinal molecular side chain that is attached to the opsin protein by a Schiff base linkage [7.122, 123]. The retinal, although 20 Å long, has a molecular weight of 284, and therefore is puny in comparison with the opsin protein. When light is absorbed by the chromophore in the retinal, several intermediate species are quickly formed. The first two new intermediates are called prelumirhodopsin and lumirhodopsin – *Yoshizawa* [7.124] has also proposed a third short-lived intermediate which he calls hypsorhodopsin. All the newly formed species can be differentiated from rhodopsin by their absorption spectra. The absorption spectrum of rhodopsin is shown in Fig. 7.20 with its main absorption band in the visible at about 500 nm. Prelumirhodopsin, on the other hand, has at 543 nm a strong absorption band whose wing is quite strong at 561 nm where rhodopsin has much less absorption [7.116]. As we shall see, this difference in absorption spectra is the key to the picosecond measurements on these species by *Busch* et al. [7.125]. It should be noted that the peak absorption of the prelumirhodopsin molecule is stronger than that of rhodopsin itself. The reason for

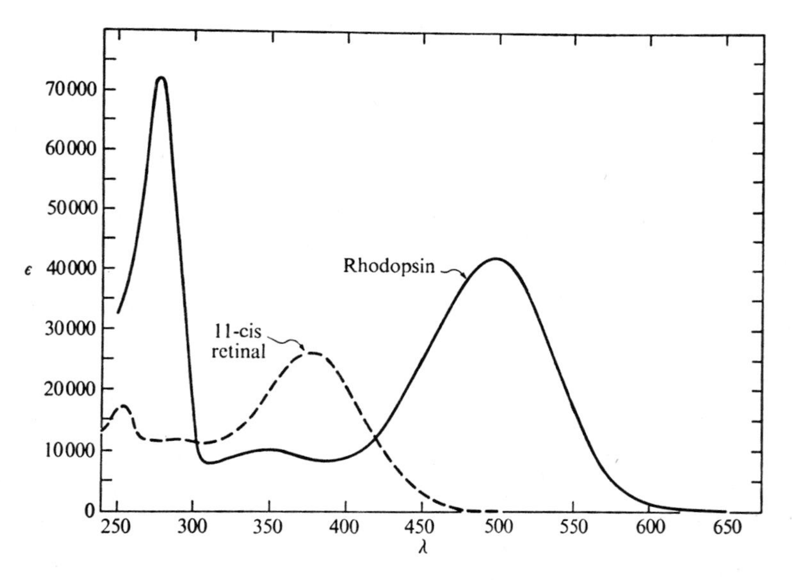

Fig. 7.20. Absorption spectra of the chromophore of rhodopsin, 11-cis retinal, and of rhodopsin. Main absorption band of chromophore shifts from 380 nm in solution to 500 nm in the rhodopsin molecule and further toward the red in prelumirhodopsin (Figure from *Ebrey* and *Honig* [7.126])

the absorption band shift in prelumirhodopsin lies mainly with the isomerization of the retinal chain attached to the opsin molecule. As has been shown by *Wald* and coworkers [7.118], the first step in the visual process after the absorption of a photon is the transformation of the 11-cis bond in the retinal substructure into a 11-trans bond. Such a transformation is diagrammatically depicted in Fig. 7.21 (for a more detailed description of different isomers of retinal, the reader should see *Ebrey* and *Honig* [7.126]). Minor structural changes, such as the type of isomerization depicted, account for the first few steps in the visual process, e.g., for the formation of prelumi- und lumirhodopsin. Because of the rapidity of some of these transformations, many studies of these species are performed at low temperatures where they can be observed more easily.

The first studies of a vision molecule using picosecond techniques were made by *Busch* et al. [7.125]. These studies also represent the first measurements in the biological area that utilized picosecond pulses. The new techniques allowed Busch et al. to study the formation and decay of prelumirhodopsin at room temperature, which is very near physiological temperatures. Because of the different time scales involved, their studies were essentially divided into two parts – measurement of the formation of prelumirhodopsin and measurement of its decay.

To measure the formation time of prelumirhodopsin, a pulse from a modelocked Nd:glass laser train was extracted, amplified, and then frequency doubled to a wavelength of 530 nm via second-harmonic generation. The green pulse was then sent by means of beam splitters along two separate paths

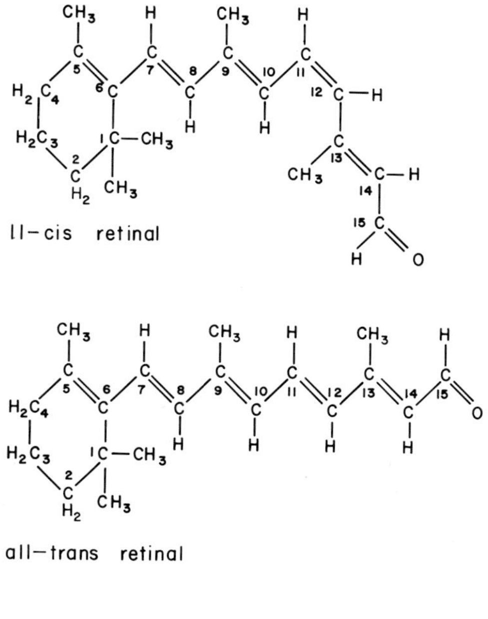

11-cis retinal

all-trans retinal

Fig. 7.21. Typical conformational change occurring after absorption of quantum by retinal molecule. Figure shows isomerization of 11-cis retinal to all-trans retinal. Conformational changes like that depicted occur when rhodopsin changes into prelumirhodopsin, and can take place on the picosecond scale. For a more detailed discussion of the many types of isomerization prozesses see *Ebrey* and *Honig* [7.126]

that intersected at the sample. The pulse traveling along the first path was allowed to excite the rhodopsin sample, while the pulse traveling along the second path passed through a cell of benzene with sufficient intensity to generate a new pulse by stimulated Raman scattering at 561 nm. The 561 nm pulse was reflected off an echelon producing a set of pulses spaced 2 ps apart. The 530 nm pulse excited the rhodopsin sample obtained from bovine retina and the set of 561 nm pulses probed the sample at set time intervals to determine how quickly prelumirhodopsin is formed, as determined by the increase of absorption at 561 nm. An experimental result showing the rapidity of prelumirhodopsin formation is shown in Fig. 7.22. The risetime for the formation of prelumirhodopsin was less than 6 ps, the laser pulse duration. This result is consistent with the expected rapidity of an isomerization process thought to be responsible for the formation of prelumirhodopsin, and prelumirhodopsin therefore arises as a direct consequence of the relaxation of electronically excited rhodopsin. Also, the absorption of the 561 nm pulse was greater when the polarizations of the 530 and 561 nm pulses were parallel than when they were perpendicular to each other.

To measure the much longer decay times for prelumirhodopsin a different experimental arrangement was required because the optical delays must be much greater. A train of pulses was used with the polarization of one pulse rotated 90° to the remainder of the pulses. Only the single pulse polarized at 90° generated the 530 nm second harmonic in a properly oriented KDP crystal, and this 530 nm pulse was directed along one path to the sample. The remaining pulses were reflected along a second path, which contained a KDP crystal and a cell of benzene where stimulated Raman scattering at 561 nm was again generated. The first 530 nm pulse excited the sample, while a comb of 561 nm pulses traversed a second path and probed the absorption as a function

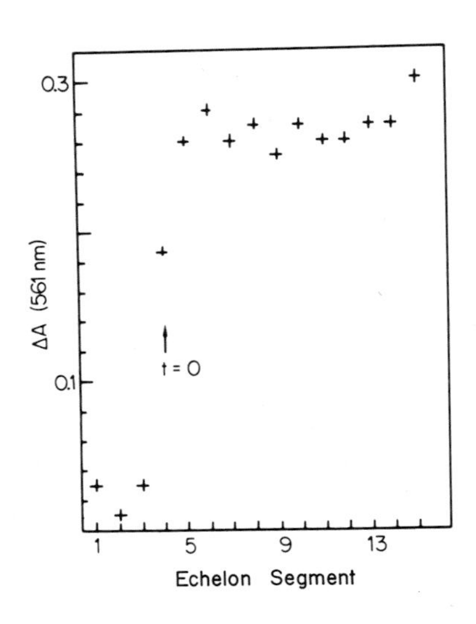

Fig. 7.22. Change in absorbance of rhodopsin sample at 561 nm is plotted as function of time. Each echelon segment represents a time delay of 20 ps. An intense 530 nm pulse excites the sample coincident with the 561 nm light produced by the fourth echelon segment. An abrupt rise in the absorbance demonstrates the ultrafast creation time for prelumirhodopsin (from *Busch* et al. [7.125])

of delay time. The time interval between the 561 nm pulses is equivalent to the roundtrip time of the laser and so can be many nanoseconds long. Experiment revealed that the newly induced absorption decays in about 30 ns. This decay time is surmised to be that of prelumirhodopsin, because no other transient is known to exist in this region of the spectrum.

In another recent experiment by *Bensasson* et al. [7.127] transient absorption changes were observed over the range 300 to 600 nm in bovine rod outer segments. The only chromophore located in these segments is rhodopsin. At 550, 450, and 330 nm transient absorptions were found that decayed in ~125 ns; at 470 nm a second transient was found. Bensasson et al. have suggested a charge transfer is taking place in rhodopsin and the charge transfer complex may be the primary intermediate in the visual process.

Bacteriorhodopsin, another molecule which may be related to rhodopsin, has received much attention lately, mainly because this molecule is found in the purple membrane of *Halobium Halobacterium* where it converts light directly into chemical energy [7.128]. The absorption of light within the purple membrane results in the transfer of a proton across the membrane to generate an electrical potential that can then be used to synthesize ATP directly. This rather amazing conversion process, which does not involve chlorophyll, is of great interest because it may give us insight into both the function and the solar-energy conversion processes of these membranes and because it may help explain the visual processes. Like rhodopsin, bacteriorhodopsin contains retinal as a chromophore and is attached to an opsin-like molecular protein via a Schiff-base linkage. The absorption spectrum of bacteriorhodopsin, which has broad featureless bands at 280 and 570 nm, is in many ways similar to that of rhodopsin, as is the photochemistry. Bacteriorhodopsin has a molecular weight of about 26000 and is found in a rather rigid membrane, whereas the visual pigments are found in fluid membranes. Despite the similarity in names, the nomenclature is not as yet based upon the structural properties of the opsin protein. In the case of bacteriorhodopsin one would hope that by measuring how long certain states last, one could define the sequence of steps leading from light absorption to the generation of a proton gradient.

The first steps in the primary photochemical processes in bacteriorhodopsin have been studied recently by *Kaufmann* et al. [7.129]. They excited the purple membrane in the light-adapted state with an intense single pulse at 530 nm and observed a rapid increase of absorption (<6–10 ps) at a monitoring wavelength of 635 nm, and a fast transient decrease of the absorption at 580 nm with a decay time of ~15 ps. The increase of absorption at 635 nm is interpreted as due to the formation of an early intermediate, bK_{590}, which had previously been detected with microsecond resolution. The decay time of this absorbance was measured to be greater than 300 ps, consistent with an earlier microsecond lifetime estimation for this state. At present, these experimenters believe that the fast transient observed near 580 nm is due to a photoproduct of bacteriorhodopsin, an excited singlet, which would extend the formation time of bK_{590} to about 15 ps. Because bK_{590} is similar in its spectral characteristics and its kinetics to rhodopsin, they suggest that the formation of the first intermediate state in bovine rhodopsin does not involve a complete cis-trans isomerization. As the authors point out, other possible interpretations arise such as absorption of the pulse by bK_{590}, because of the high intensity of their excitation pulse (greater than 10^{16} photons/cm^2).

Alfano et al. [7.130] have attempted to study the initial steps after light absorption by the purple membrane by using an optical-gate detection scheme. Although they were unable to resolve the fluorescence lifetime at room temperature, they were able to estimate the lifetime is about 3 ps and measured a lifetime of 40 ps at liquid-nitrogen temperature. Because of the short lifetime observed, they have postulated that the process responsible is isomerization. More recent work by *Hirsch* et al. [7.131] indicates a lifetime of 15 ± 3 ps at room temperature. According to these workers, the short lifetimes observed by Alfano et al. are a consequence of singlet-singlet annihilation and are not due to an intrinsic property of the purple membrane system. The purple membrane is composed of 21 molecules held together as trimers; it is arranged in a two-dimensional array with the molecules close enough for exciton coupling to occur. The intensities quoted by Alfano et al. are 6×10^{14} photons/cm^2, sufficient for singlet-singlet annihilation, whereas Hirsch et al. use a somewhat lower figure, 3×10^{14} photons/cm^2. Arguing against this hypothesis is the fact that Alfano et al. obtain a similar lifetime as in the membrane at low temperature when the membrane is broken up by a detergent into individual molecules. Note that phase fluorometry measurements yield a lifetime of about 100 ps, but phase fluorometry uses continuous light and the intermediates are formed immediately, again making interpretation difficult. Because both Alfano et al. and Hirsch et al. used modelocked laser "trains", it is in our opinion certainly desirable to definitively exclude the possibility that intermediates are responsible for the short lifetimes observed.

7.4 Deoxyribonucleic Acid (DNA)

7.4.1 Possible Applications in DNA

The structure of deoxyribonucleic acid (DNA) is well known [7.132, 133] and consists of two helical chains coiled around an axis. The two chains are held together by bases that are arranged perpendicular to that axis. Only specific pairs of bases bond to each other because of restrictions placed by hydrogen bonding; adenine binds with thymine and guanine with cytosine. The backbone consists of a regular ordering of sugar and phosphate groups, with one of the four bases attaching to each sugar. Each monomeric unit consisting of a phosphate, sugar, and base is called a mononucleotide. Two such units connected together form a dinucleotide, and multiply connected units are referred to as polynucleotides. Any sequence of bases can be fit in the sugar-phosphate structure. However, one chain is the complement of the other because of the specific pairing of the bases, and biological duplication of the molecule takes place when the hydrogen bonding between the bases is broken. The double helix must unwind to accomplish this duplication. A typical DNA chain has a molecular weight of one to two million.

Processes which may take place on a picosecond time scale in DNA include energy migration along the molecular chain and deactivation of the electronic states. Also the fluorescence decay of the mononucleotide and dinucleotide subunits is expected to be in the picosecond range. However, no firm evidence is available that indicates singlet migration occurs along the DNA chain, either at room temperature or at low temperature, although triplet exciton migration is well established. Evidence for triplet migration in poly A is fairly strong [7.134]. *Bersohn* and *Isenberg* [7.135] have shown that the triplet emission is strongly quenched by adding small quantities of certain metal ions, and measurements indicate that the diffusion length may be over one hundred adenine residues. *Somer* and *Jortner* [7.136] have estimated the triplet jump time to be 10^{-9} s, whereas *Eisinger* and *Lamola* [7.137] estimated a jump time of 10^{-7} s.

Electronic energy migration in DNA is of interest as a possible probe of structural and energetic changes induced by the introduction of chemical agents or by exposure to damaging uv radiation. If migration times can be determined using picosecond techniques and are found to vary under such changes, much new information can be deduced. Answers to questions as to why certain dye molecules are known to induce powerful mutagenic activity on particular bacteriophages may consequently be uncovered. Other possibilities include studying the formation of dimers on the DNA chain upon exposure to uv radiation or studying breaks in a chain that can be produced upon exposure to high-energy particles or photons. It is well known that the formation of pyrimidine dimers is hindered by certain dyes that bind to DNA; the long-range nature of the hindrance has been attributed to the interception of the electronic energy migration by the dimer which acts as an acceptor, the excited singlet migrating down the DNA until it reaches a pyrimidine dimer site. Picosecond studies offer the possibility of studying coherent and incoherent states in DNA and other biological molecules, and, in principle, offer the opportunity of examining dynamic properties when the double helix is untwisting during different stages of cell development, and of examining for possible differences in tumor cells.

The main absorption band in DNA is broad and featureless and is centered near 265 nm and can be conveniently excited by the fourth harmonic of the Nd:glass laser. The fluorescence and phosphorescence from DNA are strongly quenched at room temperature, so that the remaining weak emission has not been well characterized. It has been speculated that the short singlet lifetimes act as a protection mechanism against photo-induced damage. The quantum yield of the fluorescence at room temperature has been estimated to be of the order of a few times 10^{-4}, which implies that the nonradiative decay times are less than 1 ps, and similar conclusions can be reached for the individual mononucleotides at room temperature. Therefore, much of the information on fluorescence properties of the nucleic acids is obtained at liquid-nitrogen temperatures, usually in ethylene glycol-water glasses. Figure 7.23 shows the fluorescence from native DNA. At 77 K a broad and featureless fluorescence band emerges near 355 nm and a phosphorescence emission is detected at about 450 nm. It is not known exactly why the fluorescence intensity is such a strong function of temperature.

Although the absorption spectra of the mononucleotides are quite similar to those of the polynucleotides at 77 K, their emission spectra are considerably different. The fluorescence peaks of the mononucleotides are near 325 nm and are considerably narrower than those of the polynucleotides. The fluorescence from the dinucleotides and from the polynucleotides are both similarly red shifted from the mononucleotides. *Eisinger* et al. [7.138] have interpreted this red shift as due to exciplex formation in the polynucleotides; that is, the bases in their excited states interact much more strongly with neighboring bases than when in their ground states. The singlet-to-triplet yields of the individual monophosphates are of the order of 15% at 77 K.

Some peculiarities of fluorescence emission of DNA and of mononucleotides are not well understood. The decay rate from the first excited singlet state can be estimated from a knowledge of the radiative lifetime as calculated from the oscillator strengths and the measured quantum efficiency via the

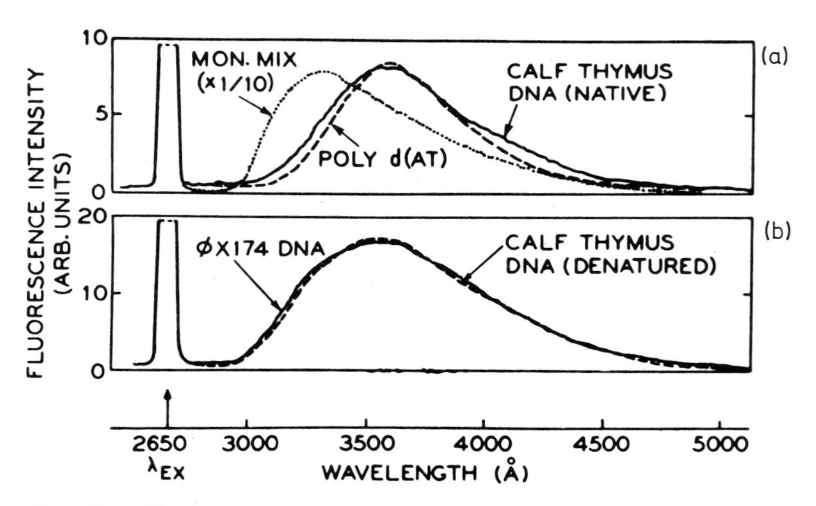

Fig. 7.23. (a) Fluorescence spectra of double stranded DNA, and poly dAT in an ethylene glycol-water glass at 85 K compared with fluorescence spectrum of an equimolar mixture of the monophosphates of adenine, guanine, thymine, and cytosine. DNA fluorescence is quenched and shifted to the red relative to the equimolar mixture. Excimer formation is believed to be responsible for these changes. (b) Spectra for heat denatured DNA and for the single strand $\phi(X)174$ DNA. Emission becomes broader and quantum yield is larger for single stranded samples (from *Eisinger* et al. [7.138])

relationship $\tau = \phi\tau_r$. Since for DNA τ_r is of the order of 1 to 2 ns, and the quantum yield at low temperatures is about 1%, such an estimate yields a value of 10 to 20 ps. However, the measured fluorescence lifetime is several nanoseconds [7.139]. Similar discrepancies have also been noted with the mononucleotides; calculated lifetimes range from 10 ps to 1.6 ns and measured lifetimes range from 2.8 to 5 ns. *Eisinger* and *Shulman* [7.140] hypothesize that the observed fluorescence comes from a lower-lying state, not observed in absorption because of a weak transition moment but which appears in emission with a correspondingly long lifetime. The main absorption band in the mononucleotides must correspond to the excitation of a π electron into a previously empty π orbital because the absorption is strong; the weak absorption that has been postulated could possibly originate from the excitation of an n electron into a π orbital. *Eisinger* and *Lamola* [7.141] have estimated that the excited singlet states of the mononucleotides in deuterated water range from 1.3 to 23 ps from measurements of transfer of the excitation energy to a europium 3+ ion which emits in the red.

No lifetimes have as yet been published for the dinucleotides; however, because the mononucleotides have lifetimes in the nanosecond range, their formation times must be very fast, much less than a nanosecond, as indicated by a spectral shift. The spectra of poly A and DNA resemble much more closely those of the dinucleotides than those of the mononucleotides. Similar fluorescence-emission spectra are also observed from DNA that has been denatured by heating. The fluorescence quantum yield for the single strands increases several-fold over that of native DNA.

Förster's theory, discussed earlier in Subsection 7.1.2, should also apply to the excitation transfer between the bases in a DNA chain. Although Förster's model takes into account only the dipole-dipole interactions, which go inversely as R^6, a simple calculation shows that the exchange interactions due to the overlap of the wavefunctions of neighboring bases are much weaker than dipole-dipole interactions. The neighboring bases are spaced only 3.5 Å apart, and the transfer time between them is therefore by any calculation exceedingly fast. Due to the R^{-6} dependence only transfer between neighboring bases is usually considered because transfer to even the second-nearest neighbor is expected to be at least 64 times slower (the exchange interaction only produces a steeper dependence with distance). In fact, the transfer may be so rapid that it precedes vibrational relaxation in the excited state, which is thought to be in the range of 0.1 to 1 ps. The very weak coupling case of *Förster* [7.22] should apply, but (7.13) may be inaccurate because it assumes that the nucleotide has vibrationally relaxed in the donor prior to transfer to the acceptor. The appropriate spectrum for the donor may be the emission spectrum prior to relaxation. If so, the transfer time should be wavelength dependent. There is much better overlap between the donor emission spectrum and the acceptor absorption spectrum before vibrational relaxation than after. Förster's theory gives a pair transfer time for DNA of about 10^{-14} s. This value must be reduced somewhat, to 2.5×10^{-14} s, because the spacing of the bases is comparable to the extension of the excited electronic systems. On the other hand, after vibrational relaxation, the transfer time is about 1 ps. Thus, over most of the excitation lifetime, the pairwise transfer time might be about 1 ps.

Gueron et al. [7.142] have calculated the excitation-transfer rates between the different bases of DNA at liquid nitrogen temperature after vibrational relaxation. The Förster radius R_0 was at most the distance between several bases in the most favorable case. The distance between bases is about 3.5 Å in the Watson-Crick model. In the case of transfer between two adenine bases, R_0 was calculated to be 0.9 times the distance which means that the molecule will more likely relax to the ground state rather than transfer its energy to the neighboring adenine. These calculations are summarized by *Eisinger* and *Lamola* (Ref. [7.137] pp. 175, 176) as well as their own calculations of transfer between the different bases before vibrational relaxation. The results "before relaxation" range from a shortest time of 2×10^{-14} s between cytosine and guanine to a longest time of 3 ps between thymine and adenine. These latter calculations were made on the basis that the emission spectrum of the donor has the same shape as the absorption spectrum and that the bases are excited at the peak of their absorption bands. Because many of the mononucleotides have very broad structureless spectra, it is possible that vibrational relaxation can compete against some of these very fast transfer processes.

Because of the rapidity of many of the transfer processes in DNA, little picosecond work has been attempted. However, the fluorescence from DNA at liquid-nitrogen temperature in ethylene glycol-water glasses is quite easily detected by a streak camera with picosecond pumping at 265 nm, the fourth harmonic of the Nd:glass laser [7.143].

7.4.2 Transfer to Intercalated Dye Molecules

A more profitable method of studying energy transfer in macromolecules by picosecond methods may be to add dye molecules at certain sites and then to excite the macromolecule at one site and observe the emission from the dye molecule at another site. By determining the transfer time much can be learned about energy migration or, in some cases, about the molecular structure itself. This method is effective in DNA; by choosing different dye molecules and by varying their concentration, some control of conditions is obtained. *Lerman* [7.144] has shown by structural considerations that certain dyes combine with DNA, most probably by intercalation. According to this theory, the double helix untwists sufficiently to provide enough space for the acridines to enter the helices, leaving the hydrogen-bond pairing of the nucleotides intact. *Lerman* [7.145], on the basis of fluorescence and dichroism experiments, has further shown that the bound acridine is more nearly perpendicular than tangent to the helix axis, and that the perpendicularity of the base pairs to the helix axis is not significantly altered. It is well known that the addition of acridine dyes to T-even phages has a potent mutagenic effect.

Many optical properties of acridine dyes bound to DNA were studied by *Weill* and *Calvin* [7.146] in 1963. By investigating the quantum yield for the fluorescence from the bound acridine as a function of dye concentration on the polymer, they were able to find some evidence for both energy transfer from acridine dye to acridine dye and from pyrimidine bases to the dye. The strongest evidence for excitation transfer from the bases to the dye is offered by the observation that the quantum yield for dye fluorescence increases under 260 nm pumping, as opposed to visible pumping, for higher polymer-to-dye ratios. The best available estimates seem to indicate a path length of migration from base to base of ten to twenty base pairs; after that, the energy must disappear in the form of heat by vibrational decay mechanisms. Such exciton-phonon interactions are probably also the main cause for the weak fluorescence in pure DNA at room temperature.

Another dye which intercalates between the base pairs of DNA and is also frequently used in energy transfer studies is ethidium bromide [7.147]. *Lepecq* and *Paoletti* [7.148] have shown that uv radiation, when absorbed by DNA in an ethidium bromide-DNA solution, sensitizes the fluorescence of ethidium bromide thus implying energy transfer. In the time domain this type of transfer has been measured by *Weill* [7.149] and by *Burns* [7.150]. Weill, by using a hydrogen flashlamp, showed that the time of transfer from the DNA to the bound dye proflavine was, after deconvolution, less than 0.1 ns; whereas Burns estimated a transfer time of less than 1 ns for transfer to the dye ethidium bromide. *Sutherland* and *Sutherland* [7.151] found further evidence for a Förster mechanism of transfer in DNA.

7.4.3 Picosecond Studies

At present the only direct time-domain measurements for studying rapid energy-transfer processes in DNA-dye complexes by using picosecond techniques are by *Shapiro* et al. [7.152]. They excited a room temperature acridine-orange DNA complex with a 10 ps pulse at 265 nm generated by repeated second-harmonic generation from a modelocked Nd:glass laser. The second KDP crystal was only 3 mm long to ensure that the 265 nm pulses were not temporally stretched beyond 10 ps due to group dispersion. Filters ensured that only the 265 nm pulses excited the sample. Fluorescence emitted in the 530 to 600 nm range from the dye complex was collected onto the slit of the streak camera.

The temporal characteristics for the visible fluorescence are shown in Fig. 7.24, which is a densitometer trace of a streak photograph and clearly shows that the fluorescence rises abruptly. From calibrated densitometer traces it was estimated that the fluorescence risetime (10–90%) from acridine-orange DNA complexes with a polymer-to-dye ratio of 400:1 was 30 ps. After deconvolution of the streak camera and laser pulse functions, they estimate that the risetime was less than 20 ps.

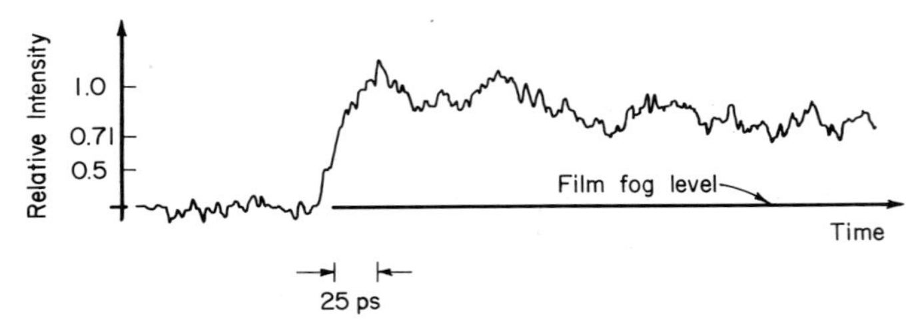

Fig. 7.24. Densitometer trace of streak camera photograph of fluorescence from a DNA-acridine orange complex with a polymer-to-dye ratio of 400:1. Excitation is with 265 nm pulse and risetime is observed to be abrupt (from *Shapiro* et al. [7.152])

Because most of the fluorescence is caused by excitations transferred to the intercalated acridine dye and because estimates of transfer distances by *Weill* and *Calvin* [7.146] and by *Lepecq* and *Paoletti* [7.148] range roughly from 5 to 10 base pairs, a measurement of the risetime <20 ps gives the period over which the excitation transfer takes place. Because DNA is a double-helix chain molecule, the diffusion is essentially in one dimension. Therefore, the exciton diffusion coefficient D in (base pairs)^2/ps units is related to the mean migration distance $(x^2)^{1/2}$ by $x^2 = D\tau$. Here τ is the lifetime whose upper limit, together with the mean migration distance of 5 to 10 base pairs, places a lower limit on the migration rate. The result is that D is at least 1 (base pair)^2/ps and may be much higher.

For reference, Förster's formula applied to transfer between adjacent bases gives for D about 100 (base pairs)^2/ps. This formula assumes the coupling strength for point dipole transition moments calculated from absorption spectra and neglects many complicating factors, such as the finite extension of electronic systems (comparable to their separation); the interplay of inter-base coherence and vibrational relaxation; the possible evolution of the excited singlet state to an exciplex [7.138]; and the coupling between bases more distant than immediate neighbors.

Many complicating effects occur during such experiments. As in photosynthetic experiments, high intensities may lead to nonlinear optical effects, especially for molecules of large size where several excitations may be present on the molecule simultaneously. Further studies are necessary to clarify the effects of a high density of singlet, triplet or exciplex states and their relationship to migration along the chain. Exciplexes should act as traps intercepting singlet migration. Similarly, dimers can be formed on thymine neighbors with a quantum efficiency of about 1% in DNA and they may act as acceptors in a Förster migration process. Further studies are also desirable with other DNA-type structures such as poly A; higher temporal resolutions may be obtained with a subpicosecond resolution streak camera or with a subpicosecond cw modelocked dye laser. The problems involved are interesting because they may ultimately lead to an understanding of the unwinding of the double helix, of radiation damage processes induced by both uv and particle excitation, and of processes related to cancer [7.153, 154].

Recently, an interesting approach for studying acridine-DNA complexes in the time domain has been introduced by *A. Andreoni, A. Longoni, C. A. Sacchi, O. Svelto,* and *G. Bottiroli*, to be published. They studied the fluorescence decay times for an acridine dye, quinacrine mustard (QM), bound within DNA samples and also within chromosomes. It had been previously known from quantum efficiency measurements that the quantum yield for QM intercalated within an AT–AT sequence had a much higher quantum yield than QM intercalated within an AT–GC or an GC–GC sequence. Andreoni et al. verified the quantum yield result in the time domain by pumping QM bound to DNA samples with a subnanosecond pulse from a dye laser at a wavelength of 419 nm. They found a fluorescence decay time of 25 ns for a PolydA–PolydT sample, whereas results from a PolydG–PolydC sample consisted of a superposition of a fast component with a lifetime of 0.5 ± 0.2 ns and a slow component with a lifetime of about 5 ns. The identification of the different lifetimes with the base-pair sequences then allowed them to study the base-pair sequences within chromosomes. When QM binds to chromosomal DNA, some parts of the chromosome fluoresce more strongly than others giving a reproducible fluorescence pattern. The brightly fluorescing regions are called chromosome bands. The increase of fluorescence might be due to either an increase in dye concentration or due to a different quantum yield of the dye. To discriminate between these two possibilities, Andreoni et al. focused the beam with a microscope down to about 1 μm allowing them to selectively excite either the bright or dim portions of the chromosome of *Vicia Faba S*. The fluorescence decay curves were different for the bright and dim regions; fluorescence emitted from the bright bands decayed slowly, whereas fluorescence outside the band decayed much more sharply. The authors were then able to reach the conclusion that they had observed the first direct indication that a larger fraction of AT–AT sequences were present within the bands than outside. This new approach shows clear promise for obtaining more detailed information on base-pair sequences within chromosomes, and this method may be useful for obtaining information on the effect of certain antitumoral agents or special drugs.

7.4.4 Picosecond Pulses and Selective Biochemical Reactions in DNA

Selective breaking of molecular bonds by infrared and ultraviolet radiation is a persistent theme in laser-isotope separation experiments and is accomplished by the use of high-power tunable lasers acting discriminately upon one vibrational or electronic level while ignoring nearby levels [7.155, 156]. These schemes have been used to separate isotopes of such "complex" molecules as sulfur hexafluoride [7.157] and boron trichloride [7.158]. *Letokhov* [7.159] has considered the possibility of selective biochemical reactions induced by laser radiation and concluded that picosecond techniques are essential.

Because, in general biophysical macromolecules have electronic absorption bands that are several hundred angstroms wide, any selective scheme must rely on vibrational excitation, and the vibronic levels must be narrow enough to permit selective absorption. Also, selectivity must be preserved after excitation in subsequent physical and chemical processes. *Letokhov* [7.159] considers two schemes: the first involves selective excitation with an infrared laser pulse followed by a visible pulse that dissociates only the vibrationally excited molecules; and the second involves multiphoton excitation with infrared pulses up the vibrational ladder leading to dissociation. In a solution there are severe requirements on these two schemes. Whereas in the gaseous phase, where these two schemes have been successful, the vibrational lifetime is relatively long, the depopulation time in the liquid phase for vibrational deactivation has been shown to be typically in the neighborhood of 10 ps [7.160]; see also Chapter 5. Therefore, to avoid nonselective deactivation of the vibrational levels by thermal deactivation, the energy must be deposited in an ultrashort pulse. Experiments with the gaseous phase are not a viable alternative because macromolecules would dissociate if they were heated to the point of vaporization. For the infrared-ultraviolet selection scheme both pulses must deposit sufficient energy before the vibrations relax. For DNA this requires powers in the neighborhood of 3×10^9 W/cm^2 for each pulse. Furthermore, the pulses must be synchronous to within the vibrational relaxation time.

For the infrared multiphoton scheme the power must be delivered onto the sample in a time shorter than the dephasing time (see Chap. 5 for definition). This is a more stringent requirement than for the infrared-ultraviolet pumping scheme because the dephasing time is usually much shorter than the depopulation time for molecules in solution.

Letokhov [7.159] points out that selective excitation may provide a method for stimulating the untwisting of the double helix of DNA. Certain vibrations such as those at the 1700 cm^{-1} and 1720 cm^{-1} correspond to the hydrogen bonding between adenine and thymine, and guanine and cystosine, respectively. These vibrations are not found in denatured DNA and can be stimulated by infrared pulses whose frequency matches the vibrational frequency. He also believes that the polynucleotide chain can be stimulated by action on the phosphate backbone. The PO_2 group has a vibrational mode at 1230 cm^{-1} so that intense radiation at 8.13 μm must be generated for this action.

One limitation to these schemes is caused by the fact that spacial localization is accurate only to within the wavelength of excitation. Thus, a molecule with reiterative elements like DNA can be broken in many locations. However, such schemes can be successful in simple molecules for providing selective chemical reactions. Thus we conclude our chapter with mention of these future possibilities; picosecond pulses should find wide application in the biological area in the future.

Acknowledgements. We thank Drs. *E.P. Ippen, R.S. Knox, W.W. Parson, C.V. Shank,* and *C.E. Swenberg* for helpful discussions of sections of the chapter; and *F. Skoberne* for editorial assistance;

and Drs. *L. Dutton, K. Kaufmann* and *P. M. Rentzepis* for providing figures and/or results prior to publication.

References

7.1 R. Emerson, W. Arnold: J. Gen. Physiol. **16**, 191 (1932)
7.2 Govindjee, R. Govindjee: Sci. Amer. **230**, 68 (1974)
7.3 Govindjee, R. Govindjee: In *Bioenergetics of Photosynthesis*, ed. by Govindjee (Academic Press, New York, San Francisco, London 1975) pp. 1–50
7.4 E. Rabinowitch, Govindjee: *Photosynthesis* (Wiley, New York 1969)
7.5 E. I. Rabinowitch: *Photosynthesis and Related Processes*, Vol. I (Interscience, New York 1945) and Vol. II (Interscience, New York 1951)
7.6 W. W. Parson: In *Chemical and Biochemical Applications of Lasers*, Vol. I, ed. by C. B. Moore (Academic Press, New York, San Francisco, London 1974) pp. 339–372
7.7 G. Wald: Sci. Amer. **201**, 92 (1959)
7.8 J. J. Katz, J. R. Norris: In *Current Topics in Bioenergetics*, Vol. 5, ed. by D. R. Sanadi, L. Packer (Academic Press, New York 1973) pp. 41–75
7.9 F. P. Zscheile, C. L. Comar: Botan. Gaz. **102**, 463 (1941)
7.10 K. D. Philipson, K. Sauer: Biochemistry **11**, 1880 (1972)
7.11 D. W. Reed: J. Biol. Chem. **244**, 4936 (1968)
7.12 P.-S. Song, T. A. Moore, M. Sun: In *The Chemistry of Plant Pigments*, ed. by C. O. Chichester (Academic Press, New York, London 1972) pp. 33–74
7.13 A. J. Campillo, R. C. Hyer, V. H. Kollman, S. L. Shapiro, H. D. Sutphin: Biochim. Biophys. Acta **387**, 533 (1975)
7.14 V. H. Kollman, S. L. Shapiro, A. J. Campillo: Biochem. Biophys. Res. Commun. **63**, 917 (1975)
7.15 S. L. Shapiro, V. H. Kollman, A. J. Campillo: FEBS Lett. **54**, 358 (1975)
7.16 V. Z. Paschenko, A. B. Rubin, L. B. Rubin: Kvant. Elek. (Moscow) **2**, 1336 (1975)
7.17 V. Z. Paschenko, S. P. Protasov, A. B. Rubin, K. N. Timofeev, L. M. Zamazova, L. B. Rubin: Biochim. Biophys. Acta **408**, 145 (1975)
7.18 N. K. Boardman: *Advances in Enzymology*, ed. by F. F. Nord (Interscience, New York, London, Sydney 1968) pp. 1–79
7.19 S. S. Brody, E. Rabinowitch: Science **125**, 555 (1957)
7.20 W. F. Watson, R. Livingston: J. Chem. Phys. **18**, 802 (1950)
7.21 T. Förster: Ann. Physik **2**, 55 (1948)
7.22 T. Förster: In *Modern Quantum Chemistry*, Part III, ed. by O. Sinanoglu (Academic Press, New York and London 1965) pp. 93–137
7.23 R. S. Knox: In *Primary Molecular Events in Photobiology*, ed. by A. Checcucci, R. A. Weale (Elsevier, Amsterdam, London, New York 1973) pp. 45–77
7.24 R. S. Knox: In *Bioenergetics of Photosynthesis*, ed. by Govindjee (Academic Press, New York, San Francisco, London 1975) pp. 183–221
7.25 G. Hoch, R. S. Knox: In *Photophysiology*, Vol. III, ed. by A. Giese (Academic Press, New York 1968) pp. 225–251
7.26 L. Stryer, R. P. Haugland: Proc. Nat. Acad. Sci. USA **58**, 719 (1967)
7.27 T. Förster: Z. Naturforsch. **4a**, 321 (1949)
7.28 J. C. Chang: Ph. D. Thesis, University of Rochester, Rochester, NY (1972)
7.29 D. Fröhlich, H. Mahr: Phys. Rev. **141**, 692 (1966)
7.30 I. Z. Steinberg: J. Chem. Phys. **48**, 2411 (1968)
7.31 D. Rehm, K. B. Eisenthal: Chem. Phys. Lett. **9**, 387 (1971)
7.32 T.S. Rahman, R. S. Knox: Phys. Stat. Sol. (b) **58**, 715 (1973)
7.33 B. Kok: Plant Physiol. **34**, 184 (1959)
7.34 K. Ballschmiter, K. Truesdell, J. J. Katz: Biochim. Biophys. Acta **184**, 604 (1969)
7.35 F. K. Fong: Proc. Nat. Acad. Sci. USA **71**, 3692 (1974)

7.36 F.K.Fong: Appl. Phys. **6**, 151 (1975)
7.37 G.S.Beddard, G.Porter: Nature **260**, 366 (1976)
7.38 F.W.Craver, R.S.Knox: Molec. Phys. **22**, 385 (1971)
7.39 A.Ore: J. Chem. Phys. **31**, 442 (1959)
7.40 R.S.Knox: Physica **39**, 361 (1968)
7.41 R.M.Pearlstein: Bull. Am. Phys. Soc. **17**, 31 (1972); unpublished notes
7.42 G.W.Robinson: Brookhaven Symp. Biol. **19**, 16 (1967)
7.43 J.Franck, E.Teller: J. Chem. Phys. **6**, 861 (1938)
7.44 E.W.Montroll: J. Math. Phys. **10**, 753 (1969)
7.45 R.M.Pearlstein: Brookhaven Symp. Biol. **19**, 8 (1967)
7.46 R.S.Knox: J. Theoret. Biol. **21**, 244 (1968)
7.47 K.Sauer: In *Bioenergetics of Photosynthesis,* ed. by Govindjee (Academic Press, New York, San Francisco, London 1975) pp. 115–181
7.48 J.P.Thornber: Biochemistry **9**, 2688 (1970)
7.49 J.P.Thornber: Ann. Rev. Plant Physiol. **26**, 127 (1975)
7.50 V.M.Kenkre, R.S.Knox: Phys. Rev. Lett. **33**, 803 (1974)
7.51 R.M.Pearlstein: Ph. D. Thesis, University of Maryland, College Park, Maryland (1966)
7.52 W.J.Vredenberg, L.N.M.Duysens: Nature **197**, 355 (1963)
7.53 Z.Bay, R.M.Pearlstein: Proc. Nat. Acad. Sci. USA **50**, 1071 (1963)
7.54 E.W.Montroll, K.Shuler: Advan. Chem. Phys. **1**, 361 (1958)
7.55 R.M.Pearlstein: J. Chem. Phys. **56**, 2431 (1972)
7.56 R.P.Hemenger, R.M.Pearlstein, K.Lakatos-Lindenberg: J. Math. Phys. **13**, 1056 (1972)
7.57 C.E.Swenberg, R.Dominijanno, N.E.Geacintov: Photochem. Photobiol. **24**, 601 (1976)
7.58 V.K.Shante, S.Kirkpatrick: Adv. Phys. **20**, 325 (1971)
7.59 A.Müller, R.Lumry, H.Kokubun: Rev. Sci. Instr. **36**, 1214 (1965)
7.60 A.Yu.Borisov, V.I.Godik: Bioenergetics **3**, 211 (1972)
7.61 A.Yu.Borisov, V.I.Godik: Bioenergetics **3**, 515 (1972)
7.62 A.Yu.Borisov, M.D.Il'ina: Biochim. Biophys. Acta **305**, 364 (1973)
7.63 A.Müller, R.Lumry, M.S.Walker: Photochem. Photobiol. **9**, 113 (1969)
7.64 H.Merkelo, S.R.Hartman, T.Mar, G.S.Singhal, Govindjee: Science **164**, 301 (1969)
7.65 H.Merkelo, J.H.Hammond, S.R.Hartman, Z.I.Derzko: J. Luminescence **1/2**, 502 (1970)
7.66 M.A.Duguay, J.W.Hansen: Appl. Phys. Lett. **15**, 192 (1969)
7.67 M.Seibert, R.R.Alfano, S.L.Shapiro: Biochim. Biophys. Acta **292**, 493 (1973)
7.68 M.Seibert, R.R.Alfano: Biophys. J. **14**, 269 (1974)
7.69 W.Yu, P.Ho, R.R.Alfano, M.Seibert: Biochim. Biophys. Acta **387**, 159 (1975)
7.70 A.J.Campillo, S.L.Shapiro, V.H.Kollman, K.R.Winn, R.C.Hyer: Biophys. J. **16**, 93 (1976)
7.71 A.J.Campillo, S.L.Shapiro: unpublished
7.72 G.S.Beddard, G.Porter, C.J.Tredwell, J.Barber: Nature **258**, 166 (1975)
7.73 D.Mauzerall: Biophys. J. **16**, 87 (1976)
7.74 R.C.Hyer, H.D.Sutphin, K.R.Winn: Rev. Sci. Instr. **46**, 1333 (1975)
7.75 M.Pope, H.Kallman, J.Giachino: J. Chem. Phys. **42**, 2540 (1965)
7.76 M.Pope, J.Burgos: Molec. Cryst. **3**, 215 (1967)
7.77 A. J. Campillo, R. C. Hyer, T. Monger, W. W. Parson, S. L. Shapiro: Proc. Nat. Acad. Sci. USA, in press
7.78 W.B.Goad, V.Gutschick: To be published
7.79 C.E.Swenberg, N.E.Geacintov, M.Pope: Biophys. J. **16**, 1447 (1976)
7.80 A.J.Campillo, V.H.Kollman, S.L.Shapiro: Science **193**, 227 (1976)
7.81 C.E.Swenberg: Private communication
7.82 J.Breton, N.Geacintov: FEBS. Lett. **69**, 86 (1976)
7.83 J.Breton, P.Mathis: Compt. Rend. Acad. Sci. (Paris) **271**, 1094 (1970)
7.84 W.T.Stacy, T.Mar, C.Swenberg, Govindjee: Photochem. Photobiol. **14**, 197 (1971)
7.85 R.K.Clayton: Ann. Rev. Biophys. Bioeng. **2**, 131 (1973)
7.86 W.W.Parson, R.J.Cogdell: Biochim. Biophys. Acta **416**, 105 (1975)

7.87 W. W. Parson: Ann. Rev. Microbiol. **28**, 41 (1974)
7.88 W. W. Parson, R. K. Clayton, R. J. Cogdell: Biochim. Biophys. Acta **387**, 265 (1975)
7.89 K. J. Kaufmann, P. L. Dutton, T. L. Netzel, J. S. Leigh, P. M. Rentzepis: Science **188**, 1301 (1975)
7.90 M. G. Rockley, M. W. Windsor, R. J. Cogdell, W. W. Parson: Proc. Nat. Acad. Sci. USA **72**, 2251 (1975)
7.91 T. L. Netzel, P. M. Rentzepis, J. Leigh: Science **182**, 238 (1973)
7.92 R. R. Alfano, S. L. Shapiro: Phys. Rev. Lett. **24**, 584 (1970)
7.93 R. R. Alfano, S. L. Shapiro: Phys. Rev. Lett. **24**, 592 (1970)
7.94 R. R. Alfano, S. L. Shapiro: Phys. Rev. Lett. **24**, 1217 (1970)
7.95 R. R. Alfano, S. L. Shapiro: Chem. Phys. Lett. **8**, 631 (1971)
7.96 F. Shimizu: Phys. Rev. Lett. **19**, 1097 (1967)
7.97 P. L. Dutton, K. J. Kaufmann, B. Chance, P. M. Rentzepis: FEBS Lett. **60**, 275 (1975)
7.98 J. Fajer, D. C. Brune, M. S. Davis, A. Forman, L. D. Spaulding: Proc. Nat. Acad. Sci. USA **72**, 4956 (1975)
7.99 J. S. Leigh, T. L. Netzel, P. L. Dutton, P. M. Rentzepis: FEBS Lett. **48**, 136 (1974)
7.100 K. J. Kaufmann, K. M. Petty, P. L. Dutton, P. M. Rentzepis: Biochem. Biophys. Res. Commun. **70**, 839 (1976)
7.101 A. Müller, J. Schulz-Hennig, H. Tashiro: Paper Q-10, IXth Intern. Conf. Quantum Electronics, Amsterdam, 1976
7.102 H. Muirhead, M. F. Perutz: Nature **199**, 633 (1963)
7.103 M. F. Perutz: Nature **194**, 914 (1962)
7.104 M. F. Perutz: Sci. Amer. **211**, 64 (1964)
7.105 F. Hoppe-Seyler: Z. physiol. Chem. **3**, 339 (1879)
7.106 F. Hoppe-Seyler: Z. physiol. Chem. **4**, 193 (1880)
7.107 G. S. Marks: *Heme and Chlorophyll* (Van Nostrand Co. Ltd., London, Princeton NJ, Toronto, Melbourne 1969)
7.108 A. E. Sidwell, R. H. Munch, E. S. G. Barron, T. R. Hogness: J. Biol. Chem. **123**, 335 (1938)
7.109 E. Gordy, D. L. Drabkin: J. Biol. Chem. **227**, 285 (1957)
7.110 R. W. Noble, M. Brunori, J. Wyman, E. Antonini: Biochemistry **6**, 1216 (1967)
7.111 J. Haldane, J. Lorrain-Smith: J. Physiol. **20**, 497 (1895)
7.112 L. Stryer: Rad. Res. Suppl. **2**, 432 (1960)
7.113 T. Bücher, E. Negelein: Biochem. Z. **311**, 163 (1952)
7.114 C. V. Shank, E. Ippen, R. Bersohn: Science **193**, 50 (1976)
7.115 G. Wald: Nature **132**, 316 (1933)
7.116 G. Wald: J. Gen. Physiol. **18**, 905 (1934–35)
7.117 W. A. Hagins, W. H. Jennings: Discuss. Faraday Soc. **27**, 180 (1960)
7.118 G. Wald: Science **162**, 230 (1968)
7.119 W. G. Marks, W. H. Dobelle, E. F. MacNichol: Science **143**, 1181 (1964)
7.120 P. K. Brown, G. Wald: Science **144**, 45 (1964)
7.121 E. F. MacNichol: Sci. Amer. **211**, 48 (1964)
7.122 F. D. Collins: Nature **171**, 469 (1953)
7.123 R. A. Morton, G. A. J. Pitt: Biochem. J. **59**, 128 (1955)
7.124 T. Yoshizawa: In *Handbook in Sensory Physiology*, Vol. VII/I, ed. by H. J. A. Dartnall (Springer Berlin, Heidelberg, New York 1972) pp. 146–179
7.125 G. E. Busch, M. L. Applebury, A. A. Lamola, P. M. Rentzepis: Proc. Nat. Acad. Sci. USA **69**, 2802 (1972)
7.126 T. G. Ebrey, B. Honig: Quart. Rev. Biophys. **8**, 129 (1975)
7.127 R. Bensasson, E. J. Land, T. G. Truscott: Nature **258**, 768 (1975)
7.128 W. Stoeckenius: Sci. Amer. **234**, 38 (1976)
7.129 K. J. Kaufmann, P. M. Rentzepis, W. Stoeckenius, A. Lewis: Biochem. Biophys. Res. Commun. **68**, 1109 (1976)
7.130 R. R. Alfano, W. Yu, R. Govindjee, B. Becher, T. G. Ebrey: Biophys. J. **16**, 541 (1976)
7.131 M. D. Hirsch, M. A. Marcus, A. Lewis, H. Mahr, N. Frigo: Biophys. J. **16**, 1399 (1976)
7.132 J. D. Watson, F. H. C. Crick: Nature **171**, 737 (1953)

7.133 J.D.Watson, F.H.C.Crick: Nature **171**, 964 (1953)
7.134 J.Eisinger, R.G.Shulman: Proc. Nat. Acad. Sci. USA **55**, 1387 (1966)
7.135 R.Bersohn, I.Isenberg: J. Chem. Phys. **40**, 3175 (1964)
7.136 R.S.Somer, J.Jortner: J. Chem. Phys. **49**, 3919 (1968)
7.137 J.Eisinger, A.A.Lamola: In *Excited State of Proteins and Nucleic Acids*, ed. by R.F.Steiner, I.Weinryb (Plenum Press, New York, London 1971) pp. 107–198
7.138 J.Eisinger, M.Guéron, R.G.Shulman, T.Yamane: Proc. Nat. Acad. Sci. USA **55**, 1015 (1966)
7.139 W.E.Blumberg, J.Eisinger, G.Navon: Biophysical J. (abstr.) **8A**, 4106 (1968)
7.140 J.Eisinger, R.G.Shulman: Science **161**, 1311 (1968)
7.141 J.Eisinger, A.A.Lamola: Biochim. Biophys. Acta **240**, 299 and 313 (1971)
7.142 M.Guéron, J.Eisinger, R.G.Shulman: J. Chem. Phys. **47**, 4077 (1967)
7.143 A.J.Campillo, V.H.Kollman, S.L.Shapiro: unpublished
7.144 L.S.Lerman: J. Molec. Biol. **3**, 18 (1961)
7.145 L.S.Lerman: Proc. Nat. Acad. Sci. USA **49**, 94 (1963)
7.146 G.Weill, M.Calvin: Biopolymers **1**, 401 (1963)
7.147 M.J.Waring: J. Mol. Biol. **13**, 269 (1965)
7.148 J.-B.Lepecq, C.Paoletti: J. Mol. Biol. **27**, 87 (1967)
7.149 G.Weill: Biopolymers **3**, 567 (1965)
7.150 V.W.Burns: Biophys. Soc. Annu. Meet. Abstr. **A**, 172 (1969)
7.151 B.M.Sutherland, J.C.Sutherland: Biophys. J. **9**, 1045 (1969)
7.152 S.L.Shapiro, A.J.Campillo, V.H.Kollman, W.B.Goad: Opt. Commun. **15**, 308 (1975)
7.153 M.R.Alvarez: Cancer Res. **35**, 93 (1975)
7.154 S.H.Tomson, E.A.Emmett, S.H.Fox: Cancer Res. **34**, 3124 (1974)
7.155 V.S.Letokhov, R.V.Ambartzumian: IEEE J. QE-**7**, 305 (1971)
7.156 R.V.Ambartzumian, V.S.Letokhov: Chem. Phys. Lett. **13**, 446 (1972)
7.157 R.V.Ambartzumian, Y.A.Gorokhov, V.S.Letokhov, G.N.Makarov: JETP Lett. **21**, 171 (1975)
7.158 S.D.Rockwood, S.W.Rabideau: IEEE J. QE-**10**, 789 (1974)
7.159 V.S.Letokhov: J. Photochem. **4**, 185 (1975)
7.160 A.Laubereau, D. von der Linde, W.Kaiser: Phys. Rev. Lett. **28**, 1162 (1972)

Additional References with Titles

R. W. Anderson, R. M. Hochstrasser: Dynamics of photodissociation in solution using picosecond spectroscopy. J. Phys. Chem. **80**, 2155 (1976)

R. W. Anderson, R. M. Hochstrasser, H. J. Pownall: Picosecond absorption spectra of the second excited singlet state of a molecule in the condensed phase: xanthione. Chem. Phys. Lett. **43**, 224 (1976)

S. Arnold, R. R. Alfano, M. Pope, W. Yu, P. Ho, R. Selsby, J. Tharrats, C. E. Swenberg: Triplet exciton caging in two dimensions. J. Chem. Phys. **64**, 5104 (1976)

G. S. Beddard, S. E. Carlin, G. Porter: Concentration quenching of chlorophyll fluorescence in bilayer lipid vesicles and liposomes. Chem. Phys. Lett. **43**, 27 (1976)

E. S. Bliss, J. T. Hunt, P. A. Renard, G. E. Sommargren, H. J. Weaver: Effects of nonlinear propagation on laser focusing properties. IEEE J. Quant. Electron. **12**, 402 (1976)

J. L. Bocher, J. C. Griesemann, M. Louis-Jacquet, M. Decroisette: Self-phase modulation in the breakdown of gases by ultra-short pulses. Opt. Commun. **16**, 262 (1976)

A. P. Bruckner: Some applications of picosecond-optical range gating. Proc. Soc. Photo-opt. Instr. Engrs. **94**, 41 (1976)

G. E. Busch, P. M. Rentzepis: Picosecond chemistry. Science **194**, 276 (1976)

A. J. Campillo, S. L. Shapiro: Use of picosecond lasers for studying photosynthesis. Proc. Soc. Photo-opt. Instr. Engrs. **94**, 89 (1976)

L. W. Coleman: Ultrafast X-ray diagnostics for laser experiments. Proc. Soc. Photo-opt. Instr. Engrs. **94**, 19 (1976)

B. Couillaud, A. Ducasse: Synchronous mode locking of two cw dye lasers sharing a common amplifier medium. Appl. Phys. Lett. **29**, 665 (1976)

J. Covey: Use of picosecond lasers in measuring ultrafast molecular processes. Proc. Soc. Photo-opt. Instr. Engrs. **94**, 107 (1976)

J.-C. Diels: Two-photon coherent propagation transmission of 90° phase-shifted pulses, and application to isotope separation. Opt. Quant. Electron. **8**, 513 (1976)

M. A. Duguay: The ultrafast optical Kerr shutter. In: *Progress in Optics*, Vol. 14, ed. by E. Wolf (North-Holland, Amsterdam 1976)

M. A. Duguay, M. A. Palmer, R. E. Palmer: Laser driven subnanosecond blast shutter. Proc. Soc. Photo-opt. Instr. Engrs. **94**, 2 (1976)

M. D. Feit, J. A. Fleck: Amplitude and phase modulation accompanying laser beam trapping in plasmas. Appl. Phys. Lett. **29**, 234 (1976)

E. Fill, K. Hohla, G. T. Schappert, R. Volk: 100-ps pulse generation and amplification in the iodine laser. Appl. Phys. Lett. **29**, 805 (1976)

G. R. Fleming, I. R. Harrowfield, A. E. W. Knight, J. M. Morris, R. J. Robbins, G. W. Robinson: Properties of single picosecond pulses from Neodymium: phosphate glass. Opt. Commun. **20**, 36 (1977)

L. S. Goldberg, J. N. Bradford: Passive mode-locking and picosecond pulse generation in Nd: lanthanum beryllate. Appl. Phys. Lett. **29**, 585 (1976)

L. Harris, G. Porter, J. A. Synowiec, C. J. Tredwell, J. Barber: Fluorescence lifetimes of chlorella pyrenoidosa. Biochim. Biophys. Acta **449**, 329 (1976)

F. Heisel, J. A. Miehe, B. Sipp, M. Schott: Experimental evidence for the influence of exciton interactions on the temporal decay of crystalline tetracene fluorescence. Chem. Phys. Lett. **43**, 534 (1976)

J.P.Heritage, A.Penzkofer: Relaxation dynamics of the first excited electronic singlet state of azulene in solution. Chem. Phys. Lett. **44**, 76 (1976)
Y.Hirata, I.Tanaka: Intersystem crossing to the lowest triplet state of phenazine following singlet excitation with a picosecond pulse. Chem. Phys. Lett. **43**, 568 (1976)
Y.Hirata, I.Tanaka: Buildup of *T-T* absorption of acridine following singlet excitation with a picosecond pulse. Chem. Phys. Lett. **41**, 336 (1976)
R.M.Hochstrasser, A.C.Nelson: A study of energy transfer between electronically excited states using a picosecond laser pulse. Opt. Commun. **18**, 361 (1976)
R.M.Hochstrasser, D.L.Narva, A.C.Nelson: Picosecond photophysics of *trans*-retinal. Chem. Phys. Lett. **43**, 15 (1976)
R.M.Hochstrasser, D.S.King, A.C.Nelson: Subnanosecond dynamics of the fluorescence and singlet absorption of s-tetrazine. Chem. Phys. Lett. **42**, 8 (1976)
D.Holten, M.Gouterman, W.W.Parson, M.W.Windsor, M.G.Rockley: Electron transfer from photoexcited singlet and triplet bacteriopheophytin. Photochem. Photobiol. **23**, 415 (1976).
L.Holz, K.Kneipp, A.Lau, W.Werncke: Stimulated Raman scattering of picosecond laser pulses by polaritons in $LiIO_3$ and $LiNbO_3$ single crystals. Phys. stat. sol. **36**, K5 (1976)
J.F.Holzrichter, D.R.Speck: Laser focusing limitations from nonlinear beam instabilities. J. Appl. Phys. **47**, 2459 (1976)
D.Huppert, P.M.Rentzepis: Picosecond kinetics. In *Molecular Energy Transfer*, ed. by R.Levine and J.Jortner (John Wiley and Sons, New York 1976) p. 270
D.Huppert, P.M.Rentzepis, G.Tollin: Picosecond kinetics of chlorophyll and chlorophyll/quinone solutions in ethanol. Biochim. Biophys. Acta **440**, 356 (1976)
E.P.Ippen, C.V.Shank, R.L.Woerner: Picosecond dynamics of azulene. Chem. Phys. Lett. (to be published)
J.Jaraudias, P.Goujon, J.C.Mialocq: Picosecond absorption relaxation of DODCI using single pulse excitation. Chem. Phys. Lett. **45**, 107 (1977)
G.I.Kachen, W.H.Lowdermilk: Self-induced gain and loss modulation in coherent, transient Raman pulse propagation. Phys. Rev. A **14**, 1472 (1976)
C.J.Kennedy: Mode-locked and frequency-doubled laser efficiencies. Appl. Opt. **15**, 2955 (1976)
T.Kobayashi, S.Nagakura: Picosecond time-resolved spectroscopy and the intersystem crossing rates of anthrone and fluorenone. Chem. Phys. Lett. **43**, 429 (1976)
P.Kolodner, C.Winterfeld, E.Yablonovitch: Molecular dissociation of SF_6 by ultra-short CO_2 laser pulses. Opt. Commun. **20**, 119 (1977)
H.S.Kwok, E.Yablonovitch: 30-psec CO_2 laser pulses generated by optical free induction decay. Appl. Phys. Lett. **30**, 158 (1977)
J.R.Lalanne, B.Martin, B.Pouligny, S.Kielich: Fast picosecond reorientation in the isotropic phase of nematogens. Opt. Commun. **19**, 440 (1976)
Chi H.Lee: Picosecond optoelectronic switching in GaAs. Appl. Phys. Lett. **30**, 84 (1977)
M.Lequime, J.Mlynek, J.P.Hermann: Intensity dependent total reflection in carbon disulfide with picosecond pulses. Opt. Commun. **19**, 423 (1976)
V.S.Letokhov: Use of laser radiation in field-electron and field-ion microscopy for observation of biological molecules. Sov. J. Quant. Electron. **5**, 506 (1975)
A.J.Lieber, H.D.Sutphin, C.B.Webb: Subpicosecond proximity-focused streak camera for X-ray and visible light. Proc. Soc. Photo-opt. Industr. Engrs. **94**, 7 (1976)
S.H.Lin: Theory of vibrational relaxation and infrared absorption in condensed media. J. Chem. Phys. **65**, 1053 (1976)
V.N.Lugovoi, V.N.Strel'tsov: Possibility of generation of femtosecond pulses by stimulating Raman scattering in optical resonators. Sov. J. Quant. Electron. **6**, 971 (1976)
H.Mahr: Tunable two-wavelength modelocking of the cw dye laser. J. Quant. Electron. **12**, 554 (1976)
M.Ojima, T.Kushida, Y.Tanaka, S.Shionoya: Picosecond time-resolved luminescence spectra of excitonic molecules in CuCl. Solid State Commun. **20**, 847 (1976)
G.L.Olson, G.E.Busch: Picosecond photodynamics of intense pulse interactions with molecular populations. J. Chem. Phys. **66**, 1183 (1977)

A. Penzkofer, W. Falkenstein: Photoinduced dichroism and vibronic relaxation in rhodamine dyes. Chem. Phys. Lett. **44**, 547 (1976)

A. Penzkofer, W. Falkenstein, W. Kaiser: Vibronic relaxation in the S_1 state of rhodamine dye solutions. Chem. Phys. Lett. **44**, 82 (1976)

H. Puell, K. Spanner, W. Falkenstein, W. Kaiser, C. R. Vidal: Third-harmonic generation of mode locked Nd: glass laser pulse in phased-matched Rb-Xe mixtures. Phys. Rev. A **14**, 2240 (1976)

J. Reintjes, R. C. Eckardt: Efficient harmonic generation from 532 to 266 nm in ADP and KD*P. Appl. Phys. Lett. **30**, 91 (1977)

J. Reintjes, R. C. Eckardt, C. Y. She, N. E. Karangelen, R. C. Elton, R. A. Andrews: Generation of coherent radiation at 53.2 nm by fifth-harmonic conversion. Phys. Rev. Lett. **37**, 1540 (1976)

J. F. Reintjes, T. N. Lee, R. C. Eckardt, R. A. Andrews: Interferometric study of laser-produced plasmas. J. Appl. Phys. **47**, 4457 (1976)

J. Reintjes, C. Y. She, R. C. Eckardt, N. E. Karangelen, R. C. Elton, R. A. Andrews: Generation of coherent radiation at 38 and 53 nm using high-order optical nonlinearities. J. Opt. Soc. Am. **67**, 251 (1977)

H. Saito, A. Kuroiwa, S. Kuribayashi, Y. Aogaki, S. Shionoya: Optical formation and luminescence of excitonic molecules in CdS. J. Luminesc. **12**, 575 (1976)

H. Salzmann, H. Strohwald: Single picosecond dye laser pulses by resonator transients. Phys. Lett. A **57**, 41 (1976)

A. Scavennec: Mismatch effects in synchronous pumping of the continuously operated mode-locked dye laser. Opt. Commun. **17**, 14 (1976)

C. V. Shank, E. P. Ippen, O. Teschke: Sub-picosecond relaxation of large organic molecules in solution. Chem. Phys. Lett. (to be published)

D. K. Sharma, R. W. Yip, D. F. Williams, S. E. Sugamori, L. L. T. Bradley: Generation of an intense picosecond continuum in D_2O by a single picosecond 1.06 μ pulse. Chem. Phys. Lett. **41**, 460 (1976)

H. Shizuka, K. Matsui, Y. Hirata, I. Tanaka: Direct measurement of intramolecular proton transfer in the excited state of 2,4-bis(dimethylamino)-6-2-hydroxy-5-methylphenyl)-s-triazine with picosecond pulses. J. Phys. Chem. **80**, 2070 (1976)

A. E. Siegman: Proposed picosecond excited-state measurement method using a tunable-laser-induced grating. Appl. Phys. Lett. **30**, 21 (1977)

W. L. Smith, J. H. Bechtel: A simple technique for individual picosecond laser pulse duration measurements. J. Appl. Phys. **47**, 1065 (1976)

W. L. Smith, P. Liu, N. Bloembergen: Superbroadening in H_2O and D_2O by self-focused picosecond pulses from a YAlG:Nd laser. Phys. Rev. A (to be published)

K. Spanner, A. Laubereau, W. Kaiser: Vibrational energy redistribution of polyatomic molecules in liquids after ultrashort infrared excitation. Chem. Phys. Lett. **44**, 88 (1976)

H. Tashiro, T. Yajima: Direct measurement of blue fluorescence lifetimes in polymethine dyes using a picosecond laser. Chem. Phys. Lett. **42**, 553 (1976)

I. V. Tomov, R. Fedosejevs, M. C. Richardson: Generation of single synchronizable picosecond 1.06 μm pulses. Appl. Phys. Lett. **30**, 164 (1977)

I. V. Tomov, R. Fedosejevs, M. C. Richardson, W. J. Orr: Synchronizable actively mode-locked Nd:glass laser. Appl. Phys. Lett. **29**, 193 (1976)

I. V. Tomov, R. Fedosejevs, M. C. Richardson, W. J. Sarjeant, A. J. Alcock, K. E. Leopold: Picosecond XeF amplified laser pulses. Appl. Phys. Lett. **30**, 146 (1977)

M. R. Topp: Biphotonic excitation of the fluorescence of benzophenone ketyl. Chem. Phys. Lett. **39**, 423 (1976)

H. M. van Driel, A. Elci, J. S. Bessey, M. O. Scully: Photoluminescence spectra of germanium at high excitation intensities. Solid State Commun. **20**, 837 (1976)

A. C. Walker, A. J. Alcock: Picosecond resolution, real-time linear detection system for 10-μm radiation. Rev. Sci. Instr. **47**, 915 (1976)

R. B. Weisman, S. A. Rice: Tunable infrared ultrashort pulses from a mode-locked parametric oscillator. Opt. Commun. **19**, 28 (1976)

M. W. Windsor: Picosecond flash photolysis studies of dyes, inorganic complexes, biological pigments, and photosynthetic systems. J. Phys. Chem. **80**, 2278 (1976)

P. Wirth, S. Schneider, F. Dörr: S_1-lifetimes of triphenylmethane and indigo dyes determined by the two-photon-fluorescence technique. Opt. Commun. **20**, 155 (1977)

P. Wirth, S. Schneider, F. Dörr: Ultrafast electronic relaxation in the S_1 state of azulene and some of its derivatives. Chem. Phys. Lett. **42**, 482 (1976)

Z. A. Yasa, A. Dienes, J. R. Whinnery: Subpicosecond pulses from a cw double mode-locked dye laser. Appl. Phys. Lett. **30**, 24 (1977)

Z. A. Yasa, O. Teschke, L. W. Braverman: Formation of ultrashort pulses in dye lasers with saturable absorbers. J. Appl. Phys. **47**, 174 (1976)

Subject Index